Spectral Methods in Geodesy and Geophysics

Cover

The globes depict progressively greater resolution of the Earth's gravitational field as expressed in a truncated spherical harmonic series excluding the main field associated with an approximating ellipsoid. The top left globe starts with a field to harmonic degree 4 and subsequent globes increase this to degrees 8, 16, 32, and 64. In each case the north pole is at the top tilted forward such that the view is of the Himalayas at the nadir. Yellows are highs and blues are lows.

Spectral Methods in Geodesy and Geophysics

Christopher Jekeli

Division of Geodetic Science
School of Earth Sciences
Ohio State University
Columbus, Ohio, USA

CRC Press
Taylor & Francis Group
Boca Raton London New York

CRC Press is an imprint of the
Taylor & Francis Group, an **informa** business
A SCIENCE PUBLISHERS BOOK

CRC Press
Taylor & Francis Group
6000 Broken Sound Parkway NW, Suite 300
Boca Raton, FL 33487-2742

First issued in paperback 2021

Version Date: 20170509

ISBN 13: 978-0-367-78182-8 (pbk)
ISBN 13: 978-1-4822-4525-7 (hbk)

Library of Congress Cataloging-in-Publication Data

Names: Jekeli, Christopher, 1953-
Title: Spectral methods in geodesy and geophysics / Christopher Jekeli, Division of Geodetic Science, School of Earth Sciences, Ohio State University, Columbus, Ohio, USA.
Description: Boca Raton, FL : CRC Press, 2017. | "A science publishers book." | Includes bibliographical references and index.
Identifiers: LCCN 2017018988| ISBN 9781482245257 (hardback : alk. paper) | ISBN 9781482245264 (e-book : alk. paper)
Subjects: LCSH: Fourier transform spectroscopy. | Spectrum analysis. | Fourier analysis. | Geodesy. | Geophysics.
Classification: LCC QC454.F7 J45 2017 | DDC 526/.101517222--dc23
LC record available at https://lccn.loc.gov/2017018988

Visit the Taylor & Francis Web site at
http://www.taylorandfrancis.com

and the CRC Press Web site at
http://www.crcpress.com

To Roberta

Preface

The genesis of this book is a course first taught to students in Geodetic Science at the Ohio State University soon after the Fast Fourier Transform was introduced to spherical harmonic analysis of the Earth's gravity field. That was in the 1980s and only a few text books existed that treated the subject from the geophysical viewpoint, specifically with the spatial coordinates as independent variables in two dimensions. The majority of books on Fourier analysis were written strictly with time as the independent variable because spectral analysis originated with and still primarily concerns communications theory and electrical engineering. In geodesy and geophysics the coordinate domain usually is approximated by a sphere, although much can be done by further approximating this domain locally by a plane. Both Cartesian and spherical domains are treated here with equal emphasis. Ellipsoidal spectral analysis, accounting for Earth's equatorial eccentricity, is not covered since it quickly becomes difficult, if not intractable, for most of the common applicable analytical tools and because the methodology under the spherical approximation is still viable and quite adequate. On the other hand, a basic understanding of concepts for the sphere is most easily obtained by first considering the Cartesian one-dimensional case; and thus each chapter begins in this domain.

The text develops the principal aspects of applied Fourier analysis and methodology with the main goal to inculcate a different way of perceiving global and regional geodetic and geophysical data, namely from the perspective of the frequency, or spectral, domain rather than the spatial domain. As the word "methods" in the title of the book suggests, the transformation of a geophysical signal into the spectral domain has applications beyond an analysis. Indeed, the use of Fourier transforms became feasible in physical geodesy at a time when the increase in global and regional data started to overwhelm computational capability. That situation is still in effect today, though to a lesser extent. The mathematician may be disappointed that the subject is not developed in the most rigorous and comprehensive way for Lebesgue-integrable measurable functions. Rather the approach is concerned more with the ultimate applications to ordinary continuous functions that represent spatial geophysical signals, such as geopotential fields. The text is written for graduate students; however, Chapters 1 through 4 and parts of 5 can also benefit undergraduates who have a solid and fluent knowledge of integral and differential calculus, have some statistical background, and are not uncomfortable with complex numbers. Sections 6.2.4 and 6.5.1 require a graduate level knowledge of physical geodesy. Much of the mathematics is derived in full, but several details along the way are left as exercises with hints to the reader, who in any case will extract the most value from the text by following the maxim that mathematics is not a spectator sport.

The concepts are developed by starting from the one-dimensional domain and working up to the spherical domain, which because of the spherical topology is not a simple generalization and not all the analytical tools of the Cartesian case carry forward without some sacrifices. Yet there is a strong similarity, and comprehension is straightforward with this background. Several concepts in Chapters 3, 5, and 6 are illustrated graphically with actual geophysical data, although some are contrived to some extent, in that they are selected specifically to demonstrate the essential points. The reader is cautioned that the world is not always so neat and tidy and only experience and experimentation may decide whether particular spectral analytical tools perform as advertised.

Of paramount importance in any text like this is the symbology, specifically the notation used to represent a function, its spectral counterpart, and their independent variables. It is attempted to adhere to notation that is as consistent and clear as possible throughout the chapters, even as rather diverse topics are encountered. It cannot be avoided that the present notation is at variance with notation found elsewhere, chiefly because the time variable is not the focus of the application. Nevertheless, it is hoped that the notation is transparent and uncomplicated yet precise.

Numerous individuals have contributed directly or indirectly to this text, starting with my graduate advisor, Prof. Richard H. Rapp, at Ohio State University, who formally introduced a course on spectral methods in 1990 and laid the groundwork for my offering of the course since 1994. Working as a geodesist at the Air Force Geophysics Laboratory in Bedford, MA (prior to joining the OSU faculty in 1993) I interacted with top engineers at contracting companies, Drs. Warren Heller, Jacob Goldstein, Jim White, among others, at The Analytic Sciences Corporation, and Dr. Stan Jordan at Geospace Systems Corp., who instilled in me a sober and practical approach to theoretical Fourier analysis. I am indebted, as well, to the many students who cheerfully suffered through my offerings of the Spectral Methods course at OSU based principally on notes always in a state of flux and on disconnected excerpts from other texts. Their patience, participation, and aptitude in finding errors in my notes were always appreciated. And, of course, my home department, the School of Earth Sciences at the Ohio State University, under the directorship of Prof. W. Berry Lyons, deserves my thanks for allowing me to undertake this project, in part with a sabbatical leave of absence. In addition, I am grateful to Prof. Jakob Flury and the Institute of Geodesy at the University of Hannover, Germany, for the opportunity to spend a summer working with their students and faculty, but also "to do what I wished." And then, there is that unfathomable entity called the World Wide Web that today contains an inexhaustible amount of easily accessible material on this (and every other imaginable) topic. From the many, mostly anonymous, authors of Wikipedia and those who post their questions and answers, or simply short passages,

I found insight both to the optimal and less transparent ways of presenting the material. Finally, and most of all, my eternal gratitude goes to my wife who endured my obsession with finishing this in a timely manner, who never doubted it (though I did more than once), and who, throughout my career and my focus at times away from family, has always supported me in countless ways. To her, this book is dedicated.

Contents

Chapter 1

Introduction

1.1 Definitions and Notations

Data analysis is the process of extracting useful information from observed or measured signals and phenomena. Signals are interpreted here in the mathematical sense of functions that depend on one or more independent variables in a given domain. Specifically, the present text is concerned with signals that play important roles in physical geodesy and particular geophysical investigations. In this emphasis, the domain of the independent variable typically is two-dimensional, either the plane or the sphere, and is described in terms of spatial, respectively angular coordinates. This contrasts with the usual one-dimensional time domain for which spectral analysis (defined below) was originally developed and that is still the typical domain for the independent variable in most texts on the subject. On the other hand, it is convenient initially to draw extensively on a one-dimensional exposition as a way to introduce and elucidate the theory and practice that then readily extends to higher dimensions. The advantages of this approach to the development are simpler notation and easier conceptual visualization. In addition, however, the *spherical* domain, being of particular relevance, is treated here in detail and in comparison to the Cartesian case. This generalization from the one-dimensional analysis to the spherical analysis is not trivial due to the special topology of the sphere. Still, for many local applications, the plane is a good approximation of the spherical surface, and thus the Cartesian two-dimensional analysis is equally important.

It is a truism, if not a tautology, that spatial data collected for analysis create their first impact on the analyst in the domain in which they are observed—the *space domain*. A two-dimensional map or a one-dimensional profile of gravity acceleration on the Earth's surface immediately identifies highs and lows that may be interpreted as generated by subsurface mass-density anomalies (after accounting for topographic variations). However, what may be less obvious from such a depiction are the magnitudes of the gravity variations at different horizontal scales of the data. This kind of analysis, important to discern and help constrain the depths and sizes of the density anomalies, requires that the data are transformed to a domain of scales or resolutions, what more precisely is called the *frequency* or *spectral domain* comprising frequencies (or wavelengths) of the data. These concepts are fundamental to the theory of spectral analysis, which, as noted above, was fully developed initially for time-varying signals that contain significant amplitudes at certain frequencies or sub-domains ("bands") of frequencies, particularly the signals associated with electromagnetic transmissions.

A wave belonging to a signal in the time domain has a duration in time called its *period*, or wavelength, where "length" refers to a time interval. Typically, in the case of electromagnetic radiation and for ease of visualization, the period of the wave is multiplied by the speed of signal propagation (e.g., the speed of light in a vacuum) and the wavelength then has units of distance. For example, the L1 carrier wave of the GPS (Global Positioning System) signal has a wavelength of about 19 cm. With the speed of light ($c = 2.998 \times 10^8$ m/s) this means that an L1 wave passes through an ideal receiver circuit (no electronic delays) every 0.64 nanoseconds. The term *frequency* is associated generally with time signals and indicates the number of waves that can be packed end-to-end into a unit of time. Formally, frequency is the inverse of period or of wavelength (implicitly assuming a propagation speed). Indeed, because the term, period, can relate more generally to signals comprising many different waves, one considers frequency and wavelength to be fundamental reciprocal quantities. Therefore, high frequency implies short wavelength and low frequency means long wavelength. The units of measure of frequency depend on a somewhat more precise definition of wavelength. Thinking of a single sine wave for illustration, if the wavelength is identified with an angle, such as the time to complete 2π radians in rotation, then frequency has units of radian per time unit. If wavelength is associated with an interval (the peak-to-peak interval of the sine wave), or a cycle, then frequency may be thought to have units of cycles per time unit. It is usually a matter of personal preference or convention within a discipline, and one type can be converted to the other by noting that 1 cycle is 2π radians. The usual notational convention is f for cyclical frequency and ω for radian frequency, with $\omega = 2\pi f$.

The cyclical frequency convention is adopted here in order to instill more emphatically the notion of frequency as the reciprocal of wavelength, and, in addition, because it is directly related to *wave number* that is used also for periodic signals (and to free up the use of ω to denote other quantities). This means that if "cycle" is thought to be a unit, then formally wavelength has units of length (or other quantity) per cycle. For example, temporal frequency typically has units of cycle per second, or Hz (*Hertz*)—the L1 GPS carrier has a frequency of 1575.42 MHz. A spatial frequency has units such as cycles per meter, corresponding to a wavelength in units of meters per cycle. On the other hand, under the System of International Units (SI) (Taylor and Thompson 2008, pp.25–27), "cycle" is not an official unit and frequency, f, simply has units, 1/s (which also defines the unit, Hz). The term "cycle" is used only to imply the context in which frequency is used. A radian, however, is a unit of angle, and thus, ω definitely has units of radian per second (or, per meter, for spatial radian frequency). These conventions notwithstanding, the units of cyclical frequency in this text are annotated by "cy/s" or "cy/m" to distinguish it from radian frequency.

The decomposition, or the *transformation*, of a signal in terms of its wavelengths, or equivalently, in terms of its frequencies, is known as its *spectrum*. This is completely analogous to a decomposition of a vector into its coordinate elements. A vector may be viewed as a directed magnitude, but a more analytical description is in terms of coordinates with respect to a defined set of basis vectors in independent directions. For signals or functions, the spectrum can be interpreted as a set of its "coordinates" with respect to a defined set of waves, or basis functions at different frequencies. The spectral domain consists of all the possible frequencies of these constituents. *Spectral analysis*, then, refers to the analysis of a signal from the viewpoint of its spectrum.

A one-dimensional signal is indicated in this text by a function, $g(x)$, where the independent variable, x, generally refers to distance. This notation is used for the independent variable rather than t, the usual variable for time-varying signals, in order to emphasize the general application in the space domain. It is assumed that the domain of x is the entire real line, although it may also be a finite interval on the line. The domain in this case is continuous. If a function (or, signal) is sampled at discrete and uniformly distributed points on the real line, which corresponds to a typical acquisition of data in practice, then the function is more aptly called a sequence, denoted by g_ℓ, where the index, ℓ, is an integer. In this case, the domain is discrete. The sampling interval, Δx, is a constant that carries the units of the space domain (in many texts it is set equal to one for convenience, but with the loss of explicit units). Independent variables, $x_1, x_2, \ldots$, are used for higher Cartesian dimensions, again with the implicit assumption that each is a real number. These may be combined in a vector, $\boldsymbol{x} = (x_1 \; x_2 \; \cdots)^T$, where its dimension is obvious from the context in which it is used. The corresponding higher-dimensioned discrete domain is denoted by integers, $\ell_1, \ell_2, \ldots$.

Frequency in one dimension is denoted, f, and in multiple (Cartesian) dimensions, $f_1, f_2, \ldots$, or as a vector, $\boldsymbol{f} = (f_1 \; f_2 \; \cdots)^T$. The spectral domain corresponding to the Cartesian spatial domain is, likewise, a possibly infinite and continuous interval of the entire real line, or plane, etc. For a discrete subset of the spectral domain, the independent variable is indicated by an integer index, such as k. Again, the interval between adjacent frequencies of the discrete domain is a constant, Δf. Corresponding functions and sequences on the spectral domain, i.e., the *transforms* of the space-domain functions, are denoted by $G(f)$ and G_k, respectively.

The two types of functions on the Cartesian space domain, $g(x)$ and g_ℓ, are further divided into periodic and non-periodic functions and sequences, giving four types in total. Similarly, the spectral domain functions and sequences may be periodic or non-periodic. Specific additional assumptions for these functions are noted when appropriate; see also Section 1.3. Functions on the continuous domain form the historical, theoretical framework for spectral analysis, while the discrete analogues come closer to practical applications, since our data invariably are samples from a continuous domain. That is, the geodetic and geophysical signals of interest here generally are defined on a continuum and a solid underpinning of the theory for functions on such a domain is necessary in order to understand the consequences of sampling.

General space domain functions are denoted by lower-case roman letters and their transforms in the spectral domain by corresponding upper case letters. Exceptions occur for specific geophysical signals whose conventional notation is adopted where possible. Periodic functions or sequences include the over-script "~", as in $\tilde{g}$. Functions on the continuous domain are distinguished, unless the context makes it clear otherwise, by inclusion of the independent variable in parentheses, such as $g(x)$; while a sequence is denoted by the letter of the parent function with appropriate integer subscript, for example, g_ℓ. The same conventions hold for the frequency domain. Thus, in the space domain one has $g(x)$, $\tilde{g}(x)$, g_ℓ, and $\tilde{g}_\ell$; and, in the frequency domain, $G(f)$, G_k, $\tilde{G}(f)$, and $\tilde{G}_k$. As may be surmised already there are different types of transforms, even for just the Fourier kind, from the space domain to the spectral domain, depending on the nature of the function (the indicated correspondence of the transforms above is correct as proved in Chapters 2 and 4). When necessary the generic notations for the direct and inverse Fourier transforms are

$$G = \mathcal{F}(g), \; g = \mathcal{F}^{-1}(G). \tag{1.1}$$

For the spherical domain, spherical polar coordinates, θ and λ, are used and known geographically as co-latitude and longitude (Figure 1.1). In one dimension, on a great circle of the sphere, like a meridian, the coordinate is θ (also, ψ). The domains of these variables conventionally are $0 \leq \theta \leq \pi$ and $0 \leq \lambda \leq 2\pi$, but may be extended to infinite domains under appropriate interpretations. Assuming a unit radius for the sphere, the corresponding global Cartesian coordinates are

$$\begin{aligned} x &= \sin\theta\cos\lambda \\ y &= \sin\theta\sin\lambda \\ z &= \cos\theta \end{aligned} \tag{1.2}$$

Local Cartesian coordinates x_j, $j = 1, 2, 3$, are defined in the east-north-up directions (Figure 1.1).

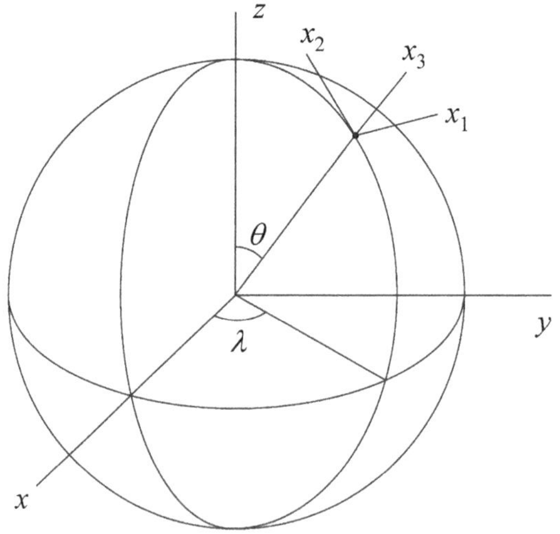

Figure 1.1: Spherical polar coordinates (θ, λ), co-latitude and longitude, for a point on a sphere. Positive coordinate axes, x and z, indicate the origins of the spherical coordinates. Local Cartesian coordinates are indicated by x_1, x_2, x_3 in the east, north, and up directions.

Implicitly, the letters $\{j, k, \ell, m, n, J, K, L, M, N\}$, are integers, unless otherwise noted; all other letters refer to real-valued or complex-valued quantities, again, unless indicated differently. Additional notation is introduced and defined as needed. Certain conventions regarding Fourier series and transforms are followed, always accompanied by clear definitions, because while they may be common they are not necessarily uniform in the literature.

1.2 Geophysical Motivation

It is worthwhile to reinforce the concepts of frequency and wavelength with a specific geophysical example before delving from a more theoretical viewpoint into the mathematics of spectral analysis. Although the accent here is on data in the spatial domain, time-domain signals abound in geophysics and geodesy, as well; and, they illustrate the power of spectral analysis for signals that intuitively contain amplitudes at specific frequencies. Thus, consider the motion of the Earth's spin axis relative to the Earth's crust as predicted in theory and as derived from astronomic observations of coordinates at terrestrial points. Assume for the moment that the spin axis is fixed in direction relative to the stars—it is not, but actually precesses, like a toy top, due to the gravitational torques exerted by the sun, moon, and planets on the equatorial bulge of the Earth that extends out of the plane of its orbit. The motion of the spin axis inside a rigid body is predicted by Euler's rotation equations (Moritz and Mueller 1987, Lambeck 1988). Therefore, astronomically determined latitude at a point fixed to the Earth, referenced to the spin axis (see Figure 1.1), changes in time as the spin axis moves relative to the Earth (which, of course, is not a desirable trait for a coordinate of a point that might serve as geodetic control for positioning). Conversely, knowledge of the constant latitude of a point in an Earth-fixed coordinate system then yields the motion of the spin axis in that coordinate system from the same type of observations. Defining local orthogonal axes, x_p, y_p, tangent to a sphere and with origin at the north pole, the coordinates of the spin axis describe *polar motion* as shown in Figure 1.2. (These are conventional coordinate notations designating directions, respectively, along the prime meridian (origin for longitudes) and 90° due west, or the x-axis and $-y$-axis in Figure 1.1. Moreover, the polar motion coordinates conventionally are given as angles in units of arcsecond, which may be converted to linear distance using Earth's polar radius of curvature, 6400 km.) Clearly visible is a periodic signal with period, or peak-to-peak interval, of about 1.2 years or 435 days. This "Chandler period", discovered by Seth Carlo Chandler in 1891 (Carter and Carter 2002) and geophysically explained by Simon Newcomb shortly thereafter, disagrees with the theoretical value of 304 days based on Euler's equations for a rigid Earth and thus gave proof to the fact that the Earth's interior is not rigid.

If the Chandler variation is removed from the total signal, the residual is still periodic, but now with a period of 1 year (peak-to-peak interval, Figure 1.2). This annual component is not immediately visible in the time-domain data, but as shown later it is distinct once the data are transformed to the spectral domain. The example thus points towards the utility of spectral analysis that in this case not only proved the non-rigidity of the Earth, but identifies also the main contributing components,

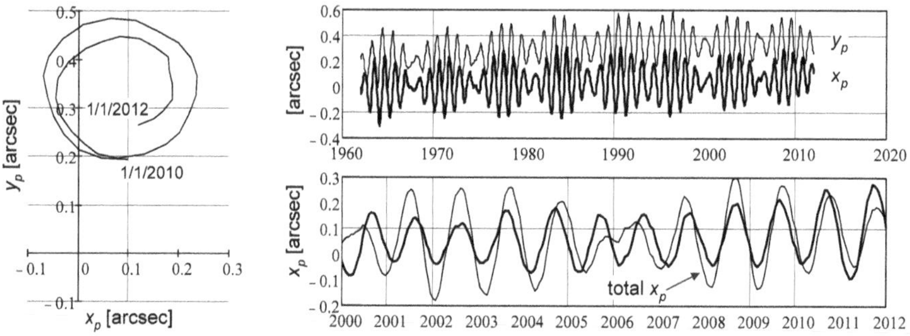

Figure 1.2: Polar motion coordinates for 2010.00–2012.00 (left), 2000.00–2012.00 (bottom right), and 1962.00–2012.00 (top right). The abscissas on the right represent calendar years. On the bottom right, for the x_p-coordinate over the interval 2000–2012, the light curve is the total polar motion; whereas, the dark curve has the Chandler component removed. One second of arc on the Earth's surface is about 30 m.

which can then be studied more easily to determine their cause (fluid outer core, oceans, and atmosphere); see (Wahr 1988, Cross 2000).

There are many other examples in geophysics that illustrate the time-variable, periodic components in natural phenomena. Perhaps the largest class of such signals comprises the elastic waves that travel through the Earth as a result of an earthquake. The field of seismology deals with these waves, as well as those induced by controlled explosions, to infer the geologic substructure of the Earth. Many other naturally occurring periodic signals are associated in one way or another with Earth's rotation and orbital motion around the sun (e.g., tidal signals and water cycles in large river basins); others are linked to the solar magnetic activity cycle; and still others are less predicable but still recurring, such as climate effects associated with ocean circulation.

Static signals on the spatial domain, such as the mean Earth's gravitational field (with all time-variable parts removed) usually do not exhibit predominant periods, or resonances, with respect to the spatial coordinates, but their analysis in the corresponding spectral domain is nonetheless informative. As the number of measurement systems increases, and with them the heterogeneity of data sets that provide a more comprehensive view of the Earth, it is important to relate the resolution and accuracy of the data to the scales of the signals being measured. Automated remote sensing system, such as radar altimetry, and *in situ* sensing systems, such as gravity gradiometry, on aircraft and satellites provide different data sets that ultimately might be combined or correlated in order to obtain a more complete representation of the Earth. A solid understanding of the spectral magnitudes of the various signals not only helps in designing the required level of accuracy and resolution in the data collection systems, but allows a more consistent combination of different types of data for a synergistic modeling. For example, it makes little sense to measure gravity at intervals of 10 m if the signal to be modeled, such as the geoid height (Chapter 6), has no significant features or amplitudes at that scale. Conversely, the data from a satellite orbiting the Earth at a speed of 7 km/s and providing a measurement of the gravity gradient every 10 s cannot resolve subsurface geologic structures with significant amplitudes at 1 km scales, no matter how accurate the data are. In this case, it is necessary to supplement the satellite data with gravity surveys at finer local detail. Although these examples may seem intuitively obvious, it is the intent of this book to formalize the analysis from the perspective of a signal's spectrum. A fluency in spectral thinking goes a long way to designing projects and interpreting data.

There is an additional, albeit rather mundane, benefit to transforming data and signals into their corresponding spectra. A significant computational advantage results when processing data in a convolution integral, ubiquitous in geodesy and geophysics (Chapter 6), by using the corresponding frequency-domain formulation. Although computational capability may still follow Moore's law of exponential growth (Moore 1965, Wikipedia 2014), meaning that convolutions are becoming easier to handle in the spatial domain, the amount of data is increasing as well ("big data"), and efficiencies are still needed in these types of computations.

1.3 Mathematical Preliminaries

Spectral analysis has its foundations with the French mathematicians Jean Baptiste Joseph Fourier (1768–1830) and Augustin-Louis Cauchy (1789–1857) (Titchmarsh 1948), and many more since then who have developed a solid mathematical theory that now finds application in all the sciences and beyond. The *Fourier transform*, explicitly defined in subsequent chapters, decomposes a function into its spectrum using sines and cosines, viewed as basis functions. The basis functions for the spherical domain are more complicated because of the particular geometry of the sphere, but the essential ideas of spectral analysis are the same, which can also be applied to functions on hypersurfaces or manifolds of higher dimensions. Moreover, the theory of Fourier analysis is applicable to a broad range of function spaces. It hinges on two main aspects—the transformation of a function to another function (the "transform"), and the representation of the original function by its transform, i.e., that the *inverse* transform exists and is equal to the original function. The first aspect does not always guarantee the second.

The mathematical treatment here is very much limited, with some exceptions, to well-behaved functions that are equal to their inverse transforms and we may assume that physical signals and data in geodesy and geophysics have their foundation in such functions. Moreover, it is no particular restriction to assume that our functions are real-valued, even though the analysis with complex-valued functions is straightforward. The spectrum for the Cartesian domain, on the other hand, is most conveniently expressed with complex numbers, even if the function is real-valued. It means that the basis functions also are complex, specifically the exponentials with imaginary argument, thus combining both sine and cosine basis functions, according to Euler's formula (equation (1.5), below). In transforming back, the complex spectrum combines appropriately with the complex basis function to yield the original real-valued function. A real-valued spectrum corresponding to sine and cosine basis functions can always be determined unambiguously from the complex-valued spectrum. For the spherical domain both real-valued and complex spectra have specific advantages, but the former is the usual representation. Both are developed in Section 2.6.2.

It is assumed that the reader has some familiarity with complex numbers; a very brief introduction follows. A complex number is the algebraic sum,

$$z = x + iy, \tag{1.3}$$

with "real" and "imaginary" parts, $x = \mathrm{Re}(z)$ and $y = \mathrm{Im}(z)$, that are real numbers, and where the prefix of y symbolizes the imaginary unit, $i = \sqrt{-1}$. It should be noted that there is nothing particularly mysterious about i. It simply separates real and imaginary parts of a complex number, but it is manipulated algebraically like any number and it multiplies the imaginary part in the usual sense of multiplication. For example, $i^2 = -1$, $i^3 = -i$, $1/i = -i$, etc. However, addition and subtraction are always done separately for the real and imaginary parts. The *complex conjugate*, denoted by z^*, is defined by replacing *every* occurrence of i in an expression of complex numbers by $-i$. Clearly, if z is a real number then $z = z^*$, and this is one way to verify that a more complicated expression is real. The square of the magnitude of a complex number is defined by

$$|z|^2 = z \cdot z^* = (x + iy)\,(x - iy) = x^2 - i^2y^2 = x^2 + y^2. \tag{1.4}$$

The reason that the Fourier spectrum is more easily expressed with complex numbers is embodied in *Euler's formula*,

$$e^{i\beta} = \cos\beta + i\sin\beta, \tag{1.5}$$

for any real number, β. This vastly simplifies formulas in Fourier spectral analysis by combining both the cosine and sine transforms, which most physical signals possess. The inverse to Euler's formula is given by

$$\cos\beta = \frac{1}{2}\left(e^{-i\beta} + e^{i\beta}\right) \quad \text{and} \quad \sin\beta = \frac{i}{2}\left(e^{-i\beta} - e^{i\beta}\right). \tag{1.6}$$

Equations (1.5) and (1.6) are readily proved by considering the infinite, uniformly converging Taylor series for the exponential and for the sine and cosine,

$$e^x = 1 + x + \frac{1}{2!}x^2 + \frac{1}{3!}x^3 + \frac{1}{4!}x^4 + \frac{1}{5!}x^5 + \ldots, \tag{1.7}$$

$$\sin x = x - \frac{1}{3!}x^3 + \frac{1}{5!}x^5 - \frac{1}{7!}x^7 + \ldots, \tag{1.8}$$

$$\cos x = 1 - \frac{1}{2!}x^2 + \frac{1}{4!}x^4 - \frac{1}{6!}x^6 + \ldots. \tag{1.9}$$

Note that from equation (1.5),

$$e^{i\pi k} = (-1)^k, \quad \text{for any integer, } k. \tag{1.10}$$

It is assumed throughout this text that the functions representing geodetic and geophysical signals are benign, or well-behaved, in the sense that they are generally continuous, although we may allow step discontinuities. Where needed for purposes of analysis, it is assumed that they are differentiable. Of course, one may argue that not all physical signals can satisfy these criteria; consider, for example, the topographic elevation at meter-level or smaller scales. But, our observations of geophysical signals are still at a level that permits their *models* to have the necessary properties that allow spectral analyses to proceed without additional mathematical justification. One notable exception concerns the interpretation, in Section 5.6, of a geophysical signal as the realization of a stochastic (random) process. Formally this requires the consideration of stochastic equivalents of the classical concepts of continuity, differentiation, and integration (Priestley 1981, Chapters 3 and 4). These are briefly noted where needed, but Fourier transforms of random processes are mostly restricted here to harmonic processes whose realizations are amenable to the usual operations in Riemanian calculus. The other exception concerns the Dirac delta function that should perhaps best be handled with the theory of distributions (e.g., Lighthill 1958). Again, this mathematical formality is omitted with the idea that such a function and operations with it can be viewed as a limiting process of an ordinary function. Aside from these complications the main mathematical difficulties of Fourier transforms arise because they involve infinite series and integrals with infinite limits, which always raise questions on convergence.

Nevertheless, rather simple mathematical conditions on functions exist that lead to their representation in terms of Fourier transforms. The conditions under which a function, $g(x)$, defined on a finite interval, $[a,b]$, may be represented by a convergent series of sines and cosines were formally established with a theorem due to P.G. Dirichlet (Boas 1966, p.294; Stein and Shakarchi 2003, p.128), known now as *Dirichlet's conditions*,

a) $g(x)$ is absolutely integrable,

$$\int_a^b \left| g(x) \right| dx < \infty; \tag{1.11}$$

b) $g(x)$ is piecewise continuous (having a finite number of jump- or step-discontinuities);

c) $g(x)$ is bounded; and,

d) $g(x)$ has a finite number of maxima and minima.

If a function has *bounded variation* (Papoulis 1977, p.94) or *limited total fluctuation*, that is, it is bounded and it does not oscillate too much, then it satisfies Dirichlet's conditions b), c), and d). For example, sin $(1/x)$ does not have bounded variation for $0 < x \le 1$. Dirichlet's theorem for the representation of a function by its Fourier series was recast by C. Jordan in terms of bounded variation (Titchmarsh 1939, p.406; Butzer and Nessel 1971, p.52; see also Lakatos 1976, p.148). Without limiting the applications, it may be assumed that Dirichlet's conditions hold for all geophysical signals.

While one can go far with absolutely integrable functions, it happens that square-integrability adds useful properties, especially when considering correlation of signals. A function, $g(x)$, is square-integrable on the finite interval, $[a,b]$, if

$$\int_a^b \left| g(x) \right|^2 dx < \infty. \tag{1.12}$$

On a finite interval, such as this, the function space of square-integrable functions is not as large as the space of absolutely integrable functions; for example, $g(x) = 1/\sqrt{x}$ defined on $(0,1)$ is absolutely integrable, but not square-integrable. However, there is no loss of applicability by assuming that geophysical signals are not only absolutely integrable, but also square integrable. This also allows us to view signals in terms of their *energy*, which is defined by an integration of squares.

A key aspect of Fourier transforms is the decomposition of a function into basis functions that are *orthogonal*. Analogous to the orthogonality of two vectors, manifested by their vanishing inner product, two complex functions are orthogonal on an interval, $[a,b]$, if the integral of their product is zero. Let $\{\eta_j(x)\}, j = 1,\ldots,\infty$, be a set of possibly complex functions in the space of square-integrable functions that satisfy

$$\langle \eta_j, \eta_k \rangle = \int_a^b (\eta_j(x))^* \, \eta_k(x) dx = \delta_{j-k}, \tag{1.13}$$

where $\langle \cdot,\cdot \rangle$ is another notation for "inner product" of functions, and

$$\delta_m = \begin{cases} 0, & m \neq 0 \\ 1, & m = 0 \end{cases} \tag{1.14}$$

is the *Kronecker delta*. Then $\{\eta_j(x)\}$ is a set of orthogonal functions; indeed, they are *orthonormal*, since their magnitude, $\sqrt{\langle \eta_j, \eta_j \rangle} = 1$. They also form a *complete basis* for square-integrable functions, $g(x)$, on $[a,b]$ if the partial sums, $\sum_{j=1}^{n} c_j \eta_j(x)$, converge to $g(x)$ in the sense of a "limit in the mean,"

$$\lim_{n \to \infty} \int_a^b \left| g(x) - \sum_{j=1}^{n} c_j \eta_j(x) \right|^2 dx = 0, \tag{1.15}$$

where

$$c_j = \int_a^b \eta_j^*(x) \, g(x) dx \tag{1.16}$$

(e.g., Cushing 1975, p.142). For example, the sines and cosines of multiple angles, $k\beta$, or written according to equations (1.6) as polynomials in the variable, $e^{i\beta}$, are orthogonal function on an interval (Section 2.2). And, it turns out that they constitute one option for a complete basis of square-integrable functions on $[a,b]$. Indeed, equation (1.16) is the Fourier transform of a function if the basis functions, $\eta_j(x)$, are the sines and cosines, or the complex exponentials, equation (1.5).

For infinite intervals, $(-\infty, \infty)$, the situation in some sense is more difficult, as the basis for the function space is not countable, i.e., its elements are not associated with an integer. However, sufficient conditions for functions to be representable by correspondingly defined Fourier transforms are, again, Dirichlet's conditions, plus absolute integrability (e.g., Boas 1966, p.602),

$$\int_{-\infty}^{\infty} \left| g(x) \right| dx < \infty. \tag{1.17}$$

This is generally a more stringent requirement than being integrable over a finite domain and means that $\lim_{x \to \pm\infty} g(x) = 0$. The condition of square-integrability,

$$\int_{-\infty}^{\infty} \left| g(x) \right|^2 dx < \infty, \tag{1.18}$$

unlike the case for finite intervals, is a weaker condition than the inequality (1.17) (e.g., $g(x) = 1/x$ and $g(x) = \sin(x)/x$ are square integrable, but not absolutely integrable on $(-\infty, \infty)$). And, the theory of Fourier transforms can be built on this condition, as well, attributed to M. Plancherel (Titchmarsh 1948).

The mathematical theory of the Fourier transform of functions defined on finite or infinite intervals $(-\infty, \infty)$ is extensive, and the interested reader is referred to such classic texts as (Titchmarsh 1948, Butzer and Nessel 1971, Priestley 1981), and (Brillinger 2001); a brief review is also found in (Chui 1992, Ch.2). The general development of Fourier theory from first principles is not pursued here and it is assumed that the function representing a geophysical signal has bounded variation and satisfies *both* integrability conditions (1.17) and (1.18). However, certain individual important functions appear that do not satisfy these conditions, and yet Fourier transforms can be found that also represent them. Chapter 2 gives a rudimentary exposition of Fourier transforms for functions on a continuous domain, whether finite or infinite.

When dealing with sums and integrals having infinite limits, it is often necessary in deriving certain relationships to interchange the order of combinations of such sums and integrals, or to interchange limits and integrals or sums. The following theorems are standard material in most text books on real analysis (e.g., Goldberg 1964, Berberian 1999). For iterated integrals, the result goes variously by the name of *Fubini's theorem* or the *Fubini-Tonelli theorem* and states that

$$\iint_{U\times V} \left| g(u,v) \right| d(u,v) < \infty \Rightarrow \iint_{U\times V} g(u,v)\, d(u,v) = \int_U \left(\int_V g(u,v)\, dv \right) du = \int_V \left(\int_U g(u,v)\, du \right) dv, \tag{1.19}$$

where U and V are the possibly infinite domains of the variables, u and v, respectively. For infinite sums, there is also

$$\sum_{j=-\infty}^{\infty} \sum_{k=-\infty}^{\infty} \left| g_j h_k \right| < \infty \Rightarrow \sum_{j=-\infty}^{\infty} \sum_{k=-\infty}^{\infty} g_j h_k = \sum_{j=-\infty}^{\infty} g_j \sum_{k=-\infty}^{\infty} h_k = \sum_{k=-\infty}^{\infty} h_k \sum_{j=-\infty}^{\infty} g_j. \tag{1.20}$$

Similar statements hold for mixed sums and integrals. In addition, it is often necessary to take infinite limits inside infinite integrals (or sums). This is possible with the *Lebesgue Dominated Convergence Theorem*,

$$\lim_{n\to\infty}\int_{-\infty}^{\infty} g_n(x)\,dx = \int_{-\infty}^{\infty} \lim_{n\to\infty} g_n(x)dx, \tag{1.21}$$

assuming the limits and integrals exist.

Another powerful result for infinite integrals and sums is *Schwarz's inequality* (Kaplan 1973), which states that for functions, g and h, square-integrable on the domain, U (finite or infinite),

$$\left|\int_U g(u)\,h(u)\,du\right|^2 \le \int_U |g(u)|^2\,du \int_U |h(u)|^2\,du. \tag{1.22}$$

Again, the analogous result for summations is

$$\left|\sum_j g_j h_j\right|^2 \le \sum_j |g_j|^2 \sum_j |h_j|^2. \tag{1.23}$$

where the summation index may range over a finite or the infinite set of integers.

1.4 Summary

This text develops the background of Fourier analysis, considering first periodic and non-periodic functions on the line and on the plane, then moving to the more practical aspects of dealing with corresponding discrete functions, or sequences, both infinitely and finitely extended. Special emphasis is devoted to functions on the sphere, the natural domain for global applications in geodesy and geophysics. The independent variable almost exclusively is the spatial coordinate, rather than time that is commonly used in most textbooks on spectral analysis. This is not to say that temporal signals are uncommon in geophysics (or geodesy), but is meant to distinguish the present text from such excellent books that focus on the temporal aspects, especially in seismology (e.g., Bath 1974, Meskó 1984, Buttkus 2000). As such the principal direction of the theory is toward applications in geopotential fields, which also have a temporal component, but are viewed primarily as static signals varying in space.

It is assumed that all relevant geophysical signals satisfy the needed properties that allow a Fourier spectral analysis. There are exceptions to these adopted precepts for particular functions and they are noted and dealt with at the appropriate juncture in the discussions. Although the theory of Fourier analysis is well developed for Lebesgue-integrable functions, the classical Riemann integrals are assumed here, following Chui's (1992, p.23) assurance that the "sacrifice is small" in doing so for our applications.

Chapter 2

Fourier Transforms of Functions on the Continuous Domain

2.1 Introduction

The basic concepts of spectral analysis through Fourier transforms typically are developed for functions on a one-dimensional domain where the independent variable is time. On the other hand, the domain of many geodetic and geophysical signals is the sphere that approximates the Earth, which in some applications may also be approximated locally by a plane. In order to extend the spectral concepts to these more general domains, this chapter first describes the Fourier transform in one dimension, using the independent variable, x, to emphasize and prepare for a multi-dimensional Cartesian domain. Once firmly established in one dimension, the Fourier analysis generalizes with no effort to higher dimensions, specifically the plane. The adaptation to the spherical domain is less trivial and requires a change in basis functions from sines and cosines to Legendre and sinusoidal functions. The general assumptions of geophysical signals that allow a spectral analysis on the plane and sphere, discussed in Section 1.3, are reiterated in subsequent sections in a more specific context.

2.2 Fourier Series

Consider first a real function defined on the finite interval, $0 \le x < P$, on the real line and assume that it is extended periodically for $-\infty < x < \infty$. Thus, let $\tilde{g}(x)$ be a real periodic function, with period, P,

$$\tilde{g}(x \pm nP) = \tilde{g}(x), \quad n = 0,1,2,\ldots . \tag{2.1}$$

It is assumed that $\tilde{g}(x)$ is absolutely integrable, equation (1.11), so that one may form the constants,

$$a_k = \frac{2}{P}\int_0^P \tilde{g}(x)\cos\left(\frac{2\pi}{P}kx\right)dx, \qquad b_k = \frac{2}{P}\int_0^P \tilde{g}(x)\sin\left(\frac{2\pi}{P}kx\right)dx, \qquad k \geq 0. \tag{2.2}$$

Under certain conditions the sum of all these coefficients, each multiplied by a corresponding sine or cosine function, reproduces the function, $\tilde{g}(x)$. Specifically, if $\tilde{g}(x)$ is continuous and has bounded variation on the interval, $[0, P]$ (Section 1.3), then

$$\tilde{g}(x) = \frac{1}{2}a_0 + \sum_{k=1}^{\infty} a_k \cos\left(\frac{2\pi}{P}kx\right) + \sum_{k=1}^{\infty} b_k \sin\left(\frac{2\pi}{P}kx\right), \quad -\infty < x < \infty. \tag{2.3}$$

Thus, not only does the infinite series converge, it converges uniformly to $\tilde{g}(x)$. It is also possible to allow piecewise continuity of $\tilde{g}(x)$ (e.g., $\tilde{g}(P) \neq \tilde{g}(0)$ may be allowed), but then the series converge to the average of $\tilde{g}(x)$ on either side of a step discontinuity (Titchmarsh 1939, p.406; Sneddon 1961, p.26; Cushing 1975, p.149),

$$\frac{1}{2}\lim_{\delta x \to 0+}\left(\tilde{g}(x-\delta x) + \tilde{g}(x+\delta x)\right) = \frac{1}{2}a_0 + \sum_{k=1}^{\infty} a_k \cos\left(\frac{2\pi}{P}kx\right) + \sum_{k=1}^{\infty} b_k \sin\left(\frac{2\pi}{P}kx\right). \tag{2.4}$$

where "$\delta x \to 0+$" means that the limit to zero is approached with δx always positive. More general convergence theorems for more general functions have been developed, but for practical purposes, we may limit the present exposition to the case corresponding to the stated assumptions (see also Chapter 1).

Equation (2.3) (or (2.4)) for $\tilde{g}(x)$ is called its *Fourier series* representation and a_k, b_k are called *Fourier coefficients*. Later they are also identified as the spectrum of the function, $\tilde{g}(x)$. The sinusoidal functions are orthogonal over the interval, $[0, P]$, which means that for integers, $k, \ell \geq 0$,

$$\int_0^P \sin\left(\frac{2\pi k}{P}x\right)\cos\left(\frac{2\pi \ell}{P}x\right)dx = 0; \tag{2.5}$$

$$\int_0^P \sin\left(\frac{2\pi k}{P}x\right)\sin\left(\frac{2\pi \ell}{P}x\right)dx = \frac{P}{2}\left(\delta_{k-\ell} - \delta_{k+\ell}\right); \tag{2.6}$$

$$\int_0^P \cos\left(\frac{2\pi k}{P}x\right)\cos\left(\frac{2\pi \ell}{P}x\right)dx = \frac{P}{2}\left(\delta_{k-\ell} + \delta_{k+\ell}\right); \tag{2.7}$$

where $\delta_{k-\ell}$ is the Kronecker delta, equation (1.14). Equations (2.5) through (2.7) can be proved with integration by parts, or more easily using the complex exponential forms, equations (1.6); see also equation (2.12). Multiplying both sides of the series expression, equation (2.3), by either $\cos(2\pi\ell x/P)$ or $\sin(2\pi\ell x/P)$, interchanging summation and integration (permitted because the series converge uniformly for all x), and making use of their orthogonality, the coefficients, a_k, b_k, are again given by equations (2.2).

An example of a periodic function represented by a Fourier series is illustrated in Figure 2.1. In this case the series are finite (all but a finite number of the coefficients are zero); and, it is noted that necessarily, $\tilde{g}(0) = \tilde{g}(P)$, because a finite series of continuous functions is continuous. In the case of the saw-tooth function, $\tilde{g}(x) = 0.5\ \mathrm{mod}_P(x)$ (Figure 2.2), where $\tilde{g}(0) \neq \lim_{x \to P} \tilde{g}(x)$, the series converges in accordance with equation (2.4) to half its maximum amplitude at $x = nP$. This is easily checked by computing the Fourier coefficients and evaluating the series (2.3), for example, at $x = P$, which turns out to be $0.5P/2 = 5$ (Exercise 2.1).

The arguments of the sines and cosines in the series expression (2.3) for $\tilde{g}(x)$ are proportional to the quantities, k/P. These are the harmonic *frequencies* of the function, where the *fundamental frequency* is $1/P$ ($k = 1$). The term "frequency", usually encountered with time-varying signals, is used here in connection with

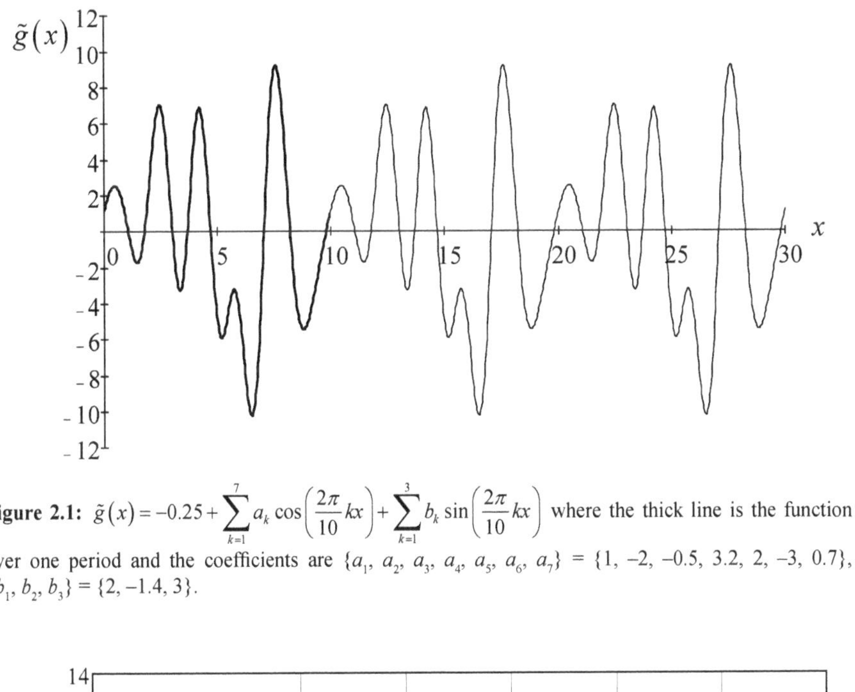

Figure 2.1: $\tilde{g}(x) = -0.25 + \sum_{k=1}^{7} a_k \cos\left(\frac{2\pi}{10}kx\right) + \sum_{k=1}^{3} b_k \sin\left(\frac{2\pi}{10}kx\right)$ where the thick line is the function over one period and the coefficients are $\{a_1, a_2, a_3, a_4, a_5, a_6, a_7\} = \{1, -2, -0.5, 3.2, 2, -3, 0.7\}$, $\{b_1, b_2, b_3\} = \{2, -1.4, 3\}$.

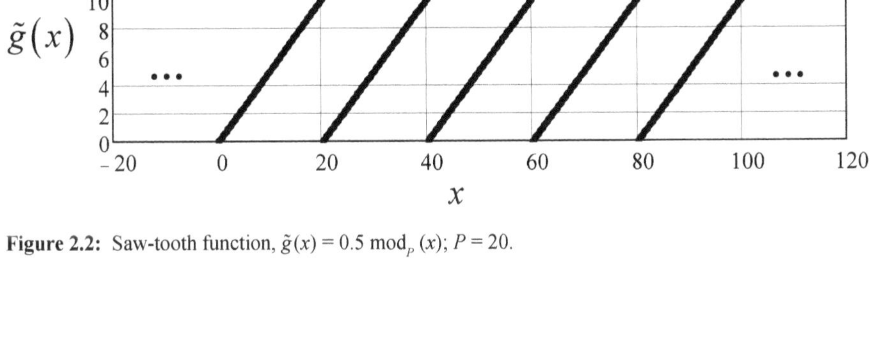

Figure 2.2: Saw-tooth function, $\tilde{g}(x) = 0.5\ \mathrm{mod}_P(x)$; $P = 20$.

functions depending on spatial variables, and in geodetic or geophysical texts it is also called "spatial frequency". Another way to identify a constituent of the series is by *wave number* (referring to the integer, k). Here, frequency and wave number always refer to functions on a spatial domain unless otherwise indicated.

It is clear from the preceding development that the use of sines and cosines adds a level of inefficiency in notation that may be alleviated by combining both sinusoids into a single function using Euler's formula, equation (1.5). And, indeed, one may write the Fourier series of $\tilde{g}(x)$, equation (2.3), as

$$\tilde{g}(x)=\frac{1}{P}\sum_{k=-\infty}^{\infty}G_k e^{i\frac{2\pi}{P}kx}, \tag{2.8}$$

where the Fourier coefficients, G_k, in general are complex numbers. This expansion holds also for complex-valued periodic functions, but for real functions, as assumed here, the complex Fourier coefficients combine with the complex basis functions, $e^{i2\pi kx/P}$, to produce real values on the left side. At points of discontinuity, the series converges as indicated in equation (2.4),

$$\frac{1}{2}\lim_{\delta x\to 0+}\left(\tilde{g}(x-\delta x)+\tilde{g}(x+\delta x)\right)=\frac{1}{P}\sum_{k=-\infty}^{\infty}G_k e^{i\frac{2\pi}{P}kx}. \tag{2.9}$$

The fact that the series requires both positive and negative frequencies, or wave numbers, comes from the combination of the sine and cosine coefficients and should not cause undue concern. Later it is shown that for real functions, the coefficient with negative wave number is simply the complex conjugate of the coefficient with positive wave number. From equations (1.5) and (1.6), it is straightforward to find the corresponding relationship between the coefficients, $\{a_k, b_k\}$, in equation (2.3) and the coefficients, $\{G_k\}$, in equation (2.8),

$$G_k=\frac{P}{2}\begin{cases} a_k-ib_k, & k>0\\ a_0, & k=0\\ a_{-k}+ib_{-k}, & k<0\end{cases} \tag{2.10}$$

The inverse relationship is given by

$$\begin{aligned} a_k&=\frac{1}{P}\left(G_k+G_{-k}\right), \quad k\ge 0\\ b_k&=\frac{i}{P}\left(G_k-G_{-k}\right), \quad k>0\end{aligned} \tag{2.11}$$

The orthogonality of the sinusoids, equations (2.5) through (2.7), also condenses to

$$\int_0^P e^{i\frac{2\pi}{P}(k-\ell)x}\,dx=P\delta_{k-\ell}, \tag{2.12}$$

which is easily proved by substituting equation (1.5) and integrating. The antecedent factor, $1/P$, in equation (2.8) makes the units of G_k consistent with later transforms of non-periodic functions. Thus, the units of G_k are the units of $\tilde{g}(t)$ per frequency unit, since the units of P are inverse to those of frequency.

The spectral domain in this case of periodic functions constitutes the set of frequencies, k/P, or wave numbers, k. It is a *discrete*, but countably infinite set, which is another way of saying that the set of basis functions is discrete and countably infinite, or that the function space of $\tilde{g}$ is *separable*. The set of coefficients, $\{a_k, b_k\}$ or $\{G_k\}$, is known more precisely as the *Fourier spectrum* of $\tilde{g}$. It is also the *Fourier transform* of $\tilde{g}$. The Fourier transform terminology is used generally for other types of functions, as well (e.g., non-periodic functions). Thus, to be more specific in this case, or where it could lead to confusion, one should say that it is the Fourier *series* transform. Although the frequency domain is discrete, this Fourier transform should not be confused with the *discrete* Fourier transform that refers to discrete periodic functions for which the frequency domain is also finite (Section 4.3.1).

If $\tilde{g}$ is given, its spectrum formally is obtained using the orthogonality, equation (2.12). Multiplying equation (2.8) on both sides by $e^{-i2\pi \ell x/P}$ and integrating yields

$$G_k = \int_0^P \tilde{g}(x) e^{-i\frac{2\pi}{P}kx} dx. \tag{2.13}$$

Since $e^{-i2\pi kx/P}$ is periodic with period, P, these coefficients could be determined over *any* interval of length, P, which is easily proved with a change in the integration variable, $x \to x + a$, for any real constant, a. The Fourier spectrum may also be written in polar form in terms of amplitude and phase,

$$G_k = A_k e^{i\phi_k}, \tag{2.14}$$

where the *amplitude spectrum* is

$$A_k = |G_k|, \tag{2.15}$$

and the *phase spectrum* is

$$\phi_k = \tan^{-1} \frac{\mathrm{Im}(G_k)}{\mathrm{Re}(G_k)}. \tag{2.16}$$

These components are discussed further for the continuous spectrum in Section 2.3.

In summary, a function, $\tilde{g}(x)$, is given on a finite interval defined without loss in generality by $[0, P)$ and satisfies the Dirichlet conditions. Considering the entire real line, it is assumed that $\tilde{g}(x)$ is extended periodically on $(-\infty, \infty)$ so that $\tilde{g}(x+mP) = \tilde{g}(x)$ for $x \in [0, P)$, where m is any integer. Then one can find its Fourier spectrum according to equation (2.13). Conversely, given its spectrum, G_k, the function, $\tilde{g}(x)$, is obtained by equation (2.8) (except at points of discontinuity). Determining the Fourier coefficients is also termed an *analysis* of $\tilde{g}(x)$; and, if the spectrum is given then the reconstruction of the function is called the *synthesis* of $\tilde{g}(x)$. If the function is continuous over its period, with $\tilde{g}(0) = \tilde{g}(P)$, then it and its Fourier spectrum are *dual* representations of the same information—one is equivalent to the other. The spectrum of a function displays the same information, but in a different domain—the *frequency domain*, where it is often more useful than in the space- (or time-) domain. The two relationships, (2.13) and (2.8), constitute a *Fourier (series) transform pair* for continuous periodic functions, denoted as

$$\tilde{g}(x) \leftrightarrow G_k. \tag{2.17}$$

2.2.1 Properties of the Fourier Series Transform

The Fourier series transform pair (2.17) may be manipulated using several linear operations whose results are summarized with the following (non-exhaustive) list of properties. They can be proved with relative ease (Exercise 2.2) from the basic transform pair, repeated here for convenience,

$$\tilde{g}(x) = \frac{1}{P}\sum_{k=-\infty}^{\infty} G_k e^{i\frac{2\pi}{P}kx}, \quad G_k = \int_0^P \tilde{g}(x) e^{-i\frac{2\pi}{P}kx}\, dx. \tag{2.18}$$

The generalization for points of discontinuity, equation (2.9), may be used when needed. It is noted that while we assume only real-valued functions, $\tilde{g}$, all definitions and properties hold equally for complex-valued functions.

1. Fourier transform pair: $\tilde{g}(x) \leftrightarrow G_k$; (2.19)
2. Proportionality: $a\tilde{g}(x) \leftrightarrow aG_k$, a is any constant; (2.20)
3. Superposition: $\tilde{g}_1(x) + \tilde{g}_2(x) \leftrightarrow (G_1)_k + (G_2)_k$, provided $\tilde{g}_1$ and $\tilde{g}_2$ have the same period, P; (2.21)
4. Symmetry: $\tilde{g}(-x) \leftrightarrow G_{-k}$; (2.22)
5. Translation in x: $\tilde{g}(x + x_0) \leftrightarrow G_k e^{i\frac{2\pi}{P}kx_0}$; (2.23)
6. Translation in k: $\tilde{g}(x) e^{-i\frac{2\pi}{P}k_0 x} \leftrightarrow G_{k+k_0}$; (2.24)
7. Differentiation: $\frac{d^p}{dx^p}\tilde{g}(x) \leftrightarrow \left(i\frac{2\pi k}{P}\right)^p G_k$, provided $\tilde{g}$ is differentiable at x up to order, p; (2.25)

If $\tilde{g}$ is an even function, then so is its Fourier spectrum, and vice versa,

$$\tilde{g}(-x) = \tilde{g}(x) \quad \text{if and only if} \quad G_{-k} = G_k, \tag{2.26}$$

which follows from Fourier transform pairs (2.19) and (2.22). If $\tilde{g}$ is a real-valued function (as in all our applications), then its spectrum for negative wave numbers is the conjugate mirror image of its spectrum for positive wave numbers (the spectrum is *Hermitian*); the converse is also true:

$$\tilde{g}(x) = \tilde{g}^*(x) \quad \text{if and only if} \quad G_{-k} = G_k^*. \tag{2.27}$$

Similarly, Hermitian functions have real-valued spectra,

$$\tilde{g}(-x) = \tilde{g}^*(x) \quad \text{if and only if} \quad G_k = G_k^*. \tag{2.28}$$

Properties (2.27) and (2.28) are evident immediately by taking the complex conjugate of equations (2.18) (Exercise 2.3). Combining properties (2.26) and (2.27), if $\tilde{g}$ is both real and even, then so is its spectrum, and conversely,

$$\tilde{g}(x) = \tilde{g}^*(x) \text{ and } \tilde{g}(-x) = \tilde{g}(x) \quad \text{if and only if} \quad G_k = G_k^* \text{ and } G_{-k} = G_k. \tag{2.29}$$

From equation (2.13), one sees that the spectrum at zero frequency is proportional to the average of the function over one period,

$$G_0 = \int_0^P \tilde{g}(x)dx. \tag{2.30}$$

Since the Fourier series converges, it is also evident from the property of convergent power series that the spectrum must attenuate to zero as the frequency or wave number increases,

$$\lim_{k\to\pm\infty} |G_k| = 0. \tag{2.31}$$

This is also known as the *Riemann-Lebesgue Theorem* (e.g., Titchmarsh 1939, p.403). If the derivative, $d\tilde{g}(x)/dx$, exists then its Fourier spectrum, from property (2.25), is $i2\pi kG_k/P$, which must attenuate to zero if the derivative is absolutely integrable. In this case, the Fourier coefficients, G_k, attenuate sufficiently quickly so that also the series of coefficients is absolutely summable,

$$\sum_{k=-\infty}^{\infty} |G_k| < \infty; \tag{2.32}$$

that is, the Fourier series converges absolutely (Davis 1975, p.293).

We assume also, as indicated in Section 1.3, that the function, $\tilde{g}(x)$, is square-integrable on the interval, $[0, P]$,

$$\int_0^P |g(x)|^2\, dx < \infty. \tag{2.33}$$

Then, the functions, $e^{i2\pi kx/P}$, $-\infty < k < \infty$, form a complete basis for such functions. In fact, the partial sum,

$$\hat{\tilde{g}}(x) = \frac{1}{P}\sum_{k=-N/2}^{N/2-1} G_k e^{i\frac{2\pi}{P}kx}, \tag{2.34}$$

is the best linear approximation of $\tilde{g}(x)$ in terms of these basis functions. To prove this, let the *inner product* of functions, $g(x)$ and $h(x)$, be denoted (as in equation (1.13)) by

$$\langle \tilde{g}, \tilde{h} \rangle = \int_0^P \tilde{g}(x) \tilde{h}^*(x)\, dx, \tag{2.35}$$

and the "length", or *norm*, of a function, analogous to the length of a vector, by

$$\|\tilde{g}(x)\| = \sqrt{\langle \tilde{g}, \tilde{g} \rangle}. \tag{2.36}$$

The inner product is linear in the sense that

$$\langle \tilde{g}_1 + a\tilde{g}_2, \tilde{h} \rangle = \langle \tilde{g}_1, \tilde{h} \rangle + a\langle \tilde{g}_2, \tilde{h} \rangle, \tag{2.37}$$

for any constant, a. Equations (2.13) and (2.12) with the inner product notation become

$$G_k = \left\langle \tilde{g}(x), e^{i\frac{2\pi}{P}kx} \right\rangle, \ \left\langle e^{i\frac{2\pi}{P}kx}, e^{i\frac{2\pi}{P}k'x} \right\rangle = P\delta_{k-k'}. \tag{2.38}$$

For a finite linear combination of the basis functions with arbitrary coefficients, $\hat{G}_k$, viewed as an approximation of $\tilde{g}$,

$$\hat{g}(x) = \frac{1}{P} \sum_{k=-N/2}^{N/2-1} \hat{G}_k e^{i\frac{2\pi}{P}kx}, \tag{2.39}$$

it is desired to determine the coefficients, $\hat{G}_k$, that minimize the error of approximation, $\hat{g}(x) - \tilde{g}(x)$. The square of the norm of this error is given by

$$\begin{aligned}
\|\hat{g}(x) - \tilde{g}(x)\|^2 &= \left\langle \frac{1}{P} \sum_{k=-N/2}^{N/2-1} \hat{G}_k e^{i\frac{2\pi}{P}kx} - \tilde{g}(x), \frac{1}{P} \sum_{k=-N/2}^{N/2-1} \hat{G}_k e^{i\frac{2\pi}{P}kx} - \tilde{g}(x) \right\rangle \\
&= \frac{1}{P^2} \sum_{k=-N/2}^{N/2-1} \sum_{k'=-N/2}^{N/2-1} \hat{G}_k \hat{G}_{k'}^* \left\langle e^{i\frac{2\pi}{P}kx}, e^{i\frac{2\pi}{P}k'x} \right\rangle - \frac{1}{P} \sum_{k=-N/2}^{N/2-1} \hat{G}_k \left\langle e^{i\frac{2\pi}{P}kx}, \tilde{g}(x) \right\rangle \\
&\quad - \frac{1}{P} \sum_{k=-N/2}^{N/2-1} \hat{G}_k^* \left\langle \tilde{g}(x), e^{i\frac{2\pi}{P}kx} \right\rangle + \langle \tilde{g}(x), \tilde{g}(x) \rangle
\end{aligned} \tag{2.40}$$

where the linearity of the inner product, equation (2.37), is used. Now with equations (2.38) and $\langle \tilde{g}(x), \tilde{h}(x) \rangle = \langle \tilde{h}(x), \tilde{g}(x) \rangle^*$, one finds

$$\|\hat{g}(x) - \tilde{g}(x)\|^2 = \frac{1}{P} \sum_{k=-N/2}^{N/2-1} |\hat{G}_k|^2 - \frac{1}{P} \sum_{k=-N/2}^{N/2-1} \hat{G}_k G_k^* - \frac{1}{P} \sum_{k=-N/2}^{N/2-1} \hat{G}_k^* G_k + \langle \tilde{g}(x), \tilde{g}(x) \rangle. \tag{2.41}$$

Adding and subtracting $\frac{1}{P}\sum_{k=-N/2}^{N/2-1} G_k^* G_k$ yields,

$$\left\|\tilde{g}(x)-\breve{g}(x)\right\|^2 = \left\langle \tilde{g}(x),\tilde{g}(x)\right\rangle - \frac{1}{P}\sum_{k=-N/2}^{N/2-1} G_k^* G_k + \frac{1}{P}\sum_{k=-N/2}^{N/2-1}\left|G_k-\widehat{G}_k\right|^2. \tag{2.42}$$

Only the last term depends on the unknown coefficients; and, the error is least only if one sets this term to zero, which happens only if

$$\widehat{G}_k = G_k. \tag{2.43}$$

Therefore, the best approximation of a function on a finite interval in terms of a finite series of Fourier basis functions is the series with the Fourier series transform for its coefficients. A similar result holds for any orthogonal set of basis functions (Davis 1975).

Since with equations (2.43) the approximation error goes to zero as $N \to \infty$, equation (2.42) results in a form of *Parseval's theorem*, which holds in general for square-integrable complex-valued functions (Titchmarsh 1939, p.424),

$$\int_0^P \left|\tilde{g}(x)\right|^2 dx = \frac{1}{P}\sum_{k=-\infty}^{\infty}\left|G_k\right|^2. \tag{2.44}$$

This also shows that, in view of equation (1.12),

$$\sum_{k=-\infty}^{\infty}\left|G_k\right|^2 < \infty. \tag{2.45}$$

The importance of Parseval's theorem lies in the interpretation of the left side as the total *energy* of the function (over one period), which according to the right side is also distributed over its spectrum. For example, the energy at frequency, $|k|/P$, is $(|G_k|^2 + |G_{-k}|^2)/P$. It is another consequence of the fact that the (continuous) function over the period, P, and its Fourier spectrum contain identical information, only expressed in different domains. If $\tilde{g}_1$ and $\tilde{g}_2$ have the same period, P, Parseval's theorem may be generalized (sometimes known as *Plancherel's theorem*),

$$\int_0^P \tilde{g}_1(x)\tilde{g}_2^*(x)\,dx = \frac{1}{P}\sum_{k=-\infty}^{\infty}(G_1)_k (G_2)_k^*, \tag{2.46}$$

obtained by substituting equation (2.8) on the left side,

$$\frac{1}{P^2}\int_0^P\left(\sum_{k=-\infty}^{\infty}(G_1)_k e^{-i\frac{2\pi}{P}kx}\right)\left(\sum_{k'=-\infty}^{\infty}(G_2)_{k'}^* e^{i\frac{2\pi}{P}k'x}\right)dx = \frac{1}{P^2}\sum_{k=-\infty}^{\infty}\sum_{k'=-\infty}^{\infty}(G_1)_k (G_2)_{k'}^*\int_0^P e^{i\frac{2\pi}{P}(k'-k)x}\,dx. \tag{2.47}$$

Equation (2.46) then follows immediately by the orthogonality, equation (2.12), and by noting that the sum converges on account of Schwarz's inequality (1.23),

$$\left|\sum_{k=-\infty}^{\infty}(G_1)_k(G_2)_k^*\right|^2 \le \sum_{k=-\infty}^{\infty}\left|(G_1)_k\right|^2\sum_{k=-\infty}^{\infty}\left|(G_2)_k\right|^2 < \infty. \tag{2.48}$$

Parseval's theorem or its generalized form comes in many flavors depending on the type of function. The universal terminology of "Parseval's theorem" is adopted in this text for each of these.

2.3 The Fourier Integral

For many geophysical applications, the signals in the spatial domain are not endowed with a fundamental period and the preceding development of Fourier series has perhaps only limited intuitive appeal. On the other hand, it should be noted that if the domain of definition of the function is a finite interval, as may be dictated in practice, then it could be represented as a Fourier series by assuming simply that the length of the domain is the period. Complications arise, however, when dealing with multiple signals that are defined on domains of different lengths. It is common, therefore, as in time-domain applications, to expand the theory to non-periodic functions where the domain is the entire real line (or plane, Section 2.4.2).

One may think of extending the primary interval for periodic functions, say $[-P/2, P/2)$, to the domain, $(-\infty, \infty)$. Again, we assume that $g(x)$ is both absolutely and square-integrable,

$$\int_{-\infty}^{\infty}\left|g(x)\right|dx < \infty, \tag{2.49}$$

$$\int_{-\infty}^{\infty}\left|g(x)\right|^2 dx < \infty. \tag{2.50}$$

However, certain individual functions are also considered that do not satisfy these conditions if Fourier transforms can be found that represent them. The most pathological case is the Dirac delta function (Section 2.3.4), which is not bounded, but can be represented by a Fourier transform. On the other hand, at this stage, periodic functions are excluded, since they clearly are not absolutely nor square-integrable on $(-\infty, \infty)$; although, one can get around this restriction in a formal sense (see below).

From equations (2.8) and (2.13) for the equivalent interval, $[-P/2, P/2)$, one may write

$$\tilde{g}(x) = \frac{1}{P}\sum_{k=-\infty}^{\infty}\int_{-P/2}^{P/2}\tilde{g}(x')e^{-i\frac{2\pi}{P}k(x'-x)}dx'. \tag{2.51}$$

Setting $f = k/P$ with $1/P = \delta f$ and taking the limit, $P \rightarrow \infty$, the frequency, f, changes from being countably infinite to continuous over the entire real line, and the sum becomes an integral (in the Riemannian sense),

$$g(x) = \int_{-\infty}^{\infty}\left(\int_{-\infty}^{\infty} g(x')e^{-i2\pi f x'}dx'\right)e^{i2\pi f x}df, \tag{2.52}$$

which is also known as the *Fourier integral equation* or *theorem* (Titchmarsh 1948, Cushing 1975). A single-integral form is obtained firstly by changing the integral on f to the limit as $f_0 \rightarrow \infty$ of a definite integral over the interval, $(-f_0, f_0)$, and interchanging integrations,

$$g(x) = \lim_{f_0\rightarrow\infty} \int_{-\infty}^{\infty} g(x') \int_{-f_0}^{f_0} e^{i2\pi f(x-x')}df\,dx'. \tag{2.53}$$

Then, with equation (1.5) and noting that the integral of the sine function, being an odd function, is zero, one obtains,

$$g(x) = \lim_{f_0\rightarrow\infty} \frac{1}{\pi}\int_{-\infty}^{\infty} g(x')\frac{\sin\left(2\pi f_0\left(x-x'\right)\right)}{x-x'}dx', \tag{2.54}$$

where the integrand is also well defined if $x' = x$. The corresponding Fourier integral at points of discontinuity of $g(x)$ follows in the same way starting with equation (2.9),

$$\frac{1}{2}\lim_{\delta x\rightarrow 0+}\left(g(x-\delta x)+g(x+\delta x)\right) = \lim_{f_0\rightarrow\infty}\frac{1}{\pi}\int_{-\infty}^{\infty} g(x')\frac{\sin\left(2\pi f_0\left(x-x'\right)\right)}{x-x'}dx'. \tag{2.55}$$

The *Fourier transform* of $g(x)$ is defined by

$$\mathcal{F}(g) \equiv G(f) = \int_{-\infty}^{\infty} g(x)e^{-i2\pi f x}dx; \tag{2.56}$$

and, equation (2.52) then shows that the *inverse Fourier transform* converges to $g(x)$ at points of continuity,

$$\mathcal{F}^{-1}(G) \equiv g(x) = \int_{-\infty}^{\infty} G(f)e^{i2\pi f x}df, \tag{2.57}$$

and to the average of $g(x)$ on either side of a step-discontinuity,

$$\frac{1}{2}\lim_{\delta x\to 0+}\left(g\left(x-\delta x\right)+g\left(x+\delta x\right)\right)=\int_{-\infty}^{\infty}G\left(f\right)e^{i2\pi f x}df. \tag{2.58}$$

Analogous to the Fourier coefficients, equation (2.13), $G(f)$ is known as the *Fourier spectrum* (or *spectral density*) of g. It is a function of continuous frequency, f, which, as before is defined here as *cyclical frequency* and, just like k/P, has units of cycle per unit of x. The spectral domain in this case is continuous and infinite. Like its discrete cousin, $G(f)$ generally is complex even if $g(x)$ is a real function and it can be decomposed into its *amplitude spectrum*,

$$A(f)=\left((\mathrm{Re}\,G(f))^2+(\mathrm{Im}\;G(f))^2\right)^{1/2}, \tag{2.59}$$

and its *phase spectrum*,

$$\phi\left(f\right)=\tan^{-1}\frac{\mathrm{Im}\,G\left(f\right)}{\mathrm{Re}\,G\left(f\right)}. \tag{2.60}$$

Amplitude and phase together yield the spectrum in the form,

$$G(f)=A(f)e^{i\phi(f)}. \tag{2.61}$$

The amplitude often is displayed as the proxy characterization of the complex spectrum, mainly to identify particular resonances or spectral trends of a signal (see the example, below). The phase spectrum plays a role in characterizing the performance of filters as illustrated in Section 3.5.1. The units of $G(f)$ are the units of $g(x)$ per unit of frequency, whence its alternative name as spectral density.

A *band-limited* function, $g_{f_0}(x)$, is one whose non-zero spectrum is limited to a finite spectral domain or a finite band of frequencies. For example, usually it is implied that the spectrum vanishes for all frequencies greater (in magnitude) than some f_0,

$$G_{f_0}(f)=0,\quad |f|>f_0; \tag{2.62}$$

The frequency, f_0, is also called the *cutoff frequency*, although, this may not always define a definite boundary between non-zero and zero spectral content. There may be a transition zone, sometimes rather wide, that separates the essential spectral domain of the function and its near-zero part. The *bandwidth* of the function is defined here by the domain of frequencies bounded by the cutoff frequency,

$$W(g)=2f_0. \tag{2.63}$$

The definition of the cutoff frequency can take various forms if the function is not exactly band-limited and this is elaborated in Section 3.5.2. Substituting equation (2.56) into equation (2.57), one obtains as in equation (2.54),

$$
\begin{aligned}
g_{f_0}(x) &= \int_{-\infty}^{\infty} g(x')\left(\int_{-f_0}^{f_0} e^{-i2\pi f(x'-x)}\,df\right)dx' \\
&= \frac{1}{\pi}\int_{-\infty}^{\infty} g(x')\frac{\sin\left(2\pi f_0(x-x')\right)}{x-x'}\,dx'
\end{aligned}
\tag{2.64}
$$

which shows that band-limiting a function according to equation (2.62) can create a considerable distortion of its original form (Gibbs's effect, Section 2.3.2).

Analogously, a *space-limited* function, $g_T(x)$, vanishes outside a sub-domain in space, for example,

$$
g_T(x) = 0, \quad |x| > \frac{T}{2}. \tag{2.65}
$$

(If time is the independent variable, these are known as time-limited functions.) Note that a space- (or, time-) limited function is not the same as a periodic function since its specific definition of zero outside an interval precludes periodicity over the real line.

An extremely important result is that no function, other than the zero function, can be simultaneously band-limited and space-limited (Slepian 1983). It is proved by noting that a band-limited function is analytic everywhere when extended to the complex plane; i.e., all its derivatives exist and its Taylor series converges everywhere. As such, if it vanishes on some interval (i.e., it is space-limited), then its Taylor series must also vanish everywhere, leaving only the trivial, zero-valued signal as being both band-limited and space-limited (see also Papoulis 1977, p.188). There is a reciprocal tradeoff between limited frequency and limited space domains, which is expressed by an *uncertainty principle* analogous to the one introduced by Heisenberg in quantum mechanics. It is elaborated further in Section 3.5.2 in connection with the bandwidth of filters and window functions.

This section closes with the continuation of the polar motion example of Chapter 1 (Figure 1.2). The amplitude spectrum for the x-coordinate, shown in Figure 2.3,

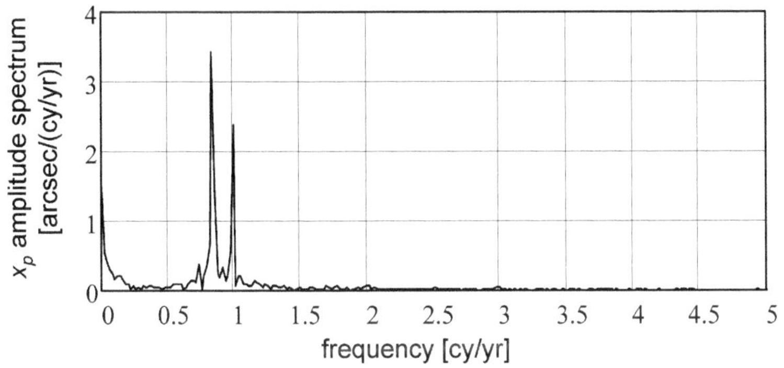

Figure 2.3: Amplitude spectrum of the x-coordinate of polar motion.

identifies much more clearly the frequencies corresponding to the periods embedded in the signal. A similar result (not shown) may be obtained for the y-coordinate. The infinite integral in equation (2.56) is truncated of practical necessity and discretized to a finite sum, and the spectral values (for discrete frequencies) on the plot are connected by lines. These approximations are discussed in detail in Chapter 4. The definite resonances of polar motion at frequencies, $f = 0.84$ cy/yr and $f = 1.0$ cy/yr, corresponding to periods of $P = 1.19$ yr $= 435$ days (the Chandler period) and $P = 1.0$ yr (annual period), are much easier to discern in Figure 2.3 than in Figure 1.2. The large value at zero frequency is proportional to the average value of the x-coordinate of polar motion over the interval defined by the truncation of the signal. The other minor resonances near the principal ones are spectral leakage errors discussed in Section 3.6.1.

Finally, it is noted that the spectra of spatial geophysical signals (e.g., topographic elevation of Earth's surface), in contrast, generally have energy distributed more evenly, though with attenuating amplitude, over all frequencies, as illustrated in Figure 2.19. This leads then also to their interpretations in the space and frequency domains from a stochastic or statistical viewpoint, which is the main topic of Chapter 5.

2.3.1 Properties of the Fourier Integral Transform

Many of the properties of the Fourier series transform pair (Section 2.2.1) carry over directly to the Fourier integral transform pair, equations (2.56) and (2.57), repeated here,

$$g(x) = \int_{-\infty}^{\infty} G(f) e^{i2\pi fx} df, \quad G(f) = \int_{-\infty}^{\infty} g(x) e^{-i2\pi fx} dx, \tag{2.66}$$

where convergence of the first integral is given by equation (2.58) at points with step discontinuities in $g(x)$. It can be shown that $G(f)$ is *continuous* (Butzer and Nessel 1971, p.189). Heuristically, even if a function has a finite number of step-discontinuities, its integral is continuous. The transform is also bounded, according to equation (2.49),

$$|G(f)| = \left| \int_{-\infty}^{\infty} g(x) e^{-i2\pi fx} dx \right| \le \int_{-\infty}^{\infty} \left| g(x) e^{-i2\pi fx} \right| dx = \int_{-\infty}^{\infty} |g(x)|\, dx < \infty. \tag{2.67}$$

Moreover, the Riemann-Lebesgue theorem (e.g., Cushing 1975, p.158) says that the Fourier transform of an absolutely integrable function attenuates to zero as the absolute frequency increases,

$$\lim_{f \to \pm\infty} G(f) = 0. \tag{2.68}$$

Additional properties beyond those for the Fourier series transform ensue if $g(x)$ is also differentiable. For example, the nth derivative of $g(x)$ is square integrable if and only if $(2\pi f)^n\, G(f)$ is square integrable (Daubechies 1992, p.xii), $n = 0,1,2,\ldots$ (see also property (2.75) and Parseval's theorem, equation (2.87)).

While our applications generally involve real-valued signals, the following properties hold also for complex-valued functions.

1. Fourier transform pair:	$g(x) \leftrightarrow G(f);$	(2.69)
2. Proportionality:	$ag(x) \leftrightarrow aG(f);$	(2.70)
3. Superposition:	$g_1(x) + g_2(x) \leftrightarrow G_1(f) + G_2(f);$	(2.71)
4. Symmetry:	$g(-x) \leftrightarrow G(-f);$	(2.72)
5. Translation in x:	$g(x + x_0) \leftrightarrow G(f)e^{i2\pi f x_0};$	(2.73)
6. Translation in f:	$g(x)\, e^{-i2\pi f_0 x} \leftrightarrow G(f + f_0);$	(2.74)
7. Differentiation in space (time):	$\frac{d^p}{dx^p} g(x) \leftrightarrow (i2\pi f)^p\, G(f);$	(2.75)
8. Differentiation in frequency:	$(-i2\pi x)^p\, g(x) \leftrightarrow \frac{d^p}{df^p} G(f);$	(2.76)
9. Duality:	$G(x) \leftrightarrow g(-f);$	(2.77)
10. Similarity (scaling):	$g(ax) \leftrightarrow \frac{1}{\lvert a \rvert} G\left(\frac{f}{a}\right),\ a \neq 0;$	(2.78)

The proofs of transform pairs (2.70) through (2.76) are left to the reader (Exercise 2.4). It is assumed in properties (2.75) and (2.76) that $g(x)$, respectively $G(f)$, is differentiable up to order, p. Property (2.77) assumes that G as a function of x is absolutely integrable and has bounded variation. Its Fourier transform is given by

$$\mathcal{F}\left(G(x)\right) = \int_{-\infty}^{\infty} G(x)\, e^{-i2\pi f x}\, dx = g(-f). \tag{2.79}$$

The proof of equation (2.78) follows from a change in integration variable, $x' = ax$,

$$\begin{aligned} \mathcal{F}\left(g(ax)\right) &= \int_{-\infty}^{\infty} g(ax)\, e^{-i2\pi f x}\, dx = \frac{1}{\lvert a \rvert} \int_{-\infty}^{\infty} g(x')\, e^{-i2\pi f \frac{x'}{a}}\, dx' \\ &= \frac{1}{\lvert a \rvert} G\left(\frac{f}{a}\right) \end{aligned} \tag{2.80}$$

The antecedent absolute value combines the separate cases, $a > 0$ and $a < 0$. As an example of the utility of this property, consider the change of units of a signal from temporal to spatial units, as in the case that involves the velocity of a sensor system. The sensor may yield quantities depending on time (t), but it is desired to transform these to quantities depending on distance (x). Assuming a constant velocity, v, the change of variables is given by

$$x = vt. \tag{2.81}$$

Let the spectrum of the time signal, $g_t(t)$, be $G_t(f_t)$, where f_t is temporal frequency; and let $g_x(x) = g_t(x/v)$. Then with $a = 1/v$ the spectrum of $g_x(x)$ is $G_x(f_x) = |v|G_t(vf_x)$, where f_x is spatial frequency. For example, suppose that the sensor is a gravity meter on board an aircraft. If g_t has units [mGal] and t has units [s], then $G_t(f_t) = G_t(vf_x)$ has units [mGal/(cy/s)]. Now if v has units [m/s], then x has units [m], f_x has units [cy/m], $g_x(x)$ still has units [mGal], and its spectrum, $G_x(f_x)$, has units [mGal/(cy/m)].

Aside from this rather proletarian utility, the similarity or scaling property demonstrates a deeper and fundamental relationship between a function and its spectrum. Written as

$$|a|^{1/2} g(ax) \leftrightarrow \frac{1}{|a|^{1/2}} G\left(\frac{f}{a}\right), \tag{2.82}$$

a scale increase in the function implies a corresponding scale decrease in the spectrum; and, an expansion of the spatial domain implies a contraction of the frequency domain. The latter is analogous to the "uncertainty principle" noted previously in Section 2.3 by which a function and its spectrum cannot both be of limited extent. This is important in spectral estimation from a finite interval of data and is discussed in more detail in Section 5.7.2.

Just like periodic functions, there are certain symmetries in the Fourier transform pairs if g is even, real, or Hermitian. Analogous to results (2.26) through (2.29), one has,

$$g(-x) = g(x) \quad \text{if and only if} \quad G(-f) = G(f); \tag{2.83}$$

$$g(x) = g^*(x) \quad \text{if and only if} \quad G(-f) = G^*(f); \tag{2.84}$$

$$g(-x) = g^*(x) \quad \text{if and only if} \quad G(f) = G^*(f); \tag{2.85}$$

$$g(x) = g^*(x) \text{ and } g(-x) = g(x) \quad \text{if and only if} \quad G(f) = G^*(f) \text{ and } G(-f) = G(f). \tag{2.86}$$

For square-integrable functions, Parseval's theorem states that

$$\int_{-\infty}^{\infty} g_1^*(x) g_2(x)\, dx = \int_{-\infty}^{\infty} G_1^*(f) G_2(f)\, df, \tag{2.87}$$

which is proved simply by substituting the inverse Fourier transform for g_1 on the left side and interchanging integrals,

$$\int_{-\infty}^{\infty}\left(\int_{-\infty}^{\infty} G_1^*(f)e^{-i2\pi f x}df\right)g_2(x)dx = \int_{-\infty}^{\infty} G_1^*(f)\left(\int_{-\infty}^{\infty} g_2(x)e^{-i2\pi f x}dx\right)df = \int_{-\infty}^{\infty} G_1^*(f)G_2(f)df. \tag{2.88}$$

If $g_1(x) = g(x) = g_2(x)$, and equation (2.50) holds, then g has *finite energy*, and Parseval's theorem shows that the total energy of a function is equal to the total energy of the spectrum,

$$\int_{-\infty}^{\infty}|g(x)|^2\,dx = \int_{-\infty}^{\infty}|G(f)|^2\,df. \tag{2.89}$$

The squared magnitude of the spectrum, $|G(f)|^2$, is also called the *energy spectral density* (Section 5.2).

2.3.2 Rectangle Function

The rectangle function is one of the more useful functions in the study of spectral analysis. It will surface repeatedly in the discussion of filters and spectral density estimation, as well in any other application of practical data analysis since observations of a geophysical signal are available only in a limited spatial domain (or, for a limited duration in time). This practical reality is represented by the product of the rectangle function and the data function, theoretically defined over the entire domain. The basic form of the rectangle function, shown in Figure 2.4, is defined by

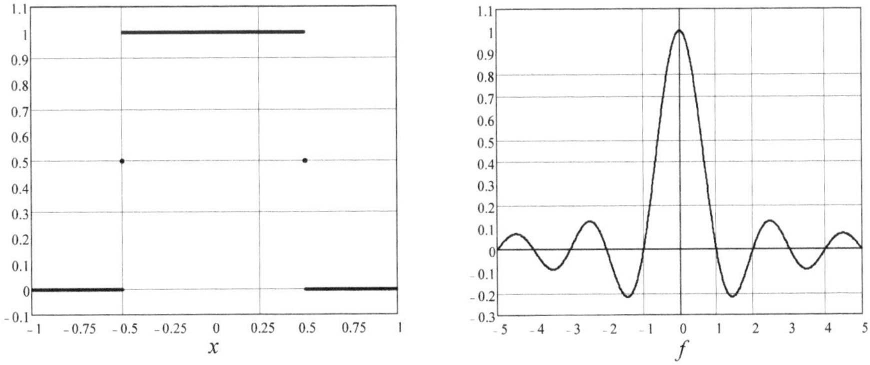

Figure 2.4: Rectangle function (left panel) and its Fourier transform, the sinc function (right panel).

$$b(x)=\begin{cases}1, & |x|<\frac{1}{2}\\ 0.5, & |x|=\frac{1}{2}\\ 0, & |x|>\frac{1}{2}\end{cases} \tag{2.90}$$

Alternative names for the rectangle function are *box-car* function and *top-hat* function, both evident from the figure with a little imagination. The definition of $b(\pm 1/2)$ is not universal, but is illustrated here primarily so that the inverse Fourier transform converges to $b(x)$ for all x, as in equation (2.58). However, for simplicity and where the convergence question is not an issue, we may define $b(\pm 1/2)=1$.

The Fourier transform, according to equation (2.56), is given by

$$\begin{aligned}B(f)=\mathcal{F}\left(b(x)\right)&=\int_{-\infty}^{\infty} b(x)e^{-i2\pi f x}dx=\int_{-1/2}^{1/2} e^{-i2\pi f x}dx=\int_{-1/2}^{1/2}\cos(2\pi f x)\,dx\\ &=\frac{\sin(\pi f)}{\pi f}\end{aligned} \tag{2.91}$$

or,

$$B(f)=\mathrm{sinc}(f), \tag{2.92}$$

where the so-called “sinc” function of f (formally the *cardinal sine* function),

$$\mathrm{sinc}(f)=\frac{\sin(\pi f)}{\pi f}, \tag{2.93}$$

has this special name and notation because it finds frequent usage. An alternative definition also appears in the literature, $\mathrm{sinc}(x)=(\sin x)/x$; however, equation (2.93) is the defining notation used here. In either case, the apparent singularity at the origin disappears,

$$\mathrm{sinc}(0)=1, \tag{2.94}$$

which is proved mathematically by l’Hôpital’s rule for the limit as $x\to 0$, and is evident from its graph (Figure 2.4). Furthermore, the sinc function is continuous and differentiable everywhere (Exercise 2.5).

The inverse Fourier transform at $x=\pm 1/2$ is

$$\mathcal{F}^{-1}\left(B(f)\right)\Big|_{x=\pm 1/2}=\int_{-\infty}^{\infty} B(f)e^{i2\pi f\left(\pm\frac{1}{2}\right)}df=2\int_{0}^{\infty}\frac{\sin(\pi f)}{\pi f}\cos(\pi f)\,df=\frac{1}{2}, \tag{2.95}$$

using equation (1.5) and noting that $(1/f)\sin^2(\pi f)$ is an odd function and integrates to zero. The final integral may be determined by consulting a standard table of integrals (e.g., Gradshteyn and Ryzhik 1980, p.405). Thus, there is agreement with the given values of the rectangle function at its points of discontinuity. Equation (2.58), in this case, reduces to equation (2.57); and, one may write,

$$b(x) = \int_{-\infty}^{\infty} \operatorname{sinc}(f)\, e^{i2\pi fx}\, df. \tag{2.96}$$

Evaluated at $x = 0$, this yields

$$\int_{-\infty}^{\infty} \operatorname{sinc}(f)\, df = 1; \tag{2.97}$$

that is, the area enclosed by the sinc function in the frequency domain is also unity, as it is for the rectangle function in the space domain.

The zeros of the sinc function occur at $f = \pm n,\ n \geq 1$. It is not absolutely integrable because the areas of its "side lobes" do not decay fast enough; indeed, one has (Exercise 2.6)

$$\int_{-\infty}^{\infty} \left|\operatorname{sinc}(f)\right| df > \frac{4}{\pi^2} \sum_{k=1}^{\infty} \frac{1}{k} \to \infty. \tag{2.98}$$

Therefore, one cannot a priori claim that there is a Fourier integral for the sinc function with point-wise convergence. Nevertheless, it can be shown (Exercise 2.7) that

$$\mathcal{F}(\operatorname{sinc}(x)) = b(f), \tag{2.99}$$

which is also consistent with the duality property (2.77).

Consider the more general rectangle function, $b(x/T)/T$, for $T > 0$, that is non-zero over the interval, $(-T/2,\ T/2)$, but scaled to preserve unit area. By the properties of similarity, equation (2.78), and proportionality, equation (2.70), there is,

$$b^{(T)}(x) = \frac{1}{T} b\left(\frac{x}{T}\right) \quad \leftrightarrow \quad B^{(T)}(f) = \frac{1}{T} T \cdot B(Tf) = \frac{\sin(\pi Tf)}{\pi Tf} = \operatorname{sinc}(Tf). \tag{2.100}$$

Thus, as the rectangle function's base shrinks (T decreases), the main "lobes" of the sinc function expand (the zeros of $B(Tf)$ are at $f = \pm\, n/T$); the opposite clearly holds: as T increases, the lobes of the sinc function become narrower. In all cases, the Fourier transform is unity at zero frequency, which is important, as shown in Section 3.5.1, when considered as a filter applied to a geophysical signal that should preserve its average.

The rectangle function is the mathematical tool that truncates, or space-limits, a function to a finite domain. That is, a function, $g(x)$, essentially limited to a finite domain, $(-T/2, T/2)$, by the definition

$$g_T(x) = \begin{cases} g(x), & -T/2 < x < T/2 \\ g(x)/2, & x = \pm T/2 \\ 0, & \text{otherwise} \end{cases} \tag{2.101}$$

is the same as

$$g_T(x) = b\left(\frac{x}{T}\right) g(x). \tag{2.102}$$

Thus, for example, the effect of truncation can be studied by analyzing the spectral properties of the rectangle function. Equation (2.102) is usually acceptable in the analysis also if $g_T(\pm T/2) = g(\pm T/2)$.

Equation (2.57) shows that generally the entire spectrum is needed to capture the complete function. If the function is smooth then it is reasonable to assume that it is well represented with a partial spectrum that excludes the high-frequency components that would otherwise reflect the roughness in the function. However, at points of discontinuity the convergence of the Fourier integral to the function is slow; and, since the intermediate integrals of the converging transform, equation (2.54), are continuous they must oscillate strongly near the discontinuity in order accommodate the step in the function. This is called *Gibbs's effect* (also, *Gibbs's phenomenon*) that is again demonstrated by equation (2.64), where a band-limited approximation to $g(x)$ is given by

$$g_{f_0}(x) = 2f_0 \int_{-\infty}^{\infty} g(x')\operatorname{sinc}\left(2f_0\left(x - x'\right)\right) dx'. \tag{2.103}$$

In this case the rectangle function is applied in the spectral domain and the approximation is affected by the oscillations of its inverse Fourier transform, the sinc function. For example, the discontinuities of the rectangle function, $b(x)$, cause a ripple effect in its band-limited approximation, $b_{f_0}(x)$, as shown in Figure 2.5 for cutoff frequencies, $f_0 = 2$ and $f_0 = 10$.

Gibbs's effect is present in any rendition of a general function by its band-limited approximation if the function varies more rapidly than can be accommodated by the limited spectrum. This is illustrated also in connection with filtered functions in Section 3.5.1, since a filter is nothing more than a band-limiting operation.

This section concludes with another example of a function that like the sinc function is not absolutely integrable but has a Fourier transform. This is the reciprocal function, $r(x) = 1/x$, $x \neq 0$, that is integrable on $(-\infty, \infty)$ as a Cauchy principal value. It can be shown (Exercise 2.8) that

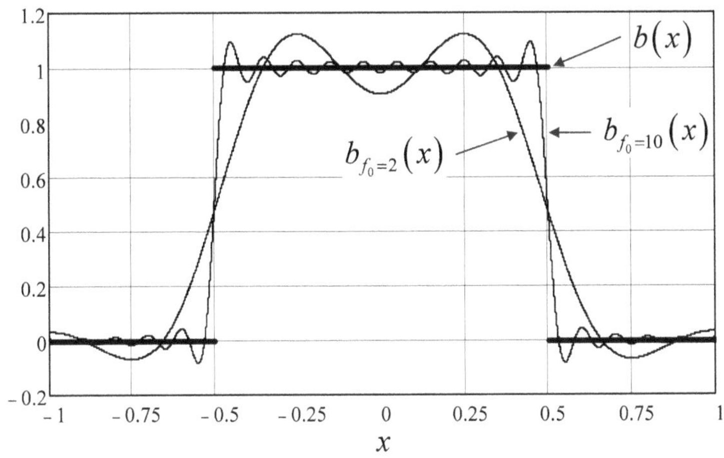

Figure 2.5: Demonstration of Gibbs's effect at the discontinuities of the rectangle function. Approximations, $b_{f_0}(x)$, approach $b(x)$ as $f_0 \to \infty$.

$$\mathcal{F}(r(x)) \equiv R(f) = \begin{Bmatrix} -i\pi, & f > 0 \\ 0, & f = 0 \\ i\pi, & f < 0 \end{Bmatrix} = -i\pi \operatorname{sgn}(f), \tag{2.104}$$

where the sign function is zero if $f = 0$. This is a step function (constant with a step, $i2\pi$, at $f = 0$), which is not absolutely integrable (but has a Cauchy principal value). The step function also has a Fourier transform, which is the reciprocal function (Section 2.3.4).

2.3.3 Gaussian Function

Unlike the incongruity between the shape of the rectangle function and its Fourier transform, the *Gaussian function* and its transform are virtually identical. Defined by

$$\gamma(x) = e^{-\pi x^2}, \tag{2.105}$$

its Fourier transform is, again, a Gaussian function (Figure 2.6),

$$\Gamma(f) = e^{-\pi f^2}, \tag{2.106}$$

The proof of this is left to Exercise 2.9. The area under both $\gamma(x)$ and $\Gamma(f)$ is unity. For scaled widths, but still unit area in the space domain, one defines

$$\gamma^{(\beta)}(x) = \frac{1}{\beta} e^{-\pi(x/\beta)^2}, \quad \beta > 0, \tag{2.107}$$

with Fourier transform, using the similarly property (2.78), given by

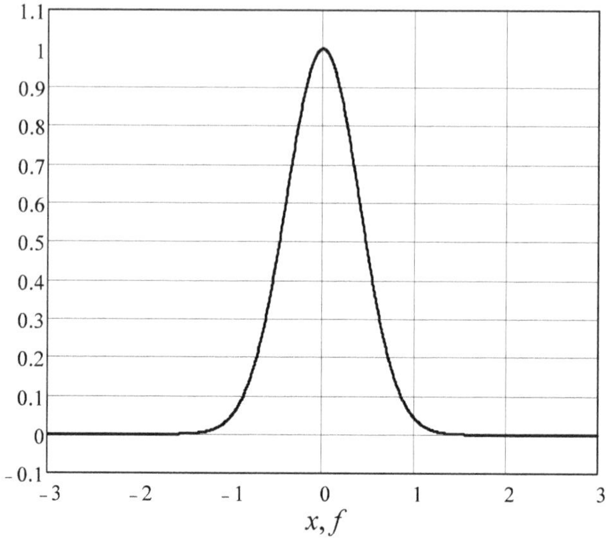

Figure 2.6: Gaussian function, $\gamma(x)$, and its Fourier transform, $\Gamma(f)$.

$$\Gamma^{(\beta)}(f) = e^{-\pi(\beta f)^2}. \tag{2.108}$$

The area under $\Gamma^{(\beta)}$ is $1/\beta$. The width of $\gamma^{(\beta)}$, may be characterized by the span between its inflection points, $x = \pm\,\beta/\sqrt{2\pi}$; and, with the same characterization, the width of its Fourier transform is $2/(\sqrt{2\pi}\beta)$. As the width of the Gaussian function increases, the width of its Fourier transform decreases; and, vice versa; thus, again illustrating the reciprocal tradeoff between space-limited and band- (or, frequency-) limited functions, even in an approximate sense, since mathematically the Gaussian function is neither space- nor frequency-limited.

2.3.4 Dirac Delta Function

Consider again the rectangle functions, $b(x/T)/T$, with $0 < T \le 1$ (Figure 2.7), recalling that the area under any of these scaled functions is unity. As T approaches zero, the magnitude of $b^{(T)}(x) = b(x/T)/T$ at $x = 0$ approaches infinity, while it approaches zero everywhere else. Mathematically, one defines

$$\delta(x) = \lim_{T\to 0} b^{(T)}(x), \tag{2.109}$$

where

$$\delta(x) = 0, \quad \text{for all } x \neq 0, \tag{2.110}$$

and it is infinite at $x = 0$. $\delta(x)$ is called the *Dirac delta* function, also simply the *delta* function, or the *impulse* function. Because it does not have a finite value at the origin, it is sometimes said that $\delta(x)$ is not a legitimate function. However, the infinite limit

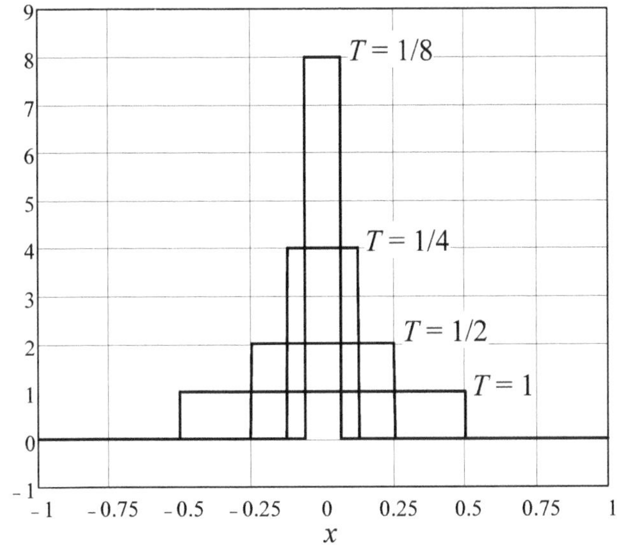

Figure 2.7: The functions $b(x/T)/T$ with $T \rightarrow 0$. The vertical lines connecting the function values at $x = \pm T/2$ only emphasize the area and do not define the function at these points.

is not arbitrary, but defined in the sense of equation (2.109) as the limit of a function that has unit area; and, for this reason, or with this understanding, $\delta(x)$ may be called a function. It could also be defined as the limit of the Gaussian function, $\gamma^{(\beta)}(x)$, equation (2.107), as $\beta \rightarrow 0$. It is noteworthy that although $\delta(x)$ clearly does not satisfy Dirichlet's conditions (it is not bounded) the Fourier transform of $\delta(x)$ exists. Indeed, from equations (2.100) and (2.94),

$$\begin{aligned}\lim_{T\to 0} B(Tf) &= \lim_{T\to 0}\left(\text{sinc}(Tf)\right) \\ &= 1, \quad \text{for all } f\end{aligned} \tag{2.111}$$

and, one may formally write

$$\mathcal{F}(\delta(x)) \equiv \Delta(f) = 1. \tag{2.112}$$

The algebraic manipulations with $\delta(t)$ are viewed generally in the sense of distribution functions (Papoulis 1977, p.96), which, however, are outside the scope of this text. With careful application of the limiting process, as in equation (2.109), one may continue to work with the Dirac delta function and take advantage of a few important results. For example, from the definition that it is the limit of rectangle functions all having unit area, one has,

$$\int_{-\infty}^{\infty} \delta(x)dx = 1; \tag{2.113}$$

which can be proved mathematically by changing the infinite integral into the limit of a proper integral and substituting equation (2.109),

$$\lim_{x_0\to\infty}\int_{-x_0}^{x_0}\lim_{T\to 0}\frac{1}{T}b\left(\frac{x}{T}\right)dx=\lim_{T\to 0}\left(\frac{1}{T}\lim_{x_0\to\infty}\int_{-x_0}^{x_0}b\left(\frac{x}{T}\right)dx\right)=\lim_{T\to 0}\frac{1}{T}\int_{-T/2}^{T/2}dx=\lim_{T\to 0}(1)=1. \quad (2.114)$$

Furthermore, it is shown in Chapter 3 (equation (3.23)) that

$$\int_{-\infty}^{\infty}\delta(x-x_0)g(x)\,dx=g(x_0), \quad (2.115)$$

which is known, for evident reasons, as the *reproducing property* of the Dirac delta function. Since the rectangle function is an even function, so is the delta function,

$$\delta(-x)=\delta(x). \quad (2.116)$$

According to equation (2.113), the units of $\delta(x)$ are 1/(units of x).

A formal verification that the inverse Fourier transform of $\Delta(f)$, equation (2.112), is the Dirac delta function is achieved, again, by re-formulating the improper integral of equation (2.57) for $\Delta(f)$ and substituting equation (1.5),

$$\begin{aligned}\int_{-\infty}^{\infty}e^{i2\pi fx}df&=\lim_{f_0\to\infty}\int_{-f_0}^{f_0}\left(\cos(2\pi fx)+i\sin(2\pi fx)\right)df\\&=2\lim_{f_0\to\infty}\int_{0}^{f_0}\cos(2\pi fx)\,df\end{aligned} \quad (2.117)$$

where the second equality derives from the facts that the sine and cosine are odd and even functions, respectively (with respect to f). The limiting process is now modified for $x\neq 0$ as

$$\lim_{f_0\to\infty}\int_{0}^{f_0}\cos(2\pi fx)\,df=\lim_{n_0\to\infty}\sum_{n=0}^{n_0}\int_{nP}^{(n+1)P}\cos(2\pi fx)\,df,\quad P=\frac{1}{x}, \quad (2.118)$$

where each integral of the sum is over one period of the cosine. These integrals thus vanish, hence, so does the limit; and,

$$\int_{-\infty}^{\infty}e^{i2\pi fx}df=0,\quad x\neq 0. \quad (2.119)$$

If $x = 0$, then the left side of equation (2.117) clearly is infinite. Therefore, insofar as the Fourier transform of $\delta(x)$ exists and equals unity, we may also write formally for the inverse Fourier transform,

$$\delta(x) = \int_{-\infty}^{\infty} e^{i2\pi f x} df = \int_{-\infty}^{\infty} e^{-i2\pi f x} df, \tag{2.120}$$

where the second equality is justified by equation (2.116). This also shows that, formally,

$$\mathcal{F}(1) = \int_{-\infty}^{\infty} 1 \cdot e^{-i2\pi f x} dx = \delta(f). \tag{2.121}$$

The function whose derivative is the Dirac delta function is the Heaviside step function,

$$h(x) = \left\{ \begin{matrix} 0, & x < 0 \\ 0.5, & x = 0 \\ 1, & x > 0 \end{matrix} \right\} = \frac{1}{2} + \frac{1}{2}\operatorname{sgn}(x); \tag{2.122}$$

that is, formally,

$$\frac{d}{dx} h(x) = \delta(x). \tag{2.123}$$

This is *not* to say that the integral of the Dirac delta function is the step function (we already have equation (2.113)); however, equation (2.123) could also be used to *define* the delta function. The Fourier transform of the Heaviside step function is (Exercise 2.10)

$$H(f) = \frac{1}{2}\left(\delta(f) - \frac{i}{\pi f} \right). \tag{2.124}$$

This allows now the determination of the Fourier transform of the step function, equation (2.104), which in terms of x is $-i\pi \operatorname{sgn}(x) = -i\pi(2h(x)-1)$. Indeed, using equations (2.124) and (2.121),

$$\mathcal{F}(-i\pi \operatorname{sgn}(x)) = -\frac{1}{f} \tag{2.125}$$

which is the reciprocal function in the frequency domain.

2.3.5 Fourier Transforms Using the Delta Function

Fourier integral transforms of periodic functions were specifically excluded since they are not absolutely integrable over the real line. Formally, however, it is possible to define such a transform for periodic, as well as discrete functions, or sequences, with appropriate use of the Dirac delta function. With a change in variables, first replacing x by k/T; subsequently replacing f by x, the second equation (2.120) becomes

$$\int_{-\infty}^{\infty} e^{-i\frac{2\pi}{T}kx}\,dx = \delta\left(\frac{k}{T}\right). \tag{2.126}$$

Substituting the Fourier series, equation (2.8), into the Fourier transform, equation (2.56), one finds,

$$\begin{aligned}\mathcal{F}\left(\tilde{g}(x)\right) &= \int_{-\infty}^{\infty}\left(\frac{1}{P}\sum_{k=-\infty}^{\infty} G_k e^{i\frac{2\pi}{P}kx}\right)e^{-i2\pi f x}\,dx = \frac{1}{P}\sum_{k=-\infty}^{\infty} G_k \int_{-\infty}^{\infty} e^{-i2\pi\left(f-\frac{k}{P}\right)x}\,dx \\ &= \frac{1}{P}\sum_{k=-\infty}^{\infty} G_k \delta\left(f-\frac{k}{P}\right)\end{aligned} \tag{2.127}$$

That is, the Fourier integral transform of a periodic function with Fourier coefficients G_k is an infinite sequence of impulses scaled by G_k/P and spaced along the frequency axis at the *discrete* frequencies k/P (Figure 2.8). The units of the Dirac delta function are the inverse of frequency units in this case; they cancel the units of $1/P$.

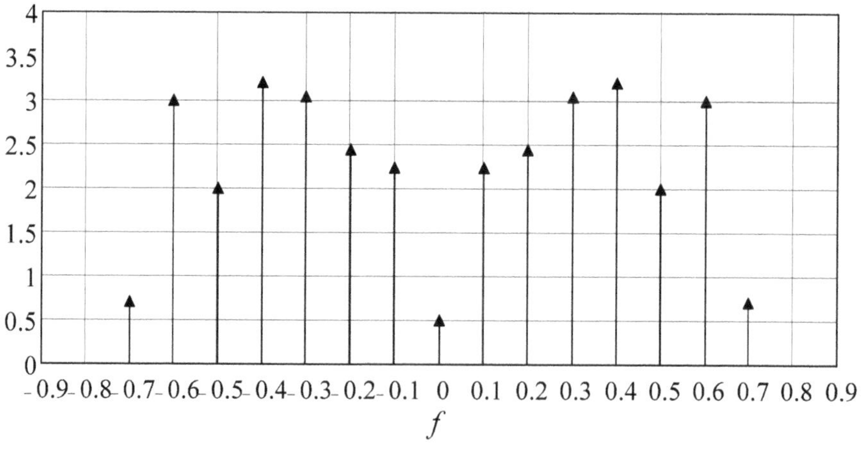

Figure 2.8: Fourier transform (amplitude spectrum) of $\tilde{g}(x)$, shown in Figure 2.1, where $\{G_{\pm7}, G_{\pm6}, G_{\pm5}, G_{\pm4}, G_{\pm3}, G_{\pm2}, G_{\pm1}, G_0\} = \{0.7, -3, 2, 3.2, -0.5 \mp i3, -2 \pm i1.4, 1 \mp i2, -0.5\}$.

The Fourier transform of a periodic function, formulated as equation (2.127), makes it consistent with the Fourier integral transform of non-periodic, absolutely integrable functions of bounded variation. It is insightful perhaps only in that sense, and serves toward a unified mathematical development of the Fourier transform for a larger class of functions. However, equations (2.8) and (2.13) are the preferred Fourier (series) transform pair for periodic functions, as they avoid the use of the Dirac delta function.

Secondly, consider the periodic function that is an infinite sequence of identical rectangle functions, each having T as its base,

$$\tilde{d}(x) = \frac{1}{T}\sum_{k=-\infty}^{\infty} b\left(\frac{x - k\Delta x}{T}\right), \tag{2.128}$$

where $\Delta x > T$ is the spacing between their centers (Figure 2.9, left) and, hence, the period of $\tilde{d}(x)$. Defining $s(x)$ to be the limit of $\tilde{d}(x)$ as $T \to 0$ while each rectangle retains unit area, it becomes a train of impulses, as given by equation (2.109) (see also Figure 2.9, right),

$$s(x) = \lim_{T\to 0} \tilde{d}(x) = \sum_{k=-\infty}^{\infty} \delta(x - k\Delta x). \tag{2.129}$$

This function, $s(x)$, (again in a formal setting) is called the *sampling function*, or *Dirac comb*, because if multiplied with some arbitrary function, it samples the latter using impulses, which has some utility when developing the Fourier transform of discrete functions (Section 4.2.1).

Being periodic, with period, Δx, $\tilde{d}(x)$, given by equation (2.128), can be represented as a Fourier series, where the Fourier coefficients are given by equation (2.13),

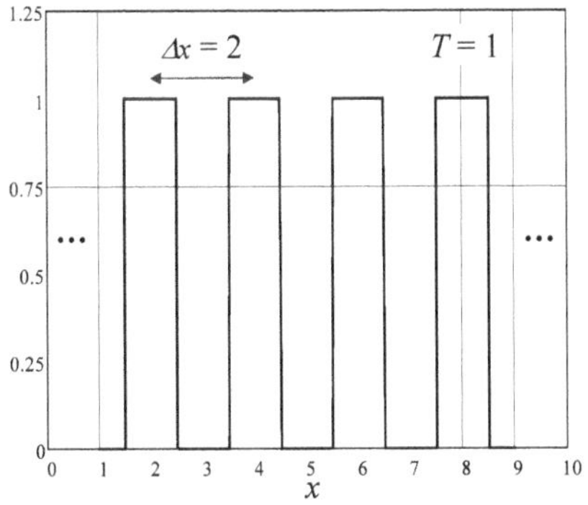

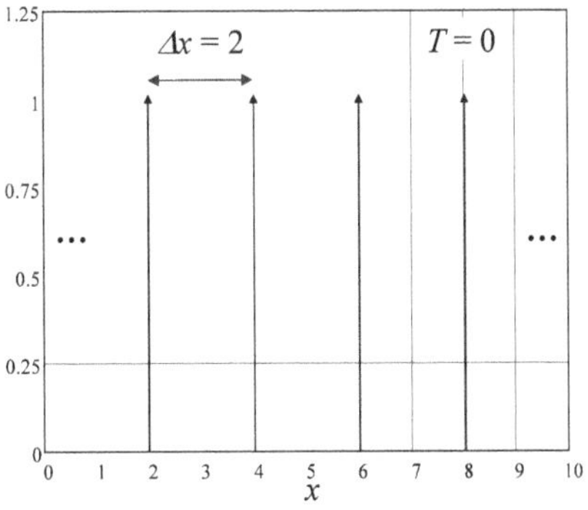

Figure 2.9: Left: Infinite sequence of rectangle functions, $\tilde{d}(x)$, where vertical lines are included only to emphasize the area under each rectangle. Right: Sampling function, $s(x)$.

which is an integral over any period of $\tilde{d}(x)$. Choosing the interval, $[-\Delta x/2, \Delta x/2]$, which covers only the $k = 0$ term in equation (2.128), yields

$$D_k = \frac{1}{T} \int_{-\Delta x/2}^{\Delta x/2} b\left(\frac{x}{T}\right) e^{-i\frac{2\pi}{\Delta x}kx} dx. \tag{2.130}$$

Now, since $\Delta x > T$, the rectangle function, $b(x/T)$, is zero outside the integration interval; and, the integration can be extended to $\pm\infty$ without changing its value. With a suitable change of integration variable ($x' = x/T$), and using the result (2.91) and the definition (2.93), one obtains,

$$\begin{aligned} D_k &= \frac{1}{T} \int_{-\infty}^{\infty} b\left(\frac{x}{T}\right) e^{-i\frac{2\pi}{\Delta x}kx} dx = \int_{-\infty}^{\infty} b(x') e^{-i\frac{2\pi}{\Delta x}kTx'} dx' \\ &= \text{sinc}\left(\frac{kT}{\Delta x}\right) \end{aligned} \tag{2.131}$$

Formally, the Fourier integral transform of $\tilde{d}(x)$, according to equation (2.127), is an infinite sequence of impulses having magnitudes, $D_k/\Delta x$. Using equation (2.127) for $\tilde{d}(x)$ and substituting equation (2.131) into the right side yields

$$\mathcal{F}\left(\tilde{d}(x)\right) = \frac{1}{\Delta x} \sum_{k=-\infty}^{\infty} \text{sinc}\left(\frac{kT}{\Delta x}\right) \delta\left(f - \frac{k}{\Delta x}\right). \tag{2.132}$$

In the limit as $T \to 0$, and with equations (2.129) and (2.94), one has the result,

$$\mathcal{F}\left(\sum_{k=-\infty}^{\infty} \delta(x - k\Delta x)\right) = \frac{1}{\Delta x} \sum_{k=-\infty}^{\infty} \delta\left(f - \frac{k}{\Delta x}\right). \tag{2.133}$$

The Fourier transform of the sampling function is, again, a sampling function, but now in the frequency domain.

2.4 Two-Dimensional Transforms in Cartesian Space

Geodetic and geophysical data may be collected in one spatial dimension, such as topography or potential field quantities measured on profiles using airborne sensor systems, and can certainly be analyzed in that single dimension. However, it is more typical that such data ultimately are amassed to represent a signal whose domain is more naturally the plane, or the sphere, or even three-dimensional space. All the concepts of Fourier series and integral transforms carry over into these higher dimensions with no effort if the underlying coordinate system is Cartesian.

The reason for this facility is the *separability* of the basis functions of these spaces, the complex exponentials, into products of univariate functions corresponding to the individual coordinates, $x_1, x_2, \ldots$. Therefore, the following can be stated without much further theoretical discussion, where it is mostly a straightforward matter of including an additional coordinate, summation, and integral.

2.4.1 Two-Dimensional Fourier Series Transform

Periodic functions in two Cartesian dimensions are periodic with respect to each coordinate with possibly different periods, P_1 and P_2,

$$\tilde{g}(x_1 \pm n_1 P_1, x_2 \pm n_2 P_2) = \tilde{g}(x_1, x_2), \text{ for any integers, } n_1 \text{ and } n_2. \tag{2.134}$$

The Dirichlet conditions are assumed for $\tilde{g}(x_1, x_2)$ with appropriate modifications; for example, absolute integrability means that

$$\int_0^{P_1}\int_0^{P_2} |\tilde{g}(x_1, x_2)|\, dx_1 dx_2 < \infty. \tag{2.135}$$

Furthermore, as before, we assume that $\tilde{g}(x_1, x_2)$ is also square-integrable and that is has bounded variation in both dimensions. The basis functions for the space of these periodic signals are $e^{i2\pi(x_1 k_1/P_1 + x_2 k_2/P_2)}$, which satisfy the orthogonality relationship (cf. equation (2.12)),

$$\int_0^{P_1}\int_0^{P_2} e^{i2\pi\left(\frac{k_1-\ell_1}{P_1}x_1 + \frac{k_2-\ell_2}{P_2}x_2\right)} dx_2 dx_1 = P_1 P_2 \delta_{k_1-\ell_1} \delta_{k_2-\ell_2}, \qquad P_1 \neq 0, P_2 \neq 0, \tag{2.136}$$

for all integers, k_1, k_2, ℓ_1, ℓ_2. Then, the Fourier coefficients are defined by

$$G_{k_1,k_2} = \int_0^{P_1}\int_0^{P_2} \tilde{g}(x_1, x_2) e^{-i2\pi\left(\frac{k_1}{P_1}x_1 + \frac{k_2}{P_2}x_2\right)} dx_1 dx_2, \quad -\infty < k_1, k_2 < \infty, \tag{2.137}$$

where the frequencies in the two coordinate directions are k_1/P_1 and k_2/P_2, respectively. The units of G_{k_1,k_2} are the units of $\tilde{g}(x_1, x_2)$ divided by the units of both frequencies. For continuous functions, the corresponding uniformly convergent Fourier series is

$$\tilde{g}(x_1, x_2) = \frac{1}{P_1 P_2} \sum_{k_1=-\infty}^{\infty} \sum_{k_2=-\infty}^{\infty} G_{k_1,k_2} e^{i2\pi\left(\frac{k_1}{P_1}x_1 + \frac{k_2}{P_2}x_2\right)}, \quad -\infty < x_1, x_2 < \infty. \tag{2.138}$$

Allowing for a finite number of step-discontinuities, the series converges to

$$\frac{1}{4}\lim_{\delta x_1\to 0+}\lim_{\delta x_2\to 0+}\begin{pmatrix}\tilde{g}(x_1-\delta x_1,x_2-\delta x_2)+\tilde{g}(x_1+\delta x_1,x_2-\delta x_2)+\\ \tilde{g}(x_1-\delta x_1,x_2+\delta x_2)+\tilde{g}(x_1+\delta x_1,x_2+\delta x_2)\end{pmatrix}$$

$$=\frac{1}{P_1P_2}\sum_{k_1=-\infty}^{\infty}\sum_{k_2=-\infty}^{\infty}G_{k_1,k_2}e^{i2\pi\left(\frac{k_1}{P_1}x_1+\frac{k_2}{P_2}x_2\right)}. \tag{2.139}$$

Periodic functions defined on higher-dimensioned Cartesian space have analogous series expansions and Fourier series transforms with obvious generalization.

The properties of proportionality, superposition, translation, and differentiation (properties (2.20), (2.21), (2.23), and (2.25)) for two- (or higher-) dimensioned functions are completely analogous and need not be repeated. The symmetry property in two dimensions is

$$\tilde{g}(-x_1,-x_2)\leftrightarrow G_{-k_1,-k_2},\quad \tilde{g}(-x_1,x_2)\leftrightarrow G_{-k_1,k_2},\quad \tilde{g}(x_1,-x_2)\leftrightarrow G_{k_1,-k_2}, \tag{2.140}$$

where the proofs are identical to those for the transform pair (2.22) (see Exercise 2.2). Results similar to properties (2.26) through (2.28) are the following (Exercise 2.11). Directly from the symmetry property,

$$\tilde{g}(\pm x_1,\pm x_2)=\tilde{g}(x_1,x_2)\quad\text{if and only if}\quad G_{\pm k_1,\pm k_2}=G_{k_1,k_2}, \tag{2.141}$$

where the signs of the space variables on the left match those of the wave numbers on the right. For real-valued functions,

$$\tilde{g}(x_1,x_2)=\tilde{g}^*(x_1,x_2)\quad\text{if and only if}\quad G_{-k_1,k_2}=G^*_{k_1,-k_2}; \tag{2.142}$$

and, for real-valued spectra,

$$\tilde{g}(-x_1,x_2)=\tilde{g}^*(x_1,-x_2)\quad\text{if and only if}\quad G_{k_1,k_2}=G^*_{k_1,k_2}. \tag{2.143}$$

Finally, combining statements (2.141) and (2.142), one obtains

$$\tilde{g}(x_1,x_2)=\tilde{g}^*(x_1,x_2)\text{ and }\tilde{g}(\pm x_1,\pm x_2)=\tilde{g}(x_1,x_2)\quad\text{if and only if}$$

$$G_{-k_1,k_2}=G^*_{k_1,-k_2}\text{ and }G_{\pm k_1,\pm k_2}=G_{k_1,k_2}. \tag{2.144}$$

again, where the options on "±" must match on both sides of the mutual implications. The spectrum is real only if the signs are negative for both dimensions; that is, $G_{-k_1,k_2}=G^*_{k_1,-k_2}$ and $G_{-k_1,-k_2}=G_{k_1,k_2}$ implies that $G^*_{k_1,k_2}=G^*_{-k_1,-k_2}=G_{k_1,k_2}$.

Parseval's theorem for two-dimensional, square-integrable, functions follows in the same manner as for equation (2.46),

$$\int_0^{P_1}\int_0^{P_2}\tilde{g}_1(x_1,x_2)\,\tilde{g}_2^*(x_1,x_2)\,dx_1dx_2=\frac{1}{P_1}\frac{1}{P_2}\sum_{k_1=-\infty}^{\infty}\sum_{k_2=-\infty}^{\infty}(G_1)_{k_1,k_2}(G_2)^*_{k_1,k_2}, \tag{2.145}$$

provided that both $\tilde{g}_1(x_1, x_2)$ and $\tilde{g}_2(x_1, x_2)$ have the same periods, P_1 and P_2, in their respective dimensions.

2.4.2 Two-Dimensional Fourier Integral Transform

The transition to non-periodic functions in two Cartesian dimensions now follows the lines of reasoning developed for one dimension. The Dirichlet conditions require absolute integrability over the x_1, x_2-plane,

$$\int_{-\infty}^{\infty}\int_{-\infty}^{\infty}\left|g(x_1,x_2)\right|dx_1dx_2<\infty, \tag{2.146}$$

as well as bounded variation in the two dimensions. As in one dimension, this means that the function attenuates to zero in all directions,

$$\lim_{|x_1|\to\infty \text{ or } |x_2|\to\infty} g(x_1,x_2)=0. \tag{2.147}$$

The Fourier integral equation, analogous to equation (2.52), for continuous functions then holds,

$$g(x_1,x_2)=\int_{-\infty}^{\infty}\int_{-\infty}^{\infty}\left(\int_{-\infty}^{\infty}\int_{-\infty}^{\infty}g(x_1',x_2')e^{-i2\pi(f_1x_1'+f_2x_2')}dx_1'dx_2'\right)e^{i2\pi(f_1x_1+f_2x_2)}df_1df_2. \tag{2.148}$$

Thus, the Fourier transform pair for continuous functions is

$$G(f_1,f_2)=\int_{-\infty}^{\infty}\int_{-\infty}^{\infty}g(x_1,x_2)e^{-i2\pi(f_1x_1+f_2x_2)}dx_1dx_2, \tag{2.149}$$

$$g(x_1,x_2)=\int_{-\infty}^{\infty}\int_{-\infty}^{\infty}G(f_1,f_2)e^{i2\pi(f_1x_1+f_2x_2)}df_1df_2. \tag{2.150}$$

Convergence of the left side in the case of a step discontinuity in g is given by

$$\frac{1}{4}\lim_{\delta x_1\to 0+}\lim_{\delta x_2\to 0+}\begin{pmatrix} g(x_1-\delta x_1,x_2-\delta x_2)+g(x_1+\delta x_1,x_2-\delta x_2)+ \\ g(x_1-\delta x_1,x_2+\delta x_2)+g(x_1+\delta x_1,x_2+\delta x_2) \end{pmatrix}$$

$$=\int_{-\infty}^{\infty}\int_{-\infty}^{\infty}G(f_1,f_2)e^{i2\pi(f_1x_1+f_2x_2)}df_1df_2. \tag{2.151}$$

An example of such a function is the two-dimensional rectangle function in Section 2.4.3.

The units of $G(f_1, f_2)$ are the units of g divided by both frequency units. Other properties for the two-dimensional Fourier transform pair naturally follow from the properties of the one-dimensional transform pair, equations (2.70) through (2.78). For example, the transform pairs for partial derivatives of non-negative integer orders, p_1, p_2, are

$$\frac{\partial^{p_1}}{\partial x_1^{p_1}} \frac{\partial^{p_2}}{\partial x_2^{p_2}} g(x_1, x_2) \leftrightarrow (i2\pi f_1)^{p_1} (i2\pi f_2)^{p_2} G(f_1, f_2), \tag{2.152}$$

$$(-i2\pi x_1)^{p_1} (-i2\pi x_2)^{p_2} g(x_1, x_2) \leftrightarrow \frac{\partial^{p_1}}{\partial f_1^{p_1}} \frac{\partial^{p_2}}{\partial f_2^{p_2}} G(f_1, f_2), \tag{2.153}$$

assuming that corresponding derivatives exist. As in the case for the 2-D periodic functions, it is necessary only to elaborate on the symmetry properties. Analogous to transform pairs (2.140), one has

$$g(-x_1, -x_2) \leftrightarrow G(-f_1, -f_2),\ g(-x_1, x_2) \leftrightarrow G(-f_1, f_2),\ g(x_1, -x_2) \leftrightarrow G(f_1, -f_2). \tag{2.154}$$

Similarly, for real and/or even functions, the following results are easily proved.

$$g(\pm x_1, \pm x_2) = g(x_1, x_2) \quad \text{if and only if} \quad G(\pm f_1, \pm f_2) = G(f_1, f_2); \tag{2.155}$$

$$g(x_1, x_2) = g^*(x_1, x_2) \quad \text{if and only if} \quad G(-f_1, f_2) = G^*(f_1, -f_2); \tag{2.156}$$

$$g(-x_1, x_2) = g^*(x_1, -x_2) \quad \text{if and only if} \quad G(f_1, f_2) = G^*(f_1, f_2); \tag{2.157}$$

$$\begin{gathered} g(x_1, x_2) = g^*(x_1, x_2) \text{ and } g(\pm x_1, \pm x_2) = g(x_1, x_2) \quad \text{if and only if} \\ G(-f_1, f_2) = G^*(f_1, -f_2) \text{ and } G(\pm f_1, \pm f_2) = G(f_1, f_2) \end{gathered} \tag{2.158}$$

where, in the latter case, the spectrum is real only if $g(x_1, x_2) = g^*(x_1, x_2)$ and $g(-x_1, -x_2) = g(x_1, x_2)$. Finally, for square-integrable functions,

$$\int_{-\infty}^{\infty} \int_{-\infty}^{\infty} |g(x_1, x_2)|^2 \, dx_1 dx_2 < \infty, \tag{2.159}$$

Parseval's theorem states that

$$\int_{-\infty}^{\infty} \int_{-\infty}^{\infty} g_1(x_1, x_2) g_2^*(x_1, x_2) \, dx_1 dx_2 = \int_{-\infty}^{\infty} \int_{-\infty}^{\infty} G_1(f_1, f_2) G_2^*(f_1, f_2) \, df_1 df_2. \tag{2.160}$$

2.4.3 Special Functions in Two Dimensions

The rectangle function in two dimensions is defined as

$$b(x_1,x_2)=\begin{cases}1, & |x_1|<\frac{1}{2} \text{ and } |x_2|<\frac{1}{2}\\ \frac{1}{2}, & |x_1|=\frac{1}{2} \text{ and } |x_2|\le\frac{1}{2}, \text{ or } |x_1|\le\frac{1}{2} \text{ and } |x_2|=\frac{1}{2}\\ \frac{1}{4}, & |x_1|=\frac{1}{2} \text{ and } |x_2|=\frac{1}{2}\\ 0, & |x_1|>\frac{1}{2} \text{ or } |x_2|>\frac{1}{2}\end{cases} \tag{2.161}$$

where the special values at the edges and corners of the cube (Figure 2.10) are specified in accordance with equation (2.139). The notational distinction between the one-dimensional and the two-dimensional rectangle functions is achieved here by the number of their arguments; and, no confusion is anticipated. The Fourier transform is simply the product of sinc functions (Exercise 2.12):

$$\mathcal{F}(b(x_1,x_2))=B(f_1,f_2)=\operatorname{sinc}(f_1)\operatorname{sinc}(f_2). \tag{2.162}$$

For arbitrary extents, T_1, T_2, one has the transform pair as in equation (2.100),

$$b^{(T_1,T_2)}(x_1,x_2)=\frac{1}{T_1}\frac{1}{T_2}b\left(\frac{x_1}{T_1},\frac{x_2}{T_2}\right) \quad\leftrightarrow\quad B^{(T_1,T_2)}(f_1,f_2)=\operatorname{sinc}(T_1f_1)\operatorname{sinc}(T_2f_2)=B(T_1f_1,T_2f_2). \tag{2.163}$$

Figure 2.11 shows the two-dimensional Gaussian function and its identically shaped Fourier transform,

$$\gamma(x_1,x_2)=e^{-\pi\left(x_1^2+x_2^2\right)}, \tag{2.164}$$

$$\Gamma(f_1,f_2)=e^{-\pi\left(f_1^2+f_2^2\right)}. \tag{2.165}$$

With different scaling in the two coordinate directions, they are given by

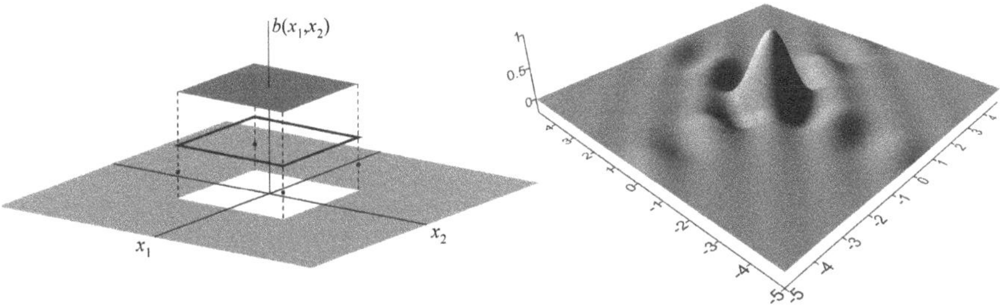

Figure 2.10: The two-dimensional rectangle function (left), equation (2.161), and its Fourier transform (right), equation (2.162).

$$\gamma^{(\beta_1,\beta_2)}\left(x_1,x_2\right)=\frac{1}{\beta_1\beta_2}e^{-\pi\left(\left(x_1/\beta_1\right)^2+\left(x_2/\beta_2\right)^2\right)},\quad \beta_1>0,\ \beta_2>0, \tag{2.166}$$

$$\Gamma^{(\beta_1,\beta_2)}\left(f_1,f_2\right)=e^{-\pi\left(\left(\beta_1 f_1\right)^2+\left(\beta_2 f_2\right)^2\right)}, \tag{2.167}$$

similar to the one-dimensional Gaussian function and its transform, equations (2.107) and (2.108). The volume enclosed by $\gamma^{(\beta_1,\beta_2)}(x_1, x_2)$ is unity for any positive values, β_1, β_2.

Finally, the generalization of the Dirac delta function to two dimensions is

$$\delta\left(x_1,x_2\right)=0,\quad \text{if}\quad x_1\neq 0\quad \text{or}\quad x_2\neq 0, \tag{2.168}$$

such that its Fourier transform is

$$\mathcal{F}\left(\delta\left(x_1,x_2\right)\right)=1,\quad \text{for all } f_1, f_2. \tag{2.169}$$

Hence, similar to equations (2.114) and (2.115), it is straightforward to derive

$$\int_{-\infty}^{\infty}\int_{-\infty}^{\infty}\delta\left(x_1,x_2\right)dx_1dx_2=1, \tag{2.170}$$

$$\int_{-\infty}^{\infty}\int_{-\infty}^{\infty}\delta\left(x_1'-x_1,x_2'-x_2\right)g\left(x_1',x_2'\right)dx_1'dx_2'=g\left(x_1,x_2\right). \tag{2.171}$$

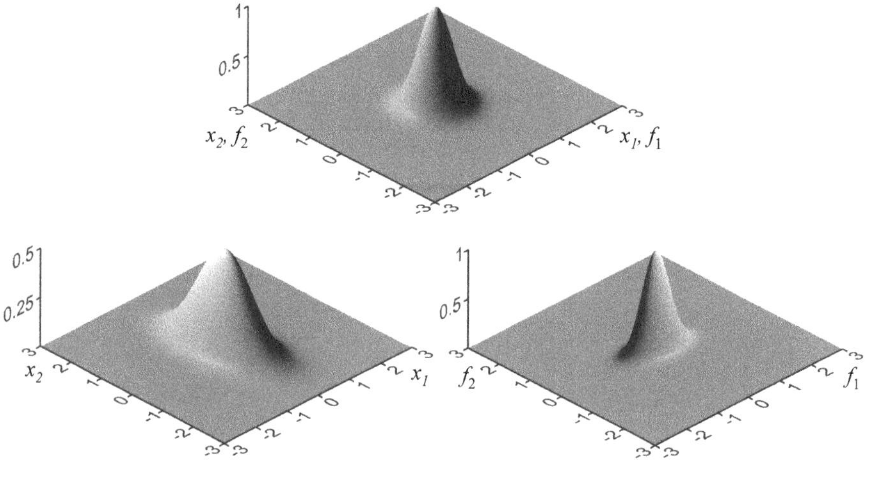

Figure 2.11: Gaussian functions, $\gamma(x_1, x_2)$ (top), $\gamma^{(\beta_1=1,\beta_2=2)}(x_1, x_2)$ (bottom, left), and their Fourier transforms, $\Gamma(f_1, f_2)$ (top), $\Gamma^{(\beta_1=1,\beta_2=2)}(f_1, f_2)$ (bottom, right).

where the latter result may be obtained as in Section 3.2.1. Following the same arguments as for equation (2.120), the inverse Fourier transform is given by

$$\delta(x_1, x_2) = \int_{-\infty}^{\infty}\int_{-\infty}^{\infty} e^{i2\pi(f_1x_1+f_2x_2)} df_1 df_2. \tag{2.172}$$

2.5 The Hankel Transform

Some problems in geodesy and geophysics are simplified significantly if a two-dimensional function on the plane or its spectrum can be approximated or modeled by an *isotropic* function, one that does not depend on direction. Certain functions, such as kernels of convolutions (Chapter 3), are defined specifically by physical laws to be isotropic. For example, the reciprocal distance function that is fundamental in potential theory is independent of direction. In these cases, the corresponding Fourier transforms on the plane simplify to a single integral.

Suppose that a function, $g(x_1, x_2)$, depends only on the distance, s, from the origin,

$$g(x_1, x_2) = g(s). \tag{2.173}$$

The appropriate two-dimensional coordinates on the plane are the polar coordinates, s, ω, related to the Cartesian coordinates by

$$x_1 = s\cos\omega, \quad x_2 = s\sin\omega, \tag{2.174}$$

where ω is an angle reckoned counter-clockwise from the x_1-axis (Figure 2.12), and

$$s = \sqrt{x_1^2 + x_2^2}. \tag{2.175}$$

The domain for these coordinates is defined by $0 \le s < \infty$ and $0 \le \omega \le 2\pi$; and, a differential area element is given by $dx_1 dx_2 = s ds d\omega$.

With a corresponding change in integration variables the Fourier transform of $g(x_1, x_2)$, equation (2.149), becomes

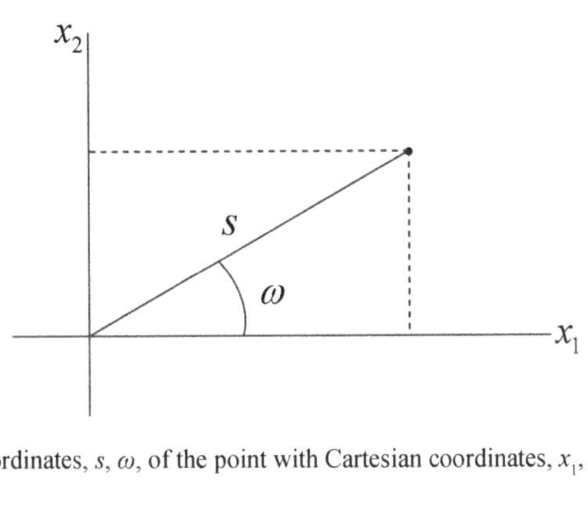

Figure 2.12: Polar coordinates, s, ω, of the point with Cartesian coordinates, x_1, x_2.

$$\mathcal{F}(g(x_1,x_2)) = \int_0^{2\pi}\int_0^{\infty} g(s)e^{-i2\pi\bar{f}s\cos(v-\omega)} s\,d\omega ds$$
$$= \int_0^{\infty} sg(s)\left(\int_0^{2\pi} e^{-i2\pi\bar{f}s\cos(v-\omega)} d\omega\right) ds \tag{2.176}$$

where

$$f_1 = \bar{f}\ \cos v,\quad f_2 = \bar{f}\ \sin v, \tag{2.177}$$

and thus,

$$\bar{f} = \sqrt{f_1^2 + f_2^2}\,. \tag{2.178}$$

The integral in the last line of equation (2.176) with respect to ω is (Gradshteyn and Ryzhik 1980, p.952; noting that the integrand is periodic)

$$\int_0^{2\pi} e^{-i2\pi\bar{f}s\cos(v-\omega)} d\omega = 2\pi J_0\left(2\pi\bar{f}s\right), \tag{2.179}$$

for *any* angle, v, where J_0 is the zero-order Bessel function of the first kind. Therefore,

$$G(f_1,f_2) = G(\bar{f}) = 2\pi\int_0^{\infty} g(s)J_0\left(2\pi\bar{f}s\right)s\,ds, \tag{2.180}$$

which is called the *Hankel transform* (of order zero) of g; also the *Fourier-Bessel transform*. If g is a real signal then since the Bessel function is also real, the Hankel transform is real. Note that although it is a function of only one frequency (the radial frequency, $\bar{f}$), the units of $G(\bar{f})$ still contain the inverse *square* of frequency units.

The Hankel transform defined above is just a special case of the two-dimensional Fourier transform, where the function being transformed has circular symmetry, given by equation (2.173). One then has also the *inverse Hankel transform*, derived from the inverse Fourier transform in exactly the same way. The result for functions defined on the infinite plane is

$$g(s) = 2\pi\int_0^{\infty} G(\bar{f})J_0\left(2\pi\bar{f}s\right)\bar{f}\,d\bar{f}. \tag{2.181}$$

If $g(s)$ has a finite number of step discontinuities, then

$$\frac{1}{2}\lim_{\delta s\to 0+}\left(g(s-\delta s)+g(s+\delta s)\right)=2\pi\int_0^\infty G(\bar{f})J_0(2\pi\bar{f}s)\bar{f}\,d\bar{f}. \tag{2.182}$$

Being a special case of the Fourier transform, the Hankel transform has similar properties. Proportionality (2.70) and superposition (2.71) are exactly the same and need not be repeated. Slight variations in the other properties are shown in the following.

1. Hankel transform pair: $g(s) \leftrightarrow G(\bar{f})$. (2.183)
2. Duality: $G(s) \leftrightarrow g(\bar{f})$; (2.184)
3. Similarity (scaling): $g(as) \leftrightarrow \frac{1}{a^2}G\left(\frac{\bar{f}}{a}\right), \quad a>0;$ (2.185)

Parseval's Theorem in this case derives directly from the corresponding theorem for functions of two independent variables, equation (2.160),

$$\int_0^\infty g_1(s)g_2^*(s)s\,ds=\int_0^\infty G_1(\bar{f})G_2^*(\bar{f})\bar{f}\,d\bar{f} \tag{2.186}$$

It is easy to show (Exercise 2.13) that if

$$\bar{g}(s)=\frac{1}{2\pi}\int_{\omega=0}^{2\pi}g(x_1,x_2)\,d\omega, \tag{2.187}$$

then the Hankel transform of this angular average is the corresponding average of the spectrum, $G(f_1,f_2)$, in the frequency domain,

$$\bar{G}(\bar{f})=2\pi\int_{s=0}^{\infty}\bar{g}(s)sJ_0(2\pi\bar{f}s)\,ds=\frac{1}{2\pi}\int_{\nu=0}^{2\pi}G(f_1,f_2)\,d\nu. \tag{2.188}$$

The isotropic rectangle function is also called the *cylinder* function, defined to enclose unit volume,

$$b_c(s)=\begin{cases}1, & 0\le s<1/\sqrt{\pi}\\ 1/2, & s=1/\sqrt{\pi}\\ 0, & s>1/\sqrt{\pi}\end{cases} \tag{2.189}$$

The Hankel transform of $b_c(s)$ is given by

$$B_c\left(\bar{f}\right)=2\pi\int_0^{1/\sqrt{\pi}} sJ_0\left(2\pi\bar{f}s\right)ds$$

$$=\frac{2\pi}{\left(2\pi\bar{f}\right)^2}\int_{s=0}^{1/\sqrt{\pi}} 2\pi\bar{f}s\,J_0\left(2\pi\bar{f}s\right)d\left(2\pi\bar{f}s\right)=\frac{1}{\bar{f}\left(2\pi\bar{f}\right)}\left[2\pi\bar{f}sJ_1\left(2\pi\bar{f}s\right)\right]_0^{1/\sqrt{\pi}}$$

$$=\frac{1}{\bar{f}\sqrt{\pi}}J_1\left(2\sqrt{\pi}\bar{f}\right) \tag{2.190}$$

where J_1 is the first-order Bessel function of the first kind, and $\frac{d}{dx}\left(xJ_1(x)\right)=xJ_0(x)$. It is noted that $B_c(0) = 1$, since $\lim_{x\to 0}\left(J_1(x)/x\right)=1/2$. For a different cylinder radius, $s_\mu=\mu/\sqrt{\pi}$, $\mu > 0$, the similarity property (2.185) shows that

$$b_c^{(\mu)}(s) = b_c(s/\mu)/\mu^2 \tag{2.191}$$

has the Hankel transform,

$$B_c^{(\mu)}\left(\bar{f}\right)=B_c\left(\mu\bar{f}\right)=\frac{1}{\mu\bar{f}\sqrt{\pi}}J_1\left(2\sqrt{\pi}\mu\bar{f}\right)=\frac{1}{\mu\bar{f}\sqrt{\pi}}J_1\left(2\pi s_\mu\bar{f}\right). \tag{2.192}$$

Both the two-dimensional Gaussian function and Dirac delta function are isotropic and their corresponding Fourier transforms, equations (2.165) and (2.169), respectively, are also their Hankel transforms. Note that the generalized Gaussian function, equation (2.166) is isotropic only if $\beta_1=\beta_2=\beta_p$,

$$\gamma^{(\beta_p)}(s)=\frac{1}{\beta_p^2}e^{-\pi s^2/\beta_p^2} \quad\leftrightarrow\quad \Gamma^{(\beta_p)}\left(\bar{f}\right)=e^{-\pi\beta_p^2\bar{f}^2}. \tag{2.193}$$

2.6 Legendre Transforms

While Cartesian coordinates are adequate for many local analyses in geodesy and geophysics, global data sets are increasingly prevalent, especially since measurement systems on board Earth-orbiting satellites are generating voluminous amounts of data related to the solid Earth, the oceans, and the atmosphere. Because of the near-spherical shape of the Earth, it is thus advantageous to develop a spectral analysis of functions on a sphere. It is also possible to refine the spectral analysis to functions on a biaxial ellipsoid (spheroid) that is rotationally symmetric with respect to the polar axis and to functions on a tri-axial ellipsoid using appropriately defined coordinates (Hobson 1965). However, these analyses are not as far reaching in applications since there is no simple spectral treatment of a convolution (Chapter 3), thus making filtering and covariance analysis difficult, if not intractable. Spheroidal analysis certainly has a following in geodesy (Bölling and Grafarend 2005) for the simple reason that the Earth has a significant flattening along the polar axis of about 0.3%. And, ellipsoidal analysis has found application for asteroids and extraterrestrial small moons that are neither spherical nor spheroidal, and might best be approximated by a tri-axial ellipsoid (Garmier and Barriot 2001).

Nevertheless, spherical spectral analysis remains particularly appealing for terrestrial applications and is usually the first analysis performed even for irregular bodies. Therefore, the exposition here is limited to functions on the spatial domain, Ω, defined as the set of points on a sphere having spherical polar coordinates, θ, λ, representing, for the Earth, the geocentric co-latitude and longitude (Figure 1.1),

$$\Omega = \{(\theta, \lambda) \mid 0 \le \theta \le \pi,\ 0 \le \lambda \le 2\pi\}. \tag{2.194}$$

One could also use geocentric latitude, ϕ, instead, where $\phi = \pi/2 - \theta$, and there is no conventional preference. Since the spherical spatial domain is finite, the basis functions form a countable set; in other words, the corresponding function space is separable. As before, it is assumed that the functions satisfy the Dirichlet conditions, where absolute integrability means that

$$\iint_{\Omega} \left| g(\theta, \lambda) \right| d\Omega < \infty. \tag{2.195}$$

Despite the obvious periodicity (period, 2π) in longitude, we abandon the notation for periodic functions since for most applications there is no need to distinguish between the finite spherical domain and a corresponding infinite domain. However, it is possible to develop two-dimensional Fourier series for g defined on the spatial domain in the form of equation (2.138) using $x_1 \equiv \theta$ and $x_2 \equiv \lambda$, with periods, $P_1 = \pi$ and $P_2 = 2\pi$. This is done to derive some results on aliasing in Chapter 4, and we revert to the notation, $\tilde{g}$, for this particular instance. The polar regions can be awkward in this case, and one must impose the constraints, $\tilde{g}(0, \lambda) = c_N$ and $\tilde{g}(\pi, \lambda) = c_S$, for all λ, where c_N, c_S are constants. No such multiplicity occurs with the use of associated Legendre functions, $P_{n,m}(\cos\theta)$ (Section 2.6.2) and together with $\sin(m\lambda)$ and $\cos(m\lambda)$ (or $e^{im\lambda}$) they form a complete basis for functions on the sphere, which is the preferred basis for spherical spectral analysis. The nomenclature for the spectral transforms varies in the literature and a particular classification is adopted here with alternatives noted, and no confusion is anticipated.

Although the geodetic and geophysical applications of spectral analysis on the sphere clearly refer to the Earth, one may assume without loss in generality a *unit sphere*, that is, a sphere whose radius is unity (with no particular units of measure). As the need arises, it is easy to scale the sphere to any radius with defined units of measure; however, the spherical spectrum depends in the first place on the angular variation of the function. The spherical spectral analysis thus always refers to the domain, Ω, and never to a domain that includes a radial dimension; there is no frequency or wave number corresponding to distance from the center of the sphere. On the other hand, for certain three-dimensional functions, such as a field potential, there is a relationship between spherical spectra on different concentric spheres that depends on their relative radii (Section 6.2.2).

The spherical functions in geodesy that depend on just one angular coordinate invariably are kernel functions in a convolution (Chapter 3); that is, they depend on the angular distance, ψ, from some given point on the sphere. Angular "distance" refers to the length in radians of a great circle arc on the unit sphere subtended at its center by the angle, ψ; its range of values is $0 \le \psi \le \pi$. Thus, any segment on a meridian (great circle containing the poles), being also an interval in co-latitude at constant longitude, is such an angular distance. An interval of longitude at constant co-latitude, however, does not define an angular distance, or great circle arc, unless the co-latitude happens to be $\theta = \pi/2$ (the equator). Thus, spherical functions of one angular coordinate may also be denoted, $g(\theta)$, with the origin for θ at the north pole.

Imagine rotating the origin of the θ, λ-coordinate system (the north pole) to a given point, (θ, λ); then, the angular distance to any other point on the sphere is its co-latitude with respect to that new pole (Figure 2.13). More generally, the coordinates of a point, (θ', λ'), relative to a pole (origin) at (θ, λ) are defined by the spherical polar coordinates, ψ, ζ, given by

$$\cos\psi = \cos\theta\cos\theta' + \sin\theta\sin\theta'\cos(\lambda' - \lambda), \tag{2.196}$$

$$\tan\zeta = -\frac{\sin\theta'\sin(\lambda' - \lambda)}{\sin\theta\cos\theta' - \cos\theta\sin\theta'\cos(\lambda' - \lambda)}. \tag{2.197}$$

Equation (2.196) is the law of cosines and equation (2.197) comes from the formulas,

$$\sin\psi\,\sin(\pi - \zeta) = \sin\theta'\sin(\lambda' - \lambda), \tag{2.198}$$

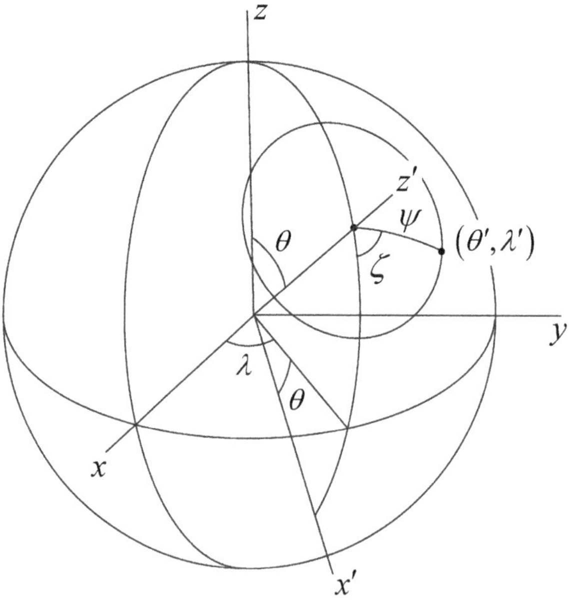

Figure 2.13: Coordinates, ψ, ζ, of the point, (θ', λ'), with respect to the pole and meridian at (θ, λ), obtained by first rotating the x-axis to the meridian plane of (θ, λ), followed by rotating the z-axis to the point, (θ, λ), resulting in the x', y', z'-system.

$$\sin\psi \cos(\pi - \zeta) = \sin\theta \cos\theta' - \cos\theta \sin\theta' \cos(\lambda - \lambda'), \tag{2.199}$$

where the first is the law of sines and second is easily derived (Exercise 2.14). Note that the angle, ζ, is like a longitude (positive counter-clockwise) with respect to the reference meridian through the pole at (θ, λ).

2.6.1 One-Dimensional Legendre Transform

A univariate function on the sphere is defined to depend solely on co-latitude, denoted in this case by $g(\psi)$ (later also by $g(\theta)$), with domain equal to a semi-circle, $0 \le \psi \le \pi$. On this interval, as noted in Section 1.3 for functions defined on a finite interval, the orthogonal polynomials, $e^{i2k\psi}$, $-\infty < k < \infty$, form a complete set of basis functions. However, in anticipation of the two-dimensional basis functions on the sphere, we use another set of orthogonal polynomials that form a complete basis for functions on the semi-circle. These are the Legendre polynomials, $P_n(y)$, $n \ge 0$, with argument, $y = \cos\psi$ $(-1 \le y \le 1)$. When generalized to the sphere they avoid the aforementioned problem with $e^{i2k\psi}$ at the poles. The argument, y, should not be confused with the global Cartesian y-coordinate.

The Legendre polynomials are solutions to Legendre's differential equation (e.g., Arfken 1970, p.384),

$$\frac{d}{dy}\left(\left(1-y^2\right)\frac{d}{dy}P_n(y)\right) + n(n+1)P_n(y) = 0, \tag{2.200}$$

and are generated, for example, by *Rodrigues's formula*,

$$P_n(y) = \frac{1}{2^n n!}\frac{d^n}{dy^n}\left(y^2 - 1\right)^n, \quad n \ge 0. \tag{2.201}$$

A more practical and numerically very stable recursion formula for $P_n(y)$ is (Abramowitz and Stegun 1972)

$$(n+1)P_{n+1}(y) = (2n+1)yP_n(y) - nP_{n-1}(y), \quad n \ge 1; \quad P_0(y) = 1, \quad P_1(y) = y. \tag{2.202}$$

Values of the Legendre polynomials at the interval endpoints, $y = \pm 1$, are

$$P_n(1) = 1, \quad P_n(-1) = (-1)^n, \quad \text{for all } n \ge 0; \tag{2.203}$$

and, the orthogonality on this interval is

$$\int_{-1}^{1} P_n(y)P_{n'}(y)\,dy = \frac{2}{2n+1}\delta_{n-n'}, \tag{2.204}$$

as proved, for example, in (Arfken 1970, p.546). Graphs of the Legendre polynomial for small n are shown in Figure 2.14.

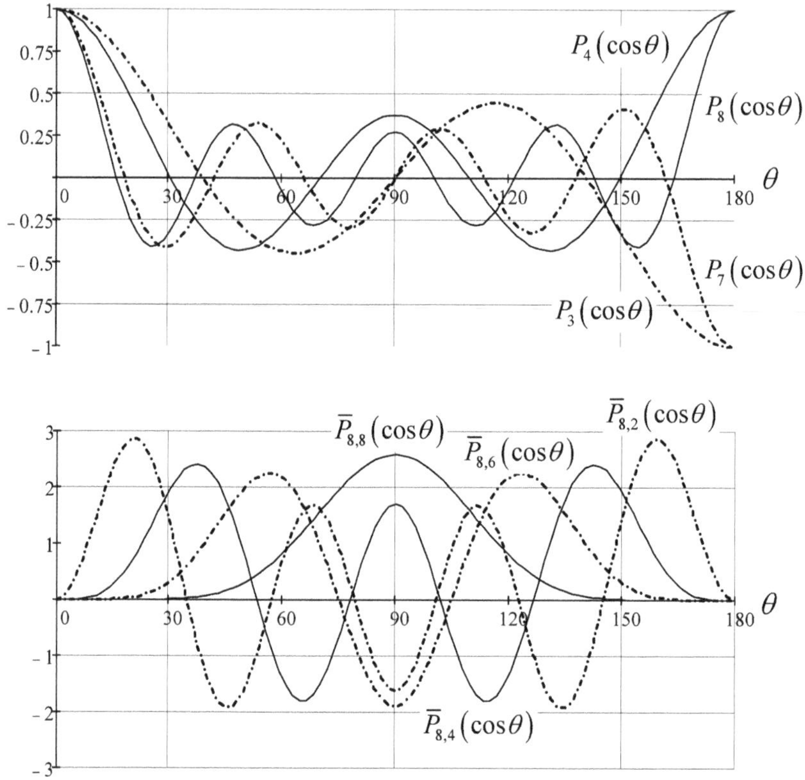

Figure 2.14: Legendre polynomials, $P_n(\cos\theta)$ (top), for selected even and odd degrees; and normalized associated Legendre functions, $\bar{P}_{n,m}(\cos\theta)$ (bottom), for degree, $n = 8$, and selected orders.

We define the *Legendre transform* of a univariate function, $g(\psi)$, by

$$G_n = \frac{1}{2}\int_0^\pi g(\psi) P_n(\cos\psi)\sin\psi\, d\psi. \tag{2.205}$$

The set, $\{G_n\}$, is also known as the *Legendre spectrum* of $g(\psi)$. If $g(\psi)$ has bounded variation and is continuous then the inverse Legendre transform converges to $g(\psi)$ for $0 \le \psi \le \pi$,

$$g(\psi) = \sum_{n=0}^{\infty}(2n+1)G_n P_n(\cos\psi). \tag{2.206}$$

At a step discontinuity in the interior of the interval, $0 < \psi < \pi$, the convergence is given by

$$\frac{1}{2}\lim_{\delta\psi\to 0+}\left(g(\psi-\delta\psi)+g(\psi+\delta\psi)\right) = \sum_{n=0}^{\infty}(2n+1)G_n P_n(\cos\psi). \tag{2.207}$$

This latter convergence depends on the behavior of the function near and at the endpoints of the interval, $(0, \pi)$, requiring that $g(\psi)\sqrt{\sin\psi}$ be integrable over the interval (Hobson 1965, p.329). An example of a piecewise continuous $g(\psi)$ is the spherical analogue of the rectangle function defined in Section 2.6.5. Together, equations (2.205) and (2.206) (or, generally, (2.207)) constitute the one-dimensional *Legendre transform pair*. The integer, n, is called the degree of the Legendre polynomial and also identifies a wave number of the Legendre spectrum. The units of the spectrum, $\{G_n\}$, in this case are the same as the units of the function, g.

The Legendre transform defined here should not be confused with the Legendre transformation (also called the Legendre transform), which is associated with a different way of encoding information about a function or signal in terms of its derivative rather than its inherent independent variable (Boas 1966, Zia et al. 2009). Since the spectral decomposition is specifically with respect to Legendre polynomials the nomenclature here, also used by Freeden et al. (1998), is natural, if not unique. Most texts simply refer to the transform as a Legendre series. To emphasize the connection to the Fourier transform Kaplan (1973) uses the term Fourier-Legendre series. The name Fourier-Legendre transform is reserved here for the two-dimensional version discussed in the next section.

Since the one-dimensional Legendre series and transform are used in geodesy and geophysics exclusively as special cases of the two-dimensional series and transform on the sphere, corresponding properties and special functions are discussed in connection with the latter.

2.6.2 Two-Dimensional Fourier-Legendre Transform

It may be assumed without loss of generality in applications that functions, g, on the sphere are defined on the finite domain, Ω (equation (2.194)), with the constraint that $g(\theta, 0) = g(\theta, 2\pi)$ for any θ. Since they may be extended periodically in λ for $-\infty < \lambda < \infty$ with period 2π, the basis functions depending on λ have the form of the cosine and sine functions, or alternatively the complex exponential function. For the co-latitude, however, the special geometry of the sphere calls for the *associated Legendre functions*, depending on $y = \cos\theta$ and generated by

$$P_{n,m}(y) = \frac{1}{2^n n!}\left(1-y^2\right)^{m/2}\frac{d^{n+m}}{dy^{n+m}}\left(y^2-1\right)^n, \tag{2.208}$$

for integers, $n \geq 0$, $0 \leq m \leq n$, called degree and order, respectively. Specifically, these are the associated Legendre functions of the first kind and solutions to the associated Legendre differential equation (Arfken 1970, p.559),

$$\frac{d}{dy}\left(\left(1-y^2\right)\frac{d}{dy}P_{n,m}(y)\right) + \left(n(n+1) - \frac{m^2}{1-y^2}\right)P_{n,m}(y) = 0. \tag{2.209}$$

Comparing equations (2.201) and (2.208), the zero-order ($m = 0$) associated Legendre functions are Legendre polynomials, $P_{n,0}(y) = P_n(y)$.

The associated Legendre functions are orthogonal on $0 \le \theta \le \pi$, or $-1 \le y \le 1$,

$$\int_{-1}^{1} P_{n,m}(y) P_{n',m}(y)\, dy = \frac{2}{2n+1} \frac{(n+m)!}{(n-m)!} \delta_{n-n'}, \tag{2.210}$$

for any $0 \le m \le n$. A suitable normalization is usually included to ease calculations and mathematical manipulation,

$$\bar{P}_{n,m}(y) = \sqrt{\frac{2n+1}{\varepsilon_m} \frac{(n-m)!}{(n+m)!}}\, P_{n,m}(y), \tag{2.211}$$

where (allowing for $m < 0$ in later formulas)

$$\varepsilon_m = \begin{cases} 1/2, & 0 < |m| \le n \\ 1, & m = 0 \end{cases} \tag{2.212}$$

Figure 2.14 illustrates the oscillatory behavior of these functions in co-latitude. Recursion relations are given by (Abramowitz and Stegun 1972),

$$\bar{P}_{n,m}(y) = \alpha_{n,m} y \bar{P}_{n-1,m}(y) - \beta_{n,m} \bar{P}_{n-2,m}(y), \quad 0 \le m \le n-2, \quad n \ge 2, \tag{2.213}$$

where

$$\alpha_{n,m} = \sqrt{\frac{(2n-1)(2n+1)}{(n-m)(n+m)}}, \quad \beta_{n,m} = \sqrt{\frac{(2n+1)(n+m-1)(n-m-1)}{(2n-3)(n+m)(n-m)}}, \tag{2.214}$$

and the starting functions are

$$\bar{P}_{n,n-1}(y) = \sqrt{2n+1}\, y \bar{P}_{n-1,n-1}(y), \quad n \ge 1; \tag{2.215}$$

$$\bar{P}_{n,n}(y) = \sqrt{\frac{2n+1}{2n}} \sqrt{1-y^2}\, \bar{P}_{n-1,n-1}(y), \quad n \ge 2; \tag{2.216}$$

$$\bar{P}_{0,0}(y) = 1, \ \bar{P}_{1,1}(y) = \sqrt{3}\sqrt{1-y^2}. \tag{2.217}$$

The recursion (2.213) becomes numerically degenerate for very high degrees, and special numerical techniques are required to obtain sufficient precision for $n > 2500$ (e.g., Fukushima 2012a).

It can be shown that the associated Legendre functions, $\bar{P}_{n,m}(\cos\theta)$, in combination with the sinusoidal functions of longitude, called the *spherical harmonic functions*,

$$\bar{Y}_{n,m}(\theta,\lambda) = \bar{P}_{n,|m|}(\cos\theta) \begin{cases} \cos m\lambda, & 0 \le m \le n \\ \sin |m|\lambda, & -n \le m < 0 \end{cases} \tag{2.218}$$

form a complete orthogonal basis for functions on the sphere (Cushing 1975, p.158). This is the form used in geodesy and geophysics. The particular notation that includes negative integers, m, is only a matter of convenience that allows a single function, $\bar{Y}_{n,m}(\theta, \lambda)$, to represent both the sine and cosine components and that avoids the use of complex functions. The normalization of the Legendre functions also yields a particularly simple orthogonality relationship. In fact, the $\bar{Y}_{n,m}(\theta, \lambda)$ are *orthonormal* on the unit sphere,

$$\frac{1}{4\pi}\iint_{\Omega}\bar{Y}_{n,m}(\theta,\lambda)\bar{Y}_{n',m'}(\theta,\lambda)\,d\Omega=\delta_{n-n'}\delta_{m-m'} \tag{2.219}$$

where $d\Omega = \sin\theta d\theta d\lambda$.

The two-dimensional *Fourier-Legendre transform* of a function, g, on the sphere is defined by

$$G_{n,m}=\frac{1}{4\pi}\iint_{\Omega}g(\theta,\lambda)\bar{Y}_{n,m}(\theta,\lambda)\,d\Omega,\quad -n\le m\le n,\quad n\ge 0; \tag{2.220}$$

and under the usual conditions of bounded variation and absolute integrability, the corresponding two-dimensional *Fourier-Legendre series* converges to g at points of continuity,

$$g(\theta,\lambda)=\sum_{n=0}^{\infty}\sum_{m=-n}^{n}G_{n,m}\bar{Y}_{n,m}(\theta,\lambda). \tag{2.221}$$

This series is sometimes called the Laplace series (Hobson 1965, which, however, should not be confused with the Laplace transform), or also the spherical surface harmonic series (Blakely 1995; Heiskanen and Moritz 1967). Since the transform and the series involve both Legendre functions and sinusoidal functions, the given nomenclature is preferred, as used also by Kaplan (1973). Note, however, that $G_{n,m}$ is *not* the Fourier series transform of g in longitude for any fixed co-latitude. The set, $\{G_{n,m}\}$, is called the (two-dimensional) *Fourier-Legendre spectrum* of $g(\theta, \lambda)$; it is real since both $g(\theta, \lambda)$ and the basis functions are real. The units of the spectrum are the units of the function.

The integers, n, m, correspond roughly to wave numbers (frequencies) in latitude and longitude. For any particular degree, n, an increase in the order, $|m|$, decreases the number of zeros, or waves, of the associated Legendre functions, $\bar{P}_{n,m}(\cos\theta)$. $\bar{P}_{n,0}(\cos\theta)$ has n zeros or about $\lfloor n/2 \rfloor$ waves; while $\bar{P}_{n,n}(\cos\theta)$ has only one wave, as it vanishes only at both poles (Figure 2.14). The spherical harmonic functions, $\bar{Y}_{n,\pm n}(\theta, \lambda)$, thus have zeros only in longitude with corresponding waves effectively dividing the sphere into sectors; whence their name, *sectorial harmonics*. The spherical harmonic functions, $\bar{Y}_{n,0}(\theta, \lambda)$, are independent of longitude and the waves in co-latitude divide the sphere into zones; they are called *zonal harmonics*. All other spherical harmonics, $\bar{Y}_{n,m}(\theta, \lambda)$, $0 < |m| < n$, tessellate the sphere with waves and are called *tesseral harmonics* (Figure 2.15).

An alternative form of the two-dimensional Fourier-Legendre transform pair may also be defined, although it is not as popular in geodesy and geophysics. If the

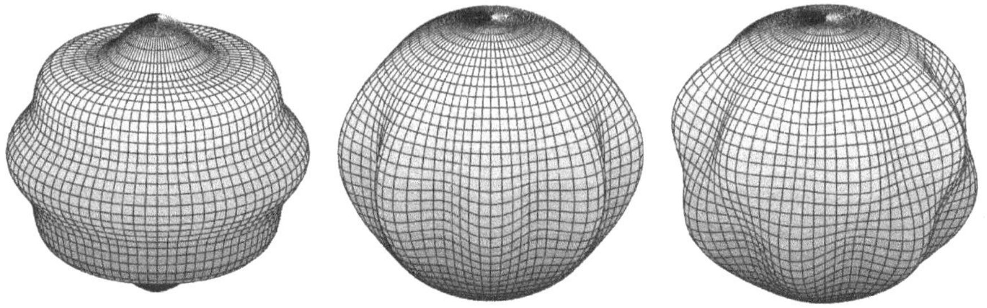

Figure 2.15: Zonal, sectorial, and tesseral spherical harmonic functions, $\bar{Y}_{8,0}(\theta, \lambda)$ (left), $\bar{Y}_{8,8}(\theta, \lambda)$ (middle), and $\bar{Y}_{8,7}(\theta, \lambda)$ (right), respectively.

basis functions combine in terms of the complex exponential in longitude, then the Fourier-Legendre transform is

$$G^c_{n,m} = \frac{1}{4\pi}\iint_{\Omega} g(\theta,\lambda)\bar{Y}^c_{n,m}(\theta,\lambda)\,d\Omega, \tag{2.222}$$

where we define

$$\bar{Y}^c_{n,m}(\theta,\lambda) = \sqrt{(2n+1)\frac{(n-|m|)!}{(n+|m|)!}}P_{n,|m|}(\cos\theta)e^{-im\lambda} = \sqrt{\varepsilon_m}\bar{P}_{n,|m|}(\cos\theta)e^{-im\lambda}. \tag{2.223}$$

The sign of the exponential, differing from definitions in other texts, such as Arfken (1970, p.571), here is consistent with the Fourier transform functions defined by equation (2.56). The complex-valued spherical harmonic functions are also orthonormal,

$$\frac{1}{4\pi}\iint_{\Omega}\left(\bar{Y}^c_{n,m}(\theta,\lambda)\right)^* \bar{Y}^c_{n',m'}(\theta,\lambda)\,d\Omega = \delta_{m-m'}\delta_{n-n'}; \tag{2.224}$$

and, the corresponding Fourier-Legendre series is

$$g(\theta,\lambda) = \sum_{n=0}^{\infty}\sum_{m=-n}^{n} G^c_{n,m}\left(\bar{Y}^c_{n,m}(\theta,\lambda)\right)^*. \tag{2.225}$$

The relationship between the real and complex spherical harmonic functions is readily seen to be

$$\begin{aligned}\bar{Y}^c_{n,m}(\theta,\lambda) &= \sqrt{\varepsilon_m}(\cos m\lambda - i\sin m\lambda)\bar{P}_{n,|m|}(\cos\theta)\\ &= \sqrt{\varepsilon_m}\begin{cases}\bar{Y}_{n,m}(\theta,\lambda) - i\bar{Y}_{n,-m}(\theta,\lambda), & 1 \le m \le n\\ \bar{Y}_{n,0}(\theta,\lambda), & m = 0\\ \bar{Y}_{n,|m|}(\theta,\lambda) + i\bar{Y}_{n,-|m|}(\theta,\lambda), & -n \le m \le -1\end{cases}\end{aligned} \tag{2.226}$$

Substituting equation (2.226) into equation (2.222) and comparing to equation (2.220) immediately yields the relationships between the coefficients, $\{G^c_{n,m}\}$ and $\{G_{n,m}\}$,

$$G^c_{n,m} = \sqrt{\varepsilon_m}\begin{cases} G_{n,m} - iG_{n,-m}, & 1 \le m \le n \\ G_{n,0}, & m = 0 \\ G_{n,|m|} + iG_{n,-|m|}, & -n \le m \le -1 \end{cases} \tag{2.227}$$

$$G_{n,m} = \frac{1}{2\sqrt{\varepsilon_m}}\begin{cases} G^c_{n,m} + G^c_{n,-m}, & 1 \le m \le n \\ 2G^c_{n,0}, & m = 0 \\ i\left(G^c_{n,|m|} - G^c_{n,-|m|}\right), & -n \le m \le -1 \end{cases} \tag{2.228}$$

It is sometimes more convenient to analyze a function or manipulate its Fourier-Legendre spectrum in terms of the complex basis functions, but the real spectrum, $\{G_{n,m}\}$, with respect to the real spherical harmonics, $\bar{Y}_{n,m}$, can always be found easily from $\{G^c_{n,m}\}$ using equation (2.228). Note that the degree and absolute order, or the wave numbers, of the complex spectrum correspond exactly to those of the real spectrum.

If the function is independent of longitude, $g(\theta,\lambda) = g(\theta)$, then by the orthogonality of $e^{-im\lambda}$, equation (2.12), the Legendre spectrum includes only zonal harmonic coefficients, $G_{n,0}$, i.e., $G_{n,m} = 0$ for $m \neq 0$; and, the series reverts to the one-dimensional Legendre series, equation (2.206). Since the Legendre polynomials generally are not normalized and, by equation (2.211), $\bar{P}_{n,0}(y) = \sqrt{2n+1}P_n(y)$, a comparison of equations (2.206) and (2.221) yields

$$G_{n,0} = \sqrt{2n+1}\,G_n. \tag{2.229}$$

One may also view $g(\theta, \lambda)$ as a function defined on the *rectangle*, $W = \{(\theta, \lambda) | 0 \le \theta < \pi, 0 \le \lambda < 2\pi\}$, and extended *periodically* over the infinite *plane*, $-\infty < \theta < \infty$, $-\infty < \lambda < \infty$. Then, distinguishing this special function as $\tilde{g}(\theta, \lambda)$, the usual Fourier series transform pair is, as in equations (2.138) and (2.137),

$$\tilde{g}(\theta,\lambda) = \frac{1}{2\pi^2}\sum_{k=-\infty}^{\infty}\sum_{m=-\infty}^{\infty} G^F_{k,m} e^{i2\pi\left(\frac{k}{\pi}\theta + \frac{m}{2\pi}\lambda\right)}, \tag{2.230}$$

$$G^F_{k,m} = \int_{\lambda=0}^{2\pi}\int_{\theta=0}^{\pi} \tilde{g}(\theta,\lambda) e^{-i2\pi\left(\frac{k}{\pi}\theta + \frac{m}{2\pi}\lambda\right)} d\theta d\lambda. \tag{2.231}$$

A relationship between the spectra, $G^F_{k,m}$ and $G_{n,m}$, is derived in Section 4.5.

Equation (2.230) holds only where $\tilde{g}(\theta, \lambda)$ is continuous, which includes the meridians, $\lambda = 0, 2\pi$, as assumed here. However, for a function that is continuous on the sphere, its planar analogue has a step discontinuity at the poles, $\theta = 0,\pi$ $(\tilde{g}(0,\lambda) \neq \tilde{g}(\pi,\lambda))$, where the Fourier series converges to $(\tilde{g}(0, \lambda) + \tilde{g}(\pi, \lambda))/2$, according to equation (2.139). Generally, any representation of a spherical geophysical signal using the basis functions, $e^{i(2k\theta+m\lambda)}$, as in equation (2.230), must ensure that the following constraints are satisfied, $\tilde{g}(0, \lambda) = c_N$, $\tilde{g}(\pi, \lambda) = c_S$ for all λ, where c_N and c_S are constants. For example, this is necessary when approximating a spherical function using tensor-product splines (Schumaker and Traas 1991). These constraints are automatically incorporated in the spherical harmonics. That is, from equation (2.208),

$$\bar{P}_{n,m}(\pm 1) = 0, \quad m \neq 0, \tag{2.232}$$

and, therefore, the spherical harmonics do not permit multiple values of $\tilde{g}(\theta, \lambda)$ at the poles for different longitudes—only the Legendre polynomials, $P_n(\cos\theta)$, contribute.

For step discontinuities of $g(\theta, \lambda)$ on the sphere, the following result is proved by Hobson (1965, Chapter 7, Section 211). The discontinuity here is assumed to be along a smooth line having "continuously turning tangent" that passes through (θ, λ) and excludes the poles. Referring to Figure 2.13, define the average of $g(\theta', \lambda')$ over a circle centered at (θ, λ) and having a great-circle arc radius, ψ,

$$\bar{g}(\psi) = \frac{1}{2\pi}\int_0^{2\pi} g(\theta', \lambda')\, d\zeta. \tag{2.233}$$

Thus, $\bar{g}(\psi)$ is an integral over a circle that for ψ in some interval, $[0, \psi_0 \ll \pi]$, crosses the line of discontinuity. If $\bar{g}(\psi)$ has bounded variation for *all* ψ in the interval, $(0, \pi)$, then the Fourier-Legendre series, equation (2.221), converges to

$$\lim_{\psi \to 0+} \bar{g}(\psi) = \sum_{n=0}^{\infty}\sum_{m=-n}^{n} G_{n,m}\bar{Y}_{n,m}(\theta, \lambda). \tag{2.234}$$

2.6.3 Properties of Fourier-Legendre Transforms

Proportionality and superposition properties for the one- and two-dimensional Fourier-Legendre transforms are analogous to corresponding properties (2.20) and (2.21) of the Fourier series transform. Many of the additional properties for Fourier series transforms also hold with respect to the longitude coordinate, λ. Thus, using the complex basis functions, the following hold, where 2. through 4. are obtained directly from properties (2.22), (2.23), and (2.25).

1. $g(\theta, \lambda) \leftrightarrow G^c_{n,m}$: (Complex) Fourier-Legendre transform pair (2.235)
2. Symmetry: $g(\theta, -\lambda) \leftrightarrow G^c_{n,-m}$; (2.236)
3. Translation in λ: $g(\theta, \lambda + \lambda_0) \leftrightarrow G^c_{n,m} e^{im\lambda_0}$; (2.237)

4. Differentiation: $$\frac{\partial^p}{\partial \lambda^p} g(\theta,\lambda) \leftrightarrow (im)^p G^c_{n,m}, \tag{2.238}$$
provided g is differentiable with respect to λ up to order, p;

5. Surface Laplacian: $$\nabla^2_{(\theta,\lambda)} g(\theta,\lambda) \leftrightarrow -n(n+1)G_{n,m}, \tag{2.239}$$

where

$$\nabla^2_{(\theta,\lambda)} = \frac{\partial^2}{\partial \theta^2} + \cot\theta \frac{\partial}{\partial \theta} + \frac{1}{\sin^2\theta} \frac{\partial^2}{\partial \lambda^2} \tag{2.240}$$

is the (scalar) *Laplace-Beltrami operator* of second derivatives on the sphere, also known as the *surface Laplacian.* From equations (2.209), (2.211), and (2.218) with $y = \cos\theta$ it is readily verified (Exercise 2.15) that

$$\nabla^2_{(\theta,\lambda)} \bar{Y}_{n,m}(\theta,\lambda) = -n(n+1)\bar{Y}_{n,m}(\theta,\lambda). \tag{2.241}$$

Applying this to equation (2.221) thus proves the transform pair (2.239).

Per degree, n, there are $2n + 1$ real and independent coefficients, $G_{n,m}$, of order, m. The $2n + 1$ complex coefficients, $G^c_{n,m}$, are not independent (if $g(\theta, \lambda)$ is real), since

$$G^c_{n,m} = \left(G^c_{n,-m}\right)^*. \tag{2.242}$$

However, having both real and imaginary parts, they carry as many independent components.

Parseval's Theorem for the spherical functions is easily derived (Exercise 2.16) using the orthogonality of the spherical harmonic functions, equation (2.219),

$$\frac{1}{4\pi} \iint_\Omega g(\theta,\lambda) h(\theta,\lambda)\, d\Omega = \sum_{n=0}^{\infty} \sum_{m=-n}^{n} G_{n,m} H_{n,m}, \tag{2.243}$$

where all quantities are real-valued; for the complex Fourier-Legendre spectrum,

$$\frac{1}{4\pi} \iint_\Omega g^*(\theta,\lambda) h(\theta,\lambda)\, d\Omega = \sum_{n=0}^{\infty} \sum_{m=-n}^{n} \left(G^c_{n,m}\right)^* H^c_{n,m}. \tag{2.244}$$

The total energy of a (generally complex) function on the sphere is then given by

$$\frac{1}{4\pi} \iint_\Omega \left|g(\theta,\lambda)\right|^2 d\Omega = \sum_{n=0}^{\infty} \sum_{m=-n}^{n} \left|G^c_{n,m}\right|^2 = \sum_{n=0}^{\infty} \sum_{m=-n}^{n} \left|G_{n,m}\right|^2. \tag{2.245}$$

Beyond these properties, the situation is more difficult due to the convergence of the meridians on the sphere. General translation and symmetry properties on the sphere devolve to properties associated with *rotations* on the sphere. From the expressions for the spherical harmonic functions, equations (2.218) and (2.208), and from equations (1.2) it is recognized that $\bar{Y}_{n,m}(\theta, \lambda)$ can be formulated as a polynomial in the global Cartesian coordinates (Müller 1966, Freeden et al. 1998),

$$\bar{Y}_{n,m}(\theta,\lambda) = \sum_{p+q+r=n} a_{p,q,r}^{(n,m)} x^p y^q z^r, \quad m = -n,\ldots,n, \tag{2.246}$$

where the $a_{p,q,r}^{(n,m)}$ are constant coefficients, p, q, r are non-negative integers, and $x^2 + y^2 + z^2 = 1$. This is a homogeneous polynomial, meaning that the powers of the coordinates in each term of the sum add to the same degree, n. It also means that any rotation of the coordinate system yields, again, a homogeneous polynomial, of the same degree, that represents a spherical harmonic function.

Consider the coordinates, ψ, ζ, for the point, (θ', λ'), that are obtained by rotating the pole and zero-meridian to (θ, λ), as in Figure 2.13. Then, a spherical harmonic function in terms of ψ, ζ may be expressed as a linear combination of the spherical harmonics, $\bar{Y}_{n,k}(\theta', \lambda')$, of the same degree,

$$\bar{Y}_{n,m}(\psi,\zeta) = \sum_{m'=-n}^{n} C_{m',m}^{n}(\alpha,\beta,\gamma)\bar{Y}_{n,m'}(\theta',\lambda'), \tag{2.247}$$

where α, β, γ are three Euler angles that describe rotations, in sequence, about the third, the resulting second, and, again, the final third axes. This relationship is derived in detail by Cushing (1975, p.595) for the complex spherical harmonics; and, it is presented there and elsewhere in the context of group theory applied to angular momentum operators in quantum mechanics. For the real spherical harmonic functions it is not difficult to translate Cushing's result to the coefficients in equation (2.247),

$$C_{m',m}^{n}(\alpha,\beta,\gamma) = \begin{cases} \dfrac{(-1)^m}{\sqrt{2\varepsilon_m}}\left((-1)^{m'} c_{m',m}^{n} + c_{-m'',m}^{n}\right), & m' > 0, \quad m \ge 0 \\ \dfrac{(-1)^m}{\sqrt{\varepsilon_m}} c_{0,m}^{n}, & m' = 0, \quad m \ge 0 \\ \dfrac{(-1)^m}{\sqrt{2\varepsilon_m}}\left((-1)^{m'} s_{-m',m}^{n} - s_{m',m}^{n}\right), & m' < 0, \quad m \ge 0 \\ (-1)^{m'} s_{m',m}^{n} + s_{-m',m}^{n}, & m' > 0, \quad m < 0 \\ \sqrt{2} s_{0,m}^{n}, & m' = 0, \quad m < 0 \\ -(-1)^{m'} c_{-m',m}^{n} + c_{m',m}^{n}, & m' < 0, \quad m < 0 \end{cases} \tag{2.248}$$

where

$$c_{m',m}^{n} = \cos(m'\alpha + m\gamma)\, d_{m',m}^{n}(\beta) \tag{2.249}$$

$$s^n_{m',m} = \sin\left(m'\alpha + m\gamma\right) d^n_{m',m}\left(\beta\right) \tag{2.250}$$

and

$$d^n_{m',m}\left(\beta\right) = \sqrt{\frac{(n+m)!(n-m)!}{(n+m')!(n-m')!}}\left(\cos\frac{\beta}{2}\right)^{-m'-m}\left(-\sin\frac{\beta}{2}\right)^{m'-m} P_{n+m}^{(m'-m,-m'-m)}\left(\cos\beta\right), \tag{2.251}$$

and where $P_\ell^{(\alpha,\beta)}(\cos\theta)$ is a Jacobi polynomial of degree, n (Abramowitz and Stegun 1972).

Thus, with reference to Figure 2.13, defining the pole and reference meridian for the coordinates, ψ, ζ, involves specifying the Euler angles, $\alpha = \lambda$, $\beta = \theta$, and $\gamma = 0$. For the special case, $m = 0$, it can be shown (Cushing 1975, pp.593–603) that

$$d^n_{m',0}\left(\theta\right) = \left(-1\right)^{m'}\sqrt{\frac{\varepsilon_{m'}}{(2n+1)}}\bar{P}_{n,m'}\left(\cos\theta\right), \tag{2.252}$$

and

$$d^n_{m',0}\left(\theta\right) = \left(-1\right)^{m'} d^n_{-m',0}\left(\theta\right). \tag{2.253}$$

Substituting these into equations (2.247) through (2.251), one obtains the important *addition theorem for spherical harmonics*,

$$P_n\left(\cos\psi\right) = \frac{1}{2n+1}\sum_{m=-n}^{n}\bar{Y}_{n,m}\left(\theta,\lambda\right)\bar{Y}_{n,m}\left(\theta',\lambda'\right); \tag{2.254}$$

for the complex spherical harmonics, equation (2.223), this is

$$P_n\left(\cos\psi\right) = \frac{1}{2n+1}\sum_{m=-n}^{n}\left(\bar{Y}^c_{n,m}\left(\theta,\lambda\right)\right)^*\bar{Y}^c_{n,m}\left(\theta',\lambda'\right). \tag{2.255}$$

If $\theta = \theta'$ and $\lambda = \lambda'$, then from equation (2.196), $\psi = 0$ and, with equation (2.203),

$$2n+1 = \sum_{m=0}^{n}\left(\bar{P}_{n,m}\left(\cos\theta\right)\right)^2. \tag{2.256}$$

If a function, $g(\theta', \lambda')$, has the Fourier-Legendre spectrum, $\left\{G_{n,m}^{(\theta',\lambda')}\right\}$, with respect to the spherical harmonic functions, $\bar{Y}_{n,m}(\theta', \lambda')$, its spectrum with respect to the spherical harmonics, $\bar{Y}_{n,m}(\psi, \zeta)$, in a coordinate system with origin rotated to (θ, λ) is obtained as follows. First express the functions, $\bar{Y}_{n,m}(\theta', \lambda')$, in terms of $\bar{Y}_{n,m'}(\psi, \zeta)$, which is achieved by setting the Euler angles, $\alpha = 0$, $\beta = -\theta$, $\gamma = -\lambda$, in equation (2.247),

$$\bar{Y}_{n,m}(\theta',\lambda') = \sum_{m'=-n}^{n} C^{n}_{m',m}(0,-\theta,-\lambda)\bar{Y}_{n,m'}(\psi,\zeta), \tag{2.257}$$

Then, insert this into equation (2.221), from which it is evident that the spectrum in the rotated coordinate system is

$$G^{(\psi,\zeta)}_{n,m'} = \sum_{m=-n}^{n} G^{(\theta',\lambda')}_{n,m} C^{n}_{m',m}(0,-\theta,-\lambda). \tag{2.258}$$

For the special case that the rotation is limited to a change in the reference meridian for the longitude, i.e., $\beta = 0$, equation (2.251) shows that $d^{n}_{m',m}(0) = 0$, for $m' \neq m$, and $d^{n}_{m,m}(0) = P^{(0,-2m)}_{n+m}(1) = 1$. Then, also $c^{n}_{m,m} = \cos(m\lambda)$ and $s^{n}_{m,m} = -\sin(m\lambda)$; and, it is readily derived that

$$G^{(\theta',\zeta)}_{n,m'} = \begin{cases} \cos(m'\lambda)G^{(\theta',\lambda')}_{n,m'} + \sin(|m'|\lambda)G^{(\theta',\lambda')}_{n,m'}, & m' > 0 \\ G^{(\theta',\lambda')}_{n,0}, & m' = 0 \\ \cos(m'\lambda)G^{(\theta',\lambda')}_{n,m'} - \sin(|m'|\lambda)G^{(\theta',\lambda')}_{n,m'}, & m' < 0 \end{cases} \tag{2.259}$$

which is the translation property (2.237) for the real spectrum, with the identifications, $\lambda = \lambda_0$ and $\lambda' = \zeta + \lambda_0$.

The Fourier-Legendre spectrum of the derivative of a function on the sphere with respect to co-latitude cannot be derived from the spectrum of the function. Mathematically the derivative of the associated Legendre function either is *not* a linear combination of associated Legendre functions of the same order (Abramowitz and Stegun 1972, p.334),

$$\sin\theta \frac{d}{d\theta}\bar{P}_{n,m}(\cos\theta) = n\cos\theta\bar{P}_{n,m}(\cos\theta) - \sqrt{\frac{(2n+1)(n+m)(n-m)}{2n-1}}\bar{P}_{n-1,m}(\cos\theta); \tag{2.260}$$

or, it is a linear combination of associated Legendre functions of *different* orders (Chapman and Bartels 1940, p.623),

$$\begin{aligned} \frac{d}{d\theta}\bar{P}_{n,m}(\cos\theta) = {} & \frac{1}{2}\sqrt{2\varepsilon_{m-1}(n+m)(n-m+1)}\bar{P}_{n,m-1}(\cos\theta) \\ & -\frac{1}{2}\sqrt{(n+m+1)(n-m)}\bar{P}_{n,m+1}(\cos\theta) \end{aligned} \tag{2.261}$$

In either case, the partial derivative, $\partial\bar{Y}_{n,m}(\theta,\lambda)/\partial\theta$, is not a linear combination of the spherical harmonics, $\bar{Y}_{p,q}(\theta,\lambda)$; consequently, the Fourier-Legendre spectra of $\partial g(\theta,\lambda)/\partial\theta$ and $g(\theta,\lambda)$ are not related analytically. Even the derivative with respect to longitude is usually scaled by $1/\sin\theta$ (e.g., see Section 5.5), which makes the differentiation property (2.238) of the transform unavailing in many applications.

2.6.4 Vector Spherical Harmonics

Although there is no direct analytic relationship between the Fourier-Legendre spectra of a function on the sphere and its horizontal derivative, one may define a basis for a *vector* of the three orthogonal derivatives whose corresponding spectrum is related to $G_{n,m}$. At a point, (θ, λ), on the sphere of radius, r, consider the unit vectors, $\boldsymbol{e}_\theta$, $\boldsymbol{e}_\lambda$, $\boldsymbol{e}_r$, rotated from the unit vectors, $\boldsymbol{e}_x = (1\ 0\ 0)^{\mathrm{T}}$, $\boldsymbol{e}_y = (0\ 1\ 0)^{\mathrm{T}}$, $\boldsymbol{e}_z = (0\ 0\ 1)^{\mathrm{T}}$, by the angles, θ and λ (Figure 2.16),

$$\boldsymbol{e}_\theta = \boldsymbol{e}_x \cos\theta\cos\lambda + \boldsymbol{e}_y \cos\theta\sin\lambda - \boldsymbol{e}_z \sin\theta, \tag{2.262}$$

$$\boldsymbol{e}_\lambda = -\boldsymbol{e}_x \sin\lambda + \boldsymbol{e}_y \cos\lambda. \tag{2.263}$$

$$\boldsymbol{e}_r = \boldsymbol{e}_x \sin\theta\cos\lambda + \boldsymbol{e}_y \sin\theta\sin\lambda + \boldsymbol{e}_z \cos\theta. \tag{2.264}$$

Clearly, they are also orthonormal,

$$|\boldsymbol{e}_\theta| = 1, \quad |\boldsymbol{e}_\lambda| = 1, \quad |\boldsymbol{e}_r| = 1; \tag{2.265}$$

$$\boldsymbol{e}_\theta \cdot \boldsymbol{e}_\lambda = 0, \quad \boldsymbol{e}_\lambda \cdot \boldsymbol{e}_r = 0, \quad \boldsymbol{e}_r \cdot \boldsymbol{e}_\theta = 0; \tag{2.266}$$

and,

$$\boldsymbol{e}_\theta \times \boldsymbol{e}_\lambda = \boldsymbol{e}_r, \quad \boldsymbol{e}_\lambda \times \boldsymbol{e}_r = \boldsymbol{e}_\theta, \quad \boldsymbol{e}_r \times \boldsymbol{e}_\theta = \boldsymbol{e}_\lambda. \tag{2.267}$$

The gradient operator with components in the directions along the unit vectors, $\boldsymbol{e}_x$, $\boldsymbol{e}_y$, $\boldsymbol{e}_z$, is

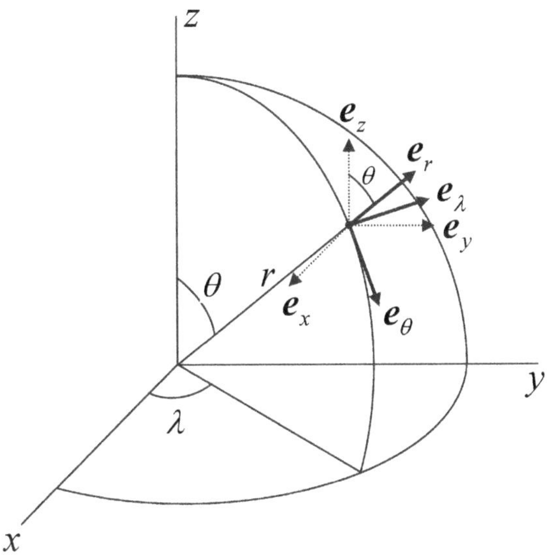

Figure 2.16: Unit vectors, $\boldsymbol{e}_\theta$, $\boldsymbol{e}_\lambda$, $\boldsymbol{e}_r$, rotated by angles, θ and λ, from $\boldsymbol{e}_x$, $\boldsymbol{e}_y$, $\boldsymbol{e}_z$.

$$\nabla = \left(\partial/\partial x \quad \partial/\partial y \quad \partial/\partial z\right)^{\mathrm{T}} = \boldsymbol{e}_x \frac{\partial}{\partial x} + \boldsymbol{e}_y \frac{\partial}{\partial y} + \boldsymbol{e}_z \frac{\partial}{\partial z}, \tag{2.268}$$

and, along the unit vectors, $\boldsymbol{e}_\theta$, $\boldsymbol{e}_\lambda$, it is

$$\nabla = \boldsymbol{e}_r \frac{\partial}{\partial r} + \boldsymbol{e}_\theta \frac{1}{r}\frac{\partial}{\partial \theta} + \boldsymbol{e}_\lambda \frac{1}{r\sin\theta}\frac{\partial}{\partial \lambda}. \tag{2.269}$$

The first derivatives of functions encountered in geodesy and geophysics are usually distinguished by vertical (normal) and horizontal (tangential) types, and one may write $\boldsymbol{\nabla} = \boldsymbol{e}_r \partial/\partial r + \boldsymbol{\nabla}_{(\theta,\lambda)}/r$, where

$$\nabla_{(\theta,\lambda)} = \boldsymbol{e}_\theta \frac{\partial}{\partial \theta} + \boldsymbol{e}_\lambda \frac{1}{\sin\theta}\frac{\partial}{\partial \lambda}. \tag{2.270}$$

It turns out that an orthogonal basis for vector functions on the sphere (that satisfy analogous integrability conditions as for scalar functions) is given by the three *vector spherical harmonics* (Barrera et al. 1985),

$$\boldsymbol{\Psi}_{n,m}^{(1)}(\theta,\lambda) = \frac{1}{\sqrt{n(n+1)}}\left(\boldsymbol{e}_\theta \frac{\partial}{\partial \theta} + \boldsymbol{e}_\lambda \frac{1}{\sin\theta}\frac{\partial}{\partial \lambda}\right) Y_{n,m}(\theta,\lambda), \tag{2.271}$$

$$\boldsymbol{\Psi}_{n,m}^{(2)}(\theta,\lambda) = \frac{1}{\sqrt{n(n+1)}}\left(\boldsymbol{e}_\lambda \frac{\partial}{\partial \theta} - \boldsymbol{e}_\theta \frac{1}{\sin\theta}\frac{\partial}{\partial \lambda}\right) Y_{n,m}(\theta,\lambda), \tag{2.272}$$

$$\boldsymbol{\Psi}_{n,m}^{(3)}(\theta,\lambda) = \boldsymbol{e}_r Y_{n,m}(\theta,\lambda); \tag{2.273}$$

see also (Hill 1954) or (Arfken 1970, p.605), who define slightly different versions. To demonstrate their mutual orthogonality, one needs two additional orthogonality properties for spherical harmonics,

$$\frac{1}{4\pi}\iint_\Omega \left(\frac{\partial \bar{Y}_{n,m}}{\partial \theta}\frac{\partial \bar{Y}_{n',m'}}{\partial \theta} + \frac{1}{\sin^2\theta}\frac{\partial \bar{Y}_{n,m}}{\partial \lambda}\frac{\partial \bar{Y}_{n',m'}}{\partial \lambda}\right) d\Omega = n(n+1)\delta_{n-n'}\delta_{m-m'}, \tag{2.274}$$

$$\frac{1}{4\pi}\iint_\Omega \frac{1}{\sin\theta}\left(\frac{\partial \bar{Y}_{n,m}}{\partial \theta}\frac{\partial \bar{Y}_{n',m'}}{\partial \lambda} - \frac{\partial \bar{Y}_{n,m}}{\partial \lambda}\frac{\partial \bar{Y}_{n',m'}}{\partial \theta}\right) d\Omega = 0, \quad \text{for all } n, n', m, m'. \tag{2.275}$$

Equation (2.274) is proved by first multiplying both sides of equation (2.241) by $Y_{n',m'}(\theta, \lambda)$ and integrating over the unit sphere, recalling the orthogonality of the spherical harmonics, equation (2.219),

$$\frac{1}{4\pi}\iint\limits_{\Omega}\nabla^2_{(\theta,\lambda)}\bar{Y}_{n,m}(\theta,\lambda)\bar{Y}_{n',m'}(\theta,\lambda)\,d\Omega = -n(n+1)\delta_{n-n'}\delta_{m-m'}. \tag{2.276}$$

Then, re-writing the surface Laplacian, the integral becomes

$$\begin{aligned}\frac{1}{4\pi}\iint\limits_{\Omega}&\left(\frac{1}{\sin\theta}\frac{\partial}{\partial\theta}\left(\sin\theta\frac{\partial\bar{Y}_{n,m}(\theta,\lambda)}{\partial\theta}\right)+\frac{1}{\sin^2\theta}\frac{\partial^2\bar{Y}_{n,m}(\theta,\lambda)}{\partial\lambda^2}\right)\bar{Y}_{n',m'}(\theta,\lambda)\,d\Omega\\ &= -\frac{1}{4\pi}\iint\limits_{\Omega}\left(\frac{\partial\bar{Y}_{n',m'}(\theta,\lambda)}{\partial\theta}\frac{\partial\bar{Y}_{n,m}(\theta,\lambda)}{\partial\theta}+\frac{1}{\sin^2\theta}\frac{\partial\bar{Y}_{n',m'}(\theta,\lambda)}{\partial\lambda}\frac{\partial\bar{Y}_{n,m}(\theta,\lambda)}{\partial\lambda}\right)d\Omega\end{aligned} \tag{2.277}$$

where the equality follows from integrations by parts, the first term with respect to θ and the second with respect to λ; and, thus proving equation (2.274) (see also Jeffreys (1955)).

The proof of equation (2.275) is immediate for any $m \neq -m'$ since the derivatives with respect to λ are

$$\frac{\partial}{\partial\lambda}\bar{Y}_{n,m}(\theta,\lambda) = -m\bar{Y}_{n,-m}(\theta,\lambda). \tag{2.278}$$

Thus, being over a full period, the integral with respect to λ, therefore, also the double integral, is zero (if either $m = 0$ or $m' = 0$, the derivative is zero, as is the double integral). If $m = -m' \neq 0$, then

$$\begin{aligned}\frac{\partial\bar{Y}_{n,m}}{\partial\theta}\frac{\partial\bar{Y}_{n',-m}}{\partial\lambda}-\frac{\partial\bar{Y}_{n,m}}{\partial\lambda}\frac{\partial\bar{Y}_{n',-m}}{\partial\theta} &= \frac{\partial\bar{Y}_{n,m}}{\partial\theta}m\bar{Y}_{n',m}+m\bar{Y}_{n,-m}\frac{\partial\bar{Y}_{n',-m}}{\partial\theta}\\ &= m\frac{d\bar{P}_{n,|m|}}{d\theta}\bar{P}_{n',|m|}\begin{cases}\cos^2 m\lambda, & 0\le m\le n\\ \sin^2|m|\lambda, & -n\le m<0\end{cases}\\ &\quad +m\bar{P}_{n,|m|}\frac{d\bar{P}_{n',|m|}}{d\theta}\begin{cases}\sin^2 m\lambda, & 0\le m\le n\\ \cos^2|m|\lambda, & -n\le m<0\end{cases}\end{aligned} \tag{2.279}$$

Hence,

$$\frac{1}{4\pi}\iint\limits_{\Omega}\frac{1}{\sin\theta}\left(\frac{\partial\bar{Y}_{n,m}}{\partial\theta}\frac{\partial\bar{Y}_{n',-m}}{\partial\lambda}-\frac{\partial\bar{Y}_{n,m}}{\partial\lambda}\frac{\partial\bar{Y}_{n',-m}}{\partial\theta}\right)d\Omega = \frac{m}{4}\int_0^{\pi}\left(\frac{d\bar{P}_{n,|m|}}{d\theta}\bar{P}_{n',|m|}+\bar{P}_{n,|m|}\frac{d\bar{P}_{n',|m|}}{d\theta}\right)d\theta, \tag{2.280}$$

since

$$\int_0^{2\pi}\cos^2(m\lambda)\,d\lambda = \int_0^{2\pi}\sin^2(m\lambda)\,d\lambda = \pi. \tag{2.281}$$

Integrating by parts and using equation (2.232), we have

$$\int_0^\pi \left(\bar{P}_{n',|m|} \frac{d\bar{P}_{n,|m|}}{d\theta} + \bar{P}_{n,|m|} \frac{d\bar{P}_{n',|m|}}{d\theta} \right) d\theta = \left(2\bar{P}_{n',|m|} \bar{P}_{n,|m|} \right) \Big|_{\theta=0}^{\pi} - \int_0^\pi \left(\bar{P}_{n,|m|} \frac{d\bar{P}_{n',|m|}}{d\theta} + \bar{P}_{n',|m|} \frac{d\bar{P}_{n,|m|}}{d\theta} \right) d\theta$$

$$= -\int_0^\pi \left(\bar{P}_{n',|m|} \frac{d\bar{P}_{n,|m|}}{d\theta} + \bar{P}_{n,|m|} \frac{d\bar{P}_{n',|m|}}{d\theta} \right) d\theta \qquad (2.282)$$

which shows that the integral, equal to its negative, must be zero, thus proving equation (2.275) also for this case. Both equations (2.274) and (2.275) hold as well for the complex spherical harmonics, $(\bar{Y}^c_{n,m})^*$ and $\bar{Y}^c_{n',m'}$.

The orthogonality of the vector spherical harmonics now follows easily. Two real-valued vector functions, $\boldsymbol{g}(\theta, \lambda)$ and $\boldsymbol{h}(\theta, \lambda)$, on the sphere are orthogonal if

$$\frac{1}{4\pi} \iint_\Omega \boldsymbol{g}(\theta,\lambda) \cdot \boldsymbol{h}(\theta,\lambda) \, d\Omega = 0. \qquad (2.283)$$

By equation (2.275),

$$\frac{1}{4\pi} \iint_\Omega \boldsymbol{\Psi}^{(1)}_{n,m}(\theta,\lambda) \cdot \boldsymbol{\Psi}^{(2)}_{n',m'}(\theta,\lambda) \, d\Omega = 0, \quad \text{for all } n, n', m, m'. \qquad (2.284)$$

And, because of equations (2.266),

$$\boldsymbol{\Psi}^{(1)}_{n,m}(\theta,\lambda) \cdot \boldsymbol{\Psi}^{(3)}_{n',m'}(\theta,\lambda) = 0, \qquad (2.285)$$

$$\boldsymbol{\Psi}^{(2)}_{n,m}(\theta,\lambda) \cdot \boldsymbol{\Psi}^{(3)}_{n',m'}(\theta,\lambda) = 0, \qquad (2.286)$$

for all n, n', m, m'; hence, their integrals vanish, as well. Finally, equations (2.274) and (2.219) yield for $\alpha = 1, 2, 3$,

$$\frac{1}{4\pi} \iint_\Omega \boldsymbol{\Psi}^{(\alpha)}_{n,m}(\theta,\lambda) \cdot \boldsymbol{\Psi}^{(\alpha)}_{n',m'}(\theta,\lambda) \, d\Omega = \delta_{n-n'} \delta_{m-m'}, \qquad (2.287)$$

so that one may write

$$\frac{1}{4\pi} \iint_\Omega \boldsymbol{\Psi}^{(\alpha)}_{n,m}(\theta,\lambda) \cdot \boldsymbol{\Psi}^{(\beta)}_{n',m'}(\theta,\lambda) \, d\Omega = \delta_{\alpha-\beta} \delta_{n-n'} \delta_{m-m'}. \qquad (2.288)$$

For future reference, it is noted and derived from equations (2.266) and (2.271) through (2.273) that

$$\left(\frac{\partial Y_{n,m}}{\partial\theta}\right)^2+\left(\frac{1}{\sin\theta}\frac{\partial Y_{n,m}}{\partial\lambda}\right)^2+(n+1)^2Y_{n,m}^2$$
$$=\frac{n(n+1)}{2}\left|\boldsymbol{\Psi}_{n,m}^{(1)}\right|^2+\frac{n(n+1)}{2}\left|\boldsymbol{\Psi}_{n,m}^{(2)}\right|^2+(n+1)^2\left|\boldsymbol{\Psi}_{n,m}^{(3)}\right|^2. \tag{2.289}$$

Then, it follows that (Exercise 2.17),

$$\frac{R^2}{4\pi(2n+1)}\iint_{\Omega}\left|\nabla\left(\left(\frac{R}{r}\right)^{n+1}\bar{Y}_{n,m}(\theta,\lambda)\right)_{r=R}\right|^2 d\Omega=n+1. \tag{2.290}$$

For a real-valued, square-integrable vector function, $\boldsymbol{g}(\theta,\lambda)$, on the sphere, define its *vector Fourier-Legendre transform* as

$$G_{n,m}^{(\alpha)}=\frac{1}{4\pi}\iint_{\Omega}\boldsymbol{g}(\theta,\lambda)\cdot\bar{\boldsymbol{\Psi}}_{n,m}^{(\alpha)}(\theta,\lambda)\,d\Omega,\quad \alpha=1,2,3. \tag{2.291}$$

If it is continuous then the vector spherical harmonic series converges to the function,

$$\boldsymbol{g}(\theta,\lambda)=\sum_{n=0}^{\infty}\sum_{m=-n}^{n}\left(G_{n,m}^{(1)}\bar{\boldsymbol{\Psi}}_{n,m}^{(1)}(\theta,\lambda)+G_{n,m}^{(2)}\bar{\boldsymbol{\Psi}}_{n,m}^{(2)}(\theta,\lambda)+G_{n,m}^{(3)}\bar{\boldsymbol{\Psi}}_{n,m}^{(3)}(\theta,\lambda)\right). \tag{2.292}$$

Note that the spectrum consists of scalar components. For example, suppose that a differentiable function, g, on the sphere has the Fourier-Legendre spectrum, $G_{n,m}$. Its horizontal gradient, given by

$$\boldsymbol{\nabla}_{(\theta,\lambda)}g(\theta,\lambda)=\boldsymbol{e}_\theta\frac{\partial g(\theta,\lambda)}{\partial\theta}+\boldsymbol{e}_\lambda\frac{\partial g(\theta,\lambda)}{\sin\theta\,\partial\lambda}, \tag{2.293}$$

has the vector Fourier-Legendre spectrum,

$$DG_{n,m}^{(\alpha)}=\frac{1}{4\pi}\iint_{\Omega}\left(\boldsymbol{\nabla}_{(\theta,\lambda)}g(\theta,\lambda)\right)\cdot\bar{\boldsymbol{\Psi}}_{n,m}^{(\alpha)}(\theta,\lambda)\,d\Omega,\quad \alpha=1,2,3, \tag{2.294}$$

For $\alpha=2$,

$$DG_{n,m}^{(2)}=\frac{1}{4\pi\sqrt{n(n+1)}}\iint_{\Omega}\left(\boldsymbol{e}_\theta\frac{\partial g}{\partial\theta}+\boldsymbol{e}_\lambda\frac{\partial g}{\sin\theta\,\partial\lambda}\right)\cdot\left(\boldsymbol{e}_\lambda\frac{\partial\bar{Y}_{n,m}}{\partial\theta}-\boldsymbol{e}_\theta\frac{1}{\sin\theta}\frac{\partial\bar{Y}_{n,m}}{\partial\lambda}\right)d\Omega$$
$$=\frac{1}{4\pi\sqrt{n(n+1)}}\iint_{\Omega}\frac{1}{\sin\theta}\left(\frac{\partial g}{\partial\lambda}\frac{\partial\bar{Y}_{n,m}}{\partial\theta}-\frac{\partial g}{\partial\theta}\frac{\partial\bar{Y}_{n,m}}{\partial\lambda}\right)d\Omega \tag{2.295}$$

Substituting the spherical harmonic expansion of g into the last integral yields a sum of integrals of the type,

$$\iint_{\Omega} \frac{1}{\sin\theta}\left(\frac{\partial \bar{Y}_{p,q}(\theta,\lambda)}{\partial\lambda}\frac{\partial \bar{Y}_{n,m}(\theta,\lambda)}{\partial\theta} - \frac{\partial \bar{Y}_{p,q}(\theta,\lambda)}{\partial\theta}\frac{\partial \bar{Y}_{n,m}(\theta,\lambda)}{\partial\lambda}\right) d\Omega, \tag{2.296}$$

which by equation (2.275) are zero for all degrees and orders. Hence, in this case, $DG_{n,m}^{(2)} = 0$. Also,

$$DG_{n,m}^{(3)} = \frac{1}{4\pi}\iint_{\Omega}\left(\boldsymbol{e}_\theta \frac{\partial g(\theta,\lambda)}{\partial\theta} + \boldsymbol{e}_\lambda \frac{\partial g(\theta,\lambda)}{\sin\theta\,\partial\lambda}\right)\cdot \boldsymbol{e}_r \bar{Y}_{n,m}(\theta,\lambda)\, d\Omega = 0 \tag{2.297}$$

That is, the horizontal gradient of a scalar function on the sphere may be expressed as a vector Fourier-Legendre series of the $\alpha = 1$ vector spherical harmonics, $\bar{\boldsymbol{\Psi}}_{n,m}^{(1)}(\theta, \lambda)$,

$$\nabla_{(\theta,\lambda)} g(\theta,\lambda) = \sum_{n=0}^{\infty}\sum_{m=-n}^{n} DG_{n,m}^{(1)} \bar{\boldsymbol{\Psi}}_{n,m}^{(1)}(\theta,\lambda). \tag{2.298}$$

Now, applying the gradient operator, equation (2.270), to

$$g(\theta,\lambda) = \sum_{n=0}^{\infty}\sum_{m=-n}^{n} G_{n,m}\bar{Y}_{n,m}(\theta,\lambda) \tag{2.299}$$

gives

$$\nabla_{(\theta,\lambda)} g(\theta,\lambda) = \sum_{n=0}^{\infty}\sum_{m=-n}^{n} G_{n,m}\nabla_{(\theta,\lambda)}\bar{Y}_{n,m}(\theta,\lambda) = \sum_{n=0}^{\infty}\sum_{m=-n}^{n} G_{n,m}\sqrt{n(n+1)}\bar{\boldsymbol{\Psi}}_{n,m}^{(1)}(\theta,\lambda). \tag{2.300}$$

Comparing equations (2.298) and (2.300),

$$DG_{n,m}^{(1)} = \sqrt{n(n+1)}G_{n,m}, \tag{2.301}$$

which gives an analytical relationship between the spectrum of a function on a sphere and the spectrum of its horizontal derivatives. Both spectra are scalars and depend on the same wave numbers. However, it must be noted that these spectra refer to functions in different spaces, one a scalar function space, the other a vector function space. In that sense they are not the same kind of spectra. On the other hand, with this caveat, certain spectral analyses may be made and relationships approximated among a function on the sphere and its horizontal derivatives (Sections 5.5 and 6.3.2).

The relationship of the Fourier-Legendre spectrum to spectra of higher than first-order horizontal derivatives requires the definition of *tensor spherical harmonics* and is outside the present scope (Rummel and vanGelderen 1992, Zerilli 1970); however, see also Section 5.5. These harmonics are fully developed with applications to angular momentum in quantum mechanics (Edmonds 1957, Ch.5), as well as more classical applications (Jones 1985) and constructive approximations on the sphere (Freeden et al. 1994).

2.6.5 Cap Function

The "rectangle" function on the sphere is defined by practical necessity in terms of just one variable, the angular distance ("distance" in radians along a great circle arc on the unit sphere) from the center of a spherical cap, $\Omega_s = \{(\psi, \zeta) \mid 0 \le \psi \le \psi_s, 0 \le \zeta \le 2\pi\}$, as

$$b_s^{(\psi_s)}(\psi) = \begin{cases} \dfrac{4\pi}{A_s}, & 0 \le \psi < \psi_s \\ \dfrac{2\pi}{A_s}, & \psi = \psi_s \\ 0, & \psi > \psi_s \end{cases} \tag{2.302}$$

where $\psi_s > 0$ is the radius of the cap that has area (see Figure 2.13),

$$A_s = \int_0^{2\pi}\int_0^{\psi_s} d\sigma = 2\pi \int_0^{\psi_s} \sin\psi \, d\psi = 2\pi\left(1 - \cos\psi_s\right). \tag{2.303}$$

Because $b_s^{(\psi_s)}(\psi)$ depends only on an angular distance, its Legendre transform is of the one-dimensional kind, equation (2.205). Substituting a formula that relates the Legendre polynomial to its derivatives (Hobson 1965, p.33),

$$(2n+1)P_n(y) = \frac{d}{dy}\left(P_{n+1}(y) - P_{n-1}(y)\right), \quad n > 0, \tag{2.304}$$

the Legendre spectrum of $b_s^{(\psi_s)}$ is,

$$\begin{aligned} B_n^{(\psi_s)} &= \frac{4\pi}{2A_s}\int_0^{\psi_s} P_n(\cos\psi)\sin\psi \, d\psi \\ &= \frac{1}{1-\cos\psi_s}\frac{1}{2n+1}\left(P_{n-1}(\cos\psi_c) - P_{n+1}(\cos\psi_s)\right), \quad n \ge 1 \end{aligned} \tag{2.305}$$

with

$$B_0^{(\psi_s)} = 1, \quad B_1^{(\psi_s)} = \frac{1}{2}\left(1 + \cos\psi_s\right). \tag{2.306}$$

The Legendre series of $b_s^{(\psi_s)}(\psi)$ is, from equation (2.206),

$$b_s^{(\psi_s)}(\psi) = \sum_{n=0}^{\infty} (2n+1) B_n^{(\psi_s)} P_n(\cos\psi), \tag{2.307}$$

which converges, according to equation (2.207), to the defined value of $b_s^{(\psi_s)}(\psi)$ at the edge of the cap. It is noted that the particular definition of $b_s^{(\psi_s)}(\psi)$ ensures that the zero-frequency value of its spectrum is unity, in agreement with the preceding definitions of the rectangle and cylinder functions. It also means that the average value of $b_s^{(\psi_s)}$ over the sphere is unity,

$$\frac{1}{4\pi} \iint_{\Omega} b_s^{(\psi_s)}(\psi)\, d\Omega = 1. \tag{2.308}$$

Recursion formulas for $B_n^{(\psi_s)}$ may be derived from corresponding recursion formulas for the Legendre polynomials, P_n. For example, substituting the formula (2.202) into the integral of equation (2.305), and using equation (2.304), one obtains

$$B_n^{(\psi_s)} = \frac{2n-1}{n+1} B_{n-1}^{(\psi_s)} \cos\psi_s - \frac{n-2}{n+1} B_{n-2}^{(\psi_s)}, \quad n \geq 2, \tag{2.309}$$

with starting values given by (2.306). The spectrum of the cap function for $\psi_s = 1°$ is shown in Figure 2.17. The first zero-crossing occurs at $n = 219$, which agrees with the first zero of the spectrum for the cylinder function, equation (2.192), $2\pi \bar{f} s_\mu = 3.8317$, assuming that the Cartesian frequency is related to harmonic degree by $\bar{f} = n/(2\pi R)$ (Section 2.7) and the cylinder radius is $s_\mu = (3.8317/219)R$. For the Earth, $R = 6371$ km, this is $s_\mu = 111.5$ km, or about 1° of arc. In fact, the spectra, $B_c^{(\sqrt{\pi}R\psi_s)}(\bar{f})$ and $B_n^{(\psi_s)}$, are asymptotically close for small ψ_s (Exercise 2.18).

2.6.6 Gaussian Function on the Sphere

The spherical analogue of the Gaussian function is

$$\gamma_s^{(\beta_s)}(\psi) = \frac{2\beta_s}{1 - e^{-2\beta_s}} e^{-\beta_s(1-\cos\psi)}, \quad 0 \leq \psi \leq \pi, \tag{2.310}$$

where $\beta_s > 0$. Again, it is defined so as to ensure unity for its zero-frequency spectral value. Also, being a function that depends only on the central angle, ψ, its Legendre transform is given by equation (2.205) that becomes with $y = \cos\psi$,

$$\Gamma_n^{(\beta_s)} = \frac{\beta_s e^{-\beta_s}}{1 - e^{-2\beta_s}} \int_{-1}^{1} e^{\beta_s y} P_n(y)\, dy. \tag{2.311}$$

Substituting the recursion formula, equation (2.304), and integrating by parts, one obtains

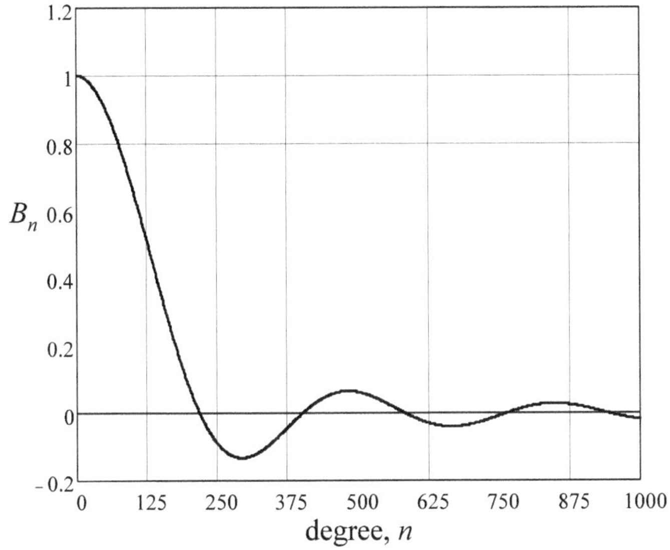

Figure 2.17: Fourier-Legendre spectrum of $b_s^{(\psi_s)}(\psi)$ for $\psi_s = 1°$.

$$\Gamma_n^{(\beta_s)} = -\frac{2n-1}{\beta_s}\Gamma_{n-1}^{(\beta_s)} + \Gamma_{n-2}^{(\beta_s)}, \quad n \geq 2, \tag{2.312}$$

with starting values

$$\Gamma_0^{(\beta_s)} = 1, \; \Gamma_1^{(\beta_s)} = \frac{1+e^{-2\beta_s}}{1-e^{-2\beta_s}} - \frac{1}{\beta_s}. \tag{2.313}$$

Since the Gaussian function is continuous for all ψ, its Legendre series converges to $\gamma_s^{(\beta_s)}(\psi)$ everywhere. Also, since $\Gamma_0^{(\beta_s)} = 1$, the average value of the Gaussian function over the sphere is unity,

$$\frac{1}{4\pi}\iint_\Omega \gamma_s^{(\beta_s)}(\psi)\,d\Omega = 1. \tag{2.314}$$

As in the Cartesian case, the parameter, β_s, defines the "width" or "significant extent" of the Gaussian curve. If significant extent is defined by the angle, ψ, for which $\gamma(\psi) = \gamma(0)e^{-1}$, then $\psi = \cos^{-1}(1 - 1/\beta_s)$. Thus, for the same extent as the cap function, one has $\beta_s = 1/(1 - \cos\psi_s)$. Other ways of defining significant extent are elaborated in Section 3.5.2. Figure 2.18 shows the spherical Gaussian function and its Legendre transform for $\beta_s = 1/(1 - \cos\psi_s)$ and $\psi_s = 1°$. The parameter, β_s, defined for the spherical Gaussian is not the same as β_p defined for the planar isotropic Gaussian function (see the transform pair (2.193)),

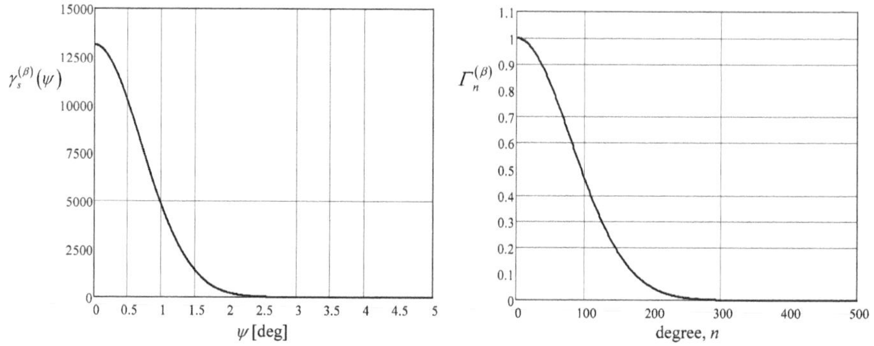

Figure 2.18: The spherical Gaussian function (left) and its Fourier-Legendre spectrum (right) for $\beta_s = 1/(1 - \cos \psi_s)$ and $\psi_s = 1°$.

$$\gamma^{(\beta_p)}(s) = \frac{1}{\beta_p^2} e^{-\pi(s/\beta_p)^2}. \tag{2.315}$$

With a planar approximation of a sphere of radius, R, that is, for small $\psi \approx s/R$ and large β_s, one has $\gamma_s^{(\beta s)}(\psi) \approx \gamma_s^{(\beta s)}(s)$, if

$$\beta_s = \frac{2\pi R^2}{\beta_p^2}. \tag{2.316}$$

where the subscript 's' on the spherical Gaussian function and the distance argument, 's', for the planar Gaussian function have entirely different meanings.

2.6.7 Spherical Dirac Delta Function

The Dirac delta function for the sphere, again, by necessity is defined only as a function of ψ and on the basis of the existence of its Legendre transform. Consider the Gaussian function on the sphere and its limit as the parameter, β_s, approaches infinity, while the significant width, $\psi_0 = \cos^{-1}(1 - 1/\beta_s)$, approaches zero. The volume of the Gaussian function, equation (2.314), remains constant at 4π. From equation (2.310), l'Hôpital's rule for limits of indeterminate expressions gives

$$\lim_{\beta_s \to \infty} \gamma_s^{(\beta_s)}(\psi) = \lim_{\beta_s \to \infty} \frac{2\beta_s}{\left(1 - e^{-2\beta_s}\right) e^{\beta_s(1-\cos\psi)}} = 0, \quad 0 < \psi \le \pi. \tag{2.317}$$

At $\psi = 0$, the limit is infinite; hence, we may define the spherical Dirac delta function on the sphere by

$$\delta_s(\psi) = \lim_{\beta_s \to \infty} \gamma_s^{(\beta_s)}(\psi), \quad 0 \le \psi \le \pi. \tag{2.318}$$

The existence of the Legendre spectrum of $\delta_s(\psi)$ then follows from equation (2.312). That is,

$$\lim_{\beta_s \to \infty} \Gamma_n^{(\beta_s)} = \lim_{\beta_s \to \infty} \Gamma_{n-2}^{(\beta_s)}, \quad n \geq 2, \tag{2.319}$$

where, from equations (2.313), $\lim_{\beta_s \to \infty} \Gamma_0^{(\beta_s)} = 1$ and $\lim_{\beta_s \to \infty} \Gamma_1^{(\beta_s)} = 1$. Thus, the spectrum of $\delta_s(\psi)$ is

$$D_n = \frac{1}{2} \int_0^\pi \delta_s(\psi) P_n(\cos\psi) \sin\psi \, d\psi = 1, \quad \text{for all } n, \tag{2.320}$$

and formally one may write

$$\delta_s(\psi) = \sum_{n=0}^{\infty} (2n+1) P_n(\cos\psi). \tag{2.321}$$

Equations (2.318) and (2.313) confirm that

$$\frac{1}{4\pi} \iint_\Omega \delta_s(\psi) \, d\Omega = \lim_{\beta_s \to \infty} \frac{1}{4\pi} \iint_\Omega \gamma_s^{(\beta_s)}(\psi) \, d\Omega = \lim_{\beta_s \to \infty} \Gamma_0^{(\beta_s)} = 1. \tag{2.322}$$

Furthermore, the corresponding reproducing property, as shown in Chapter 3, equation (3.65), is

$$\frac{1}{4\pi} \iint_{\Omega'} \delta_s(\psi) g(\theta', \lambda') \, d\Omega' = g(\theta, \lambda), \tag{2.323}$$

where the relationship between (θ, λ) and (θ', λ') is given by equations (2.196) and (2.197).

2.7 From Sphere to Plane

When analyzing a geophysical signal spectrally on the spherically approximated Earth, it is often numerically legitimate to further approximate the surface locally by a plane. If both global and local spectra of a signal have been determined with significant overlap in their corresponding spectral bands, one may wish to combine or compare them. In any of these situations, it is important to translate properly between frequency in the planar spectral domain and wave number in the spherical spectral domain.

There are a number of approaches to find such a relationship, based essentially on the asymptotic relationship between Legendre polynomials and Bessel functions (Abramowitz and Stegun 1972, p.362),

$$\lim_{n\to\infty} P_n\left(\cos\frac{x}{n}\right) = J_0(x), \quad \text{for } x > 0, \tag{2.324}$$

which is the same as

$$\lim_{n\to\infty} P_n\left(\cos\frac{x}{\sqrt{n(n+1)}}\right) = J_0(x), \quad \text{for } x > 0. \tag{2.325}$$

Equation (2.200) with $y = \cos\psi$ and $\partial y = -\sin\psi\,\partial\psi$ becomes

$$\frac{1}{\sin\psi}\frac{\partial}{\partial\psi}\left(\sin\psi\frac{\partial}{\partial\psi}P_n(\cos\psi)\right) = -n(n+1)P_n(\cos\psi). \tag{2.326}$$

On the sphere of radius, R, define a distance, $s \ll R$, by the angle, $\psi = s/R$. Then, approximately,

$$\frac{1}{s}\frac{\partial}{\partial s}\left(s\frac{\partial}{\partial s}P_n\left(\cos\frac{s}{R}\right)\right) = -\frac{n(n+1)}{R^2}P_n\left(\cos\frac{s}{R}\right). \tag{2.327}$$

Now, using the properties of the derivatives of Bessel functions, $dJ_0(x)/dx = -J_1(x)$ and $d(xJ_1(x))/dx = xJ_0(x)$, it is easily verified that

$$\frac{1}{s}\frac{\partial}{\partial s}\left(s\frac{\partial}{\partial s}J_0\left(2\pi\bar{f}s\right)\right) = -\left(2\pi\bar{f}\right)^2 J_0\left(2\pi\bar{f}s\right), \tag{2.328}$$

where the introduced frequency, $\bar{f}$, corresponds to the distance, s, as in equations (2.174) and (2.177).

Finally, let $x = 2\pi\bar{f}s$ and approximate, from equation (2.325),

$$P_n\left(\cos\frac{2\pi\bar{f}s}{\sqrt{n(n+1)}}\right) \simeq J_0\left(2\pi\bar{f}s\right). \tag{2.329}$$

Then, a comparison of equations (2.327) and (2.328) shows that

$$\bar{f} \simeq \frac{\sqrt{n(n+1)}}{2\pi R}. \tag{2.330}$$

This approximation is valid in principle only for the high degrees or high frequencies. However, it is reasonable even for the lower degrees where it is also slightly better than

$$\bar{f} \simeq \frac{n}{2\pi R}. \tag{2.331}$$

A note of caution is required here. The high-degree Fourier-Legendre spectrum of the function on the sphere represents its *global* variation at short wavelengths, whereas the determination of the high-frequency Fourier spectrum of that function over just a local area reflects is *local* short-wavelength character. Thus, while the spectral domains may be approximated by equation (2.330) at high frequencies, the spectra of a global function and of its restriction to a local area may be quite different. For example, the high-degree components of the Earth's entire topography, represented in spherical harmonics, are not the same as the high-frequency components of the topography restricted to the Colorado Rocky Mountains, nor to the topography of the northern Midwest of the U.S. (Figure 2.19). However, in nominal cases, the global and local spectra seem to match reasonably well.

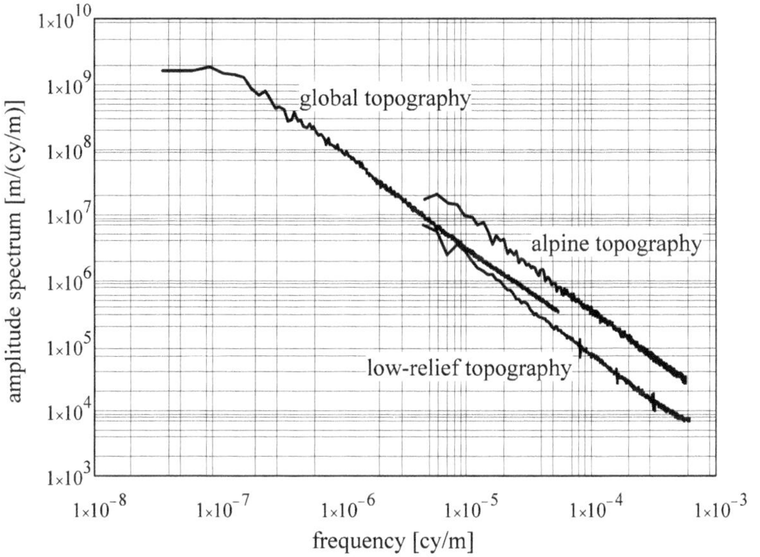

Figure 2.19: Comparison of global and local topographic spectra as azimuthally averaged amplitude values. The global spectrum is derived from a spherical harmonic series model, DTM2006 (Pavlis et al. 2012). The alpine spectrum represents terrain data of the Colorado Rocky Mountains. The lowland topography spectrum is typical of the gently rolling terrain of the northern Midwest U.S.

2.8 Examples of Fourier Transform Pairs

In this section, a few one- and two-dimensional Fourier transform pairs are listed as they are common in the applications described in Chapter 6. Most can be obtained by consulting a good Table of Integrals, e.g., Gradshteyn and Ryzhik (1980), or specialized compendia (Oberhettinger 1957). Some of these were encountered already in previous sections, and others are easily derived from these using the properties of transforms. The two-dimensional Fourier transform pairs are also Hankel transform pairs in these examples, where the following abbreviations are noted,

$$s = \sqrt{x_1^2 + x_2^2},\ \bar{f} = \sqrt{f_1^2 + f_2^2}. \tag{2.332}$$

1. $$e^{-|x|} \leftrightarrow \frac{2}{1+(2\pi f)^2}; \tag{2.333}$$

2. $$\frac{1}{\left(s^2+a^2\right)^{3/2}} \leftrightarrow \frac{2\pi}{a} e^{-2\pi a\bar{f}},\quad a>0; \tag{2.334}$$

3. $$\frac{x_1}{\left(s^2+a^2\right)^{3/2}} \leftrightarrow \frac{-i2\pi f_1}{\bar{f}} e^{-2\pi a\bar{f}},\qquad \frac{x_2}{\left(s^2+a^2\right)^{3/2}} \leftrightarrow \frac{-i2\pi f_2}{\bar{f}} e^{-2\pi a\bar{f}},\quad a \ge 0; \tag{2.335}$$

4. $$\frac{1}{s} \leftrightarrow \frac{1}{\bar{f}}; \tag{2.336}$$

5. $$\frac{1}{\left(s^2+a^2\right)^{1/2}} \leftrightarrow \frac{1}{\bar{f}} e^{-2\pi a\bar{f}},\quad a \ge 0; \tag{2.337}$$

6. $$\frac{1}{\left(s^2+a^2\right)^{5/2}} \leftrightarrow \frac{4\pi^2\bar{f}}{3a^2} e^{-2\pi a\bar{f}}\left(1+\frac{1}{2\pi a\bar{f}}\right),\quad a>0; \tag{2.338}$$

* * *

EXERCISES

Exercise 2.1

Show that the Fourier series of the saw-tooth function, $\tilde{g}(x) = c\,\mathrm{mod}_P(x)$ (Figure 2.2), converges to $cP/2$ at $x = P$. Hint: Using integration by parts, show that the Fourier coefficients, a_k, are given for $k > 0$ by

$$a_k = \frac{2c}{P}\int_0^P x\cos\left(\frac{2\pi}{P}kx\right)dx = 0, \tag{E2.1}$$

and that $a_0 = \frac{2c}{P}\int_0^P x dx = cP$. Then sum the series at $x = P$.

Exercise 2.2

Prove the Fourier series transform pairs (2.20) through (2.25). For example, for (2.22), use a change in summation index, $k = -m$,

$$\tilde{h}(x) = \tilde{g}(-x) = \frac{1}{P}\sum_{k=-\infty}^{\infty} G_k e^{-i\frac{2\pi}{P}kx} = \frac{1}{P}\sum_{m=-\infty}^{\infty} G_{-m} e^{i\frac{2\pi}{P}mx} \quad \Rightarrow \quad H_k = G_{-k}. \tag{E2.2}$$

Exercise 2.3

Prove Properties (2.27) and (2.28) using the indicated hint.

Exercise 2.4

Prove the transform pairs (2.70) through (2.76) using lines of reasoning as in Exercise 2.2.

Exercise 2.5

Show that the sinc function is differentiable at $x = 0$. First show using l'Hôpital's rule that it is continuous,

$$\lim_{x\to 0^-} \operatorname{sinc}(x) = \lim_{x\to 0^+} \operatorname{sinc}(x) = 1. \tag{E2.3}$$

Then, also show that $d(\operatorname{sinc}(x))/dx = (\pi x\cos(\pi x) - \sin(\pi x))/(\pi x^2)$ is zero at $x = 0$.

Exercise 2.6

Show that

$$\int_{-\infty}^{\infty} |\operatorname{sinc}(f)|\, df = 2\int_{0}^{\infty} |\operatorname{sinc}(f)|\, df = 2\int_{0}^{1} \left|\frac{\sin(\pi f)}{\pi f}\right| df + 2\sum_{k=1}^{\infty}\int_{k}^{k+1} \left|\frac{\sin(\pi f)}{\pi f}\right| df \tag{E2.4}$$

and consequently that inequality (2.98) holds.

Exercise 2.7

Show that the Fourier transform of the sinc function is the rectangle function, starting with

$$\mathcal{F}(\operatorname{sinc}(x)) = \int_{-\infty}^{\infty} \frac{\sin(\pi x)}{\pi x} e^{-i2\pi xf}\, dx = 2\int_{0}^{\infty} \frac{\sin(\pi x)}{\pi x} \cos(2\pi xf)\, dx, \tag{E2.5}$$

where the second equality follows from the symmetry and anti-symmetry of the components of the integrand. Show that

$$\frac{1}{x} = -\frac{1}{x} e^{-sx}\Big|_{s=0}^{\infty} = \int_{0}^{\infty} e^{-sx}\, ds, \quad x > 0, \tag{E2.6}$$

$$\sin\alpha\cos\beta = (\sin(\alpha-\beta) + \sin(\alpha+\beta))/2,$$

$$\int_0^\infty \sin(ax)e^{-sx}\,dx = \frac{a}{a^2+s^2}, \quad s>0, \qquad \int_0^\infty \frac{a}{a^2+s^2}\,ds = \frac{\pi}{2}\operatorname{sgn}(a), \tag{E2.7}$$

the latter two by consulting a standard table of integrals, and combine these to arrive at the result.

Exercise 2.8

Show that the Fourier transform of $r(x) = 1/x$, $x \neq 0$, is given by the step function in the frequency domain, equation (2.104). Hint: Note that

$$\mathcal{F}\left(\frac{1}{x}\right) = \int_{-\infty}^{\infty} \frac{1}{x}\left(\cos(2\pi fx) - i\sin(2\pi fx)\right)dx = 0 - i\int_{-\infty}^{\infty} \frac{1}{t}\sin(2\pi ft)\,dt, \tag{E2.8}$$

where the zero integral of the first term is a Cauchy principal value. Then make use of the integral of the sinc function, equation (2.97).

Exercise 2.9

Prove equation (2.106), by showing that

$$\Gamma(f) = 2\int_0^\infty e^{-\pi x^2}\cos 2\pi fx\,dx, \tag{E2.9}$$

and using a table of integrals to evaluate the integral.

Exercise 2.10

Derive the Fourier transform of the Heaviside step function. Hints: Consider the Fourier transform of $e^{-a|x|}$ for $a > 0$, showing that

$$\mathcal{F}\left(e^{-a|x|}\right) = \int_{-\infty}^{\infty} e^{-a|x|-i2\pi fx}\,dx = \int_{-\infty}^{0} e^{(a-i2\pi f)x}\,dx + \int_0^\infty e^{-(a+i2\pi f)x}\,dx = \frac{2a}{a^2+(2\pi f)^2}. \tag{E2.10}$$

Then, note that with equation (2.120),

$$\lim_{a\to 0}\frac{2a}{a^2+(2\pi f)^2} = \delta(f). \tag{E2.11}$$

Finally, prove the result using

$$\int_0^\infty e^{-(a+i2\pi f)x}\,dx = \frac{1}{a+i2\pi f} = \frac{a}{a^2+(2\pi f)^2} - \frac{i2\pi f}{a^2+(2\pi f)^2} \tag{E2.12}$$

and taking limits on both sides as $a \to 0$.

Exercise 2.11

Prove equivalences (2.141) through (2.144).

Exercise 2.12

Prove equation (2.162) using the definition of the two-dimensional Fourier transform and results from the one-dimensional case.

Exercise 2.13

Show that equations (2.187) and (2.188) constitute a Hankel transform pair. That is, show that the Hankel transform of $\bar{g}(s)$ is $\bar{G}(\bar{f})$ (and the inverse transform of $\bar{G}(\bar{f})$ is $\bar{g}(s)$).

Exercise 2.14

Derive equation (2.199) using the law of cosines for $\cos\theta'$ and substituting the law of cosines for $\cos\psi$.

Exercise 2.15

Prove equation (2.241) using the indicated procedure.

Exercise 2.16

Prove Parseval's Theorem for the sphere (equation (2.243)) using the orthogonality of the spherical harmonic functions.

Exercise 2.17

Prove equation (2.290) using equations (2.289) and (2.287).

Exercise 2.18

Using relation (2.325) show that $B_c^{(\sqrt{\pi}R\psi_s)}(\bar{f})$ and $B_n^{(\psi_s)}$ are asymptotically close for small ψ_s. Hint: Use $x = 2\pi\bar{f}s$, $\bar{f} \simeq \sqrt{n(n+1)}/(2\pi R)$ (equation (2.330)), $\psi \simeq s/R$, and $A_s R^2 \simeq s_\mu^2 \pi$, where s_μ is the radius of the cylinder function (equation (2.191)) to show first that $B_c^{(\sqrt{\pi}R\psi_s)}(\bar{f}) = \frac{2}{R^2\psi_s^2}\int_0^{R\psi_s} sJ_0(2\pi\bar{f}s)\,ds$; and, then use the first of equations (2.305) to show that this is approximately equal to $B_n^{(\psi_s)}$.

Chapter 3

Convolutions and Windows on the Continuous Domain

3.1 Introduction

In geodesy, geophysics, and many other disciplines dealing with signals of any sort, one very often operates on the signal, not from point to point, but corporately, in terms of a convolution. Many linear systems that have an input and an output can be described as a convolution. Indeed, all linear filters are convolutions. The simplest example is the moving average that assigns to every point an average (the output) of the signal (the input) in the neighborhood centered on that point. As another example, the gravitational attraction at any point (the output) due to a given distribution of mass-density (the input) is determined by summing the effects of the individual elements of the distribution using a particular weighting function derived from Newton's law of gravitation – closer elements have greater weight. The mass-density distribution is "convolved" with the weighting function (in this case called Green's function) to obtain the gravitational attraction.

The principal relevant result for spectral analysis is that convolutions of signals in the space or time domain translate into simple products of spectra in the corresponding spectral domain. This means that convolutions may be analyzed very easily if the spectra of the convolved functions are known. And, the spectral properties of a particular filter can readily be designed as the ratio of the spectrum of the desired output relative to that of the known input. There is also a computational benefit to formulating the convolution in the spectral domain. Because the Fourier transform (in its discrete form) may be computed rapidly using simple recursion formulas, the rather more cumbersome computation of the convolution for many points of the output is accomplished very efficiently by calculating the product of spectra.

The dual nature of the Fourier transform pair also means that products of functions in the space domain translate to convolutions of spectra in the spectral domain. This has particular significance in the study of truncated signals, that is, signals for which only a limited realization is available. The effect of truncation is a product of the signal and a window function, like the rectangle function; and, the consequent effect on the spectrum is then analyzed in terms of the convolution of the signal spectrum and the spectrum of the window function.

The following sections define and elaborate on these concepts for the various Cartesian and spherical domains discussed in Chapter 2. Throughout the development, it is assumed that all functions have Fourier (Legendre) transforms and that the corresponding inverse transforms converge uniformly to the functions at points of continuity and to a mean value at step discontinuities. That is, the functions are assumed to have bounded variation and to be absolutely integrable on their respective space domains. There is an important exception for various kernels, or weighting functions, that define convolutions in potential field theory and these will require special attention (Section 6.5.1) since they are not always bounded.

3.2 Convolutions of Non-Periodic Functions

Let $g(x)$ and $h(x)$ be piecewise continuous functions with bounded variation over the real line and absolutely integrable (equation (2.49)). The *convolution* of functions g and h is a function of x defined by

$$(g*h)(x) = \int_{-\infty}^{\infty} g(x')h(x-x')\,dx'. \tag{3.1}$$

It is said that g is convolved with h. This definition holds for real as well as complex-valued functions, but, as always, our principal concern is limited to real functions. The convolution may be visualized as first reflecting the function h about the ordinate axis: $h(x') \to h(-x')$, translating it to the point of evaluation, x: $h(-x') \to h(x-x')$, and then multiplying by $g(x')$ and integrating.

If the independent variable of the *convolution* is a more complicated function, $u(x)$, then one has

$$(g*h)(u(x)) = \int_{-\infty}^{\infty} g(x')h(u(x)-x')\,dx'. \tag{3.2}$$

A slightly more general notation for convolution is also used here, namely,

$$g(x)*h(x) = \int_{-\infty}^{\infty} g(x')h(x-x')\,dx', \tag{3.3}$$

that allows extra flexibility if the arguments of g and h are different functions, $u(x)$ and $v(x)$, respectively. Let $g'(x) = g(u(x))$ and $h'(x) = h(v(x))$, then,

$$g(u(x))*h(v(x)) = g'(x)*h'(x) = \int_{-\infty}^{\infty} g'(x')h'(x-x')\,dx' = \int_{-\infty}^{\infty} g(u(x'))h(v(x-x'))\,dx'. \tag{3.4}$$

It would not be easy to write this in the notation of equation (3.1). Clearly, the notation is equivalent if g and h depend directly on x; $(g*h)(x) \equiv g(x)*h(x)$.

The convolution is continuous and bounded – any step discontinuities in g or h are smoothed out by the integration; and, it is absolutely integrable (Butzer and Nessel 1971, pp.4-5). In other words, a convolution results in a "more regular" function, as illustrated by the simple example below. It also has a Fourier integral that converges everywhere to the convolution. The units of the convolution, equations (3.1) or (3.4), are the product of the units of the functions being convolved times the units of x.

As noted in the Section 3.1, one of the simplest examples of a convolution is the unweighted *moving average* of g over some given interval of length, T, and formulated as a function of x,

$$\bar{g}(x) = \frac{1}{T}\int_{x-T/2}^{x+T/2} g(x')\,dx', \tag{3.5}$$

where, in this convolution, $h(x)$ is the general, normalized, rectangle function, equation (2.100); that is,

$$\bar{g}(x) = \frac{1}{T}\int_{x-T/2}^{x+T/2} b\left(\frac{x-x'}{T}\right) g(x')\,dx' = \frac{1}{T}\int_{-\infty}^{\infty} b\left(\frac{x-x'}{T}\right) g(x')\,dx' = \frac{1}{T} g(x) * b\left(\frac{x}{T}\right). \tag{3.6}$$

Another example is the convolution of two rectangle functions that creates the *triangle function*, visualized as the outcome of sliding the two rectangle functions past each other on the x-axis, and for each relative position integrating their product. It is also formally called the *Bartlett window*. The formulation of the convolution is

$$u_{\text{Bartlett}}(x) = \int_{-\infty}^{\infty} b(x')\,b(x-x')\,dx' = \begin{cases} 1-|x|, & |x| \le 1 \\ 0, & |x| > 1 \end{cases} \tag{3.7}$$

A graph of $u_{\text{Bartlett}}(x)$ (Figure 3.1) shows the reason for its name and illustrates that the convolution of functions with step discontinuities is continuous. The moving

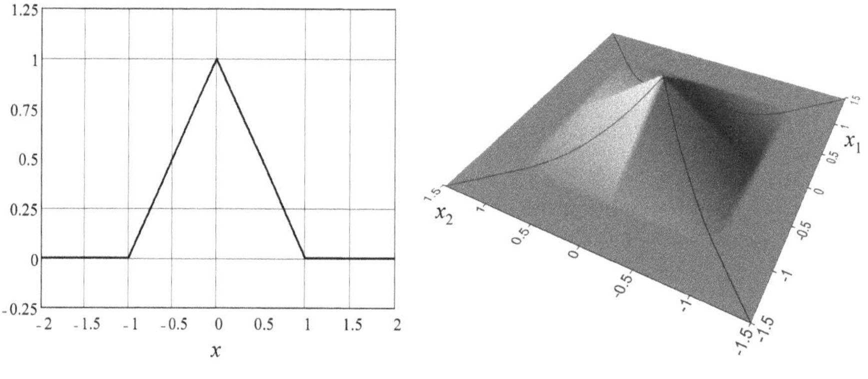

Figure 3.1: The triangle function (Bartlett window) resulting from the convolution of two rectangle functions; left: $u_{\text{Bartlett}}(x)$, equation (3.7); right: the two-dimensional version, $u_{\text{Bartlett}}(x_1, x_2)$, equation (3.25).

average, or smoothing filter, is elaborated in Section 3.5.1; whereas, window functions are discussed in Section 3.6.1.

3.2.1 Properties of the Convolution

Convolution is a linear operation and it has the following properties that are easily proved using the basic definition, equations (3.1) or (3.3). These properties also hold true for the more general definition, equation (3.4), if properly applied.

a) Proportionality: $g(x)*ah(x) = ag(x)*h(x) = a(g(x)*h(x))$, a = const.; (3.8)

b) Translation: $(g*h)(x+c) = g(x+c)*h(x) = g(x)*h(x+c)$, c = const.; (3.9)

c) Commutativity: $g(x)*h(x) = h(x)*g(x)$; (3.10)

d) Distributivity: $(g_1(x)+g_2(x))*h(x) = g_1(x)*h(x) + g_2(x)*h(x)$; (3.11)

e) Associativity: $(g_1(x)*g_2(x))*h(x) = g_1(x)*(g_2(x)*h(x))$. (3.12)

Properties (3.8) and (3.11) are obvious; and, property (3.9) follows from equations (3.2) and (3.4) with appropriate changes in integration variables. Similarly, a change in variable, $u = x - x'$, easily proves the commutativity of convolutions, equation (3.10). However, note that (Exercise 3.1)

$$(g*h)(ax+b) \neq g(ax+b)*h(x) \neq g(x)*h(ax+b) \neq g(ax+b)*h(ax+b), \tag{3.13}$$

but, $g(ax+b)*h(x) = h(x)*g(ax+b)$. Property (3.12) is proved using the definition (3.3) and interchanging integrations. Starting with

$$\begin{aligned}(g_1(x)*g_2(x))*h(x) &= \left(\int_{-\infty}^{\infty} g_1(x')g_2(x-x')\,dx'\right)*h(x)\\ &= \int_{-\infty}^{\infty}\left(\int_{-\infty}^{\infty} g_1(x')g_2(w'-x')\,dx'\right)h(x-w')\,dw'\\ &= \int_{-\infty}^{\infty} g_1(x')\left(\int_{-\infty}^{\infty} g_2(w'-x')h(x-w')\,dw'\right)dx'\end{aligned} \tag{3.14}$$

this yields, with a change in variable, $u' = w' - x'$,

$$\begin{aligned}(g_1(x)*g_2(x))*h(x) &= \int_{-\infty}^{\infty} g_1(x')\left(\int_{-\infty}^{\infty} g_2(u')h(x-x'-u')\,du'\right)dx'\\ &= g_1(x)*\left(\int_{-\infty}^{\infty} g_2(u')h(x-u')\,du'\right)\\ &= g_1(x)*(g_2(x)*h(x))\end{aligned} \tag{3.15}$$

which proves the result. Integrating both sides of equation (3.1), and changing variables as needed, we obtain the following additional property,

$$
\begin{aligned}
\int_{-\infty}^{\infty}(g*h)(x)\,dx &= \int_{-\infty}^{\infty}\int_{-\infty}^{\infty} g(x')h(x-x')\,dx'dx \\
&= \int_{-\infty}^{\infty} g(x')\left(\int_{-\infty}^{\infty} h(u)\,du\right)dx' \qquad (3.16)\\
&= \left(\int_{-\infty}^{\infty} g(x')\,dx'\right)\left(\int_{-\infty}^{\infty} h(x')\,dx'\right)
\end{aligned}
$$

That is, the area under the convolution equals the product of the areas under the functions being convolved.

The most important property of convolutions is, appropriately named, the *Convolution Theorem*. It gives the spectrum of the convolution in terms of the spectra of the convolved functions. It has many incarnations depending on the type of function; here it is given and proved for non-periodic functions.

Convolution Theorem

The Fourier transform of the convolution of two functions equals the product of their Fourier transforms; and the inverse Fourier transform of this product is the convolution. In symbols,

$$g(x) * h(x) \;\leftrightarrow\; G(f)\,H(f). \qquad (3.17)$$

The proof for the Fourier transform of the convolution follows immediately from the definition, equation (2.56), an appropriate change of integration variable for the Fourier transform of h, and an interchange of integrations,

$$
\begin{aligned}
\mathcal{F}\big(g(x)*h(x)\big)(f) &= \int_{-\infty}^{\infty}\int_{-\infty}^{\infty} g(x')h(x-x')\,dx'e^{-i2\pi f x}\,dx \\
&= \int_{-\infty}^{\infty} g(x')e^{-i2\pi f x'} \int_{-\infty}^{\infty} h(x-x')e^{-i2\pi f(x-x')}\,d(x-x')dx' \qquad (3.18)\\
&= G(f)H(f)
\end{aligned}
$$

As noted above, the convolution is absolutely integrable, bounded, and continuous; hence, the inverse Fourier transform converges to the convolution for every x, even if g or h have step discontinuities,

$$g(x) * h(x) = \mathcal{F}^{-1}(G(f)\,H(f))(x). \qquad (3.19)$$

While the numerical implementation of the convolution of two functions generally is a laborious procedure, the multiplication of their Fourier transforms by comparison is not. Since Fourier transforms can be done relatively cheaply (see Section 4.3.2), convolution by multiplication in the frequency domain is much easier. Equation (3.19) shows that a convolution (for all x) is the same as a multiplication of two spectra (for all f) and three Fourier transforms. One often encounters the problem of solving for one of the functions being convolved, say g, with both h and the convolution, $c = g*h$, known. The convolution theorem provides a relatively easy solution, by first computing $H(f)$ and $C(f)$, and then the inverse Fourier transform, $\mathcal{F}^{-1}(C(f)/H(f))$, assuming $H(f) \neq 0$. This procedure is also known as a *deconvolution*; it can take various forms depending on the problem (e.g., least-squares estimation), but essentially involves the inversion of a convolution.

By the duality of Fourier transforms, property (2.77), there is a *Dual Convolution Theorem*, also known as the *Frequency Convolution Theorem*.

Frequency (Dual) Convolution Theorem

The Fourier transform of a product of functions is the convolution of their Fourier transforms. In symbols,

$$\mathcal{F}(g(x)h(x))(f) = G(f)*H(f). \tag{3.20}$$

The proof could be obtained using the duality properties of Fourier transforms, but it is as easy to follow the steps in equations (3.18),

$$\begin{aligned}\mathcal{F}^{-1}\big(G(f)*H(f)\big)(x) &= \int_{-\infty}^{\infty}\int_{-\infty}^{\infty} G(f')H(f-f')\,df'\,e^{i2\pi f x}\,df \\ &= \int_{-\infty}^{\infty} G(f')e^{i2\pi f' x}\left(\int_{-\infty}^{\infty} H(f-f')e^{i2\pi(f-f')x}\,d(f-f')\right)df \\ &= g(x)h(x)\end{aligned} \tag{3.21}$$

Equation (3.20) then follows by applying the Fourier transform on both sides. Leaving the result as in the last of equations (3.21) assumes that the product is continuous at x. At a step discontinuity in either g or h, the left side of equation (3.21) converges to the mean of the values of $g(x)\,h(x)$ in the limit on either side of a step discontinuity, as formulated in equation (2.58). However, equation (3.20) is true irrespective of step-discontinuities in g or h since the Fourier transforms and the convolution are continuous.

From the convolution theorem, equation (3.17), the moving average, equation (3.6), for any given T can be written as

$$g(x)*\frac{1}{T}b\left(\frac{x}{T}\right) = \mathcal{F}^{-1}\big(G(f)\operatorname{sinc}(Tf)\big)(x), \tag{3.22}$$

in accordance with the Fourier transform of the rectangle function, equation (2.100). In the limit as $T \to 0$, the left side, as in equation (2.109), then yields the convolution of $g(x)$ and the Dirac delta function. On the right side, it is $g(x)$, since $\operatorname{sinc}(0) = 1$; hence,

$$\int_{-\infty}^{\infty} g(x')\delta(x-x')\,dx' = g(x). \tag{3.23}$$

This proves the reproducing property of the Dirac delta function, equation (2.115).

A final interesting example is the fact that the convolution of two Gaussian functions is again a Gaussian function (Exercise 3.2).

3.2.2 Convolutions in Higher Dimensions

The concept of convolution of non-periodic functions in Cartesian coordinates naturally extends to higher dimensions. In two Cartesian dimensions, one has

$$g(x_1,x_2) * h(x_1,x_2) = \int_{-\infty}^{\infty}\int_{-\infty}^{\infty} g(x_1',x_2')\,h(x_1-x_1',x_2-x_2')\,dx_1'dx_2'. \tag{3.24}$$

For example, the convolution of two rectangle functions, $b(x_1, x_2)$, equation (2.161), is a two-dimensional Bartlett window,

$$u_{\text{Bartlett}}(x_1,x_2) = \begin{cases} (1-|x_1|)(1-|x_2|), & |x_1| \le 1 \text{ and } |x_2| \le 1 \\ 0, & \text{otherwise} \end{cases} \tag{3.25}$$

where it is noted that the sides are not flat planes (Figure 3.1), as they would be for a pyramid.

The properties, equations (3.8) through (3.12) and (3.16), are true evidently for multi-dimensional Cartesian convolutions and need not be repeated. The convolution theorem and its dual theorem also hold for higher dimensioned Cartesian coordinates. For the convolution theorem, the multiplication of spectra is performed at each frequency pair,

$$\mathcal{F}(g*h)(f_1,f_2) = G(f_1,f_2)H(f_1,f_2). \tag{3.26}$$

The dual convolution theorem yields

$$\mathcal{F}(gh)(f_1,f_2) = G(f_1,f_2) * H(f_1,f_2). \tag{3.27}$$

A classic multi-dimensional illustration of convolution comes from potential field theory. The gravitational potential, v, due to a mass density distribution, $\rho(x_1', x_2', x_3')$, over some volume is a scalar function given by (Kellogg 1953, p.55),

$$v(x_1,x_2,x_3) = G \iiint\limits_{\text{volume}} \frac{\rho(x_1',x_2',x_3')}{\sqrt{(x_1-x_1')^2+(x_2-x_2')^2+(x_3-x_3')^2}}\,dx_1'dx_2'dx_3', \tag{3.28}$$

where G is Newton's gravitational constant. Therefore, the potential is the convolution of the density, ρ, and the reciprocal distance. It is noted that at evaluation points inside and on the boundary of the volume the integrand has a singularity as the distance between (x_1, x_2, x_3) and (x_1', x_2', x_3') becomes zero. That is, the reciprocal distance is an unbounded function for such evaluation points; and, our assumptions on the functions for the convolution are violated. However the potential for a reasonably regular (piecewise-continuous) three-dimensional density function over a finite volume is not only continuous but differentiable everywhere (Kellogg 1953, pp.150-151)—the integral is only *weakly singular* (see also Section 6.5.1.1). Usually, convolutions such as these in potential theory are evaluated for fixed x_3 such that $x_3 > x_3'$ for all x_3', for example, on the plane completely above a volume of mass density. In that case, the reciprocal distance is a bounded function and spectral analysis may proceed on the basis of the two-dimensional convolution theorem. Integrals such as equation (3.28) play an important role in geodesy and geophysics and are discussed in detail in Section 6.5.

In many cases, the convolution is an integral of a data function, $g(x_1', x_2')$, and a kernel function, $h(s)$, that is isotropic, that is, it depends only on the distance between the integration point and the evaluation point,

$$c(x_1,x_2) = \int_{-\infty}^{\infty}\int_{-\infty}^{\infty} g(x_1',x_2')\,h(s)\,dx_1'dx_2', \tag{3.29}$$

where

$$s = \sqrt{(x_1-x')^2+(x_2-x_2')^2}. \tag{3.30}$$

The Fourier transform of $h(s)$ is its Hankel transform, equation (2.180), and the convolution theorem then becomes

$$C(f_1,f_2) = G(f_1,f_2)\,H(\bar{f}), \tag{3.31}$$

with $\bar{f}$ given by (2.178).

3.3 Convolutions of Periodic Functions

For periodic functions the definition of convolution differs from equation (3.1), which holds only for functions of bounded variation and absolutely integrable, as in equation (2.49), over the entire real line. For piecewise-continuous periodic functions of bounded variation, we define a new convolution, the *cyclic* or *circular convolution*. Denoting the operation of cyclic (or periodic) convolution by "$\tilde{*}$", it is

$$\tilde{g}(x) \tilde{*} \tilde{h}(x) = \int_a^{a+P} \tilde{g}(x')\tilde{h}(x-x')\,dx', \tag{3.32}$$

where both periodic functions, $\tilde{g}$ and $\tilde{h}$, have the same period, P. As such, the cyclic convolution, $\tilde{g} \tilde{*} \tilde{h}$, is also periodic with period, P (since $\tilde{h}(x + P - x') = \tilde{h}(x - x')$). The constant, a, is arbitrary and the integral yields the same result for any such constant because the integrand is periodic; therefore, one might as well set $a = 0$,

$$\tilde{g}(x) \tilde{*} \tilde{h}(x) = \int_0^{P} \tilde{g}(x')\tilde{h}(x-x')\,dx'. \tag{3.33}$$

Just as for non-periodic functions, the cyclic convolution is continuous even if $\tilde{g}$ and $\tilde{h}$ have step discontinuities. And, the units of the cyclic convolution are, again, the product of the units of $\tilde{g}$ and $\tilde{h}$ times the units of x.

In a discussion that distinguishes between cyclic and non-cyclic convolutions, the latter is often called a *linear* convolution. However, the cyclic convolution and indeed all convolutions, by definition, are linear in the algebraic sense. Thus, the properties of the cyclic convolution also follow exactly as in equations (3.8) through (3.12) and (3.16), where the proofs that require a change in integration variable can proceed with the recognition that the convolution is periodic. For example, commutativity is proved with $u' = x - x'$ by

$$\begin{aligned}\tilde{g}(x) \tilde{*} \tilde{h}(x) &= \int_0^{P} \tilde{g}(x')\tilde{h}(x-x')\,dx' = \int_{x-P}^{x} \tilde{h}(u')\tilde{g}(x-u')\,du' = \int_0^{P} \tilde{h}(u')\tilde{g}(x-u')\,du' \\ &= \tilde{h}(x) \tilde{*} \tilde{g}(x)\end{aligned} \tag{3.34}$$

where the last equality in the first line holds according to equation (3.32) for any fixed x. The other properties are proved similarly (Exercise 3.3). Properties (3.8) and (3.11) establish the linearity of the convolution as an operator on either function.

Corresponding convolution theorems are readily obtained using equation (2.13) for the Fourier series transform. We have the following.

Cyclic Convolution Theorem

The Fourier series transform of the cyclic convolution of two periodic functions with the same period equals the product of their Fourier series transforms, and the inverse Fourier series transform is the cyclic convolution. In symbols,

$$\tilde{g}(x) \tilde{*} \tilde{h}(x) \quad \leftrightarrow \quad G_k H_k. \tag{3.35}$$

Substituting equation (3.33) into equation (2.13), the Fourier coefficient of the convolution at frequency, k/P, is

$$\int_0^P \int_0^P \tilde{g}(x')\tilde{h}(x-x')\,dx' e^{-i\frac{2\pi}{P}kx}\,dx$$

$$\begin{aligned}
&= \int_0^P \tilde{g}(x') e^{-i\frac{2\pi}{P}kx'} \left(\int_0^P \tilde{h}(x-x') e^{-i\frac{2\pi}{P}k(x-x')} dx \right) dx' \\
&= \int_0^P \tilde{g}(x') e^{-i\frac{2\pi}{P}kx'} \left(\int_{-x'}^{P-x'} \tilde{h}(u) e^{-i\frac{2\pi}{P}ku} du \right) dx' \\
&= \int_0^P \tilde{g}(x') e^{-i\frac{2\pi}{P}kx'} \left(\int_0^P \tilde{h}(u) e^{-i\frac{2\pi}{P}ku} du \right) dx'
\end{aligned} \tag{3.36}$$

The last line proves that $G_k H_k$ is the Fourier series transform of the convolution. Since $\tilde{g}(x) \tilde{*} \tilde{h}(x)$ is continuous, the Fourier series converges to the convolution for every x,

$$\tilde{g}(x) \tilde{*} \tilde{h}(x) = \frac{1}{P} \sum_{k=-\infty}^{\infty} G_k H_k e^{i\frac{2\pi}{P}kx}. \tag{3.37}$$

Cyclic Frequency (Dual) Convolution Theorem

The Fourier series transform of a product of periodic functions that have the same period is the convolution of the Fourier series transforms of the functions being multiplied. That is,

$$G_k \# H_k = \int_0^P \tilde{g}(x)\tilde{h}(x) e^{-i\frac{2\pi}{P}kx} dx, \tag{3.38}$$

where the discrete convolution, denoted by #, is defined by (see also Section 4.3.4)

$$G_k \# H_k = \frac{1}{P} \sum_{\ell=-\infty}^{\infty} G_\ell H_{k-\ell}. \tag{3.39}$$

This is proved starting, again, with the definition of Fourier series transform, equation (2.13),

$$\begin{aligned}
\int_0^P \tilde{g}(x)\tilde{h}(x) e^{-i\frac{2\pi}{P}kx} dx &= \int_0^P \frac{1}{P} \sum_{\ell=-\infty}^{\infty} G_\ell e^{i\frac{2\pi}{P}\ell x} \frac{1}{P} \sum_{k'=-\infty}^{\infty} H_{k'} e^{i\frac{2\pi}{P}k'x} e^{-i\frac{2\pi}{P}kx} dx \\
&= \frac{1}{P} \sum_{\ell=-\infty}^{\infty} G_\ell \frac{1}{P} \sum_{k'=-\infty}^{\infty} H_{k'} \int_0^P e^{i\frac{2\pi}{P}(\ell+k'-k)x} dx
\end{aligned} \tag{3.40}$$

where step discontinuities in $\tilde{g}$ or $\tilde{h}$ do not affect the integration. Now, changing the inner summation index to $p = \ell + k'$, one finds

$$\begin{aligned}\int_0^P \tilde{g}(x)\tilde{h}(x)e^{-i\frac{2\pi}{P}kx}dx &= \frac{1}{P}\sum_{\ell=-\infty}^{\infty} G_\ell \sum_{p=-\infty}^{\infty} H_{p-\ell}\frac{1}{P}\int_0^P e^{i\frac{2\pi}{P}(p-k)x}dx \\ &= \frac{1}{P}\sum_{\ell=-\infty}^{\infty} G_\ell H_{k-\ell}\end{aligned} \tag{3.41}$$

in view of the orthogonality, equation (2.12). This proves the theorem. If $\tilde{g}$ or $\tilde{h}$ have step discontinuities, then the inverse Fourier series transform of the discrete convolution, $G_k \# H_k$, converges to the average, in the limit, of the values of the product, $\tilde{g}(x)\tilde{h}(x)$, on either side of the discontinuity, in the sense of equation (2.9).

These cyclic convolution definitions, properties, and theorems can be generalized readily to periodic functions of two or more Cartesian variables, where the periodicity in each dimension is the same for each function, though it may be different for different dimensions. For example, the two-dimensional cyclic convolution is given by

$$\tilde{g}(x_1,x_2)\tilde{*}\tilde{h}(x_1,x_2) = \int_0^{P_1}\int_0^{P_2}\tilde{g}(x_1',x_2')\tilde{h}(x_1-x_1',x_2-x_2')\,dx_1'dx_2'; \tag{3.42}$$

and the convolution theorem yields

$$\tilde{g}(x_1,x_2)\tilde{*}\tilde{h}(x_1,x_2) \quad\leftrightarrow\quad G_{k_1,k_2}H_{k_1,k_2}. \tag{3.43}$$

Both functions, $\tilde{g}$ and $\tilde{h}$, and their convolution, $\tilde{g}\tilde{*}\tilde{h}$, have the same period, P_1, in x_1, and P_2 in x_2.

3.4 Convolutions on the Sphere

Most physical convolution models in geodesy and geophysics are based on a spherical formulation (Chapter 6). While the Cartesian approximation may suffice for local applications, analyses that require a global interpretation, or make use of global data, or simply require higher accuracy by accounting for the Earth's curvature, are better expressed in the spherical domain, Ω, equation (2.194). One may even consider an ellipsoidal domain, but this is outside the present scope. Not all spectral properties of the Cartesian convolution translate to the spherical case without some specializations. One-dimensional functions, depending only on θ (or, ψ) are regarded in the present context as two-dimensional functions on the sphere, but independent of λ (or, ζ). The primary focus is on the convolution of a two-dimensional function with a one-dimensional function, which readily specializes to the convolution of two one-dimensional functions on the sphere. We start, however, with the convolution of two two-dimensional functions, mostly to demonstrate the essential properties and to illustrate the difficulty of the corresponding convolution theorem. One particular specialization yields a concise result, but fortunately, in many cases there is no loss in abandoning the more general convolution in favor of the much easier convolution in which one function depends only on co-latitude, or distance along a great circle arc. Indeed, many such convolutions appear in geophysical models.

For two piecewise continuous functions of bounded variation on the unit sphere, Ω, the spherical convolution is defined as

$$g(\theta,\lambda)*h(\theta,\lambda)=\frac{1}{4\pi}\iint_{\Omega'} g(\theta',\lambda')h(\psi,\zeta)\sin\theta' d\theta' d\lambda', \tag{3.44}$$

where the integration point, (θ', λ'), is related to the evaluation point, (θ, λ), by equations (2.196) and (2.197). The coordinates, ψ, ζ, represent the "difference" between the evaluation and integration points, as shown in Figure 2.13. That is, ψ is the spherical distance (great circle arc) between the points, and ζ is the longitude of (θ', λ') with respect to the meridian through (θ, λ). The point, (θ, λ), acts as the pole and is the origin for the coordinates, ψ, ζ. Note that the definition of spherical convolution includes the scale, $1/(4\pi)$, which simplifies subsequent formulas, such as the convolution theorem. Also, the units of the convolution in this case are simply the product of the units of g and h.

The algebraic properties of convolutions in Cartesian space, equations (3.8) through (3.12), and (3.16), hold as well for the spherical convolution. The proofs for proportionality and distributivity are straightforward,

$$g(\theta,\lambda)*ah(\theta,\lambda)=ag(\theta,\lambda)*h(\theta,\lambda)=a\big(g(\theta,\lambda)*h(\theta,\lambda)\big), \tag{3.45}$$

$$\big(g_1(\theta,\lambda)+g_2(\theta,\lambda)\big)*h(\theta,\lambda)=g_1(\theta,\lambda)*h(\theta,\lambda)+g_2(\theta,\lambda)*h(\theta,\lambda). \tag{3.46}$$

One can think of the spherical convolution as the average over the unit sphere of products of g and h whose evaluation points are separated by all possible spherical distances, ψ, and all possible angles, ξ. In this view, it is evident that the commutativity property holds,

$$g(\theta,\lambda)*h(\theta,\lambda)=h(\theta,\lambda)*g(\theta,\lambda); \tag{3.47}$$

and, associativity follows under the same consideration,

$$\big(g_1(\theta,\lambda)*g_2(\theta,\lambda)\big)*h(\theta,\lambda)=g_1(\theta,\lambda)*\big(g_2(\theta,\lambda)*h(\theta,\lambda)\big). \tag{3.48}$$

The translation property, equation (3.9), is interpreted here in terms of coordinate rotations. Rotating the θ, λ-system does not change the convolution if also the θ', λ'-system in the integral is rotated in the same way. Finally, the property analogous to equation (3.16),

$$\frac{1}{4\pi}\iint_{\Omega} g(\theta,\lambda)*h(\theta,\lambda)\,d\Omega=\left(\frac{1}{4\pi}\iint_{\Omega} g(\theta,\lambda)\,d\Omega\right)\left(\frac{1}{4\pi}\iint_{\Omega} h(\theta,\lambda)\,d\Omega\right), \tag{3.49}$$

is also proved easily by substituting equation (3.44) into the left side,

$$\frac{1}{4\pi}\iint_{\Omega}\left(\frac{1}{4\pi}\iint_{\Omega'} g(\theta',\lambda')h(\psi,\zeta)d\Omega'\right)d\Omega = \left(\frac{1}{4\pi}\iint_{\Omega'}\tilde{g}(\theta,\lambda)\left(\frac{1}{4\pi}\iint_{\Omega}h(\psi,\zeta)d\Omega\right)d\Omega'\right). \tag{3.50}$$

Equation (3.49) then follows since the differential element, $d\Omega$, holds for any origin of spherical coordinates, thus making the inner integral on the right side independent of θ', λ'. Therefore, the average over the unit sphere of the spherical convolution is the product of the averages of the convolved functions.

Obtaining a convolution theorem is, however, much more difficult, and certainly not as elegant, hence likely not as useful, as for the Cartesian case. Let $\{G_{n,m}\}$ and $\{H_{n,m}\}$ be the Fourier-Legendre spectra of $g(\theta, \lambda)$ and $h(\theta, \lambda)$, respectively. The spectrum of h with respect to the origin at (θ, λ) is given by equation (2.257). Hence,

$$h(\psi,\zeta) = \sum_{n=0}^{\infty}\sum_{m'=-n}^{n}\sum_{m=-n}^{n} H_{n,m}C_{m',m}^{n}(0,-\theta,-\lambda)\bar{Y}_{n,m'}(\psi,\zeta). \tag{3.51}$$

Substituting equation (2.247) for $\bar{Y}_{n,m'}(\psi, \zeta)$, with $\alpha = \lambda$, $\beta = \theta$, and $\gamma = 0$, it is readily seen in view of the orthogonality of spherical harmonic functions that the convolution, equation (3.44), becomes

$$g(\theta,\lambda)*h(\theta,\lambda) = \sum_{p=0}^{\infty}\sum_{q=-p}^{p} G_{p,q}\sum_{m=-p}^{p} H_{p,m}\sum_{m'=-p}^{p} C_{m',m}^{p}(0,-\theta,-\lambda)C_{q,m'}^{p}(\lambda,\theta,0). \tag{3.52}$$

In order to derive a convolution theorem for this convolution it is necessary to express the last sum in equation (3.52) as a sum of spherical harmonics, $\bar{Y}_{p,m''}(\theta,\lambda), m''=-p,\ldots,p$. In other words, the right side must be expressed as a spherical harmonic series, where the coefficients then represent the Fourier-Legendre transform of the convolution. This is left to an intrepid reader (the author of this text is not aware that it has been done).

However, it is possible to design a convolution for a specialized kernel function, h, that depends on the four independent variables, θ, λ, θ', λ', and for which a convolution theorem is readily established (Han et al. 2005). Consider the convolution,

$$\bar{g}_h(\theta,\lambda) = \frac{1}{4\pi}\iint_{\Omega'} g(\theta',\lambda')h(\theta,\lambda,\theta',\lambda')\sin\theta' d\theta' d\lambda', \tag{3.53}$$

where

$$h(\theta,\lambda,\theta',\lambda') = \sum_{n=0}^{\infty}\sum_{m=-n}^{n} H_{n,m}\bar{Y}_{n,m}(\theta,\lambda)\bar{Y}_{n,m}(\theta',\lambda'). \tag{3.54}$$

By the orthogonality of the spherical harmonic functions (multiply both sides by $\bar{Y}_{p,q}(\theta, \lambda)\ \bar{Y}_{p,q}(\theta', \lambda')$ and integrate over both Ω and Ω', using equation (2.219)), the coefficients are

$$H_{n,m} = \frac{1}{16\pi^2} \iint_{\Omega} \iint_{\Omega'} h(\theta,\lambda,\theta',\lambda') \bar{Y}_{n,m}(\theta,\lambda) \bar{Y}_{n,m}(\theta',\lambda') d\Omega d\Omega'. \tag{3.55}$$

They comprise the Fourier-Legendre spectrum either of $h(\theta, \lambda, \theta', \lambda') \bar{Y}_{n,m}(\theta, \lambda)/(4\pi)$ for fixed (θ', λ'), or of $h(\theta, \lambda, \theta', \lambda') \bar{Y}_{n,m}(\theta', \lambda')/(4\pi)$ for fixed (θ, λ). Now substitute equation (3.54) into equation (3.53) to find

$$\begin{aligned} \bar{g}_h(\theta,\lambda) &= \sum_{n=0}^{\infty} \sum_{m=-n}^{n} \frac{1}{4\pi} \iint_{\Omega'} g(\theta',\lambda') \bar{Y}_{n,m}(\theta',\lambda') d\Omega' H_{n,m} \bar{Y}_{n,m}(\theta,\lambda) \\ &= \sum_{n=0}^{\infty} \sum_{m=-n}^{n} H_{n,m} G_{n,m} \bar{Y}_{n,m}(\theta,\lambda) \end{aligned} \tag{3.56}$$

Hence, the Fourier-Legendre transform of the convolution, $\bar{g}_h(\theta, \lambda)$, with h given by equation (3.54), is the product, $H_{n,m} G_{n,m}$.

If the coefficients, $H_{n,m}$, are independent of the order, m, then the kernel function depends on just one variable, ψ, in view of the addition formula (2.254),

$$h(\psi) = \sum_{n=0}^{\infty} (2n+1) H_n P_n(\cos\psi), \tag{3.57}$$

where the angle, ψ, as before, is defined by equation (2.196). We arrive, thus, at the more common convolutions in geodesy and geophysics that involve a function of two variables, θ, λ, and a function of just one variable, θ (or, ψ),

$$g(\theta,\lambda) * h(\theta) = \frac{1}{4\pi} \iint_{\Omega'} g(\theta',\lambda') h(\psi) \sin\theta' d\theta' d\lambda'. \tag{3.58}$$

The corresponding convolution theorem in this case is immediate from equation (3.56).

Convolution Theorem for the Sphere

The Fourier-Legendre spectrum of the convolution, equation (3.53), of functions defined on the sphere, where h is given by equation (3.54), is the product of their respective spectra; and, the inverse Legendre transform, equation (3.56) converges to the convolution. In symbols,

$$\bar{g}_h(\theta,\lambda) \quad \leftrightarrow \quad G_{n,m} H_{n,m}. \tag{3.59}$$

where $G_{n,m}$ is the Fourier-Legendre transform of $g(\theta, \lambda)$, equation (2.220), and $H_{n,m}$ is defined according to equation (3.55). If h depends only on θ (or, ψ), then for the convolution defined by equation (3.58),

$$g(\theta,\lambda) * h(\theta) \quad \leftrightarrow \quad G_{n,m} H_n, \tag{3.60}$$

where H_n is the Legendre transform of $h(\theta)$, equations (2.205).

An analogous dual convolution theorem for the sphere is not possible. The Fourier-Legendre spectrum of a product of functions of one and two variables, respectively, ultimately hinges on finding the value of the integral of a triple product of Legendre functions (Gaunt 1929, p.192); it is not pursued here.

The Legendre transform pair (3.60) can be specialized to functions of the single variable, θ. The convolution, equation (3.58), then is given by

$$g(\theta)*h(\theta)=\frac{1}{4\pi}\iint_{\Omega'} g(\theta')h(\psi)\sin\theta' d\theta' d\lambda', \tag{3.61}$$

where ψ is defined by $\cos\psi = \cos\theta \cos\theta' + \sin\theta \sin\theta' \cos\lambda'$ and the integration over λ' is needed for the corresponding convolution theorem (Churchill and Dolph 1954); that is, the convolution is still two-dimensional. Since it is again a function of a single variable, its Legendre spectrum is based on the Legendre polynomials, as in equation (2.205). The one-dimensional Legendre transform pair of the two-dimensional convolution on the sphere is (Exercise 3.4)

$$g(\theta)*h(\theta) \quad \leftrightarrow \quad G_n H_n. \tag{3.62}$$

Finally, consider the convolution of the spherical Gaussian function, equation (2.310), and an arbitrary function on the sphere,

$$g(\theta,\lambda)*\gamma_s^{(\beta s)}(\theta)=\frac{1}{4\pi}\iint_{\Omega'} g(\theta',\lambda')\gamma_s^{(\beta s)}(\psi)\,d\Omega', \tag{3.63}$$

By the convolution theorem (3.60),

$$\frac{1}{4\pi}\iint_{\Omega'} g(\theta',\lambda')\gamma_s^{(\beta s)}(\psi)\,d\Omega' = \sum_{n=0}^{\infty}\sum_{m=-n}^{n} \Gamma_n^{(\beta s)} G_{n,m}\bar{Y}_{n,m}(\theta,\lambda), \tag{3.64}$$

where $\Gamma_n^{(\beta s)}$ is the Legendre spectrum of $\gamma_s^{(\beta s)}$, equation (2.312). Now, in the limit as the parameter, β_s, approaches infinity, $\gamma_s^{(\beta s)}(\psi) \to \delta_s(\psi)$ (equation (2.318)) and $\Gamma_n^{(\beta s)} \to 1$, for all n, by equations (2.313). Hence, in the limit, equation (3.64) becomes

$$\frac{1}{4\pi}\iint_{\Omega'} g(\theta',\lambda')\delta_s(\psi)\,d\Omega' = g(\theta,\lambda), \tag{3.65}$$

which proves the reproducing property of the Dirac delta function for functions on the sphere, equation (2.323).

3.5 Filters on the Line, Plane, and Sphere

In introducing convolutions in Section 3.1, the analogy was made to an operation with an input and an output. In fact, the convolution is a special case of a *system*, which generally is any operation on a given input signal, $g(x)$, yielding an output, $y(x)$, written mathematically as

$$y(x) = h(g(x)), \tag{3.66}$$

where h is the *system function*. Truncation of a signal by multiplication with the rectangle function is another example of a system with $h(g(x)) = b(x/T)g(x)$. In this case the system is not a convolution, and the system function is sometimes called a "window" function, which is discussed in greater detail in Section 3.6. This section considers the class of systems that are known as *filters*. However, the terminology in the literature is not universal and filters may sometimes be considered synonymous with systems. Traditionally, a filter is a system that removes part of the spectral content of the input. More generally, and as a *definition* in the present context, a filter is any manipulation of the signal through a *convolution*,

$$y(x) = h(x)*g(x). \tag{3.67}$$

As such, the filter is *linear*, satisfying the properties of proportionality and distributivity, equations (3.8) and (3.11). Non-linear filters can also be defined, for example, in the context of estimation, such as non-linear, least-squares estimation. Our attention is restricted to linear filters; and, in that case, a filter is a convolution (and vice versa) and all properties of the convolution hold as well for the filter. It is assumed here that the function, g, is piecewise continuous; discrete or digital filters are treated briefly in Section 4.6. Filters hold a special place in signal processing and electrical engineering and some of the terminology used in those applications is adopted in geophysics and geodesy, as well.

The (linear) filter, equation (3.67), is *shift-invariant, or translation-invariant* (or, *time-invariant*, if the independent variable is time); that is,

$$y(x - x_0) = h(x)*(g(x - x_0)), \tag{3.68}$$

which follows immediately from the translation property of convolutions, equation (3.9),

$$h(x)*(g(x-x_0)) = \int_{-\infty}^{\infty} h(x')g((x-x')-x_0)\,dx' = \int_{-\infty}^{\infty} h(x')g((x-x_0)-x')\,dx' = y(x-x_0). \tag{3.69}$$

That is, the result, or output, of a filter, represented by the filter function, h, is simply translated on the x-axis by the same amount that the input is translated. In other words, there is no change in the output, relatively speaking, if there is a shift in the origin of the independent variable.

A *causal* filter is one for which the filter function satisfies

$$h(x) = 0, \quad x < 0. \tag{3.70}$$

The terminology originates with time-signal processing, where the output of a causal filter does not depend on the *future* of the input (recall that the convolution first *reflects* the filter function with respect to the origin before multiplying by the input).

All physically realizable filters in *time* are causal, because they do not use future values of the input. However, in other domains, particularly the spatial domain, filters typically are *non-causal*. For example, in optical devices, filters operate on the entire image, not just the part restricted to negative coordinates. Similarly, geodetic and geophysical data are processed in the spatial domain (or, in time, but only after all data have been collected). For periodic signals, including signals on the sphere, the "future" is known from the "past" and filters by definition are non-causal.

Assuming the Fourier transforms exist, equation (3.67) and the Convolution Theorem, equation (3.17), imply

$$Y(f) = H(f)G(f). \tag{3.71}$$

Formally, we may also consider the case when the input to the filter, g, is the Dirac delta (or, impulse) function. Then from equation (2.112), $G(f) = 1$, and the Fourier transform of the output is simply

$$Y(f) = H(f), \quad (g(x) = \delta(x)). \tag{3.72}$$

Hence, the output function in this case, i.e., the *response* of the filter to an impulse as input, is $y(x) = h(x)$; and, therefore, the filter function, $h(x)$, is also called the *impulse response*. Its Fourier transform, $H(f)$, is called variously the *frequency response*, or *transfer function*, or *filter spectrum* of the filter. In geophysics it is also called the *admittance function* (e.g., Watts 2001). In general,

$$H(f) = \frac{Y(f)}{G(f)}. \tag{3.73}$$

The frequency response is the ratio of the spectrum of the output to that of the input. Writing the Fourier transforms, $Y(f)$, $G(f)$, and $H(f)$, in the form of equation (2.61), the frequency response becomes,

$$H(f) = \frac{Y(f)}{G(f)} = \frac{|Y(f)|e^{i\varphi_Y(f)}}{|G(f)|e^{i\varphi_G(f)}} = |H(f)|e^{i(\varphi_Y(f)-\varphi_G(f))}, \tag{3.74}$$

decomposed into an *amplitude response*,

$$|H(f)| = \frac{|Y(f)|}{|G(f)|}, \tag{3.75}$$

which describes the ratio per frequency of the output to the input amplitudes, and the *phase response*,

$$\varphi_H(f) = \varphi_Y(f) - \varphi_G(f), \tag{3.76}$$

which describes how the filter changes the phase of the input signal. Amplitude response is important in designing filters as it characterizes the loss (or gain) in strength of a signal as it passes through the filter. Similarly, the phase response determines how the various spectral components of a signal are affected in phase by the filter. $\varphi_H(f) = 0$ implies no change, and the output differs from the input only in amplitude, not in phase. The simple act of unweighted averaging (filtering out high-frequency parts of a signal) has a non-zero phase response (Section 3.5.1).

The square of the amplitude response, $|H(f)|^2$, is also known as the *power transfer function*. It is often expressed in units of *decibel* (dB). These units are obtained by computing $10\log_{10}|H(f)|^2$; or $20\log_{10}|H(f)|$, which is then in dB. For example, if $|H(f_0)| = 10^{-2}$ at some frequency, f_0, then $|H(f_0)|_{dB} = -40$ dB.

Multi-dimensional filters in the Cartesian domain are multidimensional convolutions, and amplitude and phase responses are defined similarly. In two dimensions, the frequency response is

$$H(f_1, f_2) = \frac{Y(f_1, f_2)}{G(f_1, f_2)} = \frac{|Y(f_1, f_2)|e^{i\varphi_Y(f_1, f_2)}}{|G(f_1, f_2)|e^{i\varphi_G(f_1, f_2)}} = |H(f_1, f_2)|e^{i(\varphi_Y(f_1, f_2) - \varphi_G(f_1, f_2))}. \quad (3.77)$$

If the filter function is isotropic and thus depends only on distance, then the frequency response depends only on $\bar{f} = \sqrt{f_1^2 + f_2^2}$ and is real (see equation (2.180)). If it is also of one sign (either always positive or always negative, thus precluding a phase reversal), then the phase response is zero. Many physical filters (convolutions) encountered in geodesy and geophysics (Chapter 6) and approximated on the plane, in fact, have isotropic and positive frequency responses; hence, phase distortion is not a concern in those cases.

Although the term "filter" can denote any operation that results in a convolution, the classic meaning, of course, is that it removes some unwanted part of a signal. Usually those parts refer to the spectral domain, and filters are designed to pass some frequencies of the input signal and block others. For example, a smoothing filter will pass the low frequencies, or long wavelengths, and filter out the high frequencies, or short wavelengths. One speaks of a *low-pass filter* in this case. Conversely, there are *high-pass filters* and *band-pass filters* that, respectively, pass high-frequency components or only components within a certain range (or, band) of frequencies. In addition, the *band-stop filter* passes all components except those in a certain band of frequencies.

The point in the spectral domain beyond which the spectrum is passed or blocked is called the *cutoff frequency*, f_0. For example, the *ideal* low-pass filter is designed simply by setting the frequency response to unity for frequencies lower than f_0, and to zero otherwise,

$$H(f) = \begin{cases} 1, & |f| \le f_0 \\ 0, & |f| > f_0 \end{cases} \quad (3.78)$$

The bandwidth, equation (2.63) of this filter function is clear, $W(h) = 2f_0$. By the duality property (2.77), the corresponding filter function is the sinc function, equation (2.93), with appropriate scaling, equation (2.78),

$$h(x) = 2f_0 \operatorname{sinc}(2f_0 x). \quad (3.79)$$

Usually, filters are designed with less sharp frequency cutoff in order to avoid the distortion, already indicated with equation (2.103), that manifests as a ripple effect, or Gibbs's effect (Section 3.5.1). Thus, "ideal" really only refers to the truncation in the frequency domain, not to the consequent properties of the filtered function.

The ideal low-pass filter for a spherical function from the spectral viewpoint is the partial sum of Legendre polynomials,

$$h(\psi) = \sum_{n=0}^{n_0} (2n+1) P_n(\cos\psi), \tag{3.80}$$

whose Legendre spectrum is

$$H_n = \begin{cases} 1, & 0 \le n \le n_0 \\ 0, & n > n_0 \end{cases} \tag{3.81}$$

The sum in equation (3.80) has no simpler analytic form, but is similar to the sinc function. It appears again in the context of the spherical concentration problem; see Section 3.6.2 and Figure 3.15.

Filters in geodesy and geophysics typically are used to eliminate or attenuate the part of the spectrum of a signal that is irrecoverably corrupted by errors, whether due directly to measurement errors or to uncertainty propagating through various data processing steps. That is, the part of the spectrum where the "signal-to-noise" ratio is near unity or less is the object of attenuation. Filters are also used to reduce the effects of aliasing caused by under-sampling a signal (Section 4.2.3); these are specifically called *anti-aliasing* filters and invariably are low-pass filters. The filter may be a straightforward convolution with a predefined filter function, such as the filter described by simple averaging, or it may be as elaborate as a least-squares estimation of the signal with relative weights determined by the error covariances (see Sections 4.5.2 (spherical harmonic analysis) and 6.6 (least-squares collocation)). In either case, some knowledge about the spectra of the signal and the errors is assumed so as to design the appropriate frequency response of the filter. Another type of filter, the high-pass filter, is applied whenever a long-wavelength model is removed from a signal in order to isolate better the spectrum of interest or because the sampling area is of insufficient extent to determine the long wavelengths.

By removing (approximately or exactly) a part of the spectral content of an error-corrupted signal thereby improving the visualization or estimation of the signal in the space-domain, filters act as a kind of window on the frequency domain. By the spatio-frequency duality, they have spatial-domain analogues, functions that, in fact, are also called window functions. These are described in Section 3.6 and as such they are applied specifically in spectral analysis, or in the estimation of the spectrum, of a signal from data limited to a finite domain. Consequently, the filter and window functions, though operating in opposing domains, share common characteristics and can be utilized with appropriate definition in either capacity. This duality appears again in the concentration problem elaborated in Section 3.6.2. Since spectral analysis and thus window functions are the primary focus, the discussion of filters is limited here to some particular examples in the next section.

3.5.1 Filter Examples for Profiles and Spherical Signals

The moving average is perhaps the most commonly used filter in geodetic and geophysical data processing. For multiple samples of a physical quantity, such as a sequence of measurements of the gravitational or magnetic field, either in time at a single point or in space along a trajectory, the process of averaging, or smoothing, is a low-pass filter that attempts to remove high-frequency components of the data, which either are not important for the particular application or contain random measurement noise much larger than the underlying signal at those frequencies. Because it is such a ubiquitous filter, it is worth studying it in some detail, given by the convolution,

$$\bar{g}_h(x) = \int_{-\infty}^{\infty} g(x')h(x-x')\,dx'. \tag{3.82}$$

It is assumed that the frequency response of the smoothing filter is not greater than unity, $|H(f)|^2 \leq 1$; and, of course, by definition, $\lim_{f\to\pm\infty}|H(f)| = 0$. Therefore, the total energy of the average is less than the original function, since by Parseval's Theorem, equation (2.89),

$$\int_{-\infty}^{\infty} \bar{g}_h^2(x)\,dx = \int_{-\infty}^{\infty} \left|\bar{G}(f)\right|^2 df = \int_{-\infty}^{\infty} \left|G(f)H(f)\right|^2 df < \int_{-\infty}^{\infty} \left|G(f)\right|^2 df = \int_{-\infty}^{\infty} g^2(x)\,dx. \tag{3.83}$$

However, the "global average", or the zero-frequency component, of the smoothed function is preserved if the frequency response is unity at the origin,

$$\int_{-\infty}^{\infty} \bar{g}_h(x)\,dx = \bar{G}(0) = G(0)H(0) = G(0) = \int_{-\infty}^{\infty} g(x)\,dx. \tag{3.84}$$

If the filter function is the normalized rectangle function, equation (2.100),

$$h(x) = \frac{1}{T}b\left(\frac{x}{T}\right), \tag{3.85}$$

then $\bar{g}_h$ is the simple (unweighted) moving average. One might even say it is the "ideal" moving average, as it gives equal weights to $g(x)$ within, and only within, a definite interval, T, centered on x. The frequency response of this filter is the sinc function, $\mathrm{sinc}(Tf)$, equation (2.100), and by the convolution theorem (3.17),

$$\bar{G}_h(f) = G(f)\,\mathrm{sinc}(Tf). \tag{3.86}$$

On the sphere one version of the unweighted average is the convolution with the isotropic cap function, $b_s^{(\psi_s)}(\psi)$, equation (2.302),

$$\bar{g}(\theta,\lambda) = \frac{1}{A_s}\iint_{\Omega_s} \Delta g(\theta',\lambda')\,d\Omega' = \frac{1}{4\pi}\iint_{\Omega'} \Delta g(\theta',\lambda')\,b_s^{(\psi_s)}(\psi)\,d\Omega' = \Delta g(\theta,\lambda) * b_s^{(\psi_s)}(\theta). \tag{3.87}$$

The Fourier-Legendre spectrum of $\bar{g}$ is given with equations (2.305) and (3.60) by

$$\bar{G}_{n,m} = B_n^{(\psi_s)} G_{n,m}. \tag{3.88}$$

The frequency response of this filter, $b_n^{(\psi_s)}$, attenuates the high-degree (high-frequency) part of the signal much like the sinc function in Cartesian space (compare Figures 2.4 and 2.17),

Writing the sinc function in terms of an amplitude and a phase,

$$\operatorname{sinc}(Tf) = \begin{cases} |\operatorname{sinc}(Tf)| e^{i0}, & \operatorname{sinc}(Tf) \geq 0 \\ |\operatorname{sinc}(Tf)| e^{i\pi}, & \operatorname{sinc}(Tf) < 0 \end{cases} \tag{3.89}$$

the phase response is either 0 or π, depending on whether the sinc function at a particular frequency is positive or negative. Due to this phase reversal at certain frequencies, or wavelengths, this moving average inverts the input, particularly at the frequencies of the first (i.e., the largest) *side*-lobe of sinc(Tf), $1/T \leq |f| \leq 2/T$. This phase reversal is a symptom of Gibbs's effect that is demonstrated in Section 2.3.2 for functions with a step discontinuity (Figure 2.5), but is also evident where the input has large variation over intervals shorter than the cutoff wavelength. The left panel of Figure 3.2 illustrates the phase reversal for a profile of gravity anomalies that is averaged using equation (3.85). In this example, $T = 200$ km; and, waves in the input with length of about 133 km, corresponding to frequency, $f \sim 1.5/T = 0.0075$ cy/km, are inverted, e.g., near x = 2200 [km] and x = 300 [km]. This is one reason that the unweighted moving average, from the spectral viewpoint, should be avoided if possible when smoothing a signal. With an alternative, such as the Gaussian filter,

$$h(x) = \gamma^{(\beta)}(x) = \frac{1}{\beta} e^{-\pi\left(\frac{x}{\beta}\right)^2}, \quad \beta > 0, \tag{3.90}$$

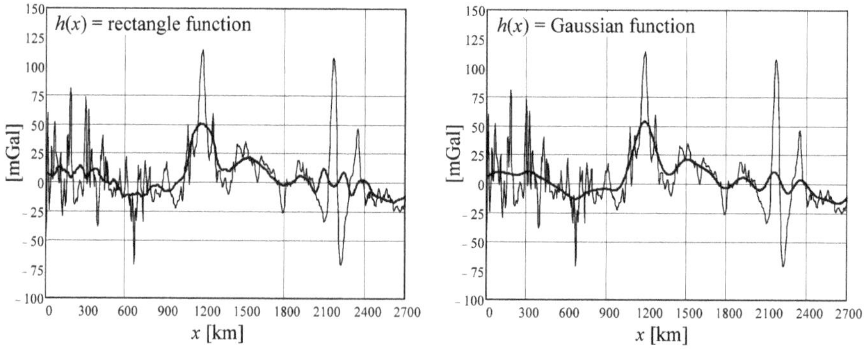

Figure 3.2: Original gravity anomaly profile (light line) and smoothed anomaly (dark line) using a simple moving average over an interval of T = 200 km (left); and using a Gaussian-weighted moving average with $\beta = T/1.33$ (equivalent bandwidth according to the –3 dB definition, which puts about 90% of the area of $\gamma^{(\beta)}(x)$ over the interval, T). Note the inversions of local extrema in the simple moving average near x = 2200 [km] and near x = 300 [km] that are absent in the Gaussian moving average.

whose spectrum is everywhere positive and real (equation 2.108), no such phase reversals occur because, clearly, the phase response is zero for all frequencies. The Gaussian moving average, shown in Figure 3.2 (right), is more faithful to the general variations of the original signal.

Since the purpose of the moving average is to remove high-frequency variations in the function, one might approach that goal from the spectral perspective and consider designing a low-pass filter directly in the spectral domain. For example, if the objective is to remove spectral components outside the band defined by $f_0 = 1/(2T)$, then one option is the "ideal" low-pass filter with frequency response, equation (3.78), and filter function, equation (3.79). But, now this also yields an unsatisfactory space-domain average, which is particularly evident on maps in the spherical domain of a truncated spherical harmonic series. Truncation at harmonic degree, n_0, represents the ideal spectral filter for spherical functions,

$$\bar{g}(\theta,\lambda) = \sum_{n=0}^{n_0} \sum_{m=-n}^{n} G_{n,m} \bar{Y}(\theta,\lambda), \tag{3.91}$$

and is equivalent to the convolution of $g(\theta, \lambda)$ with the filter function, equation (3.80), which creates a ripple effect in the output, again due to Gibbs's effect. Figure 3.3 dramatically exhibits this effect for the Earth's gravitational acceleration with cutoff degree, $n_0 = 180$, corresponding to wavelengths greater than about 220 km (equation (2.331)). In contrast, the signal smoothed by a Gaussian filter, equation (3.64),

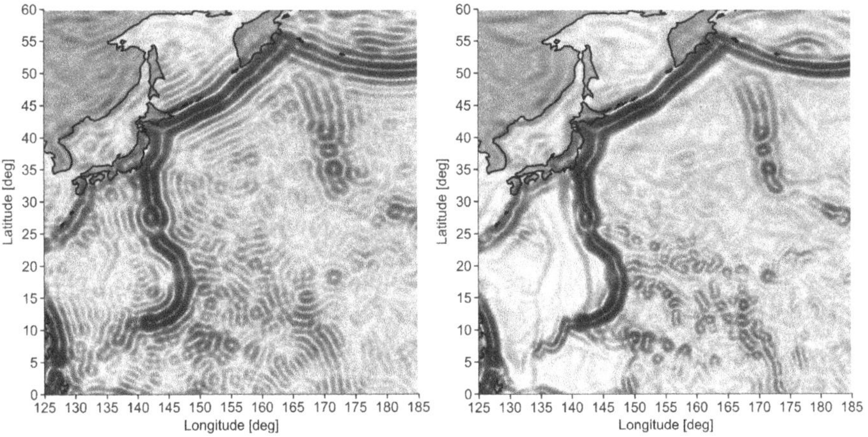

Figure 3.3: A map of the gravity disturbances (defined in Section 6.2.4) over the north Pacific trenches, smoothed using an ideal low-pass filter ($n_0 = 180$, left) and a corresponding Gaussian filter (right), and illustrating Gibbs's effect in the former. The Mariana Trench is just to the lower left of center, the Japan Trench is off the eastern coast of Japan, meeting up with the Kuril Trench off the Kuril Island chain and the Kamchatka Peninsula to the north, and the western end of the Aleutian Trench is visible at the top right of these maps. Korea is at the far left-center. The variation in the filtered gravity is between –230 mGal (dark) and +220 mGal (light) on the left, and between –150 mGal and +150 mGal on the right. Land masses to the north-west are outlined and shaded to aid in the geographical orientation.

$$\bar{g}_G(\theta,\lambda)=\sum_{n=0}^{\infty}\sum_{m=-n}^{n}\Gamma_n^{(\beta_s)}G_{n,m}\bar{Y}(\theta,\lambda), \tag{3.92}$$

for which the frequency response varies smoothly across the desired cutoff degree (Figure 2.18), exhibits no artificial oscillations. It is noted, however, that the output amplitude of the Gaussian-filtered function in this case is less than for the ideal filter because of the attenuation of the lower-degree harmonics (see also equation (3.83)). This example uses $\beta_s = (n_0 + 1)^2/2$, corresponding to the "equivalent bandwidth" as defined in association with equation (3.127) (Section 3.5.2); and, the summation is taken to sufficiently large degree for which $\Gamma_n^{(\beta)} \approx 0$.

The deleterious effect of the ideal low-pass filter becomes an issue, as well, for the ideal high-pass filter. The latter in one Cartesian dimension is defined by the frequency response,

$$H(f)=\begin{cases}0, & |f|<f_0\\ 1, & |f|\ge f_0\end{cases} \tag{3.93}$$

for some specified cut-off frequency, f_0. The corresponding filter function is (Exercise 3.5)

$$h(x)=\delta(x)-2f_0\,\mathrm{sinc}(2f_0x), \tag{3.94}$$

where $\delta(x)$ is the Dirac delta function. Applying this filter to a function, $g(x)$, leads to

$$\begin{aligned}\delta g(x)&=g(x)*h(x)\\ &=g(x)-2f_0\int_{-\infty}^{\infty}g(x')\,\mathrm{sinc}\big(2f_0(x-x')\big)\,dx'\\ &=g(x)-\int_{-f_0}^{f_0}G(f)e^{i2\pi f x}\,df\end{aligned} \tag{3.95}$$

where $\delta g(x)$ is the residual signal with respect to the low-frequency spectrum of g. Thus, if the latter (e.g., a reference model) is known, its removal from the total signal is equivalent to applying the ideal high-pass filter. Displaying a map of such residuals is subject to the same kind of ripple effect.

For geodetic and geophysical signals on the sphere, the long-wavelength model typically is a spherical harmonic series to some finite degree, n_0. Then the residual, high-pass-filtered signal is

$$\delta g(\theta,\lambda)=g(\theta,\lambda)-\sum_{n=0}^{n_0}\sum_{m=-n}^{n}G_{n,m}\bar{Y}_{n,m}(\theta,\lambda). \tag{3.96}$$

In some applications the removal of a long-wavelength model is only an intermediate step in the processing of the data, where the final step would restore that model.

In those cases, it is usually not necessary to use special filters that minimize the ripple effect. However, if the residual signal is the final product, then a more satisfactory result is obtained by applying an alternative, such as the Gaussian filter,

$$\delta g(\theta,\lambda) = g(\theta,\lambda) - \sum_{n=0}^{\infty}\sum_{m=-n}^{n} \Gamma_n^{(\beta_s)} G_{n,m} \bar{Y}_{n,m}(\theta,\lambda), \tag{3.97}$$

with an appropriate β_s for the equivalent cutoff degree. A comparison of the gravity disturbance in the north-west Pacific region according to equations (3.969) and (3.97) is given in Figure 3.4 with the same n_0 and β_s as for Figure 3.3.

Averaging data on the sphere often is performed over blocks delimited by lines of constant coordinates, θ and λ, rather than isotropically, such as by equation (3.87). Such filters are neither isotropic nor a convolution, as defined by equation (3.44), since the filter function, h, in this case also depends explicitly on the point of evaluation of the average value. Thus, approaching this kind of smoothing in the spatial domain of the sphere is difficult from the spectral viewpoint. However, as shown with equations (3.53) and (3.54), starting in the spectral domain it is possible to design filters that depend not only on harmonic degree, but also on harmonic order.

Motivated by the realization that the geopotential spectrum was determined from the satellite mission, GRACE (Tapley et al. 2004), with disproportionately better accuracy in the degree, n, than the order, m (due to the polar orbits of the satellites), and that also the global gravitational spectrum exhibited a similar pattern in its correlation with other geophysical signals, Han et al. (2005) designed a spectral response for a filter that effectively averages more in longitude than in co-latitude. This is analogous to the Cartesian case, where two parameters, T_1 and T_2, for the rectangle function, or, β_1 and β_2, for the Gaussian function, define the averaging area.

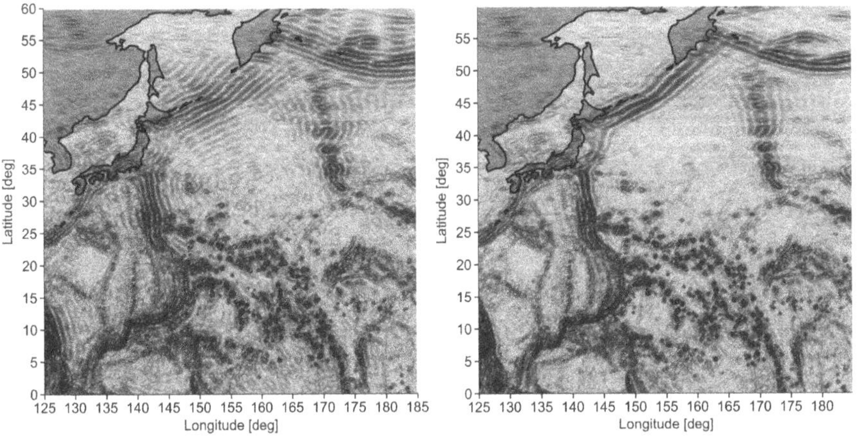

Figure 3.4: A map of the residual gravity disturbances (defined in Section 6.2.4) over the north Pacific trenches, relative to a n_0 = 180 reference field (Figure 3.3). On the left is the ideally high-pass filtered residual, and on the right is the residual as defined with a Gaussian high-pass filter. Land masses to the north-west are outlined and shaded to aid in the geographical orientation.

In the spherical case, define two horizontal distances that represent the extent of averaging in co-latitude and longitude by s_θ and s_λ. Then let

$$s^{(m_0)}(m) = \frac{m}{m_0}(s_\lambda - s_\theta) + s_\theta, \tag{3.98}$$

where m_0 is a parameter of the filter and it is assumed that $s_\lambda \geq s_\theta$. The effective averaging extent varies with order, $s_\theta \leq s^{(m_0)}(m) \leq s_\lambda$, if $0 \leq m \leq m_0$, and $s^{(m_0)}(m) > s_\lambda$ for $m > m_0$. If $s_\theta = s_\lambda$, then the averaging extent is the same for all m and the corresponding smoothing function is isotropic. The associated parameters for the cap and Gaussian smoothing functions are, respectively,

$$\psi_s^{(m_0)}(m) = \frac{s^{(m_0)}(m)}{R}, \tag{3.99}$$

$$\beta_s^{(m_0)}(m) = \frac{1}{1-\cos\left(s^{(m_0)}(m)/R\right)}, \tag{3.100}$$

which are chosen so that the frequency responses of the spherical cap and Gaussian functions enclose the same "volume" in the frequency domain, in accordance with equation (3.128) (Section 3.5.2). The frequency responses, $H_{n,m}$, for the corresponding non-isotropic smoothing filters are

$$B_{n,m} = B_n^{\left(\psi_s^{(m_0)}(m)\right)}, \quad \Gamma_{n,m} = \Gamma_n^{\left(\beta_s^{(m_0)}(m)\right)}, \tag{3.101}$$

in terms of the Legendre transforms of the spherical cap and Gaussian functions, equations (2.305) and (2.311), respectively.

Figures 3.5 and 3.6 contrast the isotropic and non-isotropic spherical cap and Gaussian function spectra for specific averaging radii, s_θ and s_λ, and $m_0 = 15$. Compared to the isotropic functions, the higher degrees of the non-isotropic functions are not as severely attenuated as the higher orders.

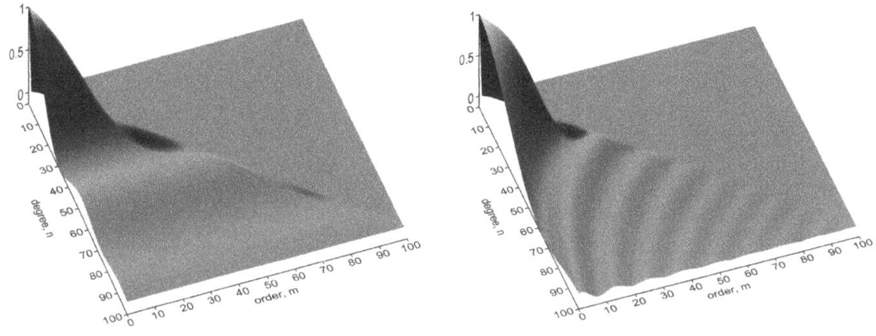

Figure 3.5: Legendre spectra of the isotropic (left) and non-isotropic cap functions. On the left the cap radius is 500 km, and on the right, $s_\theta = 500$ km, $s_\lambda = 1000$ km, with $m_0 = 15$.

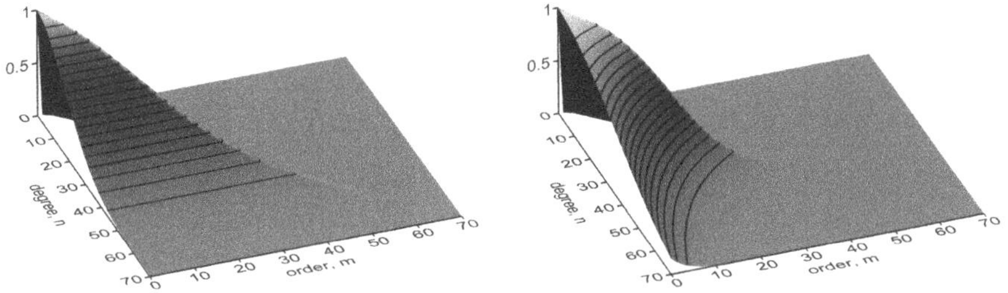

Figure 3.6: Legendre spectra of the isotropic (left) and non-isotropic Gaussian functions. On the left the equivalent extent is 250 km, and on the right, $s_\theta = 250$ km, $s_\lambda = 500$ km, with $m_0 = 15$.

3.5.2 Bandwidth and Resolution

With a smoothly attenuating response in frequency, the cutoff frequency of a filter is less definite and its specification relies on defining an *equivalent bandwidth* based on particular desirable properties. For example, if the response of a low-pass filter is normalized to unity at $f = 0$, then the effective cutoff frequency might be regarded as the –3 dB point,

$$\begin{aligned} \left|H(f_0)/H(0)\right|_{dB} = -3\text{ dB} \;&\Rightarrow\; \left|H(f_0)/H(0)\right| = 10^{-3/20} = 0.708 \approx \sqrt{2}/2 \\ &\Rightarrow\; \left|H(f_0)\right|^2 \approx \left|H(0)\right|^2/2 \end{aligned} \tag{3.102}$$

That is, the low-pass cutoff frequency, and thus the bandwidth of the filter, is defined at the point in the spectrum where the square of the amplitude response, or the energy density, is roughly half its value at the origin.

For the rectangle filter, $h(x) = b^{(T)}(x) = b(x/T)/T$, with its frequency response given by equation (2.100), the –3 dB point, f_0, is given by

$$\operatorname{sinc}^2(Tf_0) = \frac{\sin^2(\pi T f_0)}{(\pi T f_0)^2} = 0.5 \;\Rightarrow\; \sin(\pi T f_0) = \frac{\pi}{\sqrt{2}} T f_0 \;\Rightarrow\; f_0 = 0.443/T, \tag{3.103}$$

where the last equality is derived numerically. The equivalent bandwidth under this criterion is then given by equation (2.63) as

$$W\left(b^{(T)}\right) = 0.886/T. \tag{3.104}$$

Similarly, for the Gaussian filter function, $h(x) = \gamma^{(\beta)}(x) = e^{-\pi(x/\beta)^2}/\beta$, equations (2.107) and (2.108), the −3 dB cutoff frequency is

$$e^{-2\pi(\beta f_0)^2} = 0.5 \;\Rightarrow\; 2\pi(\beta f_0)^2 = \ln 2 \;\Rightarrow\; f_0 = 1/(3.01\beta); \tag{3.105}$$

and, the equivalent bandwidth is

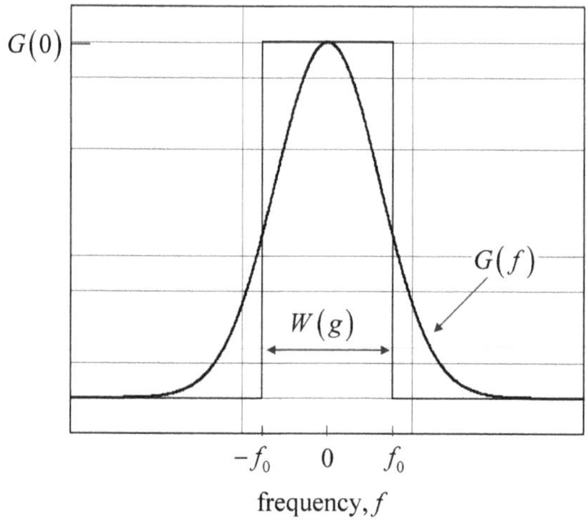

Figure 3.7: Equivalent bandwidth of g defined by the width of the rectangle in the frequency domain with area equal to that bounded by $G(f)$.

$$W\left(\gamma^{(\beta)}\right) = \frac{1}{1.505\beta}. \tag{3.106}$$

A number of alternative characterizations of equivalent bandwidth exist in the spectral analysis literature. If the frequency response is real, one option for an equivalent bandwidth of a function, g, is the width of the rectangle with height, $G(0)$, that has the same area in the frequency domain as that bounded by its Fourier transform, $G(f)$ (Figure 3.7),

$$G(0)W(g) = \int_{-\infty}^{\infty} G(f)\,df. \tag{3.107}$$

Noting that the area of the spectrum is also the value, $g(0)$ (using $x = 0$ in equation (2.57)), the equivalent bandwidth in this case is

$$W(g) = \frac{g(0)}{G(0)}. \tag{3.108}$$

Of course, this makes sense only if both $g(0) > 0$ and $G(0) > 0$. For the rectangle and Gaussian functions, the corresponding equivalent bandwidths are

$$W\left(b^{(T)}\right) = \frac{1}{T}, \tag{3.109}$$

$$W\left(\gamma^{(\beta)}\right) = \frac{1}{\beta}. \tag{3.110}$$

Another option, particularly if the frequency response is not real, is to define the equivalent bandwidth as the width of the rectangle that has the same area as the autocorrelation function of the *spectrum*. The autocorrelation function, $\overset{\circ}{\phi}$, for finite-energy functions is defined in Section 5.2, by equation (5.2). Applied to $G(f)$, it is

$$\overset{\circ}{\phi}_{G,G}(f) = \int_{-\infty}^{\infty} G^*(f')G(f'+f)\,df', \tag{3.111}$$

which is positive at the origin (inequality (5.10)) and real (equation (5.11)) if G is Hermitian, i.e., $G(-f) = G^*(f)$. This is guaranteed by property (2.84) since g is assumed to be real. Thus, analogous to equation (3.107), the equivalent bandwidth is given by

$$W(g) = \frac{\int_{-\infty}^{\infty} \overset{\circ}{\phi}_{G,G}(f)\,df}{\overset{\circ}{\phi}_{G,G}(0)} = \frac{g^2(0)}{E}, \tag{3.112}$$

where the total energy, E, of the function is given by Parseval's theorem, equation (2.89), using equation (3.111) with $f = 0$. The numerator on the right side derives from equation (5.6) and the dual convolution theorem, equation (3.20),

$$\overset{\circ}{\phi}_{G,G}(f) = G^*(-f) * G(f) = G(f) * G(f) \quad \leftrightarrow \quad g^2(x). \tag{3.113}$$

The total energy of the rectangle function, $b^{(T)}(x)$, is

$$E = \frac{1}{T^2}\int_{-\infty}^{\infty} b^2\left(\frac{x}{T}\right)dx = \frac{1}{T^2}\int_{-T/2}^{T/2} dx = \frac{1}{T}; \tag{3.114}$$

and, the equivalent bandwidth according to equation (3.112), again, is

$$W\left(b^{(T)}\right) = \frac{1/T^2}{1/T} = \frac{1}{T}. \tag{3.115}$$

The Gaussian function, $\gamma^{(\beta)}(x)$, has total energy, $E = 1/(\beta\sqrt{2})$ (Exercise 3.6), and thus, with $\gamma^{(\beta)}(0) = 1/\beta$, the equivalent bandwidth in this case is

$$W\left(\gamma^{(\beta)}\right) = \frac{1/\beta^2}{1/\left(\beta\sqrt{2}\right)} = \frac{\sqrt{2}}{\beta}. \tag{3.116}$$

The three options to define the bandwidth of the Gaussian filter, equations (3.106), (3.110), and (3.116), differ somewhat in scale, but essentially depend on the inverse of the parameter, β.

Analogously, one may define an equivalent extent in the spatial domain ("duration" in the time domain) for a function, g, that has finite energy. For example, the equivalent extent based on the area of the function is defined by

$$D(g)=\frac{\int_{-\infty}^{\infty} g(x)\,dx}{g(0)}=\frac{G(0)}{g(0)}; \tag{3.117}$$

that is, the area of g is the same as the area of the rectangle, $g(0)D(g)$. For the rectangle function, $D(b^{(T)}) = T$, as expected; and for the Gaussian function, the equivalent extent is $D(\gamma^{(\beta)}) = \beta$. The alternative characterization of extent based on the area of the correlation function, analogous to equation (3.112), is

$$D(g)=\frac{\int_{-\infty}^{\infty} \overset{\circ}{\phi}_{g,g}(x)\,dx}{\overset{\circ}{\phi}_{g,g}(0)}=\frac{G^2(0)}{E}. \tag{3.118}$$

In this case, the equivalent extent of the rectangle function is, again, T, and for the Gaussian function, it is $\beta\sqrt{2}$.

Comparing the equivalent bandwidth and equivalent extent, a kind of geometric inverse relationship between functions and their spectra is evident as it is already shown with the similarity property (2.82). As the extent decreases, the bandwidth increases, and vice versa. The product of the equivalent extent and bandwidth thus also has special significance; it is called the *space-bandwidth product* (*time-bandwidth product*, for temporal functions). One may also call it the *extent-bandwidth product*, a nomenclature that covers both space and time domains. For the definition based on the equivalent area of the spectrum or function, that product is a constant for all functions,

$$D(g)W(g)=\frac{G(0)}{g(0)}\frac{g(0)}{G(0)}=1; \tag{3.119}$$

whereas, for other definitions, it depends on the function (e.g., using the equivalent area of the correlation function it is 1 for the rectangle function and 2 for the Gaussian function).

Consider yet another definition of equivalent bandwidth (and extent),

$$\left(W(g)\right)^2=\frac{1}{E}\int_{-\infty}^{\infty} f^2\left|G(f)\right|^2 df,\quad \left(D(g)\right)^2=\frac{1}{E}\int_{-\infty}^{\infty} x^2 g^2(x)\,dx, \tag{3.120}$$

assuming the integrals exist, where, as before, E is the total energy of the function. These are like second moments if g^2/E and $|G|^2/E$ are interpretable as probability density functions, equation (5.201) (Section 5.6.2; see also (Percival and Walden

1993, p.241)). It can be shown (Exercise 3.7) that the extent-bandwidth product in this case satisfies,

$$D(g)W(g) \geq \frac{1}{4\pi}, \tag{3.121}$$

with equality holding for the Gaussian function. This is essentially *Heisenberg's uncertainty inequality* that in quantum mechanics proves the inability to determine with infinite precision both the position and momentum (velocity) of a particle. For this reason inequality (3.121) is also called a Heisenberg uncertainty principle, showing that the equivalent extent and bandwidth of a function cannot both be arbitrarily constrained.

The extent-bandwidth product, in whatever form, can be used to infer a *resolution* of a function that is either space- or band-limited. Here, we distinguish between *spectral resolution* and *spatial resolution*. According to the extent-bandwidth product, a finite bandwidth implies a finite spatial resolution, or smallest spatial feature, that is its reciprocal. With the definition of the product that leads to equation (3.119), the spatial resolution is $1/(2f_0)$, where the bandwidth is given by $W = 2f_0$. In Section 4.2.1, f_0 is also called the *Nyquist frequency*. It is noted that spatial resolution in this case also refers to the smallest half-wavelength of the band-limited function.

Conversely, a finite spectral resolution, or smallest incremental frequency, is implied by and is the reciprocal of the finite spatial extent of a function. That is, $1/D$ is the smallest difference between frequencies that could be detected. For example, the detection of the two resonances in the polar motion data (Figure 2.3), separated by about $\Delta f = 0.16$ cy/yr, requires a duration of the signal of at least 6 years (however, even this is inadequate, as discussed in Section 3.6).

These concepts of resolution are imprecise because of the different definitions of the extent-bandwidth products. However, the emphasis here is on the fact that, in principle, with finite extent or bandwidth comes finite resolution of one or the other kind. In summary, we may say that spatial resolution is defined by the reciprocal of the equivalent bandwidth and spectral resolution is defined by the reciprocal of the equivalent extent.

In higher Cartesian dimensions one could treat equivalent bandwidth and extent, and consequently spatial and spectral resolution, in isolation for each coordinate. The definitions follow in the usual straightforward manner. However, in two dimensions, if the corresponding bandwidths and extents are respectively similar, it may be more useful to describe their equivalents on the basis of isotropic, or directionally averaged, functions. Following the one-dimensional case, the conceptual starting point is the cylinder function, equation (2.191), defined on the frequency domain and with arbitrary radius. The two-dimensional equivalent bandwidth of $g(s)$, can be defined by the area in the frequency domain that, if multiplied by the Hankel transform at the origin, yields the same volume as enclosed by the transform of the function,

$$W(g) = \frac{2\pi \int_0^\infty G(\bar{f})\bar{f}d\bar{f}}{G(0)}, \tag{3.122}$$

where the frequency, $\bar{f}$, is given by equation (2.178). The frequency associated with the equivalent bandwidth is then defined by the radius of the circular area in the frequency domain, $\bar{f}_0 = \sqrt{W(g)/\pi}$. Similarly, the two-dimensional equivalent extent of $g(s)$ may be defined by the area in the space domain that, if multiplied by $g(0)$, yields the same volume as the function, itself; that is,

$$D(g) = \frac{2\pi \int_0^\infty g(s)sds}{g(0)}. \tag{3.123}$$

By equations (2.180) and (2.181), the equivalent extent-bandwidth product is

$$D(g)W(g) = \frac{G(0)\,g(0)}{g(0)\,G(0)} = 1, \tag{3.124}$$

which is not surprising since the Hankel transform is just the two-dimensional Fourier transform for an isotropic function. Other definitions of equivalent bandwidth and extent may be defined as in the one-dimensional case, but the main purpose of the foregoing is to set up the case for functions on the sphere.

Because of the spherical geometry, it is convenient, if not necessary, to consider only isotropic functions, representing functions with similar spectral and spatial characteristics in both longitude and co-latitude. The resulting definitions of equivalent extent and bandwidth are then also descriptive of and may be applied to two-dimensional functions, $g(\theta, \lambda)$. Thus, for a univariate function on the sphere, $g(\psi)$, the equivalent extent, $D(g)$, is defined as the area of a cap whose volume, $D(g)g(0)$, is the same as that bounded by g over the unit sphere. Analogous to equations (3.117) or (3.123), the equivalent extent is then given by

$$D(g) = \frac{\iint_\Omega g(\psi)\sin\psi\, d\psi\, d\zeta}{g(0)} = \frac{4\pi G_0}{g(0)}, \tag{3.125}$$

in view of equation (2.205), and with the assumptions, $g(0) > 0$ and $G_0 > 0$. For example, the cap function, $b_s^{(\psi_s)}(\psi)$, equation (2.302), analogous to the rectangle function, defines a kind of volume, height times cap area (due to the spherical curvature, this is not exactly the geometric volume), where the cap area, A_s, is given by equation (2.303),

$$A_s = 2\pi(1 - \cos\psi_s). \tag{3.126}$$

Equation (3.125) yields the (obvious) equivalent extent, $D(b_s^{(\psi_s)}) = A_s$, which for small ψ_s is $D(b_s^{(\psi_s)}) \simeq \pi\psi_s^2$. Similarly, for the Gaussian function, equation (2.310), the equivalent extent is $D(\gamma_s^{(\beta_s)}) = 4\pi\Gamma_0^{(\beta_s)}/\gamma_s^{(\beta_s)}(0) = 2\pi(1 - e^{-2\beta_s})/\beta_s$, which for large β_s is $D(\gamma_s^{(\beta_s)}) \simeq 2\pi/\beta_s$.

For the equivalent bandwidth, we note that there are $(n + 1)^2$ spectral components of a spherical function up to harmonic degree, n (for an isotropic (univariate) function the order-harmonics happen to be zero). Hence, the two-dimensional bandwidth of a function on the sphere must be interpreted in terms of the cutoff-degree as $W(g) = (n_0 + 1)^2$, or

$$n_0 = \sqrt{W(g)} - 1. \tag{3.127}$$

One option bases the cutoff-degree on the –3 dB point, that is, the degree, n_0, for which $G_{n_0}^2 = G_0^2/2$. For the cap and Gaussian functions these are approximately $n_0 = 2.865/\sqrt{A_s}$ and $n_0 = \sqrt{\beta_s \ln 2}$, respectively, derived for small extent from their asymptotic planar function equivalents; see equation (2.192) and the Hankel transform pair (2.193).

Another option for equivalent bandwidth on the sphere, similar to the one-dimensional case, is in terms of a volume in the spectral domain, defined by $\sum_{n=0}^{\infty}(2n+1)G_n$. On this basis, the equivalent bandwidth is

$$W(g) = \frac{\sum_{n=0}^{\infty}(2n+1)G_n}{G_0} = \frac{g(0)}{G_0}. \tag{3.128}$$

With equation (3.125), the extent-bandwidth product is then

$$D(g)W(g) = 4\pi. \tag{3.129}$$

For example, based on its equivalent extent, the equivalent bandwidth for the cap function is then $W(b_s^{(\psi_s)}) = 4\pi/A_s$; and, for small ψ_s, the equivalent cutoff-degree, equation (3.127), is $n_0 \simeq 2/\psi_s - 1$. Also, by equation (3.129), the equivalent bandwidth for the Gaussian function is $W\left(\gamma_s^{(\beta_s)}\right) = 2\beta_s/\left(1 - e^{-2\beta_s}\right)$; and, from equation (3.127), the approximate equivalent cutoff-degree for $\beta_s \gg 1$ is $n_0 \simeq \sqrt{2\beta_s} - 1$.

Figure 3.8 shows the frequency responses of the spherical Gaussian filter function with these two alternative bandwidths equivalent to the cap function for $\psi_s = 0.564°$. The parameter, β_s, is derived from ψ_s through the implied equivalent cutoff degree, n_0, for the cap function either on the basis of the –3 dB point, or according to the extent-bandwidth product, equation (3.129).

The spatial resolution of a spherical function that is band-limited by the maximum harmonic degree, n_0, can be interpreted as its "equivalent extent" for this bandwidth, given on the basis of equations (3.129) and (3.127) by

$$D(g) = \frac{4\pi}{(n_0 + 1)^2}. \tag{3.130}$$

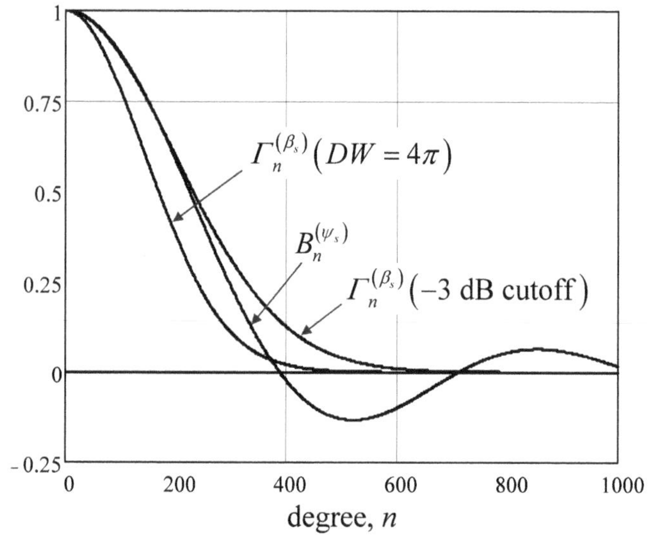

Figure 3.8: Spherical Gaussian frequency responses similar to that of the spherical cap function, $B_n^{(\psi_s)}$, for a cap radius, $\psi_s = 0.564°$. In one case, β_s is given by $\beta_s = 2.865^2/(A_s \ln 2)$ based on a common –3 dB point, and in the other case, β_s is based on a common equivalent bandwidth according to equation (3.129), $\beta_s = 2/\psi_s^2$.

In terms of an angular distance (half-wavelength), $\Delta\psi$, using $D(g) \equiv 2\pi(1 - \cos\Delta\psi) \simeq \pi\Delta\psi^2$, one obtains

$$\Delta\psi = \frac{2}{n_0 + 1} \simeq \frac{2}{n_0}, \tag{3.131}$$

where the approximation holds for large n_0. This is somewhat smaller than the often used value, $\Delta\theta = \pi/n_0$, based on the Nyquist degree limit (Section 4.5.1) and reflects a different interpretation of spatial resolution that accounts for an isotropic measure rather than one strictly with respect to co-latitude.

3.6 Window Functions

In most cases only a finite extent of the function, $g(x)$, is available in the form of data from which one wishes to determine its Fourier spectrum, $G(f)$. In view of equation (2.56), it is not generally possible to obtain the complete spectrum since this requires the entire signal. Spatial geophysical signals, such as gravitational and magnetic fields, unlike typical signals in optics or acoustics (Slepian 1976), have virtually infinite bandwidth, or at least more gradually attenuating amplitude spectra. That is, the amount of detail does not diminish with scale, which is a characteristic of fractals, classically illustrated with the shape of shorelines and topographic height profiles (Section 6.3.3). Practically determining their spectrum from a finite span of data is fraught with errors not only due to measurement noise and finite sampling (aliasing, Section 4.2.3), but because of the limit in spectral resolution dictated by the data extent. The present section introduces the basic consequences of using space-limited data for spectral analysis and alternative approaches to ameliorating the consequent effects. Section 5.7.2 elaborates on a corresponding analysis from the viewpoint of the stochastic interpretation of signals.

3.6.1 Classical Tapers

The previous Section 3.5.1 concerns filter functions with respect to their frequency response in order to ascertain the effect of their convolution with a data function. That is, different ways of band-limiting a spectrum lead to different characteristics of the filtered function. In this section, we study the method of truncating, or *windowing*, or space-limiting a function to yield particular properties of its determined spectrum. Mathematically, windowing is the multiplication of a function, $g(x)$, by a function that is space-limited (a window function), thus truncating $g(x)$ to a limited spatial domain while also possibly modifying $g(x)$ in that domain. By the duality between a function and its spectrum, the effects on the spectrum, $G(f)$, due to a window function resemble the space-domain effects on $g(x)$ due to the frequency response of a filter function. Thus, the window functions in several instances are simply the filter frequency responses formulated in the space domain, or vice versa.

A good way to start is with the rectangle function, which is the ideal window function, $b(x/T)$, in the sense that it perfectly limits and preserves the function to a given domain,

$$g_T(x) = \begin{Bmatrix} g(x), & -T/2 \leq x \leq T/2 \\ 0, & \text{otherwise} \end{Bmatrix} = b\left(\frac{x}{T}\right) g(x) \tag{3.132}$$

(The subtlety of the particular definition of the rectangle function, equation (2.90), in accordance with uniform convergence of its Fourier transform, specifically at $\pm T/2$, is ignored here as inconsequential in these applications.) Being a product rather than a convolution, windowing is not the same as filtering. On other hand, by the dual convolution theorem, equation (3.20), and with the similarity property (2.78) applied to $b(x/T)$, the spectrum of g_T is a convolution,

$$G_T(f) = T\,B(Tf) * G(f) = T\,\text{sinc}(Tf) * G(f), \tag{3.133}$$

where the spectrum, $B(f)$, is given by equation (2.92).

Analogous to equation (3.82), the spectrum, $G_T(f)$, of the truncated signal is a weighted moving average of the spectrum, $G(f)$, where the "weights" are the values of the sinc function. Thus, at any particular frequency, $G_T(f)$ is the integral of $G(f')$

multiplied by sinc($T(f - f')$) that peaks at $f' = f$ and falls off on either side of f. Hence, the truncation of a function in the space domain, equivalent to a convolution in the frequency domain, tends to smear out the true spectrum, $G(f)$. The energy of the function at a particular frequency *leaks* into neighboring parts of the spectrum via the spectral width of the main "lobe" of the sinc function, as well as its "side-lobes" (shown in Figure 2.4).

For example, restricting the polar motion data (Figure 1.2) to an interval of $T = 6$ yr means that the true spectrum is convolved with a sinc function whose first zero is at $1/T = 0.17$ cy/yr and the two actual resonances (Figure 2.3), which are separated by 0.16 cy/yr, are merged into a single large peak, as shown in Figure 3.9. To reduce this *spectral leakage* and obtain a better representation of $G(f)$ using $G_T(f)$ requires that the main lobe of the sinc function be narrower, i.e., that T become larger. Indeed, only if $T \geq 8$ yr do the two distinct resonances in the data emerge. Taking the limit of the Fourier transform pair (2.100) as $T \rightarrow \infty$, and using equation (2.121) and the duality property (2.77),

$$\lim_{T\to\infty}\left(T\,\mathrm{sinc}\left(fT\right)\right) = \lim_{T\to\infty}\mathcal{F}\left(b\left(x/T\right)\right) = \mathcal{F}\left(\lim_{T\to\infty}b\left(x/T\right)\right) = \mathcal{F}\left(1\right) = \delta\left(f\right). \qquad (3.134)$$

Thus, in the limit, the convolution, equation (3.133), is $G_T(f) \rightarrow \delta(f) * G(f) = G(f)$, by equation (3.23); and, of course, the function is no longer truncated.

In addition to the leakage resulting from the main lobe of the sinc function, the side lobes affect the computed spectrum by creating false resonances (albeit with

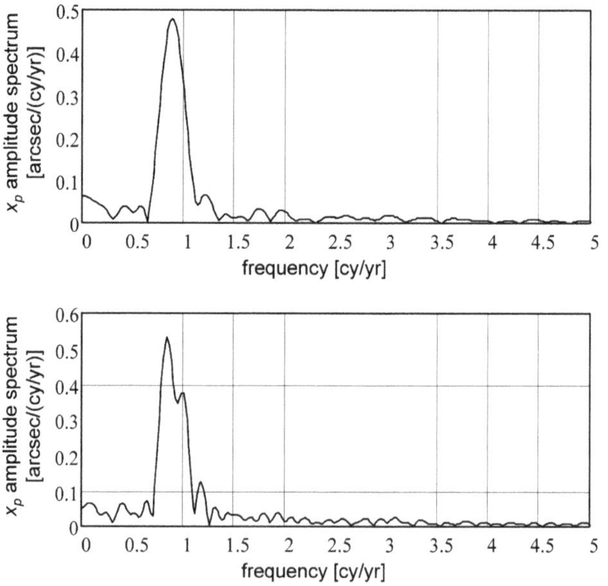

Figure 3.9: Amplitude spectrum of polar motion from data on a truncated domain of 6 years (top) and 8 years (bottom) (the visualization is improved by artificially increasing the spectral resolution with zeros added to the truncated data sequence). The amplitude at annual period is hidden within the spread out Chandler resonance (at $f = 0.84$ cy/yr) in the top graph due to spectral leakage (cf. Figure 2.3), but begins to emerge in the bottom graph as the data domain increases.

much smaller amplitudes) at frequencies near the true resonances, as also seen in Figure 3.9. Although there is no remedy to the spectral leakage caused by the main lobe except to increase the length, T, of the data interval, it is desired to reduce the spectral leakage resulting from the side-lobes. This can be accomplished, but at the cost of widening the main lobe, by alternative window functions and is analogous to modifying the frequency response of a filter to reduce Gibbs's effect. In essence, the data within the given domain, $|x| \le T/2$, are modified to create a more graceful attenuation to zero at the ends, thus eliminating the sharp discontinuity created by the rectangle function. Consequently, these window functions are also known as *data tapers*.

In general, let $u(x)$ be a window, or taper, function that vanishes outside the interval, $[-T/2, T/2]$; and, let

$$g_u(x) = u(x)\, g(x). \tag{3.135}$$

Then, the spectrum of g_u, according to equation (3.20), is

$$g_u(f) = U(f)*G(f). \tag{3.136}$$

Generally, a window function is defined to be unity at least at the origin (however, see also an alternative normalization in Section 5.72). Aside from these basic constraints, it is chosen primarily to reduce the spectral leakage due to its spectrum outside the main lobe, while also keeping the main lobe to a reasonable width.

Spectral leakage is an error given by the difference, $U(f)*G(f) - G(f)$. The first question one might ask is, which window function, $u(x)$, minimizes this error? The answer could be formulated by solving a minimization problem for $U(f)$ in terms of minimum mean-square error,

$$\int_{-\infty}^{\infty} \left|U(f)*G(f)-G(f)\right|^2 df \to \min. \tag{3.137}$$

With Parseval's theorem, equation (2.89), this is the same as solving for $u(x)$ according to

$$\int_{-\infty}^{\infty} \left|u(x)g(x)-g(x)\right|^2 dx \to \min, \tag{3.138}$$

(that is, the respective quantities in the minimization problems (3.137) and (3.138) within the absolute value symbols constitute a Fourier transform pair). Separated into three parts and noting that by definition, $u(x) = 0$ for $|x| > T/2$, this integral becomes

$$\int_{-\infty}^{\infty} \left|u(x)g(x)-g(x)\right|^2 dx = \int_{-\infty}^{-T/2} \left|g(x)\right|^2 dx + \int_{-T/2}^{T/2} \left|g(x)\right|^2 \left|u(x)-1\right|^2 dx + \int_{T/2}^{\infty} \left|g(x)\right|^2 dx. \tag{3.139}$$

It is minimized with respect to the window function if the second integral vanishes, i.e., if $u(x) = 1$, $|x| \le T/2$; thus,

$$u_{\text{optimum}}(x) = b\left(\frac{x}{T}\right). \tag{3.140}$$

That is, the rectangle window is the best "taper" from the point of view of minimizing the spectral leakage *for all* frequencies, according to the condition (3.137). On the other hand, one would like to reduce the effect of spectral leakage due to the truncation of the domain primarily for those frequencies of the spectrum that are discernible in the given finite extent of a signal. This is a more local minimization problem (in the frequency domain) than the global minimization problem (3.137). For this reason, the taper function itself is usually designed to reduce the leakage effect on the higher-frequency components of a signal. Various window functions have been discovered and designed that reduce the side lobes, as shown below. We return to the local minimization problem in Section 3.6.2.

A particular class of popular window functions in one-dimensional Cartesian space consists of cosine tapers that have the general form,

$$u_{\cos}^{(N)}(x) = \begin{cases} c_0 + 2\sum_{k=1}^{N} c_k \cos\frac{2\pi kx}{T}, & |x| \le \frac{T}{2} \\ 0, & |x| > \frac{T}{2} \end{cases} \tag{3.141}$$

where the constant coefficients, c_k, satisfy the condition,

$$c_0 + 2\sum_{k=1}^{N} c_k = 1. \tag{3.142}$$

The Fourier transform of these tapers is given by (Exercise 3.8)

$$U_{\cos}^{(N)}(f) = c_0 T \operatorname{sinc}(fT) + \sum_{k=1}^{N} c_k \left(T \operatorname{sinc}(fT - k) + T \operatorname{sinc}(fT + k)\right). \tag{3.143}$$

For $N = 0$ (i.e., $c_k = 0$, $k \ge 1$) and $c_0 = 1$, there is $u_{\cos}^{(0)}(x) = b(x/T)$. For $N \ge 1$, several have been proposed (Harris 1978); e.g., the Hann taper ($c_0 = 0.5$, $c_1 = 0.25$, $c_k = 0$, $k \ge 2$),

$$u_{\text{Hann}}(x) = \begin{cases} 0.5 + 0.5\cos\frac{2\pi x}{T}, & |x| \le \frac{T}{2} \\ 0, & |x| > \frac{T}{2} \end{cases} \tag{3.144}$$

and the Blackman taper ($c_0 = 0.42$, $c_1 = 0.25$, $c_2 = 0.04$, $c_k = 0$, $k \ge 3$),

$$u_{\text{Blackman}}(x) = \begin{cases} 0.42 + 0.50\cos\frac{2\pi x}{T} + 0.08\cos\frac{4\pi x}{T}, & |x| \le \frac{T}{2} \\ 0, & |x| > \frac{T}{2} \end{cases} \tag{3.145}$$

(coefficients for the "exact Blackman taper" differ slightly).

There are many other windows designed for particular applications or performance characteristics (Harris 1978, Wikipedia (Window Function) 2016). Among these is the Bartlett (or, triangle) window, sometimes called the Fejér window, generalized to an arbitrary interval of length, T, and obtained by utilizing equation (3.2) for the convolution in equation (3.7),

$$u_{\text{Bartlett}}^{(T)}(x) = (b*b)\left(\frac{2x}{T}\right) = \begin{cases} 1-\dfrac{2|x|}{T}, & |x| \le \dfrac{T}{2} \\ 0, & |x| > \dfrac{T}{2} \end{cases} \tag{3.146}$$

as well as the *truncated* Gaussian window, equation (2.107) modified to be unity at the origin and multiplied by the rectangle function,

$$u_{\text{Gauss}}^{(\beta)}(x) = b(x/T)e^{-\pi(x/\beta)^2}, \quad \beta > 0; \tag{3.147}$$

and, the Tukey (cosine-tapered) window function, given by

$$u_{\text{Tukey}}^{(\alpha)}(x) = \begin{cases} 1, & |x| < \alpha\dfrac{T}{2} \\ 0.5+0.5\cos\dfrac{\pi(x-\alpha T/2)}{(1-\alpha)T/2}, & \alpha\dfrac{T}{2} \le |x| \le \dfrac{T}{2} \\ 0, & |x| > \dfrac{T}{2} \end{cases} \tag{3.148}$$

where $0 \le \alpha \le 1$. Corresponding Fourier transforms are (Exercise 3.9)

$$U_{\text{Bartlett}}^{(T)}(f) = \frac{T}{2}\text{sinc}^2\left(\frac{T}{2}f\right), \tag{3.149}$$

$$U_{\text{Hann}}(f) = \frac{T}{2}\text{sinc}(fT) + \frac{T}{4}\left(\text{sinc}(fT-1)+\text{sinc}(fT+1)\right) = \frac{T}{2}\frac{\text{sinc}(fT)}{\left(1-f^2T^2\right)}, \tag{3.150}$$

$$U_{\text{Gauss}}^{(\beta)}(f) = \beta T\int_{-\infty}^{\infty}\text{sinc}(Tf')e^{-\pi(\beta(f-f'))^2}\,df', \tag{3.151}$$

$$U_{\text{Tukey}}^{(\alpha)}(f) = \frac{T}{2}\frac{\text{sinc}(fT)+\alpha\,\text{sinc}(f\alpha T)}{1-f^2(1-\alpha)^2T^2}. \tag{3.152}$$

The Tukey window is designed to taper the function only near the edges of the window, as specified by the parameter, α, and otherwise maintain full function strength. Clearly, $\alpha = 0$ reduces this to the Hann window and $\alpha = 1$ yields the rectangle

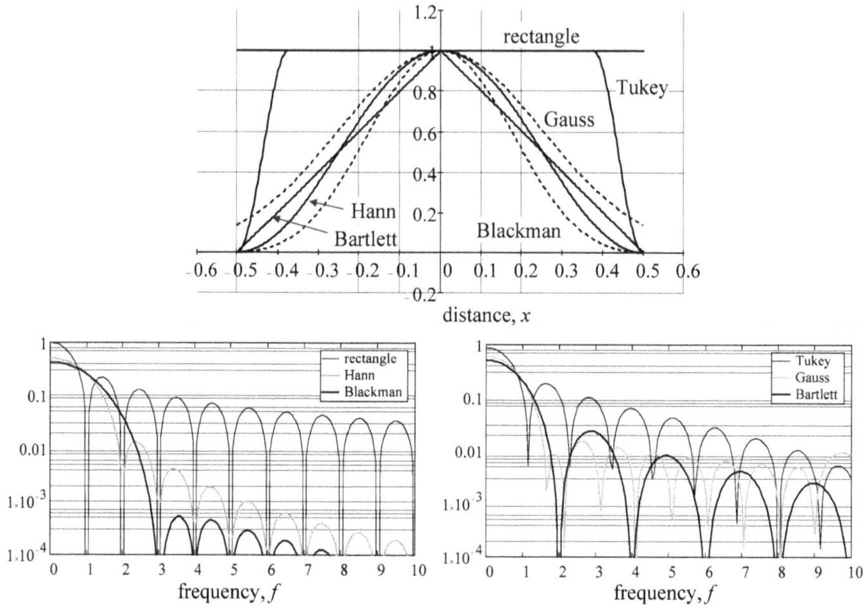

Figure 3.10: Window functions with varying taper characteristics (top) and their Fourier amplitude spectra (bottom, the legend identifies the spectra in order of magnitude at the origin). Parameters include $T = 1$ for all windows, $\beta = 0.25T\sqrt{2\pi}$ for the Gaussian window, and $\alpha = 0.75$ for the Tukey window. All amplitude spectra are symmetric with respect to the origin.

window. Note that these windows for arbitrary T do not enclose unit area, but are equal to unity at $x = 0$.

Figure 3.10 compares these windows and their Fourier transforms, showing the tradeoff between retaining the function values over the window and reducing spectral leakage. It is sometimes argued that tapering a function is altering or corrupting the function. This is true, of course, but from the point of view of spectral analysis it is not the amplitudes of the function values in the space domain that are of interest, but the locations of their salient spectral constituents in the frequency domain. And, for that purpose, tapering a function is a perfectly legitimate operation if it enhances that determination. Returning once more to the polar motion example (Figures 1.2, 2.3), application of the Hann taper to a 12-year data sequence yields the spectrum shown in Figure 3.11. The small fictitious resonances near the actual resonances (particularly at $f = 0.73$ cy/yr, implying a period of about 500 days) have vanished, compared to the rectangle window (Figure 3.9).

The easiest way to generalize these windows to higher dimensions in Cartesian space is by forming the product of the one-dimensional windows, where the scale may be different in each dimension to allow for rectangular truncation areas. By this separation of variables, the corresponding Fourier transforms are also the products of the univariate transforms. For example, the Hann taper in two dimensions is

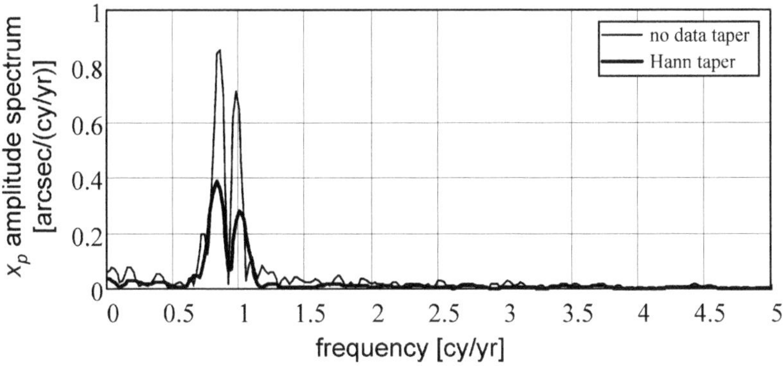

Figure 3.11: Amplitude spectrum of polar motion from data on a domain of 12 years, not tapered (thin curve) and with the Hann taper applied (thick curve).

$$u_{\text{Hann}}(x_1, x_2) = \begin{cases} \left(0.5 + 0.5\cos\dfrac{2\pi x_1}{T_1}\right)\left(0.5 + 0.5\cos\dfrac{2\pi x_2}{T_2}\right), & |x_1| \le \dfrac{T_1}{2} \text{ and } |x_2| \le \dfrac{T_2}{2} \\ 0, & \text{otherwise} \end{cases} \tag{3.153}$$

with Fourier transform

$$\begin{aligned} U_{\text{Hann}}(f_1, f_2) = &\left(\frac{T_1}{2}\operatorname{sinc}(f_1T_1) + \frac{T_1}{4}\left(\operatorname{sinc}(f_1T_1 - 1) + \operatorname{sinc}(f_1T_1 + 1)\right)\right)\cdot \\ &\left(\frac{T_2}{2}\operatorname{sinc}(f_2T_2) + \frac{T_2}{4}\left(\operatorname{sinc}(f_2T_2 - 1) + \operatorname{sinc}(f_2T_2 + 1)\right)\right) \end{aligned} \tag{3.154}$$

The spherical equivalents of these tapers could also be considered, although, like the spherical filters, the isotropic tapers are certainly easier to analyze. Moreover one might find that a planar approximation is entirely adequate. Wenzel and Arabelos (1981) describe the Hann taper on the sphere, as does Jekeli (1981a), given by

$$u_{\text{Hann}}(\psi) = \begin{cases} 0.5 + 0.5\cos\dfrac{\pi\psi}{\psi_0}, & 0 \le \psi \le \psi_0 \\ 0, & \psi_0 < \psi \le \pi \end{cases} \tag{3.155}$$

Its Legendre spectrum satisfies the recursion (correcting an error by Jekeli 1981a),

$$\left(U_{\text{Hann}}\right)_n = \frac{1}{4(2n+1)}\left(\mu_{n-1} - \mu_{n+1}\right), \quad n \ge 1, \tag{3.156}$$

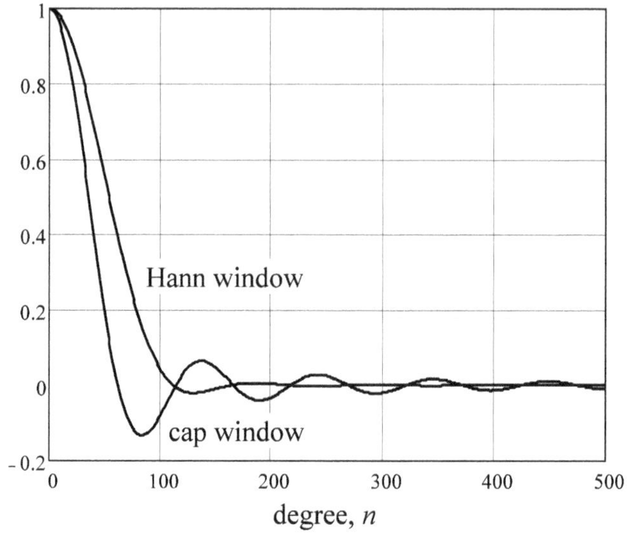

Figure 3.12: Legendre spectra of the cap and Hann windows on the sphere for $\psi_0 = 3.5°$. Both spectra are normalized to unity at the origin.

where, provided π/ψ_0 (or, $180/\psi_0$ in degrees) is not an integer and $0 < \psi_0 < \pi$,

$$\mu_{n+1} = \frac{1}{\frac{\psi_0^2}{\pi^2}(n+1)^2 - 1}\left(\left(\frac{\psi_0^2}{\pi^2}n^2 - 1\right)\mu_{n-1} - P_{n+1}(\cos\psi_0) + P_{n-1}(\cos\psi_0)\right), \tag{3.157}$$

and with starting values,

$$\left(U_{\text{Hann}}\right)_0 = \frac{1}{2}\frac{\psi_0^2 - \frac{\pi^2}{2}(1-\cos\psi_0)}{\psi_0^2 - \pi^2}, \tag{3.158}$$

$$\mu_0 = 2, \quad \mu_1 = 2\pi^2\left(\frac{\sin^2\frac{\psi_0}{2} - 1}{\psi_0^2 - \pi^2}\right). \tag{3.159}$$

If $\psi_0 = \pi/k$, then the recursion (3.157) breaks down at $n = k-1$, but it is stable enough that $\psi_0 = \pi/k + \Delta\psi$, even for $\Delta\psi = 10^{-10}$, yields numerically satisfactory results. Figure 3.12 compares the normalized spherical Hann and cap window spectra, $(U_{\text{Hann}})_n/(U_{\text{Hann}})_0$ and $B_n^{(\psi_0)}$, respectively, for $\psi_0 = 3.5°$, where as a window the cap function is unity for $\psi \le \psi_0$.

3.6.2 The Concentration Problem

A particular type of taper arises from the solution to the "concentration problem", solved famously by David Slepian (Slepian 1983). As noted in Section 2.3, and as illustrated several times in previous sections (e.g., with the Heisenberg uncertainty inequality (3.121)), a function cannot be both band-limited and space-limited. Thus, one form of the problem is to find such space-limited functions, for a given extent, whose spectral energy is as concentrated as possible to a given finite domain. As a taper or window function it would thus minimize the effects of spectral leakage. Given an extent, T, one seeks to find $g_T(x)$, equation (3.132), such that its spectrum, $G_T(f)$, for a given f_0, satisfies

$$\lambda_T(f_0) = \frac{\int_{-f_0}^{f_0} |G_T(f)|^2 \, df}{\int_{-\infty}^{\infty} |G_T(f)|^2 \, df} \to \max. \tag{3.160}$$

The numerator is the spectral energy of g_T within the desired spectral band, while the denominator is the total energy of $g_T(x)$. Since the partial energy of g_T must be less than its total energy, all values of this ratio are less than unity (and, of course, positive). The solution to the maximization problem is a space-limited function, $\hat{g}_T(x)$, that is almost band-limited. A complementary concentration problem is to find a band-limited function that maximizes its energy within a given spatial extent.

$$\mu_{f_0}(T) = \frac{\int_{-T/2}^{T/2} g_{f_0}^2(x) \, dx}{\int_{-\infty}^{\infty} g_{f_0}^2(x) \, dx} \to \max, \tag{3.161}$$

where g_{f_0} is a band-limited function (equation (2.62)). These two problems share analogous solutions by the duality between direct and inverse Fourier transforms.

The numerator in equation (3.160) is re-formulated as

$$\begin{aligned} \int_{-f_0}^{f_0} |G_T(f)|^2 \, df &= \int_{-f_0}^{f_0} \int_{-T/2}^{T/2} g_T(x) e^{i2\pi fx} dx \int_{-T/2}^{T/2} g_T(x') e^{-i2\pi fx'} dx' df \\ &= \int_{-T/2}^{T/2} \int_{-T/2}^{T/2} g_T(x) g_T(x') \left(\int_{-f_0}^{f_0} e^{i2\pi(x-x')f} df \right) dx dx' \\ &= \int_{-T/2}^{T/2} \int_{-T/2}^{T/2} g_T(x) g_T(x') \frac{\sin(2\pi f_0 (x-x'))}{\pi (x-x')} dx dx' \end{aligned} \tag{3.162}$$

noting our assumption that $g(x)$ is a real function and that the last integral in the second equation is

$$d_1(x-x';f_0) = \int_{-f_0}^{f_0} e^{i2\pi(x-x')f}\,df = \int_{-f_0}^{f_0} \cos(2\pi(x-x')f)\,df = \frac{\sin(2\pi(x-x')f_0)}{\pi(x-x')}. \tag{3.163}$$

The ratio, $\lambda_T(f_0)$, with Parseval's theorem, equation (2.87), applied to the denominator then becomes

$$\lambda_T(f_0) = \frac{\int_{-T/2}^{T/2}\int_{-T/2}^{T/2} g_T(x)\,g_T(x')\,d_1(x-x';f_0)\,dx\,dx'}{\int_{-\infty}^{\infty} g_T^2(x)\,dx}. \tag{3.164}$$

A straightforward analogous procedure applied to the ratio, $\mu_{f_0}(T)$, yields

$$\mu_{f_0}(T) = \frac{\int_{-f_0}^{f_0}\int_{-f_0}^{f_0} G_{f_0}(f)\,G_{f_0}^*(f')\,d_1(f-f';T/2)\,df\,df'}{\int_{-\infty}^{\infty} \left|G_{f_0}(f)\right|^2 df}, \tag{3.165}$$

where $d_1(f-f';T/2)$ is expressed with appropriate arguments as equation (3.163).

It is shown by the method of calculus of variations (e.g., Morse and Feshbach 1953, part II, p.1121; see also Papoulis 1977, p.218-219) that the solution to the problem of maximizing $\lambda_T(f_0)$ must also satisfy the integral equation (a Fredholm integral equation of the second kind),

$$\int_{-T/2}^{T/2} g_T(x')\,d_1(x-x';f_0)\,dx' = \lambda_T(f_0)\,g_T(x), \quad |x| \le \frac{T}{2}, \tag{3.166}$$

but not all solutions to this integral equation maximize $\lambda_T(f_0)$. With a change in variable, $y = 2x/T$, the integral equation becomes

$$\int_{-1}^{1} \psi^{(c)}(y')\,d_1\!\left(y-y';\frac{c}{2\pi}\right)dy' = \lambda^{(c)}\psi^{(c)}(y), \quad |y| \le 1, \tag{3.167}$$

where

$$c = \pi T f_0, \quad \psi^{(c)}(y) = g_T(Ty/2), \quad \lambda^{(c)} = \lambda_T(f_0), \tag{3.168}$$

and

$$\frac{T}{2}d_1\left(\frac{T}{2}(y-y');f_0\right)=d_1\left(y-y';\frac{c}{2\pi}\right)=\frac{c}{2\pi}\int_{-1}^{1}e^{ic(y-y')f}\,df. \tag{3.169}$$

Note that the left side of equation (3.167) is a *truncated* convolution of $\psi^{(c)}(y)$ with the sinc function, equation (2.93), where $d_1(y-y';c/(2\pi))=c\ \mathrm{sinc}(c(y-y')/\pi)/\pi$. Therefore, by the dual convolution theorem one can expect on the right side a function whose spectrum is approximately truncated (windowed) by the rectangle function in the frequency domain, i.e., an approximately band-limited function. By the nature of the kernel function, $d_1(y-y';c/(2\pi))$ (real, symmetric, and positive definite), there are an infinite, but countable, number of solutions to the integral equation (excluding the infinite variations in scale), corresponding to different values of the ratio, $\lambda^{(c)}$ (Courant and Hilbert 1966, pp.122-135). The solutions are *eigenfunctions*, $\psi_n^{(c)}(y)$, $n=0,1,\ldots$, with associated *eigenvalues*, $\lambda_n^{(c)}$, that are distinct and positive, and may be ordered to satisfy

$$\cdots<\lambda_n^{(c)}<\cdots<\lambda_1^{(c)}<\lambda_0^{(c)}<1. \tag{3.170}$$

The solution to the concentration problem, equation (3.160), is then the eigenfunction, $\psi_0^{(c)}(y)$, with the maximum eigenvalue, $\lambda_0^{(c)}$. The eigenvalues are also known as *concentration ratios* for the evident reason that each one quantifies the relative energy of the corresponding eigenfunction that is concentrated in the spectral band, $[-f_0,f_0]$.

Because the eigenvalues are distinct, the eigenfunctions are independent, orthogonal, and form a basis for the space of functions restricted to the domain, $[-T/2,\ T/2]$. In fact, it turns out that the numerically significant components (i.e., with significant eigenvalues) of the basis are those with $n+1<2c/\pi$ if c is large. The number, $2c/\pi=2Tf_0$, is an extent-bandwidth product (Section 3.5.1), which depends on the chosen values of T and f_0. It is known as the *Shannon number* and defines the approximate number of significant eigenfunctions needed to generate corresponding space-limited functions that are almost-band-limited (Simons 2010, p.895). In certain applications, one might even say "practically" band-limited (Slepian (1976). The transition from significant to insignificant eigenvalue is also rather abrupt; see Section 4.6.

The computation of the eigenfunctions, $\psi_n^{(c)}(y)$, is not trivial, but the key to most algorithms is the discovery (Slepian and Pollak 1961) that they also are eigenfunctions of the differential operator (Sturm-Liouville differential equation) associated with the Helmholz wave equation in terms of prolate spheroidal coordinates (Arfken 1970). The solutions to this differential equation in this form are special functions known as *prolate spheroidal wave functions* (PSWFs) whose properties lead to stable algorithms for the computation of $\psi_n^{(c)}(y)$ (Xiao et al. 2001, Moore and Cada 2004). These functions are now also known as *Slepian functions*. Figure 3.13 exhibits $\psi_n^{(c)}(y)$, $n=0,1,2,3$, for $c=2\pi$.

Similar to the Gaussian function the prolate spheroidal wave functions defined on $[-1,1]$ are their own (scaled) Fourier transforms defined on $(-\infty,\infty)$. Indeed, consider the function operators,

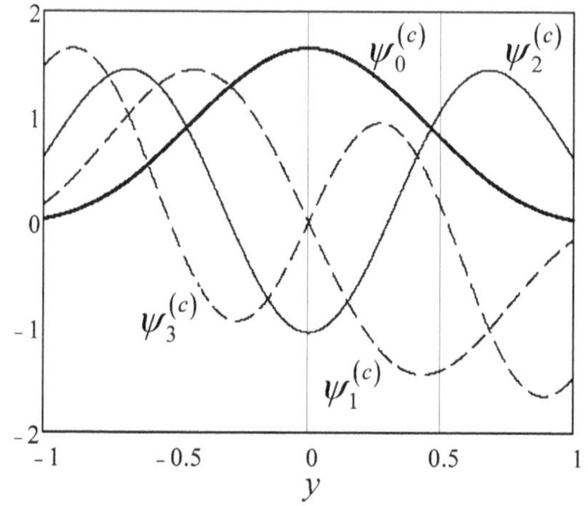

Figure 3.13: Prolate spheroidal wave functions, $\psi_n^{(c)}(y)$, for $c = 2\pi$.

$$F_c[\psi](y) = \int_{-1}^{1} \psi(y') e^{-icyy'} dy', \tag{3.171}$$

$$Q_c[\psi](y) = \int_{-1}^{1} \psi(y') d_1\left(y - y'; \frac{c}{2\pi}\right) dy', \tag{3.172}$$

which, in view of equation (3.169), are related according to

$$\begin{aligned} F_c^*\left[F_c[\psi]\right](y) &= \int_{-1}^{1}\left(\int_{-1}^{1} \psi(y') e^{-icy''y'} dy'\right) e^{icyy''} dy'' = \frac{2\pi}{c}\int_{-1}^{1} \psi(y') d_1\left(y - y'; \frac{c}{2\pi}\right) dy' \\ &= \frac{2\pi}{c} Q_c[\psi](y) \end{aligned} \tag{3.173}$$

Hence, it is easy to verify that a solution of equation (3.167), $\psi_n^{(c)}(y)$, being an eigenfunction of the operator, Q_c, with eigenvalue, $\lambda_n^{(c)}$, is also an eigenfunction of the operator, F_c, with eigenvalue, $\alpha_n^{(c)}$, where $|\alpha_n^{(c)}| = \sqrt{2\pi\lambda_n^{(c)}/c}$. Indeed, one has,

$$\lambda_n^{(c)} \psi_n^{(c)}(y) = Q_c\left[\psi_n^{(c)}\right](y) = \frac{c}{2\pi} F_c^*\left[F_c\left[\psi_n^{(c)}\right]\right](y) = \frac{c}{2\pi}\left|\alpha_n^{(c)}\right|^2 \psi_n^{(c)}(y). \tag{3.174}$$

Now, the extended functions defined by

$$\psi_n^{(T,c)}(x) = \begin{cases} \psi_n^{(c)}\left(\dfrac{2x}{T}\right), & |x| \le \dfrac{T}{2} \\ 0, & |x| > \dfrac{T}{2} \end{cases} \tag{3.175}$$

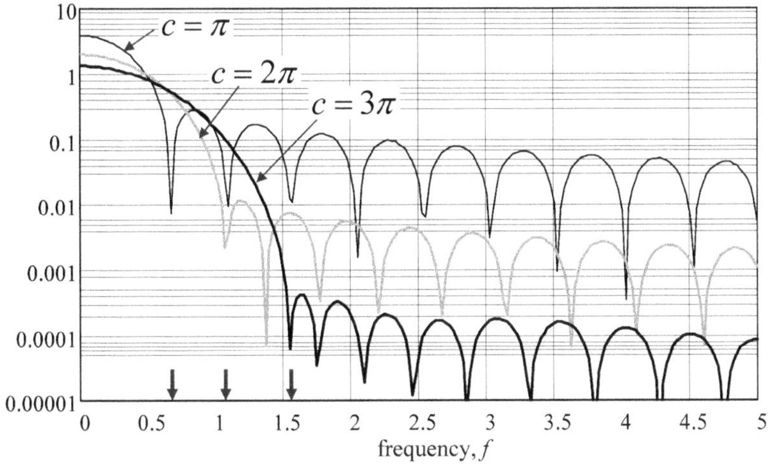

Figure 3.14: Amplitude spectra of prolate spheroidal wave functions, $\psi_0^{(T,c)}(x)$, for $T = 2$ and $c = \pi$, 2π, 3π. The thick short arrows indicate the first zero in each case.

have Fourier transforms given, with an appropriate change in integration variable, by

$$\Psi_n^{(T,c)}(f) = \int_{-\infty}^{\infty} \psi_n^{(T,c)}(x) e^{-i2\pi fx} dx = \int_{-T/2}^{T/2} \psi_n^{(c)}\left(\frac{2x}{T}\right) e^{-i2\pi fx} dx = \frac{T}{2} \alpha_n^{(c)} \psi_n^{(c)}\left(\frac{f}{f_0}\right). \quad (3.176)$$

This shows that the Fourier transform of $\psi_n^{(T,c)}(x)$ is proportional to $\psi_n^{(c)}(f/f_0)$.

The Fourier transforms of $\psi_0^{(T,c)}(x)$ for $T = 2$ and $c = \pi$, 2π, 3π ($f_0 = 0.5, 1, 1.5$) are shown in Figure 3.14. Note that the principal spectral content of $\psi_0^{(T,c)}(x)$ (delimited by the first zero) is essentially confined to the bandwidth defined by f_0, as desired in the solution to the concentration problem, and increasingly so as c (hence, f_0) increases. Moreover, the function, $\psi_0^{(T,c)}(x)$, highlighted in Figure 3.13 (where $\psi_0^{(c)}(x) = \psi_0^{(2,c)}(x)$), also serves as a taper (window function) with desirable characteristics—relatively narrow main lobe and rapidly attenuating side lobes (see also Section 5.7.2). The functions, $\psi_n^{(T,c)}(x)$, $n \geq 0$, are orthogonal, and, in turn, are successively optimally concentrated in the spectral band, $[-f_0, f_0]$ (see also Section 4.6). They form a basis for space-limited functions whose Fourier spectral content is optimally band-limited.

The concentration problems for functions on the Cartesian plane and on the sphere are set up similarly and yield analogous solutions. In fact, the concentration regions may be quite arbitrary in shape leading only to more complicated numerical algorithms to compute the eigenfunctions. For the plane, consider a function limited to a finite region, E,

$$g_E(x_1, x_2) = \begin{cases} g(x_1, x_2), & (x_1, x_2) \in E \\ 0, & \text{otherwise} \end{cases} \quad (3.177)$$

that has its spectral content limited as much as possible to the finite spectral domain, W. Such functions are determined by maximizing the ratio,

$$\lambda_E(W) = \frac{\iint\limits_W G_E^2(f_1, f_2)\,df_1 df_2}{\int\limits_{-\infty}^{\infty}\int\limits_{-\infty}^{\infty} G_E^2(f_1, f_2)\,df_1 df_2} \to \max, \tag{3.178}$$

where $G_E(f_1, f_2)$ is the Fourier transform of $g_E(x_1, x_2)$,

$$G_E(f_1, f_2) = \iint\limits_E g_E(x_1, x_2)\, e^{-i2\pi(f_1 x_1 + f_2 x_2)}\,dx_1 dx_2, \tag{3.179}$$

since $g_E(x_1, x_2) = 0$ for $(x_1, x_2) \notin E$. Substituting this into the numerator of equation (3.178), the maximization problem becomes

$$\lambda_E(W) = \frac{\iint\limits_E \iint\limits_E g(x_1, x_2)\, d_2(x_1 - x_1', x_2 - x_2'; W)\, g(x_1', x_2')\,dx_1' dx_2' dx_1 dx_2}{\iint\limits_E g^2(x_1, x_2)\,dx_1 dx_2} \to \max, \tag{3.180}$$

where

$$d_2(x_1 - x_1', x_2 - x_2'; W) = \iint\limits_W e^{-i2\pi\left(f_1(x_1 - x_1') + f_2(x_2 - x_2')\right)}\,df_1 df_2, \tag{3.181}$$

and Parseval's theorem is used in the denominator. The solution to the problem is given by the eigenfunction with maximum eigenvalue that satisfies the integral equation

$$\iint\limits_E d_2(x_1 - x_1', x_2 - x_2'; W)\, g(x_1', x_2')\,dx_1' dx_2' = \lambda_E(W)\, g(x_1, x_2). \tag{3.182}$$

If the spectral concentration region, W, is a rectangle, $W = \{(f_1, f_2) \mid |f_1| \le f_1^{(0)}, |f_2| \le f_2^{(0)}\}$, then the kernel, d_2, is given by

$$d_2\left(x_1 - x_1', x_2 - x_2'; f_1^{(0)}, f_2^{(0)}\right) = \frac{\sin\left(2\pi(x_1 - x_1') f_1^{(0)}\right)}{\pi(x_1 - x_1')} \frac{\sin\left(2\pi(x_2 - x_2') f_2^{(0)}\right)}{\pi(x_2 - x_2')}. \tag{3.183}$$

If the spectral concentration region is a disk, $W = \{(f_1, f_2) \mid 0 \le \bar{f} = \sqrt{f_1^2 + f_2^2} \le \bar{f}_0\}$, then the kernel is

$$d_2\left(x_1 - x_2, x_1' - x_2'; \bar{f}_0\right) = \int\limits_0^{2\pi}\int\limits_0^{\bar{f}_0} e^{-i2\pi \bar{f} \Delta s \cos(\nu - \omega)}\, \bar{f}\, d\bar{f}\, d\nu = 2\pi \int\limits_0^{\bar{f}_0} J_0\left(2\pi \bar{f} \Delta s\right) \bar{f}\, d\bar{f}, \tag{3.184}$$

where $\bar{f}$ is given by equations (2.177) and (2.178), $\Delta s = \sqrt{(x_1 - x_1')^2 + (x_2 - x_2')^2}$, and ω is the orientation angle of the line connecting the points, (x_1, x_2) and (x_1', x_2'). The last equality follows from equation (2.179), which in view of equation (2.190) also yields

$$d_2\left(x_1 - x_2, x_1' - x_2'; \bar{f}_0\right) = \frac{\bar{f}_0}{\Delta s} J_1\left(2\pi\Delta s\bar{f}_0\right). \tag{3.185}$$

Finally, the spectral concentration problem for functions, $g_p(\theta, \lambda)$, space-limited to a region, Ω_p, on the unit sphere, seeks to maximize the ratio,

$$\lambda_p\left(n_0\right) = \frac{\sum_{n=0}^{n_0}\sum_{m=-n}^{n}\left(G_{n,m}^{(p)}\right)^2}{\sum_{n=0}^{\infty}\sum_{m=-n}^{n}\left(G_{n,m}^{(p)}\right)^2} \rightarrow \text{max}, \tag{3.186}$$

where spectral concentration is defined here as harmonic degree and order not greater than n_0. The Fourier-Legendre transform of $g_p(\theta, \lambda)$, equation (2.220),

$$G_{n,m}^{(p)} = \frac{1}{4\pi}\iint_{\Omega_p} g_p\left(\theta, \lambda\right)\bar{Y}_{n,m}\left(\theta, \lambda\right) d\Omega, \tag{3.187}$$

substituted into equation (3.186) yields

$$\lambda_p\left(n_0\right) = \frac{\iint_{\Omega_p}\iint_{\Omega_p'} g_p\left(\theta, \lambda\right) g_p\left(\theta', \lambda'\right) d_s\left(\psi; n_0\right) d\Omega d\Omega'}{\iint_{\Omega_p} g_p^2\left(\theta, \lambda\right) d\Omega}, \tag{3.188}$$

where the denominator results from Parseval's theorem, equation (2.245), and where

$$d_s\left(\psi; n_0\right) = \frac{1}{4\pi}\sum_{n=0}^{n_0}\left(2n+1\right) P_n\left(\cos\psi\right), \tag{3.189}$$

and $\cos\psi$ is given by equation (2.196). The corresponding solution to the maximization problem is the function with the maximum eigenvalue satisfying the integral equation,

$$\iint_{\Omega_p'} g_p\left(\theta', \lambda'\right) d_s\left(\psi; n_0\right) d\Omega' = \lambda_p\left(n_0\right) g_p\left(\theta, \lambda\right). \tag{3.190}$$

The kernel function, $d_s(\psi; n_0)$, is a truncation of the Legendre series for the scaled spherical delta function, $\delta_s(\psi)/(4\pi)$, equation (2.321).

The details of computing the eigenfunctions (Slepian functions) and eigenvalues may be found in (Percival and Walden 1993, Ch.8) and in various more recent papers, e.g. (Simons 2010, Osipov and Rokhlin 2014). The computation of the eigenfunctions for the concentration problem on the plane for a disk, as well as arbitrary regions, is treated by Simons and Wang (2011). The concentration problem for the sphere is solved by Simons et al. (2006). It suffices here to note that for all the concentration problems, the kernel functions in the corresponding integral equation that define the eigenfunctions are similar, as shown in Figure 3.15. Whether the sinc function, d_1,

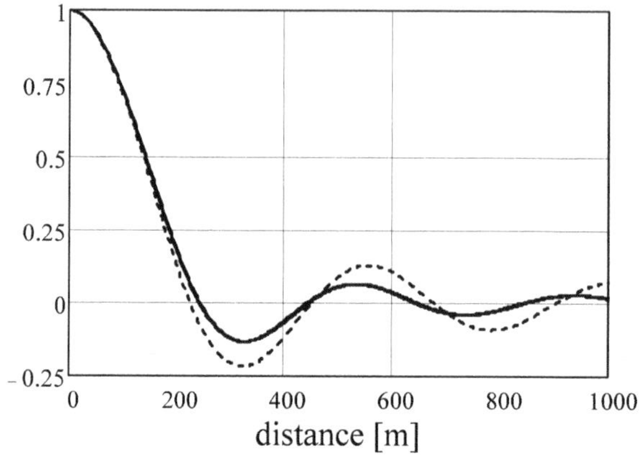

Figure 3.15: Kernel functions, $d_1(x-x';f_0)$ (dashed line), $d_2(x_1-x_2, x_1'-x_2';\bar{f}_0)$ and $d_s(\psi;n_0)$ (solid lines and indistinguishable), all normalized to unity at the origin, where $n_0=100, \bar{f}_0=n_0/(2\pi R)$ ($R=6371$ km is Earth's mean radius), and $\Delta s=R\psi; f_0=\sqrt{\pi}\bar{f}_0/2$ and $x-x'=\Delta s$ are chosen for $d_1(x-x';f_0)$.

the first-order Bessel function, d_2, or the truncated delta function series, d_s, their transforms are constants over their respective limited spectral domains.

3.6.3 Truncated Convolutions on the Sphere

Another form of optimally weighting a function in a finite domain was developed in geodesy for convolutions on the sphere, which form solutions to boundary-value problems in potential theory. The general form of these convolutions is equation (3.58), repeated here for convenience,

$$c(\theta,\lambda)=\frac{1}{4\pi}\iint\limits_{\Omega'} g(\theta',\lambda')h(\psi)\,d\Omega'. \tag{3.191}$$

The kernel function, h, as indicated, depends only on the spherical distance between the integration and evaluation points. It is also called a Green's function in boundary-value problems. Because it generally attenuates rapidly away from the origin, practical applications of the convolution make the reasonable approximation as an integral that is truncated to a neighborhood, Ω_s, of the evaluation point, typically given as a spherical cap, as for equation (2.302),

$$\hat{c}(\theta,\lambda)=\frac{1}{4\pi}\iint\limits_{\Omega_s'} g(\theta',\lambda')h(\psi)\,d\Omega'. \tag{3.192}$$

From the spectral perspective, in a rough sense, the truncated integral contributes the finer details, or high-frequency part to the convolution, while the neglected part is mostly of long-wavelength character. The truncation thus makes even more sense if a long-wavelength reference is available in the form of a finite spherical harmonic series,

$$g_M(\theta,\lambda)=\sum_{n=0}^{n_M}\sum_{m=-n}^{n}G_{n,m}\bar{Y}_{n,m}(\theta,\lambda). \tag{3.193}$$

Removing this model from the data function,

$$\delta g(\theta,\lambda)=g(\theta',\lambda')-g_M(\theta',\lambda'), \tag{3.194}$$

and restoring it in the form of the reference for the convolution, $c_M(\theta, \lambda)$, it is easily verified using the convolution theorem, equation (3.60), that

$$c(\theta,\lambda)=\frac{1}{4\pi}\iint_{\Omega'}\delta g(\theta',\lambda')h(\psi)\,d\Omega'+c_M(\theta,\lambda), \tag{3.195}$$

where, with the Legendre spectrum of h, the last term is

$$c_M(\theta,\lambda)=\sum_{n=0}^{n_M}\sum_{m=-n}^{n}H_nG_{n,m}\bar{Y}_{n,m}(\theta,\lambda). \tag{3.196}$$

Truncating the integral causes an error, but also opens the opportunity to modify the kernel so as to reduce the error. Thus, consider the expression equivalent to equation (3.195),

$$\begin{aligned}c(\theta,\lambda)=&\frac{1}{4\pi}\iint_{\Omega_s'}\delta g(\theta',\lambda')\big(h(\psi)-\delta h(\psi)\big)\,d\Omega'+\frac{1}{4\pi}\iint_{\Omega_s'}\delta g(\theta',\lambda')\delta h(\psi)\,d\Omega'\\&+\frac{1}{4\pi}\iint_{\Omega-\Omega_s'}\delta g(\theta',\lambda')h(\psi)\,d\Omega'+c_M(\theta,\lambda)\end{aligned} \tag{3.197}$$

and for the last two integrals define an error kernel that includes the modification, δh,

$$\bar{h}(\psi)=\begin{cases}\delta h(\psi), & 0\le\psi\le\psi_s\\ h(\psi), & \psi_s<\psi\le\pi\end{cases} \tag{3.198}$$

Assuming that it is square integrable on $[0, \pi]$, it has a Legendre transform, $\bar{H}_n$. The last two integrals in equation (3.197) may be combined and, with the convolution theorem, expressed as a series,

$$\begin{aligned}c(\theta,\lambda)&=\frac{1}{4\pi}\iint_{\Omega_s'}\delta g(\theta',\lambda')\big(h(\psi)-\delta h(\psi)\big)\,d\Omega'+\frac{1}{4\pi}\iint_{\Omega'}\delta g(\theta',\lambda')\bar{h}(\psi)\,d\Omega'+c_M(\theta,\lambda)\\&=\frac{1}{4\pi}\iint_{\Omega_s'}\delta g(\theta',\lambda')\big(h(\psi)-\delta h(\psi)\big)\,d\Omega'+c_M(\theta,\lambda)+\sum_{n=n_M+1}^{\infty}\sum_{m=-n}^{n}\bar{H}_nG_{n,m}\bar{Y}_{n,m}(\theta,\lambda)\end{aligned} \tag{3.199}$$

If the data function, g, is known only in the neighborhood, Ω_s, but also its spectrum, $G_{n,m}$, up to degree and order, n_M, is known, then the estimate,

$$\hat{c}(\theta,\lambda) = \frac{1}{4\pi} \iint_{\Omega_s'} \delta g(\theta',\lambda')\left(h(\psi) - \delta h(\psi)\right) d\Omega' + c_M(\theta,\lambda), \tag{3.200}$$

is in error by

$$\varepsilon_s = -\sum_{n=n_M+1}^{\infty} \sum_{m=-n}^{n} \bar{H}_n G_{n,m} \bar{Y}_{n,m}(\theta,\lambda). \tag{3.201}$$

This error is basically a high-pass filter applied to the data function. The filter function, $\bar{h}(\psi)$, equation (3.198), may be defined, through $\delta h(\psi)$, in an attempt to minimize this error. If $\delta h = 0$ (no modification to the kernel, h, over the data region), then $\bar{h}(\psi)$ has a discontinuity at $\psi = \psi_s$ and its spectrum is subject to Gibbs's effect. The resulting slow convergence tends to increase the amplitude spectrum of the error. Eliminating the discontinuity with $\delta h(\psi) = h(\psi_s)$ reduces this effect but is not optimal. Neither is the intuitively appealing modification, $\delta h(\psi) = \sum_{n=0}^{n_M} (2n+1) H_n P_n(\cos\psi)$, that imitates the removal of the reference model from the data function, g.

Molodensky (1958) had a similar idea, but designed this band-limited function by minimizing the spectrum of ε_s (see also Jekeli 1981b). To simplify the notation in the derivation let $y = \cos\psi$, and suppose that

$$\delta h(y) = \sum_{n=0}^{n_M} (2n+1) V_n P_n(y), \tag{3.202}$$

where the coefficients,

$$V_n = \frac{1}{2} \int_{-1}^{1} \delta h(y) P_n(y)\, dy, \tag{3.203}$$

are to be determined so that the error, ε_s, is minimized. From equations (3.198) and (2.205), the spectrum of $\bar{h}(y)$ is

$$\begin{aligned} \bar{H}_n &= \frac{1}{2} \int_{-1}^{y_s} h(y) P_n(y)\, dy + \frac{1}{2} \int_{y_s}^{1} \delta h(y) P_n(y)\, dy \\ &= \frac{1}{2} \int_{-1}^{y_s} \left(h(y) - \delta h(y)\right) P_n(y)\, dy + \frac{1}{2} \int_{-1}^{1} \delta h(y) P_n(y)\, dy \end{aligned} \tag{3.204}$$

where $y_s = \cos\psi_s$. For $n > n_M$, the second integral vanishes in view of equation (3.202) and the orthogonality of the Legendre polynomials, equation (2.204). Thus, the energy of the error,

$$\sum_{n=n_M+1}^{\infty}\sum_{m=-n}^{n}\left(\bar{H}_n G_{n,m}\right)^2 = \sum_{n=n_M+1}^{\infty}\bar{H}_n^2\sum_{m=-n}^{n}G_{n,m}^2, \tag{3.205}$$

is minimized by minimizing the first integral in equation (3.204),

$$\bar{H}_n = \frac{1}{2}\int_{-1}^{y_s}\left(h(y)-\delta h(y)\right)P_n(y)\,dy, \quad n > n_M. \tag{3.206}$$

In essence, the kernel function is modified optimally in order to concentrate the energy of the residual convolution within the integration area.

With a change in variable, $y = ax + a - 1$, where $a = (1 + y_s)/2$, let

$$\delta h'(x) = \delta h(ax + a - 1) \text{ and } h'(x) = h(ax + a - 1). \tag{3.207}$$

Since $-1 \leq y \leq y_s$ is equivalent to $-1 \leq x \leq 1$, they can be expanded in series of Legendre polynomials, $P_n(x)$, for instance,

$$h'(x) = \sum_{n=0}^{\infty}(2n+1)H_n'P_n(x). \tag{3.208}$$

Now, $\delta h(y)$, equation (3.202), is a polynomial of degree n_M, hence $\delta h'(x)$ is also a finite series in $P_n(x)$, $n \leq n_M$. According to the best-approximation theory for orthogonal polynomial expansions (see the conclusion after equation (2.43)), the function, $\delta h'(x)$, best approximates $h'(x)$ for $x \in [-1,1]$ if it is the partial sum of $h'(x)$,

$$\delta h'(x) = \sum_{n=0}^{n_M}(2n+1)H_n'P_n(x). \tag{3.209}$$

In this case, $\delta h(y)$ also best approximates $h(y)$ for $y \in [-1, y_s]$, thus minimizing $\bar{H}_n$ (equation (3.206)) for $n > n_M$.

Now, $P_r(x) = P_r((y + 1 - a)/a)$ is a polynomial in y that is well defined for $y \in [-1,1]$. It can be expanded as a finite linear combination of the Legendre polynomials, $P_n(y)$, $0 \leq n \leq r$, that constitute an orthogonal basis for the space of all (real) polynomials of degree, r, on $[-1,1]$,

$$P_r\left(\frac{y+1-a}{a}\right) = \sum_{n=0}^{r}(2n+1)p_{r,n}P_n(y), \tag{3.210}$$

where

$$p_{r,n} = \frac{1}{2}\int_{-1}^{1}P_r\left(\frac{y+1-a}{a}\right)P_n(y)\,dy, \quad 0 \leq n \leq r. \tag{3.211}$$

By the orthogonality of Legendre polynomials, $P_{r,n} = 0$, for $n > r$. From equation (3.208),

$$H'_r = \frac{1}{2}\int_{-1}^{1} h'(x)\,P_r(x)\,dx = \frac{1}{2a}\int_{-1}^{y_s} h(y)\,P_r\left(\frac{y+1-a}{a}\right)dy; \tag{3.212}$$

and, substituting equation (3.210), this is

$$H'_r = \frac{1}{a}\sum_{n=0}^{r}(2n+1)\,p_{r,n}Q_n, \tag{3.213}$$

where

$$Q_n = \frac{1}{2}\int_{-1}^{y_s} h(y)\,P_n(y)\,dy. \tag{3.214}$$

Finally, using equations (3.211) and (3.209) in equation (3.203), the solution for the Legendre spectrum of the optimal kernel modification, $\delta h(y)$, is given by

$$V_n = \sum_{r=n}^{n_M}(2r+1)\,H'_r p_{r,n}. \tag{3.215}$$

Also, from equations (3.206), (3.214), and (3.202)

$$\bar{H}_n = Q_n - \sum_{r=0}^{n_M}(2r+1)\,V_r q_{r,n}, \tag{3.216}$$

where

$$q_{r,n} = \frac{1}{2}\int_{-1}^{y_s} P_r(y)\,P_n(y)\,dy. \tag{3.217}$$

The integrals, $p_{r,n}$, are derived in (Molodensky 1958, p.138) (erroneously transcribed in (Jekeli 1981b)),

$$p_{r,n} = \frac{1}{a^r}\sum_{j=0}^{r-n-1}\frac{1}{j+1}\binom{r-n-1}{j}\binom{r+n}{j}(1-a)^{j+1}, \quad 0 \le n < r; \tag{3.218}$$

$$p_{r,r} = \frac{1}{(2r+1)\,a^r}, \quad r \ge 0. \tag{3.219}$$

And, the integrals, $q_{r,n}$, are obtained from (Hobson 1965, p.38), for $r \neq n$ and $r, n > 0$,

$$q_{r,n} = \frac{1}{2(r-n)(r+n+1)}\left(nP_r(y_s)\,P_{n-1}(y_s) - rP_n(y_s)\,P_{r-1}(y_s) + y_s(r-n)\,P_n(y_s)\,P_r(y_s)\right), \tag{3.220}$$

with

$$q_{r,0} = \frac{1}{2(2r+1)}\left(P_{r+1}(y_s) - P_{r-1}(y_s)\right), \quad r > 0$$

$$q_{0,0} = \frac{1+y_s}{2} \tag{3.221}$$

$$q_{r,r} = \frac{1}{2(2r+1)}\left((2r-1)q_{r-1,r-1} + y_s\left(P_r^2(y_s) + P_{r-1}^2(y_s)\right) - 2P_r(y_s)P_{r-1}(y_s)\right), \quad r > 0$$

The coefficients, Q_n, depend on the kernel, $h(\psi)$, and recursion formulas have been derived by (Hagiwara 1976) for the most common integrals in physical geodesy. If $h(\psi)$ is Stokes's kernel, then $H_n = 1/(n-1)$ (equation (6.87) with $r = R$). For this kernel, Figure 3.16 shows an example of the relative error spectrum, $\bar{H}_n$, with the Molodensky modification, equation (3.215), compared to the case, $\delta h = 0$, when $\bar{H}_n = Q_n$ (see equation (3.216)). There is significant reduction in the truncation error, equation (3.201), since $|\bar{H}_n| < |Q_n|$ for most $n > n_M$.

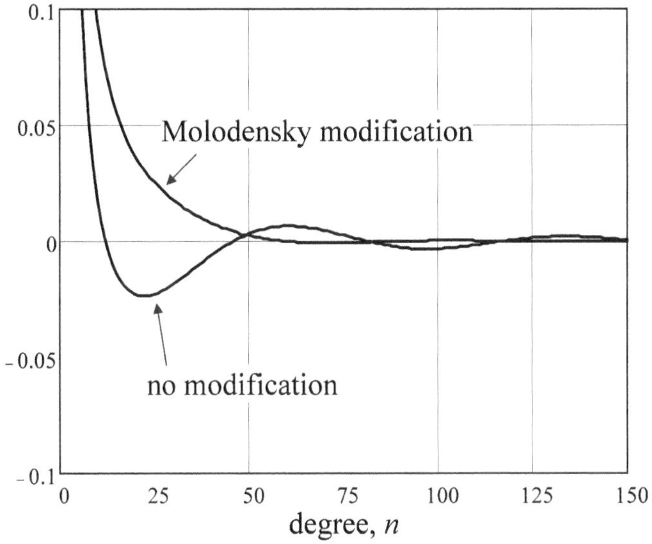

Figure 3.16: Relative error spectrum, $\bar{H}_n$, of the truncation error with Molodensky's modification, equation (3.216), and without ($\bar{H}_n = Q_n$), for the case, $\psi_s = 5°$, $n_M = 50$, and Stokes's kernel. Either coefficient enters the error only for $n > n_M$.

* * *

EXERCISES

Exercise 3.1

Using an appropriate change of integration variables in the definition of convolution, equation (3.2), show that

$$(g*h)(x+c) = g(x+c)*h(x) = g(x)*h(x+c). \tag{E3.1}$$

However, also show that, in reference to the different definitions of convolution, equations (3.2) and (3.4),

$$(g*h)(ax+b) \neq g(ax+b)*h(x) \neq g(x)*h(ax+b) \neq g(ax+b)*h(ax+b). \tag{E3.2}$$

Exercise 3.2

Show that the convolution of two Gaussian functions, equation (2.105), is again a Gaussian function. Hint: use the convolution theorem.

Exercise 3.3

Prove the following properties of the cyclic convolution,

$$(\tilde{g}\,\tilde{*}\,\tilde{h})(x+c) = \tilde{g}(x+c)\,\tilde{*}\,\tilde{h}(x) = \tilde{g}(x)\,\tilde{*}\,\tilde{h}(x+c), \tag{E3.3}$$

$$\tilde{g}(x)\,\tilde{*}\,\tilde{h}(x) = \tilde{h}(x)\,\tilde{*}\,\tilde{g}(x), \tag{E3.4}$$

$$\tilde{g}_1(x)\,\tilde{*}\left(\tilde{g}_2(x)\,\tilde{*}\,\tilde{h}(x)\right) = \left(\tilde{g}_1(x)\,\tilde{*}\,\tilde{g}_2(x)\right)\tilde{*}\,\tilde{h}(x). \tag{E3.5}$$

Exercise 3.4

Show that

$$g(\theta)*h(\theta) \quad \leftrightarrow \quad G_n H_n, \tag{E3.6}$$

where the convolution is defined by equation (3.58). Hint: Note that the convolution only depends on one variable, θ; hence, express and simplify its Legendre spectrum using the addition formula, equation (2.254), for the spectrum of *h*, integrating out λ', and finally using the orthogonality of the Legendre polynomials, equation (2.204).

Exercise 3.5

By making use of equation (2.121) and the spectrum of the delta function, show that the filter function of the ideal *high-pass* filter is given by

$$h(x) = \delta(x) - 2f_0 \operatorname{sinc}(2f_0 x). \tag{E3.7}$$

Hint: Let $H(f) = 1 - 1 + H(f)$.

Exercise 3.6

Show that the Gaussian function, $\gamma^{(\beta)}(x)$, has total energy, $E = 1/(\beta\sqrt{2})$. Hint: use the fact that the integral of the Gaussian probability density function is unity (see also equation (5.180),

$$\frac{1}{\sqrt{2\pi}\sigma}\int_{-\infty}^{\infty} e^{-\frac{1}{2}\left(\frac{x}{\sigma}\right)^2} dx = 1. \tag{E3.8}$$

Exercise 3.7

a) Show that the product, $D(g)W(g)$, of equivalent extent and equivalent bandwidth, as defined by equations (3.120), is not less than $1/(4\pi)$. Hints: Using integration by parts show that

$$\int_{-\infty}^{\infty} xg(x)\frac{dg}{dx}dx = -\frac{1}{2}E; \tag{E3.9}$$

and using appropriate properties of the Fourier transform show that

$$\int_{-\infty}^{\infty}\left(\frac{dg}{dx}\right)^2 dx = 4\pi^2\int_{-\infty}^{\infty} f^2\left|G(f)\right|^2 df. \tag{E3.10}$$

Then apply Schwarz's inequality, equation (1.22), to $\left|\int_{-\infty}^{\infty} xg(x)\frac{dg}{dx}dx\right|^2$ in order to obtain the result.

b) Show that the extent-bandwidth product equals $1/4\pi$ for the Gaussian function.

Exercise 3.8

Derive the Fourier transforms of the cosine tapers, equation (3.143). Hint: make use of the first of equations (1.6) and the translation property (2.74).

Exercise 3.9

Derive the Fourier transforms of the Bartlett, Hann, (truncated) Gaussian, and Tukey window functions, equations (3.149), (3.150), (3.151), and (3.152). Equation (3.149) is a straightforward application of the similarity property, and equation (3.150) is obtained from Exercise 3.8 and decomposing the sinc function. For equation (3.152), note that the Tukey window is a convolution,

$$u_{\text{Tukey}}^{(\alpha)}(x) = \frac{\pi}{(1-\alpha)T}v(x) * b\left(\frac{2x}{(1+\alpha)T}\right), \tag{E3.11}$$

where

$$v(x)=\begin{cases}\cos\left(\dfrac{2\pi}{(1-\alpha)T}x\right), & |x|\le\dfrac{(1-\alpha)T}{4}\\ 0, & |x|>\dfrac{(1-\alpha)T}{4}\end{cases} \tag{E3.12}$$

Find the Fourier transform of each, apply the convolution theorem, and simplify.

Chapter 4

Transforms, Convolutions, and Windows on the Discrete Domain

4.1 Introduction

The previous two chapters introduced Fourier transforms of functions of the periodic and non-periodic types on the *continuous* space domain, as well as the principal and complementary operations of convolution and truncation that have particular interpretation in the frequency domain. In practice, however, we deal with data on a *discrete*, rather than continuous, domain. That is, a geophysical signal may be defined on the basis of classical physical laws for continuous time or space, but we sample it, measure it, or realize it at discrete points in space, or at discrete instants in time. Even if an instrument responds virtually continuously to variations of a geophysical phenomenon, e.g., a tide gauge measuring sea level or an airborne magnetometer sensing the Earth's magnetic field along a flight line, the data are processed and stored with discrete time tags. Indeed, all automated recording instruments have some finite averaging time, or *integration time*, that defines the temporal resolution of the measured signal. In many cases, this integration time is not just dictated by practical recording limitations, but also by the intrinsic noise in all measurement devices, where averaging is required to reduce or filter out high frequency noise. Therefore, data in practice always constitute a *sequence* or an *array* of discrete values and a corresponding spectral analysis must be developed on the basis of this reality. This is not to imply that the foregoing mathematical developments for functions on the continuous space domain are solely an esoteric exercise to illustrate the elegance and symmetry of Fourier analysis. They do provide a foundation on which to build the practical discrete case, which is necessary for the simple reason that the data samples usually do not fully capture the information of the continuous-domain signal, nor of its spectrum. The degree to which the data represent the complete signal depends critically on the sampling interval and the significant part of spectral domain of the signal. The subject of this chapter, then, is not only to develop the transition of the Fourier and Legendre transforms and the attendant operations from the continuous to the discrete domains, but also to understand the consequent effects of sampling on the spectral analysis.

Note that only the spatial domain, e.g., the line, plane, or sphere, is discretized. The elements of the sample sequence, that is, the data, may assume any value allowed by the function that represents the signal. In the modern parlance, one usually also uses the term *digital signal* to refer to a sampled function as defined here. However, one could consider, in addition, the discretization, or *quantization*, of the values or amplitudes of the signal. Then a distinction is required between the continuous and the discrete amplitude of a signal. Clearly, the quantized-amplitude sequence is a special case of the continuous-amplitude sequence; and, the consideration of the former with respect to the latter involves primarily the incorporation of the quantization effect. This is not treated here and attention is restricted to continuous-amplitude sequences defined on a discrete domain. Moreover, the use of the word, "sequence", automatically implies a discrete domain.

The concepts of Fourier transform, convolution (filter), and window, discussed for functions on the continuous domain are readily adapted to the discrete case with one important stipulation that the spatial domain is discretized at a *constant* interval. Although the Fourier transforms can be defined for irregularly discretized domains (e.g., Marvasti 2001), the relationships between the time- or space-domain informational content of the sequence and the spectral characterization of the underlying function become complicated. For practical applications, one usually chooses to interpolate irregularly sampled data onto a regular interval or grid. However, this may also distort the spectral information that is then extracted from the data. The many ways one might approach the problem of non-uniformly sampled data is beyond the present scope and a regular distribution is assumed. For many automated data collecting systems, sampling at a constant interval is the usual mechanization.

In practice, one also has only a finitely extended sequence of data, although in some instances where the extent of the data is sufficiently great, the approximation by an infinite sequence is fully justified. For periodic functions on the line or plane and for spherical functions, the question turns on whether a full period, or the complete spherical domain, of a function is sampled.

The first consideration in this chapter is of sequences defined on the infinite line or plane. Only after obtaining an appreciation of this practical implementation of continuous-domain theory, that is, of *sampling*, do we enter the final, albeit the most useful implementation of the theory: *finitely* extended sequences. Truncation of an infinite sequence is treated, as for the continuous case, using window sequences. However, it is also convenient to interpret such finite sequences as periodic with period equal to the length of the sequence. Thus, the last type of Fourier transform to be examined is for functions that are both periodic and defined on a discrete domain, including arrays of function values covering the sphere that may be viewed also alternatively as the principal domain of a two-dimensional periodic function.

4.2 Infinite Sequences

Spectral analysis for infinite sequences can be developed without reference to a parent function from which the sequence is derived. However, since geophysical data are almost always samples of a continuous geophysical signal, for the moment, a parent function is assumed that has a Fourier integral. In one dimension, suppose that values of $g(x)$ are available at regular intervals, $x_\ell = \ell \Delta x$, where ℓ is an integer with $-\infty < \ell < \infty$. The resulting sequence of values is a set of discrete samples of g, denoted simply by the integer subscript,

$$g_\ell = g(\ell \Delta x), \quad -\infty < \ell < \infty. \tag{4.1}$$

The *sampling interval*, Δx, is constant and positive. Analogous to absolute integrability, it is assumed that the samples are absolutely summable,

$$\sum_{\ell=-\infty}^{\infty} |g_\ell| < \infty, \tag{4.2}$$

which also means that they attenuate to zero in the limits, $\ell \to \pm \infty$.

To understand sampling and the relationship between the discrete sequence and the originating continuous function one may use the concept of impulses defined by the Dirac delta function, $\delta(x)$ (Section 2.3.5). Because the parent function is not periodic, also the sequence, g_ℓ, is not periodic, leading to the expectation that the spectral domain is continuous, as it is for non-periodic functions. On further thought, however, the reader may guess that there is a limit to the frequencies that could be represented by the infinite sequence and that this depends in some way on the sampling interval. In fact, the sequence is band-limited, that is, the spectral domain is finite; and, in perfect duality with the Fourier-series transform pair, its spectrum is periodic, as seen below.

The frequency limit in the spectral domain is tied intimately to the sampling interval and defines the *spatial resolution* of the data (Section 3.5.2). The concept of resolution is all-important when designing a measurement system since it defines the tradeoffs between technical feasibility and degree of realization, or detail, of a function. Usually, high-frequency noise in an instrument constrains the upper limit of useful spectral content in the data. The constraints in data resolution then propagate to derived signals through processing operations, such as convolutions according to physical laws, as exemplified by the force fields (Chapter 6).

4.2.1 Fourier Transforms of Infinite Sequences

One approach to the Fourier transform of an infinite sequence is to use the special case of the Fourier transform for continuous functions applied to the Dirac delta function. Suppose one formally represents the infinite sequence of samples as a "function" in the continuous variable x by multiplying the parent function with the *sampling function*, (2.129):

$$g_s(x) = \sum_{k=-\infty}^{\infty} g(x)\delta(x - k\Delta x)\Delta x. \tag{4.3}$$

The factor Δx is included so that the units of $g_s(x)$ match those of $g(x)$ (since the delta function has units inverse to those of its argument). For any particular x that is not an integer multiple of Δx (i.e., $x \neq \ell\Delta x$, for any ℓ), $g_s(x) = 0$; while $g_s(\ell\Delta x)$ is a single impulse with "amplitude" $g_\ell\Delta x$. Thus, $g_s(x)$ is a train of impulses with amplitudes proportional to the sample values. As before, when dealing with the delta function, one operates on it formally, recognizing that it is not a function in the conventional meaning, and that one can always approach the result in the limit using the rectangle function with vanishing base and unit rectangular area.

If $G(f)$ is the Fourier transform of $g(x)$, then since

$$g_s(x) = g(x)\left(\sum_{k=-\infty}^{\infty} \delta(x - k\Delta x)\Delta x\right), \tag{4.4}$$

the dual convolution theorem (3.20) yields

$$\mathcal{F}(g_s(x)) = \mathcal{F}\,(g(x)) * \mathcal{F}\left(\sum_{k=-\infty}^{\infty} \delta(x - k\Delta x)\right)\Delta x. \tag{4.5}$$

It was shown with equation (2.133) that the Fourier transform of the sampling function, again, is a train of impulses,

$$\mathcal{F}\left(\sum_{k=-\infty}^{\infty} \delta(x - k\Delta x)\right) = \frac{1}{\Delta x}\sum_{k=-\infty}^{\infty} \delta\left(f - \frac{k}{\Delta x}\right). \tag{4.6}$$

Therefore, equation (4.5) becomes with the generalized convolution definition, equation (3.4),

$$\begin{aligned} \mathcal{F}(g_s(x)) &= G(f) * \sum_{k=-\infty}^{\infty} \delta\left(f - \frac{k}{\Delta x}\right) = \sum_{k=-\infty}^{\infty} G(f) * \delta\left(f - \frac{k}{\Delta x}\right) \\ &= \sum_{k=-\infty}^{\infty} \int_{-\infty}^{\infty} G(f')\delta\left(f - f' - \frac{k}{\Delta x}\right) df' \end{aligned} \tag{4.7}$$

The reproducing property of the Dirac delta function, equation (2.115) then leads to a fundamental result,

$$\tilde{G}_s(f) \equiv \mathcal{F}\,(g_s(x)) = \sum_{k=-\infty}^{\infty} G\left(f - \frac{k}{\Delta x}\right) = \sum_{k=-\infty}^{\infty} G\left(f + \frac{k}{\Delta x}\right), \tag{4.8}$$

that relates the spectrum of the sequence to the spectrum of the parent function. The last equation follows with a change in summation index, $k \to -k$.

The "~" notation is applied to the Fourier transform of the sampled signal, because $\tilde{G}_s(f)$ is *periodic* with period in frequency equal to $1/\Delta x$,

$$\tilde{G}_s\left(f+\frac{1}{\Delta x}\right)=\sum_{k=-\infty}^{\infty}G\left(f+\frac{k+1}{\Delta x}\right)=\tilde{G}_s(f). \tag{4.9}$$

Thus, the *sample spectrum*, $\tilde{G}_s(f)$, may be defined solely on the interval of frequencies (the *principal part*),

$$-\frac{1}{2\Delta x}\leq f<\frac{1}{2\Delta x}, \tag{4.10}$$

or, any other interval of length $1/\Delta x$. Periodicity implies that $\tilde{G}_s(1/(2\Delta x))=\tilde{G}_s(-1/(2\Delta x))$. The largest absolute frequency in this interval is

$$f_N=\frac{1}{2\Delta x}. \tag{4.11}$$

It is called the *Nyquist frequency* (or, folding frequency; Section 4.2.3) and is the inverse of *twice* the sampling interval.

Since $\tilde{G}_s(f)$ is a periodic function, with period, $1/\Delta x$, it can be represented as a Fourier series, equation (2.8),

$$\tilde{G}_s(f)=\Delta x\sum_{\ell=-\infty}^{\infty}c_\ell e^{-i2\pi\Delta x\ell f}. \tag{4.12}$$

There is no loss in generality in this case by choosing negative instead of positive exponents. It is noted that $\tilde{G}_s(f)$ generally is a complex function because $G(f)$ is complex. The Fourier coefficients in equation (4.12) are given by equation (2.13), with appropriate sign change in the exponential,

$$c_\ell=\int_0^{1/\Delta x}\tilde{G}_s(f)e^{i2\pi\Delta x\ell f}df=\int_{-f_N}^{f_N}\tilde{G}_s(f)e^{i2\pi\Delta x\ell f}df. \tag{4.13}$$

With the representation of $\tilde{G}_s(f)$ according to equation (4.8), one has

$$c_\ell=\int_0^{1/\Delta x}\sum_{k=-\infty}^{\infty}G\left(f+\frac{k}{\Delta x}\right)e^{i2\pi\Delta x\ell f}df=\sum_{k=-\infty}^{\infty}\int_{k/\Delta x}^{(k+1)/\Delta x}G(f')e^{i2\pi\Delta x\ell f'}df', \tag{4.14}$$

where the variable of integration is changed to $f'=f+k/\Delta x$. The sum of integrals is thus the infinite integral, and

$$c_\ell=\int_{-\infty}^{\infty}G(f')e^{i2\pi\Delta x\ell f'}df'=g(\ell\Delta x)=g_\ell. \tag{4.15}$$

That is, the "Fourier series coefficients" of the sample spectrum, $\tilde{G}(f)$, are the samples, themselves. Therefore, the *Fourier transform pair* for infinitely extended sequences is

$$\tilde{G}(f) = \Delta x \sum_{\ell=-\infty}^{\infty} g_\ell e^{-i2\pi\Delta x \ell f}, \tag{4.16}$$

$$g_\ell = \int_{-f_N}^{f_N} \tilde{G}(f) e^{i2\pi\Delta x \ell f}\, df, \tag{4.17}$$

where the subscript "*s*" is dropped because this transform pair holds in general, even if no specific reference is made to an underlying continuous function.

The delta function is used only a stepping stone to the final result that excludes this rather artificial function. On the other hand, it is instrumental in giving the fundamental relationship, equation (4.8), between the spectrum of the sequence and the Fourier transform of the parent function. Note that the definition of the Fourier transform, $\tilde{G}(f)$, agrees in terms of units with previous definitions. In the case that the sequence, g_ℓ, has no underlying parent function, a constant sampling interval is nevertheless assumed and even if it is implicit with $\Delta x = 1$, its units determine the units of frequency.

Because of the duality between the Fourier series transform pair, equations (2.18), and the transform pair for sequences, equations (4.16), (4.17), where the only difference is in the (arbitrary convention of the) sign in the exponential, all the properties of one hold for the other with corresponding domains reversed. Therefore, the properties (2.20)–(2.24) of the periodic function and its Fourier transform translate directly into the periodic spectrum and the sequence of samples. For example, the symmetry property is

$$\tilde{G}(-f) \leftrightarrow g_{-\ell}. \tag{4.18}$$

Accounting for the change in sign of the exponential, the translation properties are

$$g_\ell e^{-i2\pi\Delta x \ell f_0} \leftrightarrow \tilde{G}(f + f_0), \quad g_{\ell+\ell_0} \leftrightarrow \tilde{G}(f) e^{i2\pi\Delta x \ell_0 f}. \tag{4.19}$$

Also, the symmetry property analogous to the equivalence (2.29) is

$$g_\ell = g_\ell^* \text{ and } g_{-\ell} = g_\ell \quad \text{if and only if} \quad \tilde{G}(f) = \tilde{G}^*(f) \text{ and } \tilde{G}(-f) = \tilde{G}(f); \tag{4.20}$$

and, Parseval's theorem in this case is

$$\int_{-f_N}^{f_N} \left|\tilde{G}(f)\right|^2 df = \Delta x \sum_{\ell=-\infty}^{\infty} |g_\ell|^2. \tag{4.21}$$

The discrete version of the Dirac delta function is the Kronecker delta, equation (1.14). In order to be consistent with previous definitions in terms of units, we define a related sequence by

$$\delta_\ell^{(\Delta x)} = \begin{cases} 1/\Delta x, & \ell = 0 \\ 0, & \ell \neq 0 \end{cases} \tag{4.22}$$

From equation (4.16), its Fourier transform is then simply

$$\tilde{\Delta}(f) = 1. \tag{4.23}$$

By the duality with the Fourier series transform, one then also has a definition of the periodic delta function, $\tilde{\delta}(x)$, that is zero everywhere except at $\mathrm{mod}_P(x)$, such that

$$\int_0^P \tilde{g}(x')\tilde{\delta}(x'-x)\,dx' = \tilde{g}(x), \tag{4.24}$$

with spectrum, $\Delta_k = 1$, for all k. This also means that, using the first of equations (2.18),

$$\sum_{k=-\infty}^{\infty} e^{i\frac{2\pi}{P}kx} = P\tilde{\delta}(x). \tag{4.25}$$

The *rectangle sequence*, or discrete rectangle function is defined here (with n even) as

$$b_\ell^{(n)} = \begin{cases} 1, & -\dfrac{n}{2} \leq \ell \leq \dfrac{n}{2} - 1 \\ 0, & \text{otherwise} \end{cases} \tag{4.26}$$

where the base, $T = n\Delta x$, is associated with n intervals of length Δx. In order to find the Fourier transform of $b_\ell^{(n)}$, it is easiest first to derive an important result (Exercise 4.1),

$$D_n(x) = \sum_{k=-n}^{n} e^{ikx} = \frac{\sin\left(x(n+1/2)\right)}{\sin(x/2)}. \tag{4.27}$$

The function, $D_n(x)$, is known as the *Dirichlet kernel*, which plays a key role in the study of the convergence properties of the Fourier series transform (e.g., Davis 1975, Ch.12). The Dirichlet kernel is similar to the sinc function for small x, but it is periodic with period, 4π. The Fourier transform of $b_\ell^{(n)}$, according to equation (4.16), is

$$\tilde{B}^{(n)}(f) = \Delta x \sum_{\ell=-n/2}^{n/2-1} e^{-i2\pi\Delta x \ell f}. \tag{4.28}$$

Thus, it is related to the Dirichlet kernel as a function of f, but simplifies to (Exercise 4.2)

$$\tilde{B}^{(n)}(f) = \Delta x \frac{\sin(n\pi\Delta x f)}{\sin(\pi\Delta x f)} e^{i\pi\Delta x f}, \quad \tilde{B}^{(n)}(0) = n\Delta x. \tag{4.29}$$

The value at $f = 0$ is obtained by expanding numerator and denominator in series, equation (1.8), dividing each by f, and taking the limit as $f \to 0$. Equation (4.28) shows that $\tilde{B}^{(n)}(f)$ approaches the periodic delta function, $\tilde{\delta}(f)$, equation (4.25) (with period, $1/\Delta x$), as $n \to \infty$, demonstrating also with equation (4.29) that the periodic delta function is infinite at the origin.

Unlike the Fourier transform for the continuous rectangle function, the second of equations (2.100), $\tilde{B}^{(n)}(f)$ is complex, which results from the asymmetry in the definition of the rectangle sequence. In fact, for a symmetric rectangle sequence, the Fourier transform is (proportional to) the Dirichlet kernel. However, the definition (4.26) is preferred here since it matches the conventional form of the discrete Fourier transform, equation (4.74), thus facilitating applications with discrete data (such as windowing, Section 4.6).

The Gaussian sequence, corresponding to the Gaussian function, equation (2.107), is

$$\gamma_\ell^{(\beta,\Delta x)} = \frac{1}{\beta} e^{-\pi(\ell\Delta x/\beta)^2}, \quad \beta > 0. \tag{4.30}$$

Its Fourier transform follows directly from equation (4.16),

$$\begin{aligned} \tilde{\Gamma}^{(\beta,\Delta x)}(f) &= \frac{\Delta x}{\beta} \sum_{\ell=-\infty}^{\infty} e^{-\pi(\ell\Delta x/\beta)^2} e^{-i2\pi\Delta x \ell f} = \frac{\Delta x}{\beta}\left(1 + 2\sum_{\ell=1}^{\infty} e^{-\pi(\ell\Delta x/\beta)^2} \cos(2\pi\Delta x \ell f)\right) \\ &\simeq \frac{1}{\beta} \int_{-\infty}^{\infty} e^{-\pi(x/\beta)^2} e^{-i2\pi x f}\, dx = e^{-\pi(\beta f)^2} \end{aligned} \tag{4.31}$$

where the final equality comes from the Fourier transform of the scaled Gaussian function, equation (2.108). The integral approximation is valid for small Δx, or since $1/\beta$ scales the spatial domain, it is valid for $\beta \gg \Delta x$. In fact, for any practical utility, such as a window function, the "window parameter", β, must be significantly larger than the sampling interval, Δx, otherwise the spectrum is significantly biased at zero frequency (Figure 4.1).

A brief note on extensions to higher Cartesian dimensions closes this section. Let g_{ℓ_1,ℓ_2} be a two-dimensional, bounded, absolutely summable array of values, with constant, but not necessarily equal sample intervals, Δx_1 and Δx_2. The two-dimensional Fourier transform and its inverse are

$$\tilde{G}(f_1, f_2) = \Delta x_1 \Delta x_2 \sum_{\ell_1=-\infty}^{\infty} \sum_{\ell_2=-\infty}^{\infty} g_{\ell_1,\ell_2} e^{-i2\pi(\Delta x_1 \ell_1 f_1 + \Delta x_2 \ell_2 f_2)}, \tag{4.32}$$

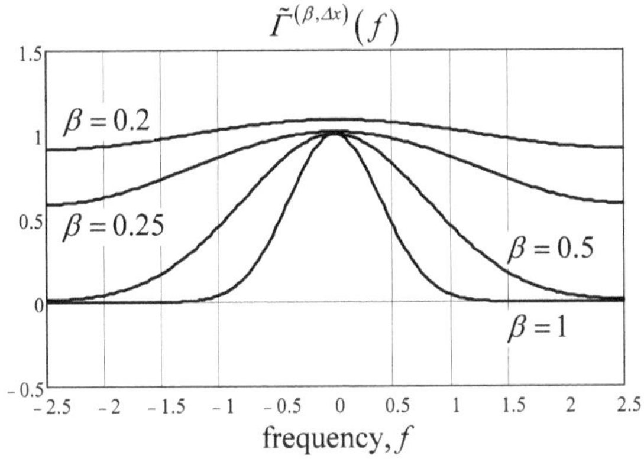

Figure 4.1: Fourier transforms of Gaussian sequences obtained with sampling interval, $\Delta x = 0.2$ (hence, $f_{\mathcal{N}} = 2.5$ and the period of $\tilde{\Gamma}^{(\beta,\Delta x)}(f)$ is 5), for various values of the parameter, β. For practical utility, one should choose $\beta \gg \Delta x$.

$$g_{\ell_1,\ell_2} = \int_{-f_{1\mathcal{N}}}^{f_{1\mathcal{N}}} \int_{-f_{2\mathcal{N}}}^{f_{2\mathcal{N}}} \tilde{G}(f_1, f_2) e^{i2\pi(\Delta x_1 \ell_1 f_1 + \Delta x_2 \ell_2 f_2)} df_1 df_2. \tag{4.33}$$

The spectrum, $\tilde{G}(f_1, f_2)$, is a periodic function with periods, $1/\Delta x_1$ and $1/\Delta x_2$, in f_1 and f_2, respectively. Corresponding Nyquist frequencies are $f_{1\mathcal{N}} = 1/(2\Delta x_1)$ and $f_{2\mathcal{N}} = 1/(2\Delta x_2)$; and the relationship between the spectra of the samples and the parent function is given, analogous to equation (4.8), by

$$\tilde{G}_s(f_1, f_2) = \sum_{k_1=-\infty}^{\infty} \sum_{k_2=-\infty}^{\infty} G\left(f_1 - \frac{k_1}{\Delta x_1}, f_2 - \frac{k_2}{\Delta x_2}\right). \tag{4.34}$$

Corresponding properties follow immediately from the duality to the two-dimensional Fourier series transform pair; see Section 2.4.1.

4.2.2 Discrete Convolutions

The concept of convolution carries over directly to the discrete domain from the continuous domain with an appropriate definition. The discrete convolution of two infinite sequences, each absolutely summable and each having the same sampling interval, Δx, is given by

$$g_\ell \# h_\ell = \Delta x \sum_{n=-\infty}^{\infty} g_n h_{\ell-n}. \tag{4.35}$$

The included antecedent factor, Δx, makes this and subsequent formulations consistent with the continuous case in terms of units; see equation (3.1). Moreover,

Δx is the reciprocal of the period of the spectra of g_ℓ and h_ℓ, and thus equations (4.35) and (3.39) are consistent.

Again, there is a convolution theorem,

$$g_\ell \# h_\ell \leftrightarrow \tilde{G}(f)\tilde{H}(f), \tag{4.36}$$

which is proved easily and analogously as in equations (3.40) and (3.41), by substituting equation (4.35) into equation (4.12) for the spectrum of the convolution (Exercise 4.3). The result is clearly similar to the Fourier transform (3.38) in the dual cyclic convolution theorem for periodic functions. On the other hand, the dual convolution theorem for sequences is analogous to the convolution theorem for periodic functions, equations (3.37) (Exercise 4.4),

$$g_\ell h_\ell \leftrightarrow \tilde{G}(f)\tilde{*}\tilde{H}(f), \tag{4.37}$$

where the cyclic convolution, $\tilde{*}$, is defined by equation (3.33). The spectra of the discrete convolution and of the discrete product both are periodic with period, $1/\Delta x$.

The discrete convolution, $g_\ell \# h_\ell$, may also be viewed in terms of the multiplication of a matrix and a vector, in this case, both having infinite dimensions,

$$\begin{pmatrix} \vdots \\ c_{\ell-1} \\ c_\ell \\ c_{\ell+1} \\ \vdots \end{pmatrix} = \Delta x \begin{pmatrix} \ddots & \ddots & \ddots & \ddots & \iddots \\ \ddots & h_{\ell-n} & h_{\ell-n-1} & h_{\ell-n-2} & \ddots \\ \ddots & h_{\ell-n+1} & h_{\ell-n} & h_{\ell-n-1} & \ddots \\ \ddots & h_{\ell-n+2} & h_{\ell-n+1} & h_{\ell-n} & \ddots \\ \iddots & \ddots & \ddots & \ddots & \ddots \end{pmatrix} \begin{pmatrix} \vdots \\ g_{n-1} \\ g_n \\ g_{n+1} \\ \vdots \end{pmatrix}. \tag{4.38}$$

Each left-to-right descending diagonal of the matrix has the same value, and the matrix is called a *Toeplitz* matrix. In essence this matrix is fully defined by a single row or column. The convolution theorem (4.36) then also immediately provides an easy way to invert this linear system of equations, provided the Fourier transform of the sequence, h_ℓ, is not zero at any frequency. That is, given c_ℓ and h_ℓ, the sequence, g_ℓ, is

$$g_\ell = \mathcal{F}^{-1}\left(\frac{\tilde{C}(f)}{\tilde{H}(f)}\right)_\ell. \tag{4.39}$$

This holds only for infinite sequences, but a similar more practical result may be derived for special finite sequences (Section 4.3.4).

The practical evaluation of a geophysical model that is a convolution of continuous functions usually involves a discretization using some form of numerical integration algorithm. However, not all such algorithms preserve the model as a convolution, which may affect the efficiency of the calculations if they rely on Fourier transform techniques (Sections 4.3.4 and 4.4). For example, applying the rectangle rule for numerical integration to the convolution,

$$(g_T * h)(x) = \int_{-T/2}^{T/2} g_T(x')h(x-x')\,dx', \tag{4.40}$$

where $g_T(x)$ is space-limited, as in equation (3.132), simply yields a discrete convolution, as in equation (4.35),

$$(g_T * h)_\ell \approx \Delta x \sum_{n=-N/2}^{N/2-1} (g_T)_n h_{\ell-n}, \quad -\frac{N}{2} \le \ell \le \frac{N}{2}-1, \tag{4.41}$$

where $(g_T)_n = g_T(n\Delta x)$. Using the trapezoid rule also yields a discrete convolution, but with additional terms,

$$\begin{aligned}(g_T * h)_\ell &\approx \frac{\Delta x}{2} \sum_{n=-N/2}^{N/2-2} \left((g_T)_n h_{\ell-n} + (g_T)_{n+1} h_{\ell-n-1}\right) \\ &= \Delta x \sum_{n=-N/2}^{N/2-1} (g_T)_n h_{\ell-n} - \frac{\Delta x}{2}\left((g_T)_{N/2-1} h_{\ell-N/2+1} + (g_T)_{-N/2} h_{\ell+N/2}\right)\end{aligned} \tag{4.42}$$

Both discrete convolutions in equations (4.41) and (4.42) can be evaluated efficiently as discussed in Section 4.3.4. However, the discretization of the integral convolution according to the well known Simpson's rule presents a greater challenge in preserving a convolution (in this case two convolutions) and is left to the reader.

In two Cartesian dimensions, the discrete convolution is

$$g_{\ell_1,\ell_2} \# h_{\ell_1,\ell_2} = \Delta x_1 \Delta x_2 \sum_{n_1=-\infty}^{\infty} \sum_{n_2=-\infty}^{\infty} g_{n_1,n_2} h_{\ell_1-n_1,\ell_2-n_2}, \tag{4.43}$$

with corresponding convolution theorems,

$$g_{\ell_1,\ell_2} \# h_{\ell_1,\ell_2} \leftrightarrow \tilde{G}(f_1,f_2)\tilde{H}(f_1,f_2), \tag{4.44}$$

$$g_{\ell_1,\ell_2} h_{\ell_1,\ell_2} \leftrightarrow \tilde{G}(f_1,f_2) \tilde{*} \tilde{H}(f_1,f_2). \tag{4.45}$$

Both infinite arrays, g_{ℓ_1,ℓ_2} and h_{ℓ_1,ℓ_2}, must have the same respective sampling intervals, Δx_1 and Δx_2; and, the spectrum of the convolution is periodic with periods equal to $1/\Delta x_1$ and $1/\Delta x_2$ in the two coordinate dimensions.

4.2.3 Aliasing of the Fourier Spectrum

Suppose that a function, $g(x)$, is band-limited (see also equation 2.62),

$$G(f) = 0, \quad \text{for} \quad f \ge f_0 \quad \text{or} \quad f < -f_0; \tag{4.46}$$

and, further suppose that it is sampled at an interval, $\Delta x = 1/(2f_0)$, so that the Nyquist frequency for the sequence, g_ℓ, is $f_N = f_0$. On the principal domain, $-f_N \le f < f_N$, its Fourier spectrum is given by equation (4.8),

$$\tilde{G}_s(f) = \sum_{k=-\infty}^{\infty} G\left(f + \frac{k}{\Delta x}\right) = \cdots + G\left(f - \frac{1}{\Delta x}\right) + G(f) + G\left(f + \frac{1}{\Delta x}\right) + \cdots. \tag{4.47}$$

Since the arguments, $f' = f \pm k/\Delta x$, $k = 1, 2, \ldots$, for $-1/(2\Delta x) \le f < 1/(2\Delta x)$ imply

$$\ldots, -\frac{5}{2\Delta x} \le f' < -\frac{3}{2\Delta x}, \quad \frac{-3}{2\Delta x} \le f' < \frac{-1}{2\Delta x}, \quad \frac{1}{2\Delta x} \le f' < \frac{3}{2\Delta x}, \quad \frac{3}{2\Delta x} \le f' < \frac{5}{2\Delta x}, \ldots, \tag{4.48}$$

it is seen that all but the $k = 0$ term of the series in equation (4.47) vanish in view of equation (4.46); and, therefore, in this case for $-f_N \le f < f_N$,

$$\tilde{G}_s(f) = G(f). \tag{4.49}$$

Here, the spectrum of the sample is denoted with the subscript "*s*" to distinguish it from the more general periodic spectrum, equation (4.16), where no specific reference to a continuous parent function is needed. Thus, according to equation (4.49), if the continuous function is band-limited to frequencies not greater than the Nyquist frequency as determined by the sampling interval, then the spectrum of the sampled sequence is equal to the spectrum of the continuous function, over the domain of the limited band of frequencies.

If $G(f) \ne 0$ for $f \ge f_N$ or $f < -f_N$ then the function, $g(x)$, has spectral content beyond the Nyquist limit; and, the spectrum, $G_s(f)$, does not equal $G(f)$ for $-f_N \le f < f_N$ due to the overlap (or, *folding*) of the parts $G(f - k/\Delta x)$, $G(f + k/\Delta x)$, etc., onto $G(f)$, as indicated in equation (4.47). This effect is known as *aliasing*, and the Nyquist frequency is also known as the *folding frequency*. That is, the spectrum of the sample from a parent function with spectral content beyond the Nyquist frequency is a corrupted version of the parent spectrum. Attempting to determine the spectrum of the parent function from the sequence of sampled data is subject to the *aliasing error*, given by

$$\tilde{G}_s(f) - G(f) = \sum_{\substack{k=-\infty \\ k \ne 0}}^{\infty} G\left(f + \frac{k}{\Delta x}\right), \quad -f_N \le f < f_N. \tag{4.50}$$

The generalization of aliasing to higher dimensions follows the usual procedure. The aliasing error in the spectrum derived from the sample of a function on the two-dimensional Cartesian domain is

$$\tilde{G}_s(f_1, f_2) - G(f) = \sum_{\substack{k_1=-\infty \\ k_1 \ne 0}}^{\infty} \sum_{\substack{k_2=-\infty \\ k_2 \ne 0}}^{\infty} G\left(f_1 + \frac{k_1}{\Delta x_1}, f_2 + \frac{k_2}{\Delta x_2}\right), \ -f_{1N} \le f_1 < f_{1N}, \ -f_{2N} \le f_2 < f_{2N} \tag{4.51}$$

Representing a continuous geophysical signal by a discrete sample introduces a *discretization error* in the space domain. That is, the representation is not complete and this misrepresentation is best understood and formulated in the frequency domain in terms of the aliasing error. This error stands in contrast to the truncation error, or windowing effect (Section 3.6), in a function that is misrepresented due to the limited spatial domain. Aliasing can affect any part of the spectrum depending on its amplitudes, most dramatically illustrated with geophysical signals that are restricted to isolated frequencies such as ocean tidal constituents (see example, below). For signals that have a continuous and attenuating spectrum, such as the Earth's gravitational field, aliasing affects primarily the higher frequencies near the Nyquist frequency as shown in the following contrived example.

Figure 4.2 shows the spectrum of a band-limited function, with $f_0 = 3$, as well as the spectrum of its sample, where the sampling interval is $\Delta x = 1/6$. Thus, the Nyquist frequency is $f_{\mathcal{N}} = 3$ and is not exceeded by the cutoff frequency of the parent function. Hence, there is no aliasing—the spectrum of the sample equals the spectrum of the parent function in the band of frequencies bounded by the Nyquist frequency. Note that the period of $\tilde{G}_s(f)$ in this case is 6. In Figure 4.3, by comparison, the sampling interval for that same continuous band-limited function is $\Delta x = 1/3$; and, consequently, the Nyquist frequency is $f_{\mathcal{N}} = 3/2 < f_0$. The period of $\tilde{G}_s(f)$ is 3, and the enveloping line in Figure 4.3 is the spectrum of the sequence of samples. It deviates at the high frequencies near f_0 from the parent spectrum due to aliasing.

Another way to view the aliasing error is to note that the samples cannot distinguish between sinusoidal components of the parent function at different frequencies, f_1, f_2, related by $f_1 = 2kf_{\mathcal{N}} \pm f_2$, where k is any integer, because the sinusoids at these frequencies pass through the same sample points of $g(x)$; that is, $e^{i2\pi f_1 x_\ell} = e^{i2\pi(2kf_{\mathcal{N}} \pm f_2)\ell\Delta x} = e^{i(2\pi k\ell \pm 2\pi f_2 x_\ell)} = e^{i2\pi f_2 x_\ell}$. One can think of the spectrum of the sampled sequence as having to accommodate the full spectrum (or, information) of the parent function values in some fashion, and it is accomplished by folding higher frequency components back onto the principal part.

The spectrum, $\tilde{G}_s(f)$, is called the *aliased spectrum* of g. Since by the Riemann-Lebesgue theorem, equation (2.77), the high-frequency spectrum of a function

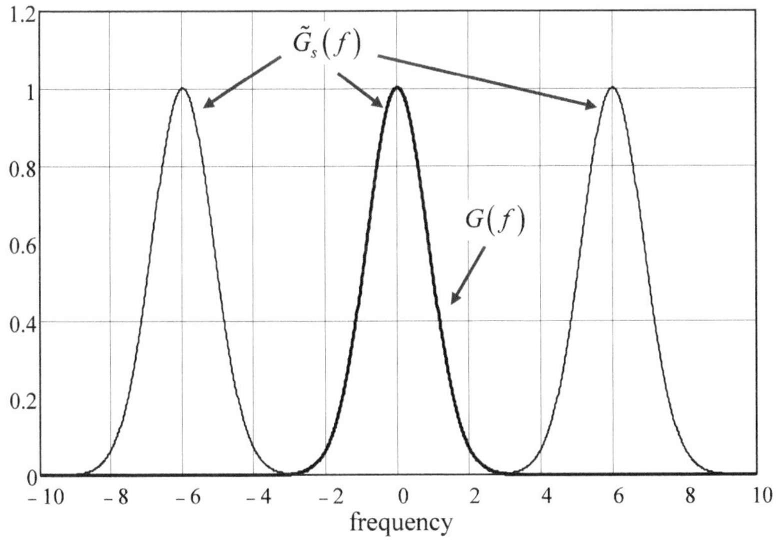

Figure 4.2: Spectra of a band-limited function (thick line) and of a sequence of its samples (thin line). The Nyquist frequency is $f_{\mathcal{N}} = 3$.

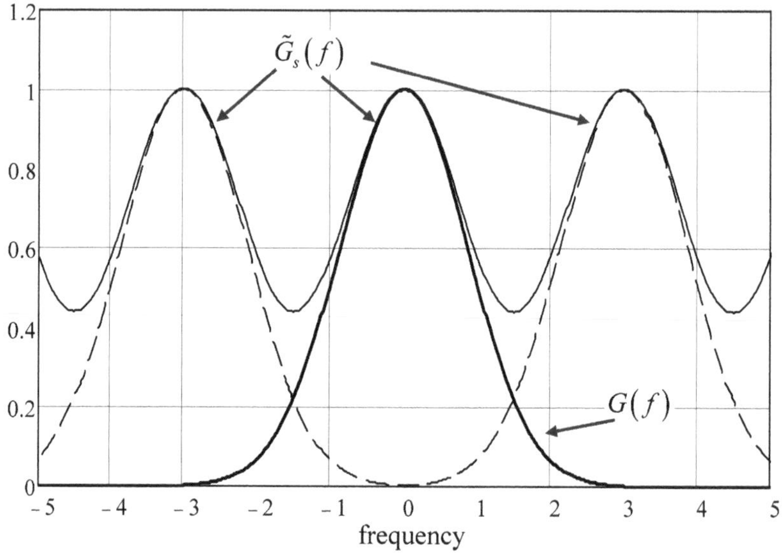

Figure 4.3: Spectrum of a band-limited function (thick line) and the aliased spectrum of a sequence of its samples (thin solid line). The Nyquist frequency is $f_N = 3/2$. The dashed lines are spectra of the band-limited function displaced by twice the Nyquist frequency, and which are terms in the sum of equation (4.47).

attenuates to zero, the estimation, or computation, of the (assumed continuous) spectrum of g based on a sampled sequence is in error primarily near and *below* the Nyquist frequency, where the spectral components are combined with components at frequencies near and *above* the Nyquist frequency, as illustrated in Figures 4.2 and 4.3 (however, see also the special case of a single sinusoid, below). Two options thus exist to reduce the aliasing error. Either, the sampled function should be *filtered* (low-pass filter, *anti-aliasing* filter) to remove the components with frequencies near and above the Nyquist frequency; or, the sampling interval should be decreased, thus increasing the Nyquist frequency to encompass a wider spectral band of the parent function.

As a special case, consider a continuous function that is a pure sinusoid (energy at only one frequency), as might be the case for an ocean tide component. The largest tide (in amplitude) observed along ocean coasts is due to the rotation of the Earth relative to the Moon, also called the "principal lunar semi-diurnal tide", or M2 tide. Its amplitude in deep water (for an *equilibrium* tide that would be subject primarily to external forcing of an elastic Earth) is about 16.8 cm with a period of 12.42 hr (Lambeck 1988, p.133); the largest solar tide, S2, has an amplitude that is about 47% that of the lunar tide with a period of 12.00 hr. In this example the independent variable, x, is time. Thus, for the M2 tide let

$$\tilde{g}(x) = \cos\left(\frac{2\pi}{P}x\right), \tag{4.52}$$

with period, P, which for the Moon is the time between consecutive zeniths of the moon over any location on the Earth. The Fourier spectrum consists of two impulses at $f_g = \pm 1/P$; specifically, equation (2.10) shows that $G_{\pm 1} = P/2$ and equation (2.127) then gives

$$G(f) = \frac{1}{2}\delta\left(f - \frac{1}{P}\right) + \frac{1}{2}\delta\left(f + \frac{1}{P}\right). \tag{4.53}$$

Clearly, this signal is band-limited, with maximum frequency, $f_0 = 1/P$.

Now, suppose one samples this signal at an interval, $\Delta x = P/c$, where c is a positive number. First, consider the case, $c \geq 2$. Then the Nyquist frequency, $f_N = 1/(2\Delta x) = c/(2P) \geq 1/P = f_0$, is greater than (or equal to) the maximum frequency of the signal; and, there is no aliasing error. But, suppose that $0 < c < 2$; that is, $\Delta x > P/2$. For example, consider an Earth-orbiting satellite system that measures the height of sea level with radar altimetry along its orbital ground track. The lunar tide signal repeats with approximately semi-diurnal period, while the time, Δx, between repeat orbits that pass over the same location on the Earth can be days or weeks, or longer, depending on the resonance of the orbital period relative to the period of Earth rotation.

The sample spectrum in this example is given by equation (4.47), which, in view of equation (4.53), is

$$\tilde{G}_s(f) = \frac{1}{2}\begin{pmatrix} \ldots + \delta\left(f - \frac{1}{P} - \frac{1}{\Delta x}\right) + \delta\left(f + \frac{1}{P} - \frac{1}{\Delta x}\right) + \\ \delta\left(f - \frac{1}{P}\right) + \delta\left(f + \frac{1}{P}\right) + \\ \delta\left(f - \frac{1}{P} + \frac{1}{\Delta x}\right) + \delta\left(f + \frac{1}{P} + \frac{1}{\Delta x}\right) + \ldots \end{pmatrix}, \quad -\frac{1}{2\Delta x} \leq f \leq \frac{1}{2\Delta x}. \tag{4.54}$$

The frequencies, f, in the specified band of the sample that contribute to its spectrum are those that give zero arguments for the delta functions in equation (4.54). The possibilities are identified by the integers, $n \geq 0$, for which (with $\Delta x = P/c$)

$$\text{either} \quad f - \left(\frac{1 \pm nc}{P}\right) = 0, \quad \text{or} \quad f + \left(\frac{1 \pm nc}{P}\right) = 0. \tag{4.55}$$

If the first equality holds for f, then the second holds for $-f$. Hence, we may restrict $f \geq 0$ and then include $-f$ as an additional frequency that contributes to the spectrum. Substituting the first of equations (4.55) into the spectral band of the sample, $0 \leq f \leq c/(2P)$, with $n \geq 0$ and $0 < c < 2$, yields

$$-\frac{1}{c} \leq \pm n \leq \frac{1}{2} - \frac{1}{c} \quad \Rightarrow \quad 0 < \frac{1}{c} - \frac{1}{2} \leq n \leq \frac{1}{c}, \tag{4.56}$$

since only $-n$ is allowed in the first inequality. Clearly, $n = 0$ is excluded and there is no contribution from the principal part in equation (4.54). The entire spectrum of the sample is an aliasing error, which has non-zero components at frequencies,

$$f_a = \pm\left(\frac{1}{P} - \frac{n}{\Delta x}\right), \tag{4.57}$$

where n is determined from the inequality (4.56), and $0 \le |f_a| \le 1/(2\Delta x)$ must also hold. The frequency, f_a, is called an *aliasing frequency*; and, the reciprocal is called an *aliasing period*, P_a. The satellite-derived signal in this case has frequency, $|f_a| \neq 1/P$. It can be shown (Exercise 4.5) that a general formula for determining the aliasing frequency when sampling a solitary sinusoid with frequency, f, is given by

$$|f_a| = |\mathrm{mod}_{2f_N}(|f| + f_N) - f_N|. \tag{4.58}$$

If $f_N \ge |f|$, then this formula gives $|f_a| = |f|$, meaning that there is no aliasing.

Table 4.1 summarizes an example in which the parent function, the M2 tide constituent, is a sinusoid with frequency, $f = 1.932$ cy/day, which is sampled at an interval such that $f_N < |f|$, hence incurring an aliasing error. The *only* non-zero component of the sample spectrum occurs at a frequency much lower than the frequency of the parent function. In the examples, the aliasing periods of the samples are $P_a = 31.0$ days and $P_a = 14.8$ days, respectively, despite the semi-diurnal period of the parent function, which, therefore, is not detected by the sample (see also (Parke et al. 1987) for additional details on aliasing related to the choice of altimeter satellite orbits).

If the function is band-limited, it can be recovered fully from the samples, provided the sampling is done at intervals corresponding to twice the highest frequency contained in the function. Indeed, suppose that $G(f) = 0$ for $|f| > f_0$; then one can write on the basis of equations (4.46) and (4.47),

$$G(f) = \tilde{G}_s(f)\,b(\Delta x f), \tag{4.59}$$

where $\Delta x = 1/(2f_0)$ and $b(\Delta x f)$ is the rectangle function, equation (2.90), now a function in the frequency domain. That is, the spectrum of the sample is truncated to just one period, and this truncated version then agrees with the total spectrum of the continuous band-limited function (Figure 4.2). Taking the inverse Fourier transform on both sides of equation (4.59) yields, with the convolution theorem (3.17),

Table 4.1: Examples of aliasing frequencies.

parent function frequency	sampling interval	Nyquist frequency	aliasing frequency
$f = 1.932$ cy/day	$\Delta x = 10$ days	$f_N = 0.05$ cy/day	$f_a = \pm 0.032$ cy/day
$f = 1.932$ cy/day	$\Delta x = 5$ days	$f_N = 0.1$ cy/day	$f_a = \pm 0.068$ cy/day

$$
\begin{aligned}
g(x) &= \mathcal{F}^{-1}(\tilde{G}_s(f)b(\Delta x f)) \\
&= \mathcal{F}^{-1}(\tilde{G}_s(f)) * \mathcal{F}^{-1}(b(\Delta x f)) \qquad (4.60) \\
&= g_s(x) * \mathcal{F}^{-1}(b(\Delta x f))
\end{aligned}
$$

where $g_s(x)$ is given by equation (4.3).

With the similarity and duality properties of the Fourier transform, equations (2.78) and (2.77), applied to the rectangle function, one obtains

$$
b(\Delta x \cdot x) \leftrightarrow \frac{1}{\Delta x}\operatorname{sinc}\left(\frac{f}{\Delta x}\right) \Rightarrow \frac{1}{\Delta x}\operatorname{sinc}\left(\frac{x}{\Delta x}\right) \leftrightarrow b(-\Delta x \cdot f) = b(\Delta x \cdot f), \qquad (4.61)
$$

where the last equality follows from the symmetry of the rectangle function with respect to the origin. Therefore, equation (4.60) becomes

$$
\begin{aligned}
g(x) &= \left(\sum_{\ell=-\infty}^{\infty} g(x)\delta(x-\ell\Delta x)\Delta x\right) * \left(\frac{1}{\Delta x}\operatorname{sinc}\left(\frac{x}{\Delta x}\right)\right) \\
&= \sum_{\ell=-\infty}^{\infty} \int_{-\infty}^{\infty} g(x')\delta(x'-\ell\Delta x)\operatorname{sinc}\left(\frac{x-x'}{\Delta x}\right)dx'
\end{aligned}
\qquad (4.62)
$$

in view of the definition of convolution, equation (3.4). Finally, with the reproducing property of the Dirac delta function, equation (2.115),

$$
\begin{aligned}
g(x) &= \sum_{\ell=-\infty}^{\infty} g(\ell\Delta x)\operatorname{sinc}\left(\frac{x-\ell\Delta x}{\Delta x}\right) \\
&= \sum_{\ell=-\infty}^{\infty} g_\ell \operatorname{sinc}\left(\frac{x}{\Delta x}-\ell\right)
\end{aligned}
\qquad (4.63)
$$

This shows that if $g(x)$ is band-limited with no frequency content for $|f| > f_0$, then the entire function can be reconstructed from its sampled values, g_ℓ, if $\Delta x = 1/(2f_0)$. This result is known as the *Whittaker-Shannon sampling theorem*. Note, however, that the entire, infinite sequence of values, g_ℓ, is needed to reconstruct the parent function. Spectral leakage occurs if one uses only a finite (truncated) subset, as discussed in Section 3.6. However, analogous to functions on the continuous domain (section 3.6.2), it is shown in Section 4.6 that an orthogonal basis exists for the index-limited sequences that minimizes spectral leakage and thus offers a very good approximation to a band-limited function.

4.3 Periodic Sequences

In addition to being discrete, data collected using a measurement or observation system usually also have finite extent in time or space. Truncation or windowing of an infinite sequence implies, according to the dual convolution theorem, that the spectrum of the finitely extended sequence is the convolution in the frequency domain of the spectrum of the infinite sequence, $\tilde{G}(f)$, with the spectrum of the discrete rectangle function, equation (4.29), or other discretized window function. This is not considered further here since it is entirely equivalent to the windowing already discussed in Section 3.6.

If, on the other hand, the parent function is periodic and the finite extent of samples covers a single period, then the samples, in fact, represent the full length of the parent function, if not its infinite resolution. Such is the case, in particular, for functions on the sphere, which may be regarded as periodic (Section 2.6). In general and almost without exception data generate a finite sequence (or array) of samples and, in order to make use of the numerical tools associated with the Fourier transform, one must often assume explicitly that they actually constitute one period from a periodic sequence. Indeed, while this seems to contradict the types of spatial geophysical or geodetic signals encountered in practice (i.e., they are patently not periodic), it is a *necessary* assumption when using *discrete* Fourier techniques, such as the fast Fourier transform (FFT, Section 4.3.2). It causes an error in the analysis, but if done properly likely not worse than using a truncated sequence without the numerical benefit of the fast techniques.

Before investigating these assumptions and their effect on operations such as convolution, the Fourier transform is developed for the fourth fundamental type of function that is periodic and defined on a discrete domain, that is, a periodic sequence (the other three fundamental cases are the periodic and non-periodic functions on the continuous domain, and the non-periodic sequence). The periodic sequence, denoted, $\tilde{g}_\ell$, may be represented by a single period, or a finite sequence. Because of this periodicity, the spectrum of the "finite" sequence, in addition to being periodic (due to a discrete domain) is, itself, *discrete*, just like the Fourier series transform of a periodic function.

4.3.1 Discrete Fourier Transform

Let $\tilde{g}_\ell = \tilde{g}(\ell\Delta x)$, $\ell = 0,\ldots, N-1$, be a finite sample from a periodic function that has bounded variation, where the sampling interval, $\Delta x > 0$, divides the period, $P = N\Delta x$; that is, $\tilde{g}_{\ell+mN} = \tilde{g}_\ell$ for any integer, m. Proceeding as for the sample of a non-periodic function, the finite periodic sequence may be represented as a periodic function of impulses (cf. equation (4.3)),

$$\tilde{g}_s(x) = \sum_{\ell=-\infty}^{\infty} \tilde{g}(x)\delta(x - \ell\Delta x)\Delta x, \tag{4.64}$$

where the infinite sum is needed to ensure periodicity on the left side. The Fourier series transform, equation (2.13), is an integral over one period, chosen for convenience to be $[-\Delta x/2, P - \Delta x/2]$,

$$G_k = \int_{-\Delta x/2}^{P-\Delta x/2} \tilde{g}_s(x) e^{-i\frac{2\pi}{P}kx}\, dx. \tag{4.65}$$

Substituting equation (4.64), we derive

$$\begin{aligned}
G_k &= \Delta x \sum_{\ell=-\infty}^{\infty} \int_{-\Delta x/2}^{P-\Delta x/2} \tilde{g}(x) e^{-i\frac{2\pi}{P}kx} \delta(x-\ell\Delta x)\,dx \\
&= \Delta x \sum_{\ell=0}^{N-1} \int_{-\Delta x/2}^{P-\Delta x/2} \tilde{g}(x) e^{-i\frac{2\pi}{P}kx} \delta(x-\ell\Delta x)\,dx \\
&= \Delta x \sum_{\ell=0}^{N-1} \tilde{g}(\ell\Delta x) e^{-i\frac{2\pi}{P}k\ell\Delta x} \\
&= \Delta x \sum_{\ell=0}^{N-1} \tilde{g}_\ell e^{-i\frac{2\pi}{N}k\ell}
\end{aligned} \tag{4.66}$$

The second equality follows from $\delta(x - \ell\Delta x) = 0$ for $-\Delta x/2 \le x \le P - \Delta x/2$, if $\ell \le -1$ or $\ell \ge N$; the third equality is a consequence of the reproducing property of $\delta(x)$, equation (2.115); and $P = N\Delta x$ yields the fourth equality, which shows that this transform is also periodic in k, and one can write $\tilde{G}_k$ for the left side.

The fundamental frequency of $\tilde{g}(x)$ is $1/P = 1/(N\Delta x)$ ($k = 1$); and, the Nyquist frequency corresponding to the sample interval, Δx, is $f_N = 1/(2\Delta x)$. Hence, assuming without loss in generality that N is even, the index, or wave number, k, counts the frequencies, $f_k = k/(N\Delta x)$, $-N/2 \le k \le N/2 - 1$; i.e., the principal part of the spectral domain for the periodic sequence comprises the finite set of discrete frequencies,

$$f_k = -\frac{1}{2\Delta x}, \frac{-N/2+1}{N\Delta x}, \ldots, -\frac{1}{N\Delta x}, 0, \frac{1}{N\Delta x}, \frac{2}{N\Delta x}, \ldots, \frac{N/2-1}{N\Delta x}. \tag{4.67}$$

If N is odd, then $-(N - 1)/2 \le k \le (N - 1)/2$. In this case, the Nyquist frequency, as defined, is not actually a constituent frequency of the principal part of the spectral domain.

The following sum can easily be verified (Exercise 4.6) and demonstrates the *orthogonality* of the discrete functions, $\exp(i2\pi k\ell/N)$, $k = 0,\ldots, N - 1$,

$$\frac{1}{N}\sum_{k=k_0}^{k_0+N-1} e^{i\frac{2\pi}{N}k(\ell-j)} = \begin{cases} 0, & \mathrm{mod}_N(\ell-j) \neq 0 \\ 1, & \mathrm{mod}_N(\ell-j) = 0 \end{cases} \tag{4.68}$$

where k_0 is any integer. Multiplying both sides of the last of equations (4.66) by $\exp(-i2\pi k\ell'/N)$, summing over all wave numbers, k, of the principal spectral domain, and making use of equation (4.68) yields

$$\begin{aligned}
\sum_{k=0}^{N-1} \tilde{G}_k e^{i\frac{2\pi}{N}k\ell'} &= \Delta x \sum_{\ell=0}^{N-1} \tilde{g}_\ell \sum_{k=0}^{N-1} e^{-i\frac{2\pi}{N}k(\ell-\ell')} \\
&= N\Delta x \tilde{g}_{\ell'}
\end{aligned} \tag{4.69}$$

Therefore, the combination of equations (4.66) and (4.69) is the *discrete Fourier transform* (DFT) pair,

$$\mathrm{DFT}\left(\tilde{g}_\ell\right)_k \equiv \tilde{G}_k = \Delta x \sum_{\ell=0}^{N-1} \tilde{g}_\ell e^{-i\frac{2\pi}{N}k\ell}, \quad 0 \le k \le N-1, \tag{4.70}$$

$$\mathrm{DFT}^{-1}\left(\tilde{G}_k\right)_\ell \equiv \tilde{g}_\ell = \frac{1}{N\Delta x} \sum_{k=0}^{N-1} \tilde{G}_k e^{i\frac{2\pi}{N}k\ell}, \quad 0 \le \ell \le N-1. \tag{4.71}$$

Both the sequence and its transform are periodic with respective periods, $N\Delta x$ in spatial distance and $1/\Delta x$ in frequency (or, period, N, in sample index and in wave number). The DFT is often given with no particular reference to Δx by simply setting it unity. Then,

$$\tilde{G}_k = \sum_{\ell=0}^{N-1} \tilde{g}_\ell e^{-i\frac{2\pi}{N}k\ell}, \quad 0 \le k \le N-1, \tag{4.72}$$

$$\tilde{g}_\ell = \frac{1}{N} \sum_{k=0}^{N-1} \tilde{G}_k e^{i\frac{2\pi}{N}k\ell}, \quad 0 \le \ell \le N-1, \tag{4.73}$$

and the user provides the domain scale if needed as in equations (4.70) and (4.71). The given steps that lead to the DFT provide an instructive connection to previous Fourier transforms, but one could as well simply recognize that the DFT pair is a mutually consistent formulation for the sets, $\{\tilde{g}_\ell\}$ and $\{\tilde{G}_k\}$, based on the orthogonality given by equation (4.68). No parent periodic function needs to enter the derivation, and thus there is no restriction on the values of $\tilde{g}_\ell$; they can be any complex numbers. However, in using the DFT it is always assumed that the finite sequence, $\tilde{g}_\ell$, continues for all integer indices with period, N. The DFT automatically implies the inverse DFT, and vice versa, unlike Fourier series or integral transforms, for example, where the inverse Fourier transform does not always converge to the function.

Since both the sequence of samples and its DFT are periodic, one may choose any N consecutive values. For example, an alternative, though less common, formulation is

$$\tilde{G}'_k = \Delta x \sum_{\ell=-N/2}^{N/2-1} \tilde{g}_\ell e^{-i\frac{2\pi}{N}k\ell}, \quad -N/2 \le k \le N/2-1, \tag{4.74}$$

$$\tilde{g}_\ell = \frac{1}{N\Delta x} \sum_{k=-N/2}^{N/2-1} \tilde{G}'_k e^{i\frac{2\pi}{N}k\ell}, \quad -N/2 \le \ell \le N/2-1, \tag{4.75}$$

which assumes that N is even. This has the advantage of more explicitly expressing the Nyquist limits and conforming more naturally to the other transform types for continuous and non-periodic functions and sequences. Note that $\tilde{G}'_k = \tilde{G}_k$, provided the same indexing is used for the sequence of samples, $\tilde{g}_\ell$, in both cases (Exercise 4.7). Other variations in the definition of the DFT exist. For example, the antecedent coefficient $1/N$ in the inverse transform might be replaced with $\sqrt{1/N}$ for both the direct and inverse transforms, equations (4.72) and (4.73), to create even greater

Table 4.2: Fourier Transform Pairs, $g \leftrightarrow G$.

	G: discrete domain	G: continuous domain
g: discrete domain	G: periodic g: periodic $\tilde{G}_k = \Delta x \sum_{\ell=0}^{N-1} \tilde{g}_\ell e^{-i\frac{2\pi}{N}k\ell}, \quad 0 \le k \le N-1$ $\tilde{g}_\ell = \frac{1}{N\Delta x} \sum_{k=0}^{N-1} \tilde{G}_k e^{i\frac{2\pi}{N}k\ell}, \quad 0 \le \ell \le N-1$	G: periodic g: non-periodic $\tilde{G}(f) = \Delta x \sum_{\ell=-\infty}^{\infty} g_\ell e^{-i2\pi\Delta x f \ell}, \quad -\frac{1}{2\Delta x} \le f \le \frac{1}{2\Delta x}$ $g_\ell = \int_{-1/(2\Delta x)}^{1/(2\Delta x)} \tilde{G}(f) e^{i2\pi\Delta x f \ell} df, \quad -\infty < \ell < \infty$
g: continuous domain	G: non-periodic g: periodic $G_k = \int_0^P \tilde{g}(x) e^{-i\frac{2\pi}{P}kx} dx, \quad -\infty < k < \infty$ $\tilde{g}(x) = \frac{1}{P} \sum_{k=-\infty}^{\infty} G_k e^{i\frac{2\pi}{P}kx}, \quad 0 \le x \le P$	G: non-periodic g: non-periodic $G(f) = \int_{-\infty}^{\infty} g(x) e^{-i2\pi f x} dx, \quad -\infty < f < \infty$ $g(x) = \int_{-\infty}^{\infty} G(f) e^{i2\pi f x} df, \quad -\infty < x < \infty$

symmetry. However, equations (4.70) and (4.71) are adopted in this text as the formal definition of the DFT pair. It has the advantage that N may be either even or odd, and inclusion of the scaling by Δx identifies the units of frequency for any practical application.

Table 4.2 summarizes and compares the Fourier transform pairs for the four types of functions in the Cartesian domain, where the domain is either discrete or continuous, and the function is either periodic or non-periodic. There is a perfect duality in these characteristics among the direct and inverse transforms: *discreteness* in the domain of either function or spectrum implies *periodicity* in its respective transform, and *continuity* in the domain of either function or spectrum implies *non-periodicity* in its respective transform. Each pair is formulated consistently in terms of units, where the spectrum has units of the function unit per frequency unit. Shown only for the singly dimensioned domain, these transform pairs extend naturally to higher Cartesian dimensions (Section 4.3.3). Also, they hold, in general, for complex functions.

4.3.2 FFT

The fast Fourier transform (FFT) is an algorithm to compute the DFT (usually as defined by equations (4.72) and (4.73)). As the name implies, it is a *fast* algorithm. There is no additional theory associated with the FFT, other than the development of the algorithm. It is a fact that

$$\text{FFT} \equiv \text{DFT}, \tag{4.76}$$

as far as the numerical results are concerned (up to a possible scaling factor). In terms of properties the two are synonymous.

It is no exaggeration to claim that the FFT nowadays is as ubiquitous and common in spectral analysis applications as the standard mathematical functions, like sine or cosine. There are many good books available (e.g., Brigham 1988) that describe the details of the algorithm. This is not done here, just as the details are not given for computing the sine or cosine of an angle. We may treat the FFT as a black box, or a library function—we know what the result, or output, should be for a given input. It is enough to know for our purposes that the FFT is the DFT; and that, when using the FFT function or subroutine from a computer software library, it is important to determine how the DFT is defined for that particular FFT algorithm in terms of the scaling factor, data indexing, and units. That is, the antecedent factor may be 1, or $1/N$, or $1/\sqrt{N}$; the data indices may start at 0, 1, or $-N/2$; and, usually Δx must be included manually to ensure the appropriate units. In some cases one must also verify the sign of the exponential function in order to distinguish between the Fourier transform and its inverse.

The speed of the algorithm depends on the prime factorization of the number, N, of sample values. The fastest computation occurs if N is a power of 2, and this was assumed for the original formulation of the FFT. The speed in this case, as measured by the number of computer multiplications, is proportional to $N \log_2 N$, compared to N^2 for the brute-force method according to the definition, equation (4.72). Thus, for example, if $N = 1024 = 2^{10}$, there is already a tremendous savings in computation time since the FFT requires only $\sim 10^4$ multiplications versus $\sim 10^6$ for the number of multiplications needed for the N sums in equation (4.72). Algorithms exist for N with arbitrary prime factorization (Singleton 1969); but, the fewer the number of different factors, the faster is the computation. Hence, when applying the FFT to a sequence of data, the greatest computational benefit is obtained if the number of data can be made equal to a power of 2, or a product of powers of primes: 2, 3, 5, 7, 11, ...; the fewer and smaller, the better.

The principal application of the FFT is, of course, in computing the spectrum of a signal, or synthesizing the signal from its spectrum (inverse transform). This holds for singly, as well as higher-dimensioned discrete signals, since corresponding algorithms simply make use of the separability of the transform in terms of the independent (Cartesian) variables. Similarly, the computation of the Legendre spectrum from data given on a regular spherical grid benefits from the FFT, but only in one dimension (Section 4.5.2). Another application of the FFT concerns the computation of convolutions, as enabled by the convolution theorem. In particular, equation (4.104) shows that the convolution can be performed in about $N + 3N \log_2 N$ multiplications (if N is a power of 2) versus N^2 multiplications using the definition (4.98). Specific applications in geodesy and geophysics abound and are the topics of Chapters 5 and 6.

4.3.3 Properties of the DFT and Higher Dimensions

The properties that were given for previous versions of the Fourier transform, such as equations (2.20) through (2.24) for the Fourier series transform, hold with corresponding modifications for the DFT, as well. They are listed here and easy to prove (Exercise 4.8).

1. Fourier transform pair: $\tilde{g}_\ell \leftrightarrow \tilde{G}_k$; (4.77)
2. Proportionality: $a\tilde{g}_\ell \leftrightarrow a\tilde{G}_k$, a is any constant; (4.78)
3. Superposition: $(\tilde{g}_1)_\ell + (\tilde{g}_2)_\ell \leftrightarrow (\tilde{G}_1)_k + (\tilde{G}_2)_k$, (4.79)
   provided Δx and N are the same for $(\tilde{g}_1)_\ell$ and $(\tilde{g}_2)_\ell$;
4. Symmetry: $\tilde{g}_{N-\ell} \leftrightarrow \tilde{G}_{N-k}$; (4.80)
5. Translation in ℓ: $\tilde{g}_{\ell+\ell_0} \leftrightarrow \tilde{G}_k e^{i\frac{2\pi}{N}k\ell_0}$; (4.81)
6. Translation in k: $\tilde{g}_\ell e^{-i\frac{2\pi}{N}k_0\ell} \leftrightarrow \tilde{G}_{k+k_0}$; (4.82)
7. Duality: $\tilde{G}_\ell \leftrightarrow \tilde{g}_{N-k}$. (4.83)

In addition, the *conjugate-symmetry* (Hermitian) property in the transform holds for real sequences (analogous to equation (2.27)) (Exercise 4.9),

$$\tilde{g}_\ell = \tilde{g}_\ell^* \quad \text{if and only if} \quad \tilde{G}_{N-k} = \tilde{G}_k^*, \tag{4.84}$$

which implies that $\tilde{G}_0$ and $\tilde{G}_{N/2}$ (N even) are both *real*, as can be seen from equation (4.70),

$$\tilde{G}_0 = \Delta x \sum_{\ell=0}^{N-1} \tilde{g}_\ell, \qquad \tilde{G}_{N/2} = \Delta x \sum_{\ell=0}^{N-1} \tilde{g}_\ell (-1)^\ell, \tag{4.85}$$

since both sums are real. By the duality property, an analogous result holds for a real-valued DFT,

$$\tilde{g}_{N-\ell} = \tilde{g}_\ell^* \quad \text{if and only if} \quad \tilde{G}_k = \tilde{G}_k^*; \tag{4.86}$$

so that, finally,

$$\tilde{g}_\ell = \tilde{g}_\ell^* \text{ and } \tilde{g}_{N-\ell} = \tilde{g}_\ell \quad \text{if and only if} \quad \tilde{G}_k = \tilde{G}_k^* \text{ and } \tilde{G}_{N-k} = \tilde{G}_k. \tag{4.87}$$

The form of the *Parseval theorem* for the DFT is

$$\Delta x^2 \sum_{\ell=0}^{N-1} (\tilde{g}_1)_\ell (\tilde{g}_2)_\ell^* = \frac{1}{N} \sum_{k=0}^{N-1} (\tilde{G}_1)_k (\tilde{G}_2)_k^*, \tag{4.88}$$

provided Δx and N are the same for $(\tilde{g}_1)_\ell$ and $(\tilde{g}_2)_\ell$. The theorem is easily proved, as for the Fourier series transform, equation (2.47), by substituting equation (4.71) on the left side,

$$\frac{1}{N^2}\sum_{\ell=0}^{N-1}\sum_{k=0}^{N-1}\left(\tilde{G}_1\right)_k e^{i\frac{2\pi}{N}k\ell}\sum_{k'=0}^{N-1}\left(\tilde{G}_2\right)_{k'}^* e^{-i\frac{2\pi}{N}k'\ell} = \frac{1}{N^2}\sum_{k=0}^{N-1}\sum_{k'=0}^{N-1}\left(\tilde{G}_1\right)_k\left(\tilde{G}_2\right)_{k'}^*\sum_{\ell=0}^{N-1} e^{i\frac{2\pi}{N}(k-k')\ell}. \tag{4.89}$$

Using the orthogonality, equation (4.68), then immediately gives the result.

Equation (4.84) shows that the number of independent spectral components (real and imaginary) is equal to the number of real sample values, which is another way of saying that the DFT contains all the information contained in the periodic sequence, and vice versa. If a real sequence is to be numerically generated (synthesized) from a discrete spectrum over a finite domain, then that spectrum must satisfy conjugate symmetry; otherwise, the sequence will not be real.

The DFT and its properties are extended naturally to higher dimensions, as for all other Fourier transforms in the Cartesian domain. In two dimensions,

$$\tilde{G}_{k_1,k_2} \equiv \mathrm{DFT}\left(\tilde{g}_{\ell_1,\ell_2}\right)_{k_1,k_2} = \Delta x_1 \Delta x_2 \sum_{\ell_1=0}^{N_1-1}\sum_{\ell_2=0}^{N_2-1}\tilde{g}_{\ell_1,\ell_2} e^{-i2\pi\left(\frac{k_1\ell_1}{N_1}+\frac{k_2\ell_2}{N_2}\right)}, \quad 0 \le k_1 \le N_1-1,\quad 0 \le k_2 \le N_2-1 \tag{4.90}$$

$$\tilde{g}_{\ell_1,\ell_2} \equiv \mathrm{DFT}^{-1}\left(\tilde{G}_{k_1,k_2}\right)_{\ell_1,\ell_2} = \frac{1}{N_1\Delta x_1 N_2 \Delta x_2}\sum_{k_1=0}^{N_1-1}\sum_{k_2=0}^{N_2-1}\tilde{G}_{k_1,k_2} e^{i2\pi\left(\frac{k_1\ell_1}{N_1}+\frac{k_2\ell_2}{N_2}\right)}, \quad 0 \le \ell_1 \le N_1-1,\quad 0 \le \ell_2 \le N_2-1 \tag{4.91}$$

The sampling interval is constant in each dimension, but Δx_1 need not equal Δx_2. The discrete signal and its transform are periodic in both dimensions, with periods, N_1 and N_2, respectively. Equivalent expressions that more clearly reflect the Nyquist limits are

$$\tilde{G}_{k_1,k_2} = \Delta x_1 \Delta x_2 \sum_{\ell_1=-N_1/2}^{N_1/2-1}\sum_{\ell_2=-N_2/2}^{N_2/2-1}\tilde{g}_{\ell_1,\ell_2} e^{-i2\pi\left(\frac{k_1\ell_1}{N_1}+\frac{k_2\ell_2}{N_2}\right)}, \quad -\frac{N_1}{2} \le k_1 \le \frac{N_1}{2}-1,\quad -\frac{N_2}{2} \le k_2 \le \frac{N_2}{2}-1 \tag{4.92}$$

$$\tilde{g}_{\ell_1,\ell_2} = \frac{1}{N_1\Delta x_1 N_2 \Delta x_2}\sum_{k_1=-N_1/2}^{N_1/2-1}\sum_{k_2=-N_2/2}^{N_2/2-1}\tilde{G}_{k_1,k_2} e^{i2\pi\left(\frac{k_1\ell_1}{N_1}+\frac{k_2\ell_2}{N_2}\right)}, \quad -\frac{N_1}{2} \le \ell_1 \le \frac{N_1}{2}-1,\quad -\frac{N_2}{2} \le \ell_2 \le \frac{N_2}{2}-1 \tag{4.93}$$

The DFT as defined by equation (4.92) is identical to equation (4.90) provided the data arrays are consistent as indicated by their indices.

Properties for the two-dimensional DFT are analogous to equations (4.78) through (4.83) with obvious extensions. Of particular note is the conjugate symmetry

(Hermitian) property of the spectrum for real signals. If and only if $\tilde{g}_{\ell_1,\ell_2}$ is *real*, then the spectrum, equation (4.90) satisfies

$$\tilde{G}_{-k_1,k_2} = \tilde{G}^*_{k_1,-k_2}, \quad \text{or also} \quad \tilde{G}_{N_1-k_1,N_2-k_2} = \tilde{G}^*_{k_1,k_2}, \quad \text{for all } k_1, k_2, \tag{4.94}$$

the latter because of its periodicity (add N_1, N_2 to the indices on the left and reverse the sign of k_2). Thus, analogous to equations (4.85), four spectral values must be real,

$$\tilde{G}_{0,0} = \tilde{G}^*_{0,0}, \quad \tilde{G}_{0,\frac{N_2}{2}} = \tilde{G}^*_{0,\frac{N_2}{2}}, \quad \tilde{G}_{\frac{N_1}{2},0} = \tilde{G}^*_{\frac{N_1}{2},0}, \quad \tilde{G}_{\frac{N_1}{2},\frac{N_2}{2}} = \tilde{G}^*_{\frac{N_1}{2},\frac{N_2}{2}}. \tag{4.95}$$

And, there is conjugate symmetry in k_1 for $k_2 = 0$ and $k_2 = N_2/2$,

$$\tilde{G}_{N_1-k_1,0} = \tilde{G}^*_{k_1,0}, \quad \tilde{G}_{N_1-k_1,\frac{N_2}{2}} = \tilde{G}^*_{k_1,\frac{N_2}{2}}; \tag{4.96}$$

and in k_2 for $k_1 = 0$ and $k_1 = N_1/2$,

$$\tilde{G}_{0,N_2-k_2} = \tilde{G}^*_{0,k_2}, \quad \tilde{G}_{\frac{N_1}{2},N_2-k_2} = \tilde{G}^*_{\frac{N_1}{2},k_2}. \tag{4.97}$$

As in the one-dimensional case, there are as many independent spectral components (real and imaginary) in the discrete spectrum, equation (4.90), as there are in the periodic sequence. If a real, two-dimensional sample is to be synthesized from a two-dimensional discrete spectrum over a finite domain, then all these symmetries must be enforced; otherwise, the sample will not be real.

4.3.4 Discrete Cyclic Convolution

The definition of convolution of periodic sequences is analogous to the definition for periodic continuous functions, equation (3.32). Let $\tilde{g}_\ell$ and $\tilde{h}_\ell$ be discrete periodic sequences each having the *same* period, N. The discrete convolution of periodic sequences is denoted "$\tilde{\#}$" and is given by

$$\tilde{c}_\ell = \tilde{g}_\ell \,\tilde{\#}\, \tilde{h}_\ell = \Delta x \sum_{n=0}^{N-1} \tilde{g}_n \tilde{h}_{\ell-n}, \quad \ell = 0, \ldots, N-1, \tag{4.98}$$

where the inclusion of Δx is consistent with previous definitions in terms of units. The convolution is periodic with period, N ($c_{\ell+mN} = c_\ell$ for any integer, m), and makes use of the fact that the sequence, $\tilde{h}_\ell$, is extended periodically beyond a single period. For example,

$$\begin{aligned} \tilde{c}_1 &= \tilde{g}_0\tilde{h}_1 + \tilde{g}_1\tilde{h}_0 + \tilde{g}_2\tilde{h}_{-1} + \tilde{g}_3\tilde{h}_{-2} + \cdots + \tilde{g}_{N-1}\tilde{h}_{-N+2} \\ &= \tilde{g}_0\tilde{h}_1 + \tilde{g}_1\tilde{h}_0 + \tilde{g}_2\tilde{h}_{N-1} + \tilde{g}_3\tilde{h}_{N-2} + \cdots + \tilde{g}_{N-1}\tilde{h}_2 \end{aligned} \tag{4.99}$$

Hence, one could also write,

$$\tilde{g}_\ell \# \tilde{h}_\ell = \Delta x \sum_{n=0}^{N-1} \tilde{g}_n \tilde{h}_{\mathrm{mod}_N(N+\ell-n)}, \quad \ell = 0,\ldots,N-1, \tag{4.100}$$

which uses only the values, $\tilde{h}_\ell$, for $\ell = 0,\ldots, N-1$. The convolution, equation (4.98) or (4.100), is known as a *discrete cyclic*, or *circular*, convolution, since it is defined specifically for periodic sequences.

As for continuous periodic functions, there is a cyclic convolution theorem and a dual cyclic convolution theorem for the discrete case.

Discrete Cyclic Convolution Theorem

The discrete Fourier transform of the discrete cyclic convolution (4.98) equals the product of the discrete Fourier transforms of the convolved periodic sequences. In symbols,

$$\tilde{g}_\ell \# \tilde{h}_\ell \quad \leftrightarrow \quad \tilde{G}_k \tilde{H}_k. \tag{4.101}$$

The proof starts with the DFT of the convolution, equation (4.98),

$$\tilde{C}_k = \mathrm{DFT}\left(\tilde{g}_\ell \# \tilde{h}_\ell\right)_k = \Delta x \sum_{\ell=0}^{N-1} \left(\Delta x \sum_{n=0}^{N-1} \tilde{g}_n \tilde{h}_{\ell-n} \right) e^{-i\frac{2\pi}{N}k\ell}. \tag{4.102}$$

Interchanging summations and changing the index, $\ell' = \ell - n$, this is

$$\begin{aligned} \tilde{C}_k &= \Delta x \sum_{n=0}^{N-1} \tilde{g}_n e^{-i\frac{2\pi}{N}nk} \Delta x \sum_{\ell'=-n}^{N-1-n} \tilde{h}_{\ell'} e^{-i\frac{2\pi}{N}\ell' k} \\ &= \Delta x \sum_{n=0}^{N-1} \tilde{g}_n e^{-i\frac{2\pi}{N}nk} \Delta x \left(\sum_{\ell'=N-n}^{N-1} \tilde{h}_{\ell'} e^{-i\frac{2\pi}{N}\ell' k} + \sum_{\ell'=0}^{N-1-n} \tilde{h}_{\ell'} e^{-i\frac{2\pi}{N}\ell' k} \right) \end{aligned} \tag{4.103}$$

where the index of the second sum is split into parts with negative and non-negative indices, ℓ', which for the negative ones are subsequently replaced according to $\ell' \to \ell' - N$, noting that both $\tilde{h}_{\ell'}$ and $\exp(-i(2\pi/N)\ell' k)$ are periodic with period, N. Combining the last two sums proves the DFT of the convolution. The inverse DFT follows automatically.

From a computational viewpoint, therefore, the discrete cyclic convolution is equivalent to

$$\begin{aligned} \tilde{g}_\ell \# \tilde{h}_\ell &= \mathrm{DFT}^{-1}\left(\tilde{G}_k \tilde{H}_k\right)_\ell \\ &= \mathrm{DFT}^{-1}\left(\mathrm{DFT}\left(\tilde{g}_{\ell'}\right)_k \mathrm{DFT}\left(\tilde{h}_{\ell'}\right)_k\right)_\ell \end{aligned} \tag{4.104}$$

It can be calculated by multiplying two discrete spectra and applying three DFTs.

Dual Discrete Cyclic Convolution Theorem

The discrete Fourier transform of the product of two periodic sequences having the same period equals the discrete convolution of their discrete Fourier transforms. In symbols,

$$\tilde{g}_\ell \tilde{h}_\ell \quad \leftrightarrow \quad \tilde{G}_k \,\tilde{\#}\, \tilde{H}_k, \tag{4.105}$$

where the convolution for discrete periodic spectra is defined by

$$\tilde{G}_k \,\tilde{\#}\, \tilde{H}_k = \frac{1}{N\Delta x}\sum_{n=0}^{N-1} \tilde{G}_n \tilde{H}_{k-n}, \quad k = 0,\ldots,N-1, \tag{4.106}$$

The proof follows easily, starting with equations (4.70) and (4.71),

$$\begin{aligned} \mathrm{DFT}\left(\tilde{g}_\ell \tilde{h}_\ell\right)_k &= \Delta x \sum_{\ell=0}^{N-1}\left(\frac{1}{N\Delta x}\sum_{n=0}^{N-1}\tilde{G}_n e^{i\frac{2\pi}{N}n\ell}\right)\tilde{h}_\ell e^{-i\frac{2\pi}{N}k\ell} \\ &= \frac{1}{N}\sum_{n=0}^{N-1}\tilde{G}_n \sum_{\ell=0}^{N-1}\tilde{h}_\ell e^{-i\frac{2\pi}{N}(k-n)\ell} \end{aligned} \tag{4.107}$$

and the result is obtained directly by using equation (4.70) for $\tilde{H}_{k-n}$. The inverse DFT follows automatically.

The convolution in the frequency domain, equation (4.106), is analogous to equation (4.98) with the recognition that the sampling interval in the frequency domain is $\Delta f = 1/(N\Delta x)$.

Because of the periodicity of $\tilde{h}_\ell$, the discrete convolution may be represented as the multiplication of a matrix and a vector (e.g., see equation (4.99)),

$$\begin{pmatrix} \tilde{c}_0 \\ \tilde{c}_1 \\ \vdots \\ \tilde{c}_{N-2} \\ \tilde{c}_{N-1} \end{pmatrix} = \begin{pmatrix} \tilde{h}_0 & \tilde{h}_{N-1} & \cdots & \tilde{h}_2 & \tilde{h}_1 \\ \tilde{h}_1 & \tilde{h}_0 & \ddots & \ddots & \tilde{h}_2 \\ \vdots & \ddots & \ddots & \ddots & \vdots \\ \tilde{h}_{N-2} & \ddots & \ddots & \tilde{h}_0 & \tilde{h}_{N-1} \\ \tilde{h}_{N-1} & \tilde{h}_{N-2} & \cdots & \tilde{h}_1 & \tilde{h}_0 \end{pmatrix} \begin{pmatrix} \tilde{g}_0 \\ \tilde{g}_1 \\ \vdots \\ \tilde{g}_{N-2} \\ \tilde{g}_{N-1} \end{pmatrix}, \tag{4.108}$$

or, in vector-matrix notation,

$$\tilde{c} = \mathbf{H}\tilde{\mathbf{g}}. \tag{4.109}$$

All elements of any left-to-right, descending diagonal of the matrix, **H**, are the same; and, as already encountered for infinite sequences, equation (4.38), such a matrix is a Toeplitz matrix. In this case, each row is the row above displaced one to the right (and each column is the column to the left displaced one down) in wrap-around fashion, and thus, **H** is more specifically a *circulant Toeplitz matrix*. Equation (4.104) is a rapid means using FFTs to calculate the product, $\mathbf{H}\tilde{\mathbf{g}}$, for circulant Toeplitz matrices (for a general Toeplitz matrix see equation (4.145)).

In most cases the problem is to solve for $\tilde{\mathbf{g}}$, given $\tilde{\mathbf{c}}$. This is a deconvolution problem (Section 3.2.1) and, again, can be done rapidly if **H** is a non-singular matrix. From the convolution theorem (4.101), there is

$$\tilde{g}_\ell = \mathrm{DFT}^{-1}\left(\frac{\tilde{C}_k}{\tilde{H}_k}\right)_\ell, \tag{4.110}$$

provided $\tilde{H}_k \neq 0$.

For two-dimensional Cartesian domains, the discrete cyclic convolution is

$$\tilde{g}_{\ell_1,\ell_2} \# \tilde{h}_{\ell_1,\ell_2} = \Delta x_1 \Delta x_2 \sum_{n_1=0}^{N_1-1}\sum_{n_2=0}^{N_2-1} \tilde{g}_{n_1,n_2}\tilde{h}_{\ell_1-n_1,\ell_2-n_2}, \quad \ell_1 = 0,\ldots,N_1-1, \quad \ell_2 = 0,\ldots,N_2-1; \tag{4.111}$$

and, corresponding convolution theorems are

$$\tilde{g}_{\ell_1,\ell_2} \# \tilde{h}_{\ell_1,\ell_2} \quad \leftrightarrow \quad \tilde{G}_{k_1,k_2}\tilde{H}_{k_1,k_2}, \tag{4.112}$$

$$\tilde{g}_{\ell_1,\ell_2}\tilde{h}_{\ell_1,\ell_2} \quad \leftrightarrow \quad \tilde{G}_{k_1,k_2} \# \tilde{H}_{k_1,k_2}, \tag{4.113}$$

where

$$\tilde{G}_{k_1,k_2} \# \tilde{H}_{k_1,k_2} = \frac{1}{N_1\Delta x_1 N_2 \Delta x_2}\sum_{n_1=0}^{N_1-1}\sum_{n_2=0}^{N_2-1}\tilde{G}_{n_1,n_2}\tilde{H}_{k_1-n_1,k_2-n_2}, \quad k_1 = 0,\ldots,N_1-1, \quad k_2 = 0,\ldots,N_2-1. \tag{4.114}$$

Explicitly, the Fourier transform pair (4.112) shows how to compute the convolution efficiently using three FFTs,

$$\tilde{g}_{\ell_1,\ell_2} \# \tilde{h}_{\ell_1,\ell_2} = \mathrm{DFT}^{-1}\left(\mathrm{DFT}\left(\tilde{g}_{\ell_1',\ell_2'}\right)_{k_1,k_2} \mathrm{DFT}\left(\tilde{h}_{\ell_1',\ell_2'}\right)_{k_1,k_2}\right)_{\ell_1,\ell_2}. \tag{4.115}$$

These results are self-evident based on the previous generalizations to higher dimensions, as discussed in Sections 3.3 and 4.2.2.

4.3.5 Aliasing of the Discrete Fourier Spectrum

Let $\tilde{g}(x)$ be a periodic function with period, P, that is band-limited,

$$G_k = 0, \quad k \le -N/2-1, \quad k \ge N/2; \tag{4.116}$$

and, suppose it is sampled at an interval, $\Delta x = P/N$. Using equation (4.74), the DFT of the samples, $\tilde{g}_\ell = \tilde{g}(\ell\Delta x)$, is

$$\begin{aligned}
\tilde{G}_k &= \Delta x \sum_{\ell=-N/2}^{N/2-1} \tilde{g}_\ell e^{-i\frac{2\pi}{N}k\ell} \\
&= \Delta x \sum_{\ell=-N/2}^{N/2-1} \left(\frac{1}{P} \sum_{k'=-N/2}^{N/2-1} G_{k'} e^{i\frac{2\pi}{P}k'\ell\Delta x} \right) e^{-i\frac{2\pi}{N}k\ell} \\
&= \frac{1}{N} \sum_{k'=-N/2}^{N/2-1} G_{k'} \sum_{\ell=-N/2}^{N/2-1} e^{i\frac{2\pi}{N}(k'-k)\ell} \\
&= G_k
\end{aligned} \tag{4.117}$$

for $k = -N/2, \ldots, N/2 - 1$, where the second equality comes from equations (2.18) and (4.116), and the last equality is due to the orthogonality, equation (4.68). Equation (2.18) then also gives

$$\tilde{g}(x) = \frac{1}{P} \sum_{k=-N/2}^{N/2-1} \tilde{G}_k e^{i\frac{2\pi}{P}kx}, \tag{4.118}$$

for all x; in other words, the periodic function, $\tilde{g}(x)$, is determined completely by its samples, $\tilde{g}(\ell\Delta x)$, provided $\Delta x = P/N$. This is another case of the Whittaker-Shannon sampling theorem.

If a periodic, *non*-band-limited signal is sampled at constant interval, Δx, then the DFT is subject to an aliasing error (Section 4.2.3). Starting with the Fourier series for $\tilde{g}(x)$, equation (2.18), the samples are given by

$$\tilde{g}_\ell = \tilde{g}(\ell\Delta x) = \frac{1}{P} \sum_{k=-\infty}^{\infty} G_k e^{i\frac{2\pi}{P}k\ell\Delta x} = \frac{1}{P} \sum_{j=-\infty}^{\infty} \sum_{k=0}^{N-1} G_{jN+k} e^{i\frac{2\pi}{P}(jN+k)\ell\Delta x}, \tag{4.119}$$

which is a more elaborate, but useful way of writing the infinite sum. Interchanging the summations, we get

$$\tilde{g}_\ell = \frac{1}{N\Delta x} \sum_{k=0}^{N-1} \sum_{j=-\infty}^{\infty} G_{jN+k} e^{i\frac{2\pi}{N}k\ell}, \tag{4.120}$$

where $P = N\Delta x$, and because $\exp(i2\pi k\ell) = 1$. A comparison of equations (4.120) and (4.75) yields the relationship between the spectrum of the sample and the spectrum of the periodic (continuous) parent function,

$$\tilde{G}_k = \sum_{j=-\infty}^{\infty} G_{jN+k}, \tag{4.121}$$

analogous to equation (4.8) for the continuous spectrum.

The aliasing error is given, as in equation (4.50), by

$$\tilde{G}_k - G_k = \sum_{\substack{j=-\infty \\ j\neq 0}}^{\infty} G_{jN+k}, \quad -N/2 \le k \le N/2-1. \tag{4.122}$$

It is zero if the periodic parent signal is band-limited by the *Nyquist wave number*, $k_N = N/2 = P/(2\Delta x)$, as shown by equation (4.118). Again, for two-dimensional domains, the analogous aliasing error in the DFT is

$$\tilde{G}_{k_1,k_2} - G_{k_1,k_2} = \sum_{\substack{j_1=-\infty \\ j_1\neq 0 \text{ and } j_2\neq 0}}^{\infty} \sum_{j=-\infty}^{\infty} G_{j_1N_1+k_1,\, j_2N_2+k_2}, \quad -N_1/2 \le k_1 \le N_1/2-1, \quad -N_2/2 \le k_2 \le N_2/2-1. \tag{4.123}$$

Throughout this section it is assumed that the period of the parent function is an integer multiple of the sampling interval and that the function is sampled over the entire period ($N\Delta x = P$). Generally, this is not a serious constraint since the total period, or interval, of an available geophysical signal is defined by a finite number of sampling intervals and $N\Delta x = P$ is automatically satisfied; this signal is then assumed to be periodic with period, P. However, it may be noted that the case, $N\Delta x < P$, represents a truncation of the domain and spectral leakage (in addition to possible aliasing) corrupts the estimation of the parent signal spectrum. If $N\Delta x > P$ and $\mathrm{mod}_{\Delta x}(P) \neq 0$, then the assumptions implicit in the application of the discrete techniques such as the DFT and the discrete cyclic convolution no longer hold. This means, again, that the parent signal has been truncated to less that a full number of periods and spectral leakage corrupts the spectrum estimated from the samples.

4.4 Cyclic Versus Linear Discrete Convolution

The concept of convolution, first introduced in Chapter 3 for (piecewise) continuous functions, is made practicable and computationally efficient using the FFT by sampling over a finite interval that is assumed to be the period of the parent function. These assumptions and restrictions cause errors when applied to actual discrete data because, in fact, geophysical models usually are based on continuous functions over domains typically larger than the data domain. Truncation of the domain causes spectral leakage and finite sampling causes aliasing. Using the FFT to compute the convolution assumes periodicity in the data (hence, in the parent function) and introduces yet another kind of error, called the *cyclic convolution error*. Spectral leakage and aliasing could be reduced, respectively, with additional data on an extended domain and finer sampling. However, the cyclic convolution error can be eliminated entirely without obtaining additional data.

To put these various errors in proper perspective, consider the typical situation where the function, g, in the convolution is the *data function* (or *signal function*) and h is a known *kernel function*, representing the convolution (or, a filter), that transforms the data in some physically meaningful way. The convolution, equation (3.3), representing the true model for the data and kernel functions is repeated here for convenience,

$$g(x)*h(x)=\int_{-\infty}^{\infty} g(x')h(x-x')\,dx'. \tag{4.124}$$

Let g_ℓ, $\ell = -N/2,\ldots, N/2-1$, be N discrete values of the data function sampled at uniform spacing, where, without loss in generality, N is even. Regarding first the finite extent of the function, define a new data function, g_T, given by

$$g_T(x)=\begin{cases} g(x), & |x|\le T/2 \\ 0, & |x|>T/2 \end{cases} \tag{4.125}$$

that reflects its truncation to a domain, $[-T/2, T/2]$. The approximation of the true convolution by the convolution with g_T, thus generates a *truncation error*, $\varepsilon_{\text{trunc}}(x)$,

$$\begin{aligned} g(x)*h(x) &= \int_{-T/2}^{T/2} g(x')h(x-x')\,dx' + \int_{|x'|>T/2} g(x')h(x-x')\,dx' \\ &= g_T(x)*h(x)-\varepsilon_{\text{trunc}}(x) \end{aligned} \tag{4.126}$$

This error is also called an edge-effect error if the kernel function attenuates away from the origin, since in that case the error is significant only when the computation point approaches the edge of the data domain and the significant part of the kernel lacks data beyond the edge (see Figure 4.4).

Assuming that the sampling interval divides the truncated domain, $T = N\Delta x$, define a data sequence, $(g_N)_\ell$,

$$(g_N)_\ell=\begin{cases} g_\ell, & -\dfrac{N}{2}\le \ell \le \dfrac{N}{2}-1 \\ 0, & \text{otherwise} \end{cases} \tag{4.127}$$

Note that $(g_N)_\ell$ is not periodic; it simply embodies the collection of available samples of the continuous function. Further approximating the continuous convolution by a discrete convolution, equation (4.126) becomes

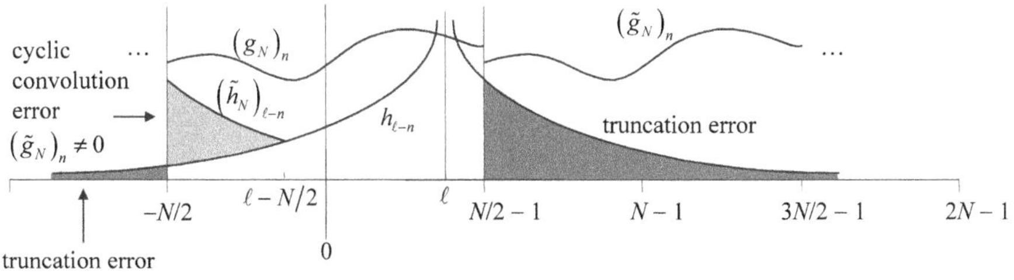

Figure 4.4: Truncation and cyclic convolution errors (shaded regions of $h_{\ell-n}$, respectively, $(\tilde{h}_N)_{\ell-n}$ multiplied by the sampled signal, $(\tilde{g}_N)_n$) for the convolution $(g_N)_\ell \# h_\ell$ evaluated at ℓ. Solid lines connect the sequence points for clarity.

$$g(x) * h(x) = (g_N)_\ell \# h_\ell - \varepsilon_{\text{discrete}}(x) - \varepsilon_{\text{trunc}}(x), \tag{4.128}$$

where the discrete convolution on the right side is defined by equation (4.35), now given by

$$(g_N)_\ell \# h_\ell = \Delta x \sum_{n=-N/2}^{N/2-1} g_n h_{\ell-n}. \tag{4.129}$$

This can be calculated for any ℓ from the data and a discretization or sampling of the known kernel function, h. With the aim of implementing the DFT (FFT), only the simple rectangle rule of numerical integration is assumed; other methods of numerical integration could be considered if they yield a convolution, such as the trapezoid rule, given by equation (4.42).

Finally, in order to evaluate the discrete convolution using the DFT, according to equation (4.104), one must assume that both the data and the kernel sequences are *periodic* with the same period, N,

$$(\tilde{g}_N)_{\ell+mN} = g_\ell, \quad -\frac{N}{2} \le \ell \le \frac{N}{2} - 1, \tag{4.130}$$

$$(\tilde{h}_N)_{\ell+mN} = h_\ell, \quad -\frac{N}{2} \le \ell \le \frac{N}{2} - 1, \tag{4.131}$$

for any integer, m. Owing to their periodic nature, the product, $(\tilde{g}_N)_n (\tilde{h}_N)_{\ell-n}$ is also periodic in n (period, N), and the discrete cyclic convolution, equation (4.98), may be written as

$$(\tilde{g}_N)_\ell \tilde{\#} (\tilde{h}_N)_\ell = \Delta x \sum_{n=-N/2}^{N/2-1} (\tilde{g}_N)_n (\tilde{h}_N)_{\ell-n}, \quad -\frac{N}{2} \le \ell \le \frac{N}{2} - 1. \tag{4.132}$$

Using this cyclic convolution instead of the *linear* convolution, equation (4.129), introduces an additional *cyclic convolution error* if one simply evaluates the convolution without modification,

$$g(x) * h(x) = (\tilde{g}_N)_\ell \tilde{\#} (\tilde{h}_N)_\ell - (\varepsilon_{\text{cyc}})_\ell - \varepsilon_{\text{discrete}}(x) - \varepsilon_{\text{trunc}}(x), \tag{4.133}$$

where the cyclic convolution error is given by

$$(\varepsilon_{\text{cyc}})_\ell = (\tilde{g}_N)_\ell \tilde{\#} (\tilde{h}_N)_\ell - (g_N)_\ell \# h_\ell, \quad -\frac{N}{2} \le \ell \le \frac{N}{2} - 1. \tag{4.134}$$

A more explicit expression for the cyclic convolution error gives further insight into its source, and offers also the opportunity to analyze its magnitude for particular applications. From the convolution definitions given by equations (4.35) and (4.132), and the definitions of the sequences, equations (4.127) and (4.130), equation (4.134) becomes

$$\left(\varepsilon_{\text{cyc}}\right)_\ell = \Delta x \sum_{n=-N/2}^{N/2-1} \left(\tilde{g}_N\right)_n \left(\tilde{h}_N\right)_{\ell-n} - \Delta x \sum_{n=-\infty}^{\infty} \left(g_N\right)_n h_{\ell-n}$$
$$= \Delta x \sum_{n=-N/2}^{N/2-1} g_n \left(\left(\tilde{h}_N\right)_{\ell-n} - h_{\ell-n} \right) \tag{4.135}$$

The error is depicted graphically in Figure 4.4. It is readily verified (Exercise 4.10) that it simplifies to

$$\left(\varepsilon_{\text{cyc}}\right)_\ell = \begin{cases} \Delta x \displaystyle\sum_{n=\ell+N/2+1}^{N/2-1} g_n \left(h_{\ell-n+N} - h_{\ell-n}\right), & -\dfrac{N}{2} \le \ell \le -2 \\ 0, & \ell = -1 \\ \Delta x \displaystyle\sum_{n=-N/2}^{-N/2+\ell} g_n \left(h_{\ell-n-N} - h_{\ell-n}\right), & 0 \le \ell \le \dfrac{N}{2} - 1 \end{cases} \tag{4.136}$$

That the error is zero at $\ell = -1$ instead of $\ell = 0$ is due to the slightly asymmetric manner in which the truncation is defined.

The summand in $(\varepsilon_{\text{cyc}})_\ell$ involves values of the kernel function potentially close to the origin. For example, when $\ell = N/2 - 1$, the error includes the product, $g_{-N/2}(h_{-1} - h_{N-1})$. Hence, if the kernel is largest near its origin (as shown in Figure 4.4) and attenuates with distance from that origin then the cyclic convolution error can be significant. Many geodetic and geophysical kernels, in fact, behave like a power of the reciprocal of the distance between the evaluation point and the integration point (Chapter 6). On the other hand, from Figure 4.4 it is evident that the truncation error and the cyclic convolution error have similar characteristics; both are largest when the computation point, ℓ, of the convolution is close to the edge of the data domain. Therefore, in avoiding the truncation error by restricting the computation point to a neighborhood of the origin of the data domain, one also tends to avoid the cyclic convolution error.

It is possible, on the other hand, to construct a discrete cyclic convolution from the given data sequence that exactly equals the discrete linear convolution. Consider the discrete cyclic convolution, $(\tilde{g}^0_{2N})_\ell \, \tilde{\#} (\tilde{h}_{2N})_\ell$, where $(\tilde{g}^0_{2N})_\ell$ is a periodic sequence whose principal part is defined over the domain, $-N \le \ell \le N - 1$, by extending (*padding*) $(\tilde{g}_N)_\ell$ with zeros on either side of its own principal interval, $-N/2 \le \ell \le N/2 - 1$,

$$\left(\tilde{g}^0_{2N}\right)_\ell = \begin{cases} g_\ell, & -\dfrac{N}{2} \le \ell \le \dfrac{N}{2} - 1 \\ 0, & -N \le \ell \le -\dfrac{N}{2} - 1 \text{ and } \dfrac{N}{2} \le \ell \le N - 1 \end{cases} \tag{4.137}$$

$$\left(\tilde{g}^0_{2N}\right)_{\ell+2mN} = \left(\tilde{g}^0_{2N}\right)_\ell, \quad -N \le \ell \le N-1, \quad m \text{ is any integer;} \tag{4.138}$$

where the second equation defines the periodic extension of $(\tilde{g}^0_{2N})_\ell$ over all integers. The extended periodic sequence, $\tilde{h}_{2N}$, is defined by

$$\left(\tilde{h}_{2N}\right)_\ell = h_\ell, \quad \left(\tilde{h}_{2N}\right)_{\ell+2mN} = \left(\tilde{h}_{2N}\right)_\ell, \quad -N \le \ell \le N-1, \quad m \text{ is any integer.} \tag{4.139}$$

Thus, while the data are extended with zeros, the kernel sequence, though assumed periodic (period $2N$), is extended naturally using *the actual known values of h.*

The following shows that the discrete cyclic convolution of these extended sequences equals the discrete linear convolution of the original finite sequences for $-N/2 \le \ell \le N/2 - 1$. First, note that

$$\left(\tilde{g}^0_{2N}\right)_n = \left(g_N\right)_n, \quad -N \le n \le N-1; \tag{4.140}$$

$$\left(\tilde{h}_{2N}\right)_{\ell-n} = \left(h_{2N}\right)_{\ell-n} = h_{\ell-n}, \quad -\frac{N}{2} \le n \le \frac{N}{2}-1 \text{ and } -\frac{N}{2} \le \ell \le \frac{N}{2}-1. \tag{4.141}$$

Hence, for $-N/2 \le \ell \le N/2 - 1$,

$$\begin{aligned}
\left(\tilde{g}^0_{2N}\right)_\ell \tilde{\#}\left(\tilde{h}_{2N}\right)_\ell &= \Delta x \sum_{n=-N}^{N-1} \left(\tilde{g}^0_{2N}\right)_n \left(\tilde{h}_{2N}\right)_{\ell-n}, && \text{from equation (4.132);} \\
&= \Delta x \sum_{n=-N}^{N-1} \left(g_N\right)_n \left(\tilde{h}_{2N}\right)_{\ell-n}, && \text{from equation (4.140);} \\
&= \Delta x \sum_{n=-N/2}^{N/2-1} \left(g_N\right)_n \left(\tilde{h}_{2N}\right)_{\ell-n}, && \text{from equation (4.127);} \\
&= \Delta x \sum_{n=-N/2}^{N/2-1} \left(g_N\right)_n h_{\ell-n}, && \text{from equation (4.141);} \\
&= \left(g_N\right)_\ell \# h_\ell, && \text{from equations (4.127) and (4.35).}
\end{aligned}$$

That is,

$$\left(g_N\right)_\ell \# h_\ell = \left(\tilde{g}^0_{2N}\right)_\ell \tilde{\#}\left(\tilde{h}_{2N}\right)_\ell, \quad -\frac{N}{2} \le \ell \le \frac{N}{2}-1, \tag{4.142}$$

and the cyclic convolution error, equation (4.134), is zero in this case. Figure 4.5 shows schematically how the cyclic convolution error vanishes for $(\tilde{g}^0_{2N})_\ell \tilde{\#}(\tilde{h}_{2N})_\ell$, provided that $-N/2 \le \ell \le N/2 - 1$. The values of the cyclic convolution of the extended sequences for other ℓ are discarded.

Equations (4.137) and (4.139) indicate precisely how the periodic sequences, $(\tilde{g}^0_{2N})_\ell$ and $(\tilde{h}_{2N})_\ell$, must be constructed so that their cyclic convolution equals the linear convolution of the original sequences. Once constructed, any period could be used in the FFT algorithm, e.g., also $0 \le \ell \le 2N - 1$.

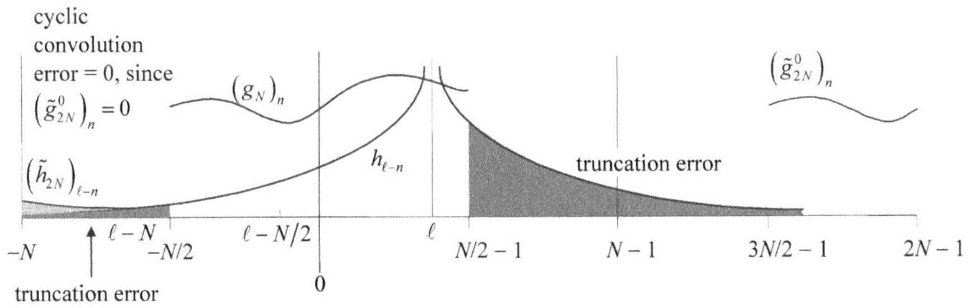

Figure 4.5: Truncation and cyclic convolution error (shaded regions of $h_{\ell-n}$, respectively, $(\tilde{h}_{2N})_{\ell-n}$ multiplied by the sampled signal, $(\tilde{g}_N)_n$) for the convolution $(g_N)_\ell \# h_\ell$ evaluated at ℓ; the cyclic convolution error is zero. Solid lines connect the sequence points for clarity.

Using this method to estimate the true convolution thus leaves only the discretization and truncation errors as indicated on the right side of equation (4.128). However, it involves twice as much computer storage space for the extended sequences (but a relatively insignificant increase in computation time). It may not be justified from a numerical standpoint to nullify the cyclic convolution error, if the truncation error, usually of similar magnitude, is already avoided by restricting the domain of the computation points of the convolution, as this also reduces at the same time the cyclic convolution error (Jekeli 1998). If the extra computer storage space is not an issue, it is prudent to eliminate the cyclic convolution error as a matter of routine practice.

The kernel function is *not* padded with zeros (as sometimes erroneously advocated), as this would cause the cyclic convolution of the extended sequences to differ from the linear convolution of the original sequences. However, if the function, h, is not known beyond some finite interval, then zero padding of this function also helps to reduce (not eliminate) the cyclic convolution error.

The elimination of the cyclic convolution error can also be applied directly to the efficient computation of a linear system of equations, written in matrix form, if the matrix is Toeplitz (cf. equation (4.108)),

$$\begin{pmatrix} c_{-N/2} \\ c_{-N/2+1} \\ \vdots \\ c_{N/2-2} \\ c_{N/2-1} \end{pmatrix} = \begin{pmatrix} h_0 & h_{-1} & \cdots & h_{-N+2} & h_{-N+1} \\ h_1 & h_0 & \ddots & \ddots & h_{-N+2} \\ \vdots & \ddots & \ddots & \ddots & \vdots \\ h_{N-2} & \ddots & \ddots & h_0 & h_{-1} \\ h_{N-1} & h_{N-2} & \cdots & h_1 & h_0 \end{pmatrix} \begin{pmatrix} g_{-N/2} \\ g_{-N/2+1} \\ \vdots \\ g_{N/2-2} \\ g_{N/2-1} \end{pmatrix}. \tag{4.143}$$

This equation is the convolution, equation (4.129),

$$c_\ell = (g_N)_\ell \# h_\ell = \sum_{n=-N/2}^{N/2-1} g_n h_{\ell-n}, \quad -\frac{N}{2} \le \ell \le \frac{N}{2} - 1. \tag{4.144}$$

From equation (4.142) and the convolution theorem (4.101), one finds immediately that

$$c_\ell = \mathrm{DFT}^{-1}\left(\mathrm{DFT}\left(\left(\tilde{g}^0_{2N}\right)_n\right)_k \mathrm{DFT}\left(\left(\tilde{h}_{2N}\right)_n\right)_k\right)_\ell, \quad -\frac{N}{2} \le \ell \le \frac{N}{2} - 1, \tag{4.145}$$

where $(\tilde{g}^0_{2N})_n$ and $(\tilde{h}_{2N})_n$ are given by equations (4.137) and (4.139), respectively. In this case, h_{-N} (not occurring in the matrix) can be arbitrary because $\ell - n = -N$ in equation (4.142) only if $N/2 \le n \le 3N/2 - 1$; and, for these n, according to equation (4.137),

$$(\tilde{g}^0_{2N})_n \, (\tilde{h}_{2N})_{-N} = 0. \tag{4.146}$$

For a *circulant*-Toeplitz matrix the solution to g_ℓ for known c_ℓ and h_ℓ is equation (4.110). The inversion of equation (4.145), however, cannot be used to find g_ℓ for a general Toeplitz matrix, since the extended sequence, c_ℓ, that yields the zero-padded sequence, $(\tilde{g}^0_{2N})_\ell$, is not known.

The two-dimensional, linear, discrete convolution of a truncated data array and a sampled kernel function is

$$\left(g_{N_1,N_2}\right)_{\ell_1,\ell_2} \# h_{\ell_1,\ell_2} = \Delta x_1 \Delta x_2 \sum_{n_1=-N_1/2}^{N_1/2-1} \sum_{n_2=-N_2/2}^{N_2/2-1} g_{n_1,n_2} h_{\ell_1-n_1,\ell_2-n_2}. \tag{4.147}$$

The corresponding discrete cyclic convolution is

$$\left(\tilde{g}_{N_1,N_2}\right)_{\ell_1,\ell_2} \tilde{\#} \left(\tilde{h}_{N_1,N_2}\right)_{\ell_1,\ell_2} = \Delta x_1 \Delta x_2 \sum_{n_1=-N_1/2}^{N_1/2-1} \sum_{n_2=-N_2/2}^{N_2/2-1} \left(\tilde{g}_{N_1,N_2}\right)_{n_1,n_2} \left(\tilde{h}_{N_1,N_2}\right)_{\ell_1-n_1,\ell_2-n_2},$$
$$-\frac{N_1}{2} \le \ell_1 \le \frac{N_1}{2} - 1, \quad -\frac{N_2}{2} \le \ell_2 \le \frac{N_2}{2} - 1 \tag{4.148}$$

They are equal over a limited domain for an appropriately zero-padded data array and the extended kernel array, analogous to equation (4.142),

$$\left(g_{N_1,N_2}\right)_{\ell_1,\ell_2} \# h_{\ell_1,\ell_2} = \left(\tilde{g}^{0,0}_{2N_1,2N_2}\right)_{\ell_1,\ell_2} \tilde{\#} \left(\tilde{h}_{2N_1,2N_2}\right)_{\ell_1,\ell_2},$$
$$-\frac{N_1}{2} \le \ell_1 \le \frac{N_1}{2} - 1, \quad -\frac{N_2}{2} \le \ell_2 \le \frac{N_2}{2} - 1, \tag{4.149}$$

where the zero padding is defined by

$$\left(\tilde{g}^{0,0}_{2N_1,2N_2}\right)_{\ell_1,\ell_2} = \begin{cases} \left(\tilde{g}_{N_1,N_2}\right)_{\ell_1,\ell_2}, & -\frac{N_1}{2} \le \ell_1 \le \frac{N_1}{2} - 1 \text{ and } -\frac{N_2}{2} \le \ell_2 \le \frac{N_2}{2} - 1 \\ 0, & -N_1 \le \ell_1 \le -\frac{N_1}{2} - 1 \text{ or } \frac{N_1}{2} \le \ell_1 \le N_1 - 1 \text{ or} \\ & -N_2 \le \ell_2 \le -\frac{N_2}{2} - 1 \text{ or } \frac{N_2}{2} \le \ell_2 \le N_2 - 1 \end{cases} \tag{4.150}$$

$$\left(\tilde{g}^{0,0}_{2N_1,2N_2}\right)_{\ell_1+2m_1N_1,\ell_2+2m_2N_2} = \left(\tilde{g}^{0,0}_{2N_1,2N_2}\right)_{\ell_1,\ell_2}, \quad -N_1 \le \ell_1 \le N_1 - 1, \quad -N_2 \le \ell_2 \le N_2 - 1 \tag{4.151}$$

where m_1, m_2 are any integers. That is, the zero-padded data array, $\left(\tilde{g}^{0,0}_{2N_1,2N_2}\right)_{\ell_1,\ell_2}$, is the original array plus a border of zeros, whose width is either $N_1/2$ or $N_2/2$, depending on the coordinate direction. This extended array is defined periodically over the entire plane. The periodic kernel array is defined, analogous to equation (4.139), by

$$\left(\tilde{h}_{2N_1,2N_2}\right)_{\ell_1,\ell_2} = h_{\ell_1,\ell_2}, \quad -N_1 \le \ell_1 \le N_1 - 1, \quad -N_2 \le \ell_2 \le N_2 - 1, \tag{4.152}$$

$$\left(\tilde{h}_{2N_1,2N_2}\right)_{\ell_1+2m_1N_1,\ell_2+2m_2N_2} = \left(\tilde{h}_{2N_1,2N_2}\right)_{\ell_1,\ell_2}, \quad m_1, m_2 \text{ are any integers,} \tag{4.153}$$

where the extension to the larger $2N_1 \times 2N_2$ grid is accomplished using the actual known values of the kernel function. The proof of equation (4.149) proceeds exactly as for the one-dimensional case (Exercise 4.11).

Most FFT algorithms assume the DFT is defined with indices starting at zero, as in equation (4.90). Since it is periodic, the cyclic convolution that is identical to the linear convolution is also in this case given by equation (4.149), but for indices, $\ell_1 = 0,\ldots, N_1 - 1$ and $\ell_2 = 0,\ldots, N_2 - 1$. The essential difference is in the padding of the *extended kernel array* prior to convolution. By shifting the index to start at zero, the extended part of the array must be such that when viewed as periodic over the plane it is still properly defined in the domain that is symmetric with respect to the origin. That is, one must always use $(\tilde{h}_{2N_1,2N_2})_{\ell_1,\ell_2}$ in its principal domain, equation (4.152), when extending it periodically over the plane. The formulas for padding the data and the kernel arrays, in this case, are

$$\left(\tilde{g}^{0,0}_{2N_1,2N_2}\right)_{\ell_1,\ell_2} = \begin{cases} g_{\ell_1,\ell_2}, & 0 \le \ell_1 \le N_1 - 1, \quad 0 \le \ell_2 \le N_2 - 1 \\ 0, & N_1 \le \ell_1 \le 2N_1 - 1, \quad 0 \le \ell_2 \le N_2 - 1 \\ 0, & 0 \le \ell_1 \le N_1 - 1, \quad N_2 \le \ell_2 \le 2N_2 - 1 \\ 0, & N_1 \le \ell_1 \le 2N_1 - 1, \quad N_2 \le \ell_2 \le 2N_2 - 1 \end{cases} \tag{4.154}$$

and

$$\left(\tilde{h}_{2N_1,2N_2}\right)_{\ell_1,\ell_2} = \begin{cases} h_{\ell_1,\ell_2}, & 0 \le \ell_1 \le N_1 - 1, \quad 0 \le \ell_2 \le N_2 - 1 \\ h_{\ell_1-2N_1,\ell_2}, & N_1 \le \ell_1 \le 2N_1 - 1, \quad 0 \le \ell_2 \le N_2 - 1 \\ h_{\ell_1,\ell_2-2N_2}, & 0 \le \ell_1 \le N_1 - 1, \quad N_2 \le \ell_2 \le 2N_2 - 1 \\ h_{\ell_1-2N_1,\ell_2-2N_2}, & N_1 \le \ell_1 \le 2N_1 - 1, \quad N_2 \le \ell_2 \le 2N_2 - 1 \end{cases} \tag{4.155}$$

Examples of this scheme are given in Section 6.5.1.

4.5 Discrete Functions on the Sphere

Samples on the sphere from a parent function, $g(\theta, \lambda)$, such as the gravitational field or the magnetic field, are usually gridded at equi-angular intervals corresponding to spherical coordinate differences (Lemoine et al. 1998, Maus et al. 2009), for example,

$$g_{j,\ell} \equiv g\left(\theta_j, \lambda_\ell\right) = g\left(\left(j+\frac{1}{2}\right)\Delta\theta, \left(\ell+\frac{1}{2}\right)\Delta\lambda\right), \quad j=0,\ldots,K-1, \quad \ell=-\frac{M}{2},\ldots,\frac{M}{2}-1, \tag{4.156}$$

where $K = \pi/\Delta\theta$, $M = 2\pi/\Delta\lambda$, and $\Delta\theta$, $\Delta\lambda$ are intervals, respectively, in co-latitude and longitude. Including the "1/2" in the definitions of θ_j and λ_ℓ places the samples of g at the center of each cell of a grid defined by the coordinate lines, thus avoiding a multiplicity at the poles.

The complex Fourier-Legendre transform pair of a continuous function on the sphere, given by equations (2.222) and equations (2.225), may be re-written using equation (2.223) as

$$G_{n,m}^c = \frac{\sqrt{\varepsilon_m}}{4\pi}\int_0^{2\pi}\int_0^{\pi} g(\theta,\lambda)\,\bar{P}_{n,|m|}(\cos\theta)\,e^{-im\lambda}\sin\theta\,d\theta\,d\lambda, \tag{4.157}$$

$$g(\theta,\lambda) = \sum_{m=-\infty}^{\infty}\sum_{n=|m|}^{\infty}\sqrt{\varepsilon_m}\,G_{n,m}^c\,\bar{P}_{n,|m|}(\cos\theta)\,e^{im\lambda}, \tag{4.158}$$

where, as before, n and m represent wave numbers, and

$$\varepsilon_m = \begin{cases} 1/2, & 0<|m|\le n \\ 1, & m=0 \end{cases} \tag{4.159}$$

The rarely utilized, but nonetheless equally valid, Fourier representation, equation (2.230), is obtained by interpreting θ, λ as Cartesian coordinates and viewing the function, $g(\theta, \lambda)$, as periodic in the plane, with respective periods, π in θ and 2π in λ. It is repeated here for convenience and with an interchange of summations,

$$\tilde{g}(\theta,\lambda) = \frac{1}{2\pi^2}\sum_{m=-\infty}^{\infty}\sum_{k=-\infty}^{\infty} G_{k,m}^F\, e^{i2\pi\left(\frac{k}{\pi}\theta+\frac{m}{2\pi}\lambda\right)}, \tag{4.160}$$

with Fourier transform, equation (2.231),

$$G_{k,m}^F = \int_{\lambda=0}^{2\pi}\int_{\theta=0}^{\pi}\tilde{g}(\theta,\lambda)\,e^{-i2\pi\left(\frac{k}{\pi}\theta+\frac{m}{2\pi}\lambda\right)}d\theta\,d\lambda. \tag{4.161}$$

Note that $\tilde{g}(\theta, \lambda) \equiv g(\theta, \lambda)$ on Ω, and $\tilde{g}(\theta, \lambda)$ is the periodic extension of $g(\theta, \lambda)$ on the infinite "plane" defined by $-\infty < \lambda < \infty$ and $-\infty < \theta < \infty$. For the sampled array, $\tilde{g}_{j,\ell} = g_{j,\ell}$, the corresponding truncated series represents the maximum recoverable resolution of $\tilde{g}(\theta, \lambda)$ according to the Whittaker-Shannon sampling theorem (cf. equation (4.118)),

$$\tilde{g}(\theta,\lambda) \approx \frac{1}{2\pi^2} \sum_{m=-M/2}^{M/2-1} \sum_{k=-K/2}^{K/2-1} \tilde{G}^F_{k,m} e^{i2\pi\left(\frac{k}{\pi}\theta + \frac{m}{2\pi}\lambda\right)}, \tag{4.162}$$

where

$$\tilde{G}^F_{k,m} = \Delta\theta\Delta\lambda \sum_{\ell=-M/2}^{M/2-1} \sum_{j=0}^{K-1} \tilde{g}_{j,\ell} e^{-i2\pi\left(\frac{jk}{K} + \frac{\ell m}{M}\right)}, \tag{4.163}$$

and where the Fourier Nyquist wave numbers are $K/2$ and $M/2$, respectively, for the coordinates, θ and λ, with corresponding Nyquist frequencies, $1/(2\Delta\theta)$ and $1/(2\Delta\lambda)$. Questions then naturally arise: what are the Nyquist wave numbers for the Fourier-Legendre spectrum and what is the optimal truncated Fourier-Legendre series in terms of maximum recoverable resolution?

The answers may be obtained by relating $G^c_{n,m}$ and $G^F_{k,m}$, since the Nyquist frequencies of the latter are already known. Ultimately, one also needs the relationship between $G^c_{n,m}$ and the discrete Fourier transform, $\tilde{G}^F_{k,m}$, to determine the maximum resolution recoverable from the samples. However, translating the Fourier formulation directly to the spherical case is complicated by the topology of the sphere on which the orthogonal basis functions, in contrast to the sinusoidal functions, do *not* create a *periodic* spectrum of the sample. That is, while the spherical harmonics, $\bar{P}_{n,|m|}(\cos\theta)e^{im\lambda}$, are periodic in longitude, with period, 2π, equal to the principal domain of λ, they are not periodic in co-latitude with period equal to its principal domain, $[0, \pi]$. Moreover, the condition, $n \geq |m|$, imposes a dependency between the maximum degree and order, even if the sample numbers, K and M, are independent.

It is more convenient to find a relationship between $G^F_{k,m}$ (or, $\tilde{G}^F_{k,m}$) and $G^c_{n,m}$ instead of $G_{n,m}$. Implications for $G_{n,m}$ then follow immediately from equation (2.228). Toward this objective, define the periodic function, with period, π,

$$\tilde{s}_{n,m}(\theta) = \bar{P}_{n,m}(\cos\theta)\sin\theta, \quad 0 \leq \theta \leq \pi, \tag{4.164}$$

such that

$$\tilde{s}_{n,m}(\theta + k\pi) = \tilde{s}_{n,m}(\theta), \text{ or any integer, } k. \tag{4.165}$$

It can be expressed as a Fourier series, equation (2.8),

$$\tilde{s}_{n,m}(\theta) = \frac{1}{\pi} \sum_{k=-\infty}^{\infty} S_k^{(n,m)} e^{i\frac{2\pi}{\pi}k\theta}, \tag{4.166}$$

with Fourier coefficients, equation (2.13),

$$S_k^{(n,m)} = \int_0^\pi \bar{P}_{n,m}(\cos\theta)\sin\theta e^{-i\frac{2\pi}{\pi}k\theta}\, d\theta, \quad -\infty < k < \infty. \tag{4.167}$$

Substituting equations (4.160) and (4.166) into equation (4.157) and re-arranging summations and integrations yields

$$G^c_{n,m} = \frac{\sqrt{\varepsilon_m}}{4\pi} \frac{1}{2\pi^2} \frac{1}{\pi} \sum_{k=-\infty}^{\infty} \sum_{m'=-\infty}^{\infty} G^F_{k,m'} \sum_{k'=-\infty}^{\infty} S^{(n,|m|)}_{-k'} \int_0^{2\pi} e^{i\lambda(m'-m)} d\lambda \int_0^{\pi} e^{i\frac{2\pi}{\pi}\theta(k-k')} d\theta, \tag{4.168}$$

where the reversal in the sign of k' is in accord with the symmetry of this summation index. By the orthogonality, equation (2.12), the integrals are $2\pi\delta_{m'-m}$ and $\pi\delta_{k'-k}$; hence,

$$G^c_{n,m} = \frac{\sqrt{\varepsilon_m}}{4\pi^2} \sum_{k=-\infty}^{\infty} G^F_{k,m} S^{(n,|m|)}_{-k}. \tag{4.169}$$

If $\tilde{g}(\theta, \lambda)$ is Fourier band-limited in the sense,

$$G^F_{k,m} = 0, \quad m < -\frac{M}{2}, \quad m > \frac{M}{2} - 1, \quad k < -\frac{K}{2}, \quad k > \frac{K}{2} - 1, \tag{4.170}$$

then clearly also $G^c_{n,m} = 0$, $m < -M/2$, $m > M/2 - 1$ ($n \geq |m|$), and $g(\theta, \lambda)$ is band-limited in the same sense with respect to the order, m, in its Fourier-Legendre spectrum. However, the Fourier band-limit in wave numbers, k, does not limit the spectrum in degrees, n. Indeed, combining equations (4.169) and (4.170), we have

$$G^c_{n,m} = \frac{\sqrt{\varepsilon_m}}{4\pi^2} \sum_{k=-K/2}^{K/2-1} G^F_{k,m} S^{(n,|m|)}_{-k}, \tag{4.171}$$

which is not zero for any degree, n, since the Fourier spectrum, $\{S^{(n,|m|)}_{-k}\}$, has at least one non-zero component for every n. Thus, the Fourier-Legendre Nyquist limit, strictly speaking, does not exist with respect to the co-latitude even if the parent function is Fourier band-limited in co-latitude (according to equation (4.170)).

Yet, it is desired to truncate the Fourier-Legendre series in an optimal way, recognizing that the samples cannot determine the entire function, or the entire spectrum, $\{G^c_{n,m}\}$. Since usually $\Delta\lambda = \Delta\theta$, let us assume for the moment that $M = 2K$. Considering the Nyquist limit in longitude, it is reasonable to truncate the Fourier-Legendre series at the Nyquist order, $m_{\mathcal{N}} = M/2$ (see Section 4.3.1) with corresponding maximum degree, $n_{\max} = K = M/2$. That this is the Nyquist limit in harmonic degree might also be argued from the viewpoint that the Legendre functions, with argument $\cos\theta$, are periodic on the principal domain, $[0, 2\pi]$. In fact, this is how the standard Fourier-Legendre series models of the potential fields of the Earth and other planets are truncated. But, this does not give the complete characterization of Nyquist limit in co-latitude. If $0 \leq |m| \leq M/2$, and $|m| \leq n \leq M/2$, then the roughly $(M/2)^2$ resulting independent spectral components do not fully represent the greater number, $MK = M^2/2$, of real-valued samples, $\tilde{g}_{j,\ell}$.

On the other hand, the discrete Fourier spectrum, $\{\tilde{G}^F_{k,m}\}$, as in equation (4.163), is the fullest representation of the function based on the samples. In order to obtain an equivalent representation in terms of the Legendre spectrum and to attempt answering the questions posed above, let $\tilde{r}_{n,m}(\theta)$ be a periodic function with period, π, defined by

$$\tilde{r}_{n,m}(\theta) = \bar{P}_{n,m}(\cos\theta), \quad 0 \le \theta \le \pi, \tag{4.172}$$

such that

$$\tilde{r}_{n,m}(\theta + k\pi) = \tilde{r}_{n,m}(\theta), \tag{4.173}$$

for any integer, k. The discrete Fourier transform pair for the samples of this function, $(\tilde{r}_{n,m})_j = \bar{P}_{n,m}(\cos\theta_j)$, $0 \le j \le K-1$, is (equations (4.70) and (4.71))

$$\tilde{R}_k^{(n,m)} = \Delta\theta \sum_{j=0}^{K-1} (\tilde{r}_{n,m})_j \, e^{-i\frac{2\pi}{K}jk}, \tag{4.174}$$

$$(\tilde{r}_{n,m})_j = \frac{1}{\pi} \sum_{k=0}^{K-1} \tilde{R}_k^{(n,m)} e^{i\frac{2\pi}{K}jk}. \tag{4.175}$$

From equations (4.158), (4.172), and (4.175), the samples, $g_{j,\ell}$, are formulated as

$$g_{j,\ell} = \sum_{m=-\infty}^{\infty} \sum_{n=|m|}^{\infty} \sqrt{\varepsilon_m} G_{n,m}^c \frac{1}{\pi} \sum_{k=0}^{K-1} \tilde{R}_k^{(n,|m|)} e^{i\frac{2\pi}{K}jk} e^{im\lambda_\ell}. \tag{4.176}$$

Substituting these into equation (4.163) yields

$$\tilde{G}_{k,m}^F = \Delta\theta\Delta\lambda \sum_{m'=-\infty}^{\infty} \sum_{n=|m'|}^{\infty} \sqrt{\varepsilon_{m'}} G_{n,m'}^c e^{i\frac{\pi}{M}m'} \frac{1}{\pi} \sum_{k'=0}^{K-1} \tilde{R}_{k'}^{(n,|m'|)} \sum_{j=0}^{K-1} e^{i\frac{2\pi}{K}j(k'-k)} \sum_{\ell=-M/2}^{M/2-1} e^{i\frac{2\pi}{M}\ell(m'-m)}, \tag{4.177}$$

where $\lambda_\ell = 2\pi(\ell + 1/2)/M$. By orthogonality, equation (4.68), the last two sums are $K\,\delta_{k'-k}$ and $M\,\delta_{\text{mod}_M(m'-m)}$, respectively; hence,

$$\tilde{G}_{k,m}^F = 2\pi \sum_{m'=-\infty}^{\infty} \sum_{n=|m'|}^{\infty} \sqrt{\varepsilon_{m'}} G_{n,m'}^c e^{i\frac{\pi}{M}m'} \tilde{R}_k^{(n,|m'|)} \delta_{\text{mod}_M(m'-m)}. \tag{4.178}$$

Now, $\text{mod}_M(m'-m) = 0$ holds for any integer, μ, such that $m' = \mu M + m$. Therefore,

$$\tilde{G}_{k,m}^F = 2\pi \sum_{\mu=-\infty}^{\infty} \sum_{n=|\mu M+m|}^{\infty} (-1)^\mu \sqrt{\varepsilon_{\mu M+m}} G_{n,\mu M+m}^c e^{i\frac{\pi}{M}m} \tilde{R}_k^{(n,|\mu M+m|)}, \tag{4.179}$$

which is the desired relationship between the *discrete* Fourier transform and the complex Fourier-Legendre transform. It is not quite the inverse to equation (4.169) since the latter relates the Fourier-Legendre transform, $G_{k,m}^c$, to the Fourier-series transform; and, the difference between the discrete Fourier transform, $\tilde{G}_{k,m}^F$, and the Fourier-series transform, $G_{k,m}^F$, is the aliasing error, equation (4.123).

To determine the resolving power of a discrete array of spherical data in terms of spherical wave numbers, n and m, consider isolating the $\mu = 0$ term in equation (4.179), and further splitting this into two terms,

$$\tilde{G}_{k,m}^{F} = 2\pi \sum_{n=|m|}^{|m|+K-1} \sqrt{\varepsilon_m} G_{n,m}^{c} e^{i\frac{\pi}{M}m} \tilde{R}_k^{(n,|m|)} + 2\pi \sum_{n=|m|+K}^{\infty} \sqrt{\varepsilon_m} G_{n,m}^{c} e^{i\frac{\pi}{M}m} \tilde{R}_k^{(n,|m|)}$$
$$+ 2\pi \sum_{\substack{\mu=-\infty \\ \mu\neq 0}}^{\infty} \sum_{n=|\mu M+m|}^{\infty} (-1)^{\mu} \sqrt{\varepsilon_{\mu M+m}} G_{n,\mu M+m}^{c} e^{i\frac{\pi}{M}m} \tilde{R}_k^{(n,|\mu M+m|)} \tag{4.180}$$

If the function is Fourier-Legendre band-limited in the sense,

$$G_{n,m}^{c} = 0, \quad \begin{cases} -\dfrac{M}{2} \leq m \leq \dfrac{M}{2} - 1 \text{ and } n \geq |m| + K \\ \left(m \leq -\dfrac{M}{2} - 1 \text{ or } m \geq \dfrac{M}{2} \right) \text{ and } n \geq |m| \end{cases} \tag{4.181}$$

then the last two sums in equation (4.180) vanish; and,

$$\tilde{G}_{k,m}^{F} = 2\pi \sum_{n=|m|}^{|m|+K-1} \sqrt{\varepsilon_m} G_{n,m}^{c} e^{i\frac{\pi}{M}m} \tilde{R}_k^{(n,|m|)}, \tag{4.182}$$

which gives a relationship between the *KM* coefficients, $\tilde{G}_{k,m}^{F}$, and the *KM* coefficients, $G_{n,m}^{c}$ (*K* degrees, *n*, for each of the *M* orders, *m*). Figure 4.6 shows the band-limited domain of the Fourier-Legendre spectrum according to equation (4.181).

The relationship, equation (4.182), is invertible, as indicated in Section 4.5.2. This means that the information contained in the samples, $\{g_{j,\ell}\}$, is fully captured either by the DFT, $\{\tilde{G}_{k,m}^{F}\}$, or by the set, $\{G_{n,m}^{c} \,|\, |m| \leq n \leq |m| + K - 1, -M/2 \leq m \leq M/2 - 1\}$. On this basis, one may define the Nyquist wave numbers that specify the limits of recoverable Fourier-Legendre spectral coefficients from the samples and that delineate the high-frequency spectrum that causes aliasing errors. Referring to Figure 4.6, the Nyquist wave numbers in longitude and co-latitude are

$$m_{\mathcal{N}} = \frac{M}{2} - 1, \tag{4.183}$$

$$n_{\mathcal{N}}(m) = |m| + K - 1, \quad 0 \leq |m| \leq m_{\mathcal{N}}, \tag{4.184}$$

where $m_{\mathcal{N}}$ is somewhat conservative since also coefficients with orders, $m = -M/2$, technically are recoverable. Figure 4.6 shows the case for $M = 2K$, but it is clear from the derivation of the Nyquist limits that *M* and *K* may be independent.

In summary, while the Nyquist limit associated with the sampling interval, $\Delta\lambda$, corresponds essentially to the Fourier (Cartesian) case, it is less definite with respect to the sampling interval, $\Delta\theta$. This also affects the defined spatial resolution (minimum half-wavelength) of a spherical harmonic series truncated at maximum *degree*, n_0. Often it is simply quoted as the reciprocal of the Nyquist limit in λ, which, moreover, is also casually identified with the Nyquist limit in θ, assuming $\Delta\theta = \Delta\lambda$. Under this adopted convention and the definition of the extent-bandwidth product, equation (3.119), the spatial resolution is stated as the reciprocal of the bandwidth,

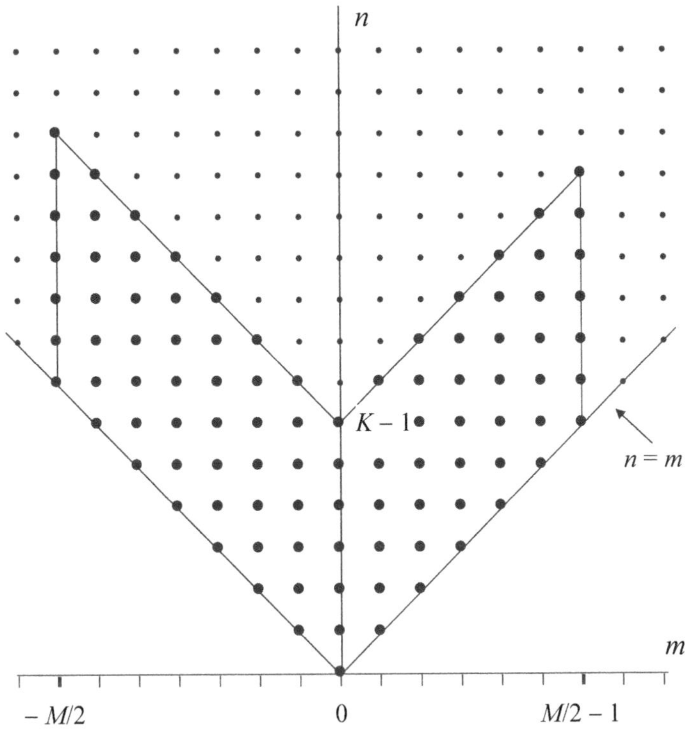

Figure 4.6: Large dots denote the domain of the Fourier-Legendre spectrum of a band-limited spherical function as defined by the limits of equation (4.181). The large and small dots together denote the entire domain, extending to $n \to \infty$ with $|m| \le n$.

$M/(2\pi) = K/\pi$, where $K \simeq n_0$, hence as $\Delta\theta = \Delta\lambda = \pi/n_0$. However, this underestimates the resolution as given perhaps more accurately by equation (3.131) that better reflects the spatial decrease in sampling interval, $\Delta\lambda$, as one approaches either pole.

4.5.1 Aliasing of the Fourier-Legendre Spectrum

For a continuous function on the sphere that is Fourier band-limited in the sense of equation (4.170), the Whittaker-Shannon sampling theorem for periodic functions, equation (4.117), states that the DFT, $\tilde{G}^F_{k,m}$, of its samples completely defines the function. Now suppose that the function is Fourier-Legendre band-limited as given by equation (4.181), and define the $K \times 1$ vectors,

$$\tilde{\boldsymbol{G}}^F_m = \left(\tilde{G}^F_{-K/2,m} \quad \cdots \quad \tilde{G}^F_{K/2-1,m} \right)^{\mathrm{T}}, \tag{4.185}$$

$$\boldsymbol{G}^c_m = \left(G^c_{|m|,m} \quad \cdots \quad G^c_{|m|+K-1,m} \right)^{\mathrm{T}}, \tag{4.186}$$

and the $K \times K$ matrix,

$$\mathbf{R}_m = 2\pi\sqrt{\varepsilon_m}\, e^{i\frac{\pi}{M}m}\begin{pmatrix} \tilde{R}_{-K/2}^{(|m|,|m|)} & \cdots & \tilde{R}_{-K/2}^{(|m|+K-1,|m|)} \\ \vdots & \ddots & \vdots \\ \tilde{R}_{K/2-1}^{(|m|,|m|)} & \cdots & \tilde{R}_{K/2-1}^{(|m|+K-1,|m|)} \end{pmatrix}. \tag{4.187}$$

Then equation (4.182), under the assumption of equation (4.181), in vector-matrix form is

$$\tilde{\boldsymbol{G}}_m^F = \mathbf{R}_m \boldsymbol{G}_m^c, \quad -\frac{M}{2} \le m \le \frac{M}{2} - 1. \tag{4.188}$$

The matrices, $\mathbf{R}_m$, defined by equations (4.187) and (4.174), are assumed to be invertible since the columns are the DFTs of uniform samples of associated Legendre functions of different degrees (and the same order) that by equation (2.210) are orthogonal. Thus, we have the Whittaker-Shannon sampling theorem for functions on the sphere. If the limited Fourier-Legendre spectrum is computed from the samples, $\tilde{g}_{j,\ell}$, of a Fourier-Legendre band-limited function (equation (4.181)) according to

$$\boldsymbol{G}_m^c = \mathbf{R}_m^{-1} \tilde{\boldsymbol{G}}_m^F, \quad -\frac{M}{2} \le m \le \frac{M}{2} - 1, \tag{4.189}$$

where the elements, $\tilde{G}_{k,m}^F$, are given by equation (4.163), then this completely determines the function for all points on the sphere,

$$\begin{aligned} g(\theta,\lambda) &= \sum_{m=-M/2}^{M/2-1} \sum_{n=|m|}^{|m|+K-1} G_{n,m}^c \left(\bar{Y}_{n,m}^c(\theta,\lambda)\right)^* \\ &= \sum_{m=-M/2}^{M/2-1} \sum_{n=|m|}^{|m|+K-1} G_{n,m} \bar{Y}_{n,m}(\theta,\lambda) \end{aligned} \tag{4.190}$$

where the second equation holds in view of the one-to-one relationship between the real and complex Legendre spectra, equations (2.227) and (2.228). It is noted that the Fourier-Legendre spectrum computed alternatively by simply discretizing the integrals of equation (2.222) according to the rectangle rule,

$$G_{n,m}^c \approx \frac{\Delta\theta\Delta\lambda}{4\pi} \sum_{j=0}^{K-1} \sum_{\ell=-M/2}^{M/2-1} g_{j,\ell} \bar{Y}_{n,m}^c\left(\theta_j, \lambda_\ell\right) \sin\theta_j, \tag{4.191}$$

does not reproduce the entire band-limited function exactly.

If the function is not band-limited, then the computation of the spectrum from the samples according to equation (4.189) is corrupted by aliasing; that is, it is corrupted by the spectral content that cannot be resolved from the finite sampling of the signal. The aliasing error may be inferred from equation (4.180). Define the semi-infinite vectors,

$$\delta\boldsymbol{G}_m^c = \begin{pmatrix} G_{|m|+K,m}^c & G_{|m|+K+1,m}^c & \cdots \end{pmatrix}^{\mathrm{T}}, \tag{4.192}$$

$$\Delta \boldsymbol{G}^{c}_{\mu,m} = \left(G^{c}_{|\mu M+m|,\mu M+m} \quad G^{c}_{|\mu M+m|+1,\mu M+m} \quad \cdots \right)^{\mathrm{T}}, \tag{4.193}$$

and the semi-infinite, $K \times \infty$, matrices,

$$\delta \mathbf{R}_{m} = 2\pi \sqrt{\varepsilon_m}\, e^{i\frac{\pi}{M}m} \begin{pmatrix} \tilde{R}^{(|m|+K,|m|)}_{-K/2} & \tilde{R}^{(|m|+K+1,|m|)}_{-K/2} & \cdots \\ \vdots & \vdots & \vdots \\ \tilde{R}^{(|m|+K,|m|)}_{K/2-1} & \tilde{R}^{(|m|+K+1,|m|)}_{K/2-1} & \cdots \end{pmatrix}, \tag{4.194}$$

$$\Delta \mathbf{R}_{\mu,m} = 2\pi (-1)^{\mu} \sqrt{\varepsilon_{\mu M+m}}\, e^{i\frac{\pi}{M}m} \begin{pmatrix} \tilde{R}^{(|\mu M+m|,|m|)}_{-K/2} & \tilde{R}^{(|\mu M+m|+1,|m|)}_{-K/2} & \cdots \\ \vdots & \vdots & \vdots \\ \tilde{R}^{(|\mu M+m|,|m|)}_{K/2-1} & \tilde{R}^{(|\mu M+m|+1,|m|)}_{K/2-1} & \cdots \end{pmatrix}. \tag{4.195}$$

Then the aliasing error, or the difference between $\boldsymbol{G}^{c}_{m}$ derived from equation (4.189) and from equation (4.180), is

$$\mathbf{R}^{-1}_{m} \tilde{\boldsymbol{G}}^{F}_{m} - \boldsymbol{G}^{c}_{m} = \mathbf{R}^{-1}_{m} \left(\delta \mathbf{R}_{m} \delta \boldsymbol{G}^{c}_{m} + \sum_{\substack{\mu=-\infty \\ \mu \neq 0}}^{\infty} \Delta \mathbf{R}_{\mu,m} \Delta \boldsymbol{G}^{c}_{\mu,m} \right), \quad -\frac{M}{2} \le m \le \frac{M}{2} - 1. \tag{4.196}$$

A less optimal aliasing error results if the spectrum is derived in other ways from the samples, for example, using equation (4.191).

4.5.2 Spectral Analysis and Synthesis on the Sphere

Solving equation (4.189) for the M vectors, $\boldsymbol{G}^{c}_{m}$, from the global array of samples, $\{g_{j,\ell}\}$, equation (4.156), is equivalent to solving for the KM coefficients, $G^{c}_{n,m}$, by inverting the linear system of KM equations,

$$g_{j,\ell} = \sum_{m=-M/2}^{M/2-1} \sum_{n=|m|}^{|m|+K-1} G^{c}_{n,m} \sqrt{\varepsilon_m} \left(\tilde{r}_{n,|m|} \right)_j e^{i\frac{2\pi}{M}m\ell}, \tag{4.197}$$

and also involves the inversion of M $K \times K$ matrices (Jekeli 1996). The spectrum is thus determined to the full extent allowed by the Nyquist limits, equations (4.183) and (4.184). However, it is common practice to compute the Fourier-Legendre spectrum from the global grid only up to the fixed degree, $K-1$. The model for the real-valued spectrum in this case is

$$g_{j,\ell} = \sum_{m=-M/2+1}^{M/2-1} \sum_{n=|m|}^{K-1} G_{n,m} \bar{Y}_{n,m}\left(\theta_j, \lambda_\ell \right), \quad j = 0,\ldots,K-1; \quad \ell = -\frac{M}{2},\ldots,\frac{M}{2}-1; \tag{4.198}$$

and, equivalently, analogous to equation (4.190), the model with complex spherical harmonics, equation (2.225), is

$$g_{j,\ell} = \sum_{m=-M/2+1}^{M/2-1} \sum_{n=|m|}^{K-1} \sqrt{\varepsilon_m} G^c_{n,m} \bar{P}_{n,|m|}\left(\cos\theta_j\right) e^{i\pi\frac{m}{M}} e^{i2\pi\frac{m\ell}{M}}, \tag{4.199}$$

where $M = 2K$ ($\Delta\lambda = \Delta\theta$), and where in each case the upper limit, $K - 1$, in the degree, n, also constrains the lower limit in the order, m, to be $-M/2 + 1$. Since the number of unknown coefficients, K^2, now is only half the number of equations, $2K^2$, they are estimated as a solution to an over-determined problem. For convenience, the following is formulated for the complex spectrum, with a transformation to the real spectrum at the end.

Restricting the vectors, $\boldsymbol{G}^c_m$, equation (4.186), to the first $K - |m|$ elements,

$$\boldsymbol{G}^c_m = \left(G^c_{|m|,m} \quad \cdots \quad G^c_{K-1,m}\right)^{\mathrm{T}}, \quad -\frac{M}{2}+1 \le m \le \frac{M}{2}-1, \tag{4.200}$$

and defining the $(K - |m|) \times 1$ vectors,

$$\boldsymbol{r}^{(j)}_m = \left(\bar{P}_{|m|,|m|}\left(\cos\theta_j\right) \quad \cdots \quad \bar{P}_{K-1,|m|}\left(\cos\theta_j\right)\right)^{\mathrm{T}}, \tag{4.201}$$

equation (4.199) becomes

$$g_{j,\ell} = \sum_{m=-M/2+1}^{M/2-1} C^{(j)}_m e^{i2\pi\frac{m\ell}{M}}, \tag{4.202}$$

where, for $j = 0,\ldots, K - 1$,

$$C^{(j)}_m = \sqrt{\varepsilon_m} e^{i\pi\frac{m}{M}} \left(\boldsymbol{r}^{(j)}_m\right)^{\mathrm{T}} \boldsymbol{G}^c_m, \quad -\frac{M}{2}+1 < m \le \frac{M}{2}-1 \tag{4.203}$$

From equations (4.74) and (4.75), it is clear that

$$C^{(j)}_m = \frac{1}{M\Delta\lambda} \mathrm{DFT}\left(g_{j,\ell}\right)_m, \tag{4.204}$$

except that the component, $C^{(j)}_{-M/2}$, thus computed, is not used in the model given by equation (4.202). $M\Delta\lambda C^{(j)}_m$ is the DFT of the data for each zone of longitudes at co-latitude, θ_j. For each m, $-M/2 + 1 \le m \le M/2 - 1$, the set of K equations (4.203) may be written in vector-matrix form as

$$\boldsymbol{C}_m = \sqrt{\varepsilon_m} e^{i\pi\frac{m}{M}} \mathbf{r}_m \boldsymbol{G}^c_m, \quad -\frac{M}{2}+1 < m \le \frac{M}{2}-1, \tag{4.205}$$

where $\boldsymbol{C}_m = (C_m^{(0)} \cdots C_m^{(K-1)})^{\mathrm{T}}$ is a $K \times 1$ vector with elements given by equation (4.204), and where, from equation (4.201),

$$\mathbf{r}_m = \begin{pmatrix} \left(\boldsymbol{r}_m^{(0)}\right)^{\mathrm{T}} \\ \vdots \\ \left(\boldsymbol{r}_m^{(K-1)}\right)^{\mathrm{T}} \end{pmatrix} = \begin{pmatrix} \bar{P}_{|m|,|m|}(\cos\theta_0) & \cdots & \bar{P}_{K-1,|m|}(\cos\theta_0) \\ \vdots & \vdots & \vdots \\ \bar{P}_{|m|,|m|}(\cos\theta_{K-1}) & \cdots & \bar{P}_{K-1,|m|}(\cos\theta_{K-1}) \end{pmatrix} \tag{4.206}$$

is a $K \times (K - |m|)$ matrix.

The set of linear equations (4.205) can be used to solve for the elements of the vectors, $\boldsymbol{G}_m^c$, from the DFT of the samples. There are K equations for any m, $-M/2 + 1 \le m \le M/2 - 1$, but only $K - |m|$ unknown coefficients, equation (4.200). Thus, except for $m = 0$, the systems of equations are over-determined and some form of constraint is needed to obtain a unique solution. Specifically, it is desired to find the solution, $\hat{\boldsymbol{G}}_m^c$, that minimizes the difference between left and right sides in the sense of the minimum squared-norm,

$$\left(\boldsymbol{C}_m - \sqrt{\varepsilon_m}\, e^{i\pi\frac{m}{M}} \mathbf{r}_m \boldsymbol{G}_m^c\right)^{\mathrm{T}} \left(\boldsymbol{C}_m - \sqrt{\varepsilon_m}\, e^{i\pi\frac{m}{M}} \mathbf{r}_m \boldsymbol{G}_m^c\right) \to \min \quad \text{with respect to } \boldsymbol{G}_m^c. \tag{4.207}$$

The solution, $\hat{\boldsymbol{G}}_m^c$, that satisfies this minimization is obtained by applying the *Moore-Penrose pseudoinverse* of $\mathbf{r}_m$ (Golub and Van Loan 1996) to equation (4.205),

$$\hat{\boldsymbol{G}}_m^c = \frac{e^{-i\pi\frac{m}{M}}}{\sqrt{\varepsilon_m}} \left(\mathbf{r}_m^{\mathrm{T}} \mathbf{r}_m\right)^{-1} \mathbf{r}_m^{\mathrm{T}} \boldsymbol{C}_m, \quad -\frac{M}{2} + 1 \le m \le \frac{M}{2} - 1, \tag{4.208}$$

where $\mathbf{r}_m^{\mathrm{T}} \mathbf{r}_m$ is a $(K - |m|) \times (K - |m|)$ matrix that is invertible because the columns of $\mathbf{r}_m$ are all assumed to be independent of each other, comprising uniform samples of orthogonal associated Legendre functions ($\mathbf{r}_m$ is assumed to have *full column rank*). From equation (2.228), and the fact that $\boldsymbol{C}_{-m} = (\boldsymbol{C}_m)^*$ (for real-valued $g_{j,\ell}$), the minimum-norm estimate of the real Fourier-Legendre spectrum is (Exercise 4.12)

$$\hat{\boldsymbol{G}}_m = \begin{cases} 2\left(\mathbf{r}_m^{\mathrm{T}} \mathbf{r}_m\right)^{-1} \mathbf{r}_m^{\mathrm{T}} \operatorname{Re}\left(e^{-i\pi\frac{m}{M}} \boldsymbol{C}_m\right), & 1 \le m \le M/2 - 1 \\ \left(\mathbf{r}_0^{\mathrm{T}} \mathbf{r}_0\right)^{-1} \mathbf{r}_0^{\mathrm{T}} \boldsymbol{C}_0, & m = 0 \\ -2\left(\mathbf{r}_m^{\mathrm{T}} \mathbf{r}_m\right)^{-1} \mathbf{r}_m^{\mathrm{T}} \operatorname{Im}\left(e^{-i\pi\frac{|m|}{M}} \boldsymbol{C}_{|m|}\right), & -M/2 + 1 \le m \le -1 \end{cases} \tag{4.209}$$

where

$$\hat{\boldsymbol{G}}_m = \begin{pmatrix} \hat{G}_{|m|,m} & \cdots & \hat{G}_{K-1,m} \end{pmatrix}^{\mathrm{T}}, \quad -\frac{M}{2}+1 \le m \le \frac{M}{2}-1. \tag{4.210}$$

The entire numerical process involves the inversion of K matrices that range in size from 1×1 ($|m| = K - 1$) to $K \times K$ ($m = 0$).

If the data errors are considered as offering weights to the data, then a corresponding least-squares solution may be formulated if the covariance matrix of the errors, $\mathbf{D}_{\varepsilon_g} = \mathbf{P}^{-1}$, is non-singular. The solution is then based on the minimization,

$$(\boldsymbol{g} - \hat{\boldsymbol{g}})^{\mathrm{T}} \mathbf{P} (\boldsymbol{g} - \hat{\boldsymbol{g}}) \to \min \quad \text{with respect to } \boldsymbol{G}_m^c, \tag{4.211}$$

where the $KM \times 1$ data vector is

$$\boldsymbol{g} = \begin{pmatrix} g_{0,-M/2} & \cdots & g_{0,M/2-1} & \cdots & g_{K-1,-M/2} & \cdots & g_{K-1,M/2-1} \end{pmatrix}^{\mathrm{T}}, \tag{4.212}$$

and the corresponding model vector, $\hat{\boldsymbol{g}}$, has elements (upon the minimization),

$$\hat{g}_{j,\ell} = \sum_{m=-M/2+1}^{M/2-1} \sqrt{\varepsilon_m}\, e^{i\pi\frac{m}{M}} \left(\boldsymbol{r}_m^{(j)}\right)^{\mathrm{T}} \hat{\boldsymbol{G}}_m^c e^{i2\pi\frac{m\ell}{M}}. \tag{4.213}$$

The least-squares solution is given by

$$\hat{\boldsymbol{G}}^c = \left(\mathbf{A}^{\mathrm{H}}\mathbf{P}\mathbf{A}\right)^{-1} \mathbf{A}^{\mathrm{H}}\mathbf{P}\boldsymbol{g}, \tag{4.214}$$

where $\mathbf{A}^{\mathrm{H}}$ is the complex transpose of the $KM \times K^2$ matrix, $\mathbf{A}$, that contains the elements,

$$a_{(j,\ell),(n,m)} = \sqrt{\varepsilon_m}\, e^{i\pi\frac{m}{M}} \bar{P}_{n,|m|}\left(\cos\theta_j\right) e^{i2\pi\frac{m\ell}{M}}. \tag{4.215}$$

In order to preserve the computational efficiency of the DFT, the error covariance matrix must be diagonal and the variances for any particular longitude must be identical; thus $\mathbf{P} = \mathrm{diag}(p_{(j,\ell)})$, where the diagonal element is $p_{(j,\ell)} = p_j$. Then an element of the $K^2 \times 1$ vector, $\boldsymbol{b} = \mathbf{A}^{\mathrm{H}}\mathbf{P}\boldsymbol{g}$, is

$$\begin{aligned} b_{(n,m)} &= \sqrt{\varepsilon_m}\, e^{i\pi\frac{m}{M}} \bar{P}_{n,|m|}\left(\cos\theta_j\right) \sum_{j=0}^{K-1} p_j \sum_{\ell=-M/2}^{M/2-1} g_{j,\ell} e^{-i2\pi\frac{m\ell}{M}} \\ &= \sqrt{\varepsilon_m}\, e^{i\pi\frac{m}{M}} \bar{P}_{n,|m|}\left(\cos\theta_j\right) \frac{1}{\Delta\lambda} \sum_{j=0}^{K-1} p_j \,\mathrm{DFT}\left(g_{j,\ell}\right)_m \end{aligned} \tag{4.216}$$

and the normal matrix, $\mathbf{A}^{\mathrm{H}}\mathbf{P}\mathbf{A}$, is block diagonal with $2K - 1$ blocks ranging in size, for $m = -M/2 + 1, \ldots, M/2 - 1$, from 1×1 to $K \times K$ and back to 1×1,

$\mathbf{a}^{(m)} =$

$$\varepsilon_m M \begin{pmatrix} \sum_{j=0}^{K-1} p_j \bar{P}_{|m|,|m|}(\cos\theta_j)\bar{P}_{|m|,|m|}(\cos\theta_j) & \cdots & \sum_{j=0}^{K-1} p_j \bar{P}_{|m|,|m|}(\cos\theta_j)\bar{P}_{K-1,|m|}(\cos\theta_j) \\ \vdots & \ddots & \vdots \\ \sum_{j=0}^{K-1} p_j \bar{P}_{K-1,|m|}(\cos\theta_j)\bar{P}_{|m|,|m|}(\cos\theta_j) & \cdots & \sum_{j=0}^{K-1} p_j \bar{P}_{K-1,|m|}(\cos\theta_j)\bar{P}_{K-1,|m|}(\cos\theta_j) \end{pmatrix}. \tag{4.217}$$

With careful attention to the indices it is readily shown that the least-squares solution reduces to

$$\hat{\boldsymbol{G}}_m^c = \frac{e^{-i\pi\frac{m}{M}}}{\sqrt{\varepsilon_m}} \left(\mathbf{r}_m^{\mathrm{T}}\mathbf{p}\mathbf{r}_m\right)^{-1} \mathbf{r}_m^{\mathrm{T}}\mathbf{p}\boldsymbol{C}_m, \quad -\frac{M}{2}+1 \le m \le \frac{M}{2}-1, \tag{4.218}$$

where $\mathbf{p} = \mathrm{diag}(p_j)$ and the vector, $\boldsymbol{C}_m$, has elements given by equation (4.204). If $\mathbf{p}$ is proportional to the identity matrix (equal weights on the data), then the least-squares estimation for the Fourier-Legendre spectrum is identical to the minimum-norm solution, equation (4.208). For variations on these methods the reader may consult (Lemoine et al. 1998, Ch.8; Colombo 1980).

The alternative, non-optimal method of spectral analysis on the sphere is based on numerical quadratures, equation (4.191) (see also Schmitz and Cain 1983, Rapp 1969), which can also be formulated in terms of the DFT of the samples, $g_{j,\ell}$. Substituting equation (2.223) into equation (4.191), an estimate of the complex spectrum is

$$\hat{G}_{n,m}^c = \sqrt{\varepsilon_m}\frac{\Delta\theta}{4\pi}\sum_{j=0}^{K-1} \mathrm{DFT}\left(g_{j,\ell}\right)_m \bar{P}_{n,|m|}(\cos\theta_j)\sin\theta_j. \tag{4.219}$$

Then, with equation (4.204) and by defining

$$\mathbf{s}_m = \begin{pmatrix} \bar{P}_{|m|,|m|}(\cos\theta_0)\sin\theta_0 & \cdots & \bar{P}_{K-1,|m|}(\cos\theta_0)\sin\theta_0 \\ \vdots & \vdots & \vdots \\ \bar{P}_{|m|,|m|}(\cos\theta_{K-1})\sin\theta_{K-1} & \cdots & \bar{P}_{K-1,|m|}(\cos\theta_{K-1})\sin\theta_{K-1} \end{pmatrix}, \tag{4.220}$$

this becomes in vector-matrix form,

$$\hat{\boldsymbol{G}}_m^c = \sqrt{\varepsilon_m}\frac{\Delta\theta}{2}\mathbf{s}_m^{\mathrm{T}}\boldsymbol{C}_m, \quad -K+1 \le m \le K-1. \tag{4.221}$$

For the real spectrum,

$$\hat{\boldsymbol{G}}_m = \begin{cases} \dfrac{\Delta\theta}{2}\mathbf{s}_m^{\mathrm{T}}\,\mathrm{Re}\left(\boldsymbol{C}_m\right), & 1 \le m \le K-1 \\ \dfrac{\Delta\theta}{2}\mathbf{s}_0^{\mathrm{T}}\boldsymbol{C}_0, & m = 0 \\ -\dfrac{\Delta\theta}{2}\mathbf{s}_{|m|}^{\mathrm{T}}\,\mathrm{Im}\left(\boldsymbol{C}_{|m|}\right), & -K+1 \le m \le -1 \end{cases} \tag{4.222}$$

where, as before, $\hat{\boldsymbol{G}}_m^c$ and $\hat{\boldsymbol{G}}_m$ are vectors whose elements constitute the Fourier-Legendre spectrum, as in equations (4.200) and (4.210). Clearly, the numerical-quadratures estimates, equation (4.221) (or, (4.222)), though not optimal, are computationally more efficient and stable for high harmonic degrees since no matrix inversion is required.

Simple integral discretizations for the analysis, such as equations (4.198) or (4.199), using data values at points of the spherical grid, may be less appropriate if the "samples" are averages over the grid cells. In this case, the right sides of these equations comprise sums of integrals of the basis functions. This is a straightforward modification that can be derived from the basic analysis equations above (Colombo 1980). Suppose that the samples on the sphere are mean values of a function over a grid cell instead of point values,

$$\bar{g}_{j,\ell} = \frac{1}{\Delta\Omega_j}\int_{\lambda_\ell-\Delta\lambda/2}^{\lambda_\ell+\Delta\lambda/2}\int_{\theta_j-\Delta\theta/2}^{\theta_j+\Delta\theta/2} g(\theta,\lambda)\sin\theta\, d\theta\, d\lambda, \tag{4.223}$$

where the area of a grid cell is

$$\Delta\Omega_j = \int_{\lambda_\ell-\Delta\lambda/2}^{\lambda_\ell+\Delta\lambda/2}\int_{\theta_j-\Delta\theta/2}^{\theta_j+\Delta\theta/2} \sin\theta\, d\theta\, d\lambda = 2\Delta\lambda\sin\left(\Delta\theta/2\right)\sin\theta_j. \tag{4.224}$$

Then, the model for spectral analysis becomes

$$\bar{g}_{j,\ell} = \sum_{m=-M/2+1}^{M/2-1}\sum_{n=|m|}^{K-1}\sqrt{\varepsilon_m}\, G_{n,m}^c\, \overline{IP}_{n,|m|}^{(j)}\, \overline{IE}_m^{(\ell)}, \tag{4.225}$$

where $\overline{IP}_{n,|m|}^{(j)}$ and $\overline{IE}_m^{(\ell)}$ are defined by the averages,

$$\overline{IP}_{n,|m|}^{(j)} = \frac{1}{2\sin\left(\Delta\theta/2\right)\sin\theta_j}\int_{\theta_j-\Delta\theta/2}^{\theta_j+\Delta\theta/2} \bar{P}_{n,|m|}\left(\cos\theta\right)\sin\theta\, d\theta, \tag{4.226}$$

$$\overline{IE}_m^{(\ell)} = \frac{1}{\Delta\lambda}\int_{\lambda_\ell-\Delta\lambda/2}^{\lambda_\ell+\Delta\lambda/2} e^{im\lambda}\, d\lambda = \mathrm{sinc}\left(\frac{m}{M}\right)e^{im\lambda_\ell}, \tag{4.227}$$

and the last equation is derived with the definition of the sinc function, equation (2.93). The integrals of the associated Legendre functions may be evaluated using a recursion algorithm (Paul 1978, Jekeli et al. 2007, Fukushima 2012b).

Proceeding as for point samples, it is not difficult to see, analogous to equations (4.204) and (4.203), that if one defines

$$\bar{C}_m^{(j)} = \sqrt{\varepsilon_m}\,\mathrm{sinc}\left(\frac{m}{M}\right) e^{i\pi\frac{m}{M}} \left(\bar{\boldsymbol{r}}_m^{(j)}\right)^{\mathrm{T}} \boldsymbol{G}_m^c, \quad -K+1 < m \le K-1, \tag{4.228}$$

where

$$\bar{\boldsymbol{r}}_m^{(j)} = \begin{pmatrix} \overline{IP}_{|m|,|m|}^{(j)} & \cdots & \overline{IP}_{K-1,|m|}^{(j)} \end{pmatrix}^{\mathrm{T}}, \tag{4.229}$$

then also,

$$\bar{C}_m^{(j)} = \frac{1}{M\Delta\lambda}\mathrm{DFT}\left(\bar{g}_{j,\ell}\right)_m. \tag{4.230}$$

Since the first zero of sinc(m/M) occurs at $m = \pm M$ (Figure 2.4) and $|m| < K < M$, equation (4.228) can be inverted as in equation (4.208) to obtain the least-squares estimates of the Fourier-Legendre spectrum, under the same conditions on the data weights,

$$\hat{\boldsymbol{G}}_m^c = \frac{e^{-i\pi\frac{m}{M}}}{\sqrt{\varepsilon_m}\,\mathrm{sinc}\left(\dfrac{m}{M}\right)} \left(\bar{\mathbf{r}}_m^{\mathrm{T}}\mathbf{p}\bar{\mathbf{r}}_m\right)^{-1} \bar{\mathbf{r}}_m^{\mathrm{T}}\mathbf{p}\bar{\boldsymbol{C}}_m, \quad -\frac{M}{2}+1 \le m \le \frac{M}{2}-1, \tag{4.231}$$

where

$$\bar{\mathbf{r}}_m = \begin{pmatrix} \left(\bar{\boldsymbol{r}}_m^{(0)}\right)^{\mathrm{T}} \\ \vdots \\ \left(\bar{\boldsymbol{r}}_m^{(K-1)}\right)^{\mathrm{T}} \end{pmatrix} = \begin{pmatrix} \overline{IP}_{|m|,|m|}^{(0)} & \cdots & \overline{IP}_{K-1,|m|}^{(0)} \\ \vdots & \vdots & \vdots \\ \overline{IP}_{|m|,|m|}^{(K-1)} & \cdots & \overline{IP}_{K-1,|m|}^{(K-1)} \end{pmatrix}. \tag{4.232}$$

The estimates for the real spectrum are analogous to equation (4.209),

$$\hat{\boldsymbol{G}}_m = \begin{cases} \dfrac{2}{\mathrm{sinc}(m/M)} \left(\bar{\mathbf{r}}_m^{\mathrm{T}}\mathbf{p}\bar{\mathbf{r}}_m\right)^{-1} \bar{\mathbf{r}}_m^{\mathrm{T}}\mathbf{p}\,\mathrm{Re}\left(e^{-i\pi\frac{m}{M}}\bar{\boldsymbol{C}}_m\right), & 1 \le m \le \dfrac{M}{2}-1 \\ \left(\bar{\mathbf{r}}_0^{\mathrm{T}}\mathbf{p}\bar{\mathbf{r}}_0\right)^{-1} \bar{\mathbf{r}}_0^{\mathrm{T}}\mathbf{p}\bar{\boldsymbol{C}}_0, & m = 0 \\ -\dfrac{2}{\mathrm{sinc}(m/M)} \left(\bar{\mathbf{r}}_m^{\mathrm{T}}\mathbf{p}\bar{\mathbf{r}}_m\right)^{-1} \bar{\mathbf{r}}_m^{\mathrm{T}}\mathbf{p}\,\mathrm{Im}\left(e^{-i\pi\frac{|m|}{M}}\bar{\boldsymbol{C}}_{|m|}\right), & -\dfrac{M}{2}+1 \le m \le -1 \end{cases} \tag{4.233}$$

Finally, the estimates of the spectrum of the *average* function from its samples and based on a simple numerical integration are obtained as in equation (4.221),

$$\hat{\bar{\boldsymbol{G}}}_m^c = \sqrt{\varepsilon_m}\frac{\Delta\theta}{2}\mathbf{s}_m^{\mathrm{T}}\bar{\boldsymbol{C}}_m, \quad -\frac{M}{2}+1 \le m \le \frac{M}{2}-1, \tag{4.234}$$

where $\bar{\boldsymbol{C}}_m$ is the vector with elements, $\bar{\boldsymbol{C}}_m^{(j)}$, given by equation (4.230). While the least-squares solution, equation (4.231), directly estimates the spectrum of the un-averaged function, converting the estimates, $\hat{\bar{\boldsymbol{G}}}_m^c$, to $\hat{\boldsymbol{G}}_m^c$ is more problematic since this requires a *deconvolution*. Using the convolution theorem (3.60) on the sphere (Section 3.4), for example, one might define a cap area corresponding to the area of an average grid cell. Then, from equation (3.88) (which holds also for the complex Fourier-Legendre spectrum),

$$\hat{G}_{n,m}^c = \frac{1}{B_n^{(\psi_s)}}\hat{\bar{G}}_{n,m}^c, \quad B_n^{(\psi_s)} \neq 0, \tag{4.235}$$

where $B_n^{(\psi_s)}$ is the Legendre spectrum, equation (2.305), of the cap function. As noted, it is assumed that this is not zero for any degree, n. However, since this frequency response oscillates similar to the sinc function, there is the possibility that $B_n^{(\psi_s)} \simeq 0$; and, empirical modifications to equation (4.235) have been implemented (e.g., Rapp and Pavlis 1990).

The *synthesis* of the function from a given Fourier-Legendre spectrum is straightforward and numerically efficient when computed on a regular grid over the entire sphere using the DFT in longitude. If the spectrum is known and band-limited and the function values are desired on a grid as in equations (4.198) or (4.199), then for the latter it is easily shown (Exercise 4.13) that

$$g_{j,\ell} = \mathrm{DFT}^{-1}\left(C_m'^{(j)}\right)_\ell, \quad j = 0,\ldots,K-1, \tag{4.236}$$

where

$$C_m'^{(j)} = \begin{cases} 0, & m = -\dfrac{M}{2} \\ \sqrt{\varepsilon_m}\, e^{i\pi\frac{m}{M}}\left(\boldsymbol{r}_m^{(j)}\right)^{\mathrm{T}}\boldsymbol{G}_m^c, & -\dfrac{M}{2}+1 \le m \le \dfrac{M}{2}-1 \end{cases} \tag{4.237}$$

and the vectors, $\boldsymbol{G}_m^c$, $\boldsymbol{r}_m^{(j)}$, are given by equations (4.200) and (4.201). For a given real-valued Fourier-Legendre spectrum, equation (2.227) yields the corresponding complex spectrum to be used in these equations. Related efficient algorithms have been developed for analysis and synthesis, for example, using the Clenshaw summation method (Gleason 1985, Holmes and Featherstone 2002).

4.5.3 DFT of Convolutions on the Sphere

Geodetic and geophysical convolutions on the sphere often involve a data function and a kernel function that depends only on a single variable, as in equation (3.58),

$$g(\theta,\lambda)*h(\theta)=\frac{1}{4\pi}\iint_{\Omega'}g(\theta',\lambda')h(\psi)\sin\theta' d\theta' d\lambda', \tag{4.238}$$

where $h(\psi)$ typically is a function of

$$\cos\psi=\cos\theta\cos\theta'+\sin\theta\sin\theta'\cos(\lambda-\lambda') \tag{4.239}$$

and attenuates to zero as the spherical distance, ψ, between evaluation and integration points increases. Then it is numerically justified that the integration region may be truncated to a neighborhood of the computation points, (θ, λ). Moreover, it is common that discrete values of the data function are distributed on a uniform grid such as defined by equation (4.156). If one could write the spherical distance, ψ, just in terms of latitude differences and longitude differences, then any convolution with such a kernel would have the form of equation (4.111) that would be amenable to fast computation using the FFT, as described by equation (4.115).

Equation (4.239) already shows that the kernel depends on the difference in longitude. For limited spherical regions, it is also possible to approximate the kernel as depending on latitude differences. Adding and subtracting $\sin\theta \sin\theta'$, one obtains

$$\begin{aligned}\cos\psi&=\cos\theta\cos\theta'+\sin\theta\sin\theta'-\sin\theta\sin\theta'+\sin\theta\sin\theta'\cos(\lambda-\lambda')\\&=\cos(\theta-\theta')-\sin\theta\sin\theta'(1-\cos(\lambda-\lambda'))\end{aligned} \tag{4.240}$$

and

$$\begin{aligned}\sin\theta\sin\theta'&=\sin\theta\sin(\theta-(\theta-\theta'))\\&=\sin^2\theta\cos(\theta-\theta')-\sin\theta\cos\theta\sin(\theta-\theta')\end{aligned} \tag{4.241}$$

Now, if the kernel is approximated by setting the co-latitudes of the evaluation points to the average co-latitude, $\theta \approx \theta_m$, of the integration region under consideration, then it depends only on $\theta - \theta'$ and $\lambda - \lambda'$.

Denoting the approximate kernel by $\hat{h}$, the convolution, equation (4.238), then becomes

$$g(\theta,\lambda)*h(\theta)\approx\frac{1}{4\pi}\iint_{\Omega_0}g(\theta',\lambda')\sin\theta'\hat{h}(\theta-\theta',\lambda-\lambda')d\theta' d\lambda', \tag{4.242}$$

where $\boldsymbol{\Omega}_0$ is the truncated spherical domain on which the approximation is numerically justified. It is noted that aside from the truncation of the integral this convolution is exact for the evaluation points with co-latitude, $\theta = \theta_m$. This suggests a compromise in computational efficiency by applying the DFT only with respect to the longitude dependence, but performing the convolution with standard numerical integration in the space domain for the co-latitude so as to avoid the error associated with the approximation, $\theta \approx \theta_m$.

Thus, one may re-write equation (4.238) as

$$c(\theta,\lambda) = g(\theta,\lambda) * h(\theta) = \frac{1}{4\pi} \iint_{\Omega} g(\theta',\lambda') h(\theta,\theta',\lambda-\lambda') \sin\theta' d\theta' d\lambda'. \tag{4.243}$$

A discretization and truncation of this integral proceeds under the assumption that the data function, g, is sampled at a uniform interval, $\Delta\lambda$, in longitude as in equation (4.156), but for any arbitrary discrete co-latitudes,

$$g_{j,\ell} = g(\theta_j, \ell\Delta\lambda), \quad j = 0,\ldots,J-1, \quad \ell = -\frac{L}{2},\ldots,\frac{L}{2}-1, \tag{4.244}$$

where, although the co-latitudes, θ_j, need not be equally spaced, we may assume a constant interval, $\Delta\theta$; and J, L delimit the area of integration. Then,

$$c_{j,\ell} = \frac{\Delta\theta\Delta\lambda}{4\pi} \sum_{j'=0}^{J-1} \sin\theta_{j'} \sum_{n=-L/2}^{L/2-1} g_{j',n} h(\theta_j,\theta_{j'})_{\ell-n}, \tag{4.245}$$

where, in order to utilize the DFT, the kernel function values must form a periodic sequence in longitude. Thus, one defines

$$h(\theta_j,\theta_{j'})_{\ell} = h(\theta_j,\theta_{j'},\ell\Delta\lambda), \quad -\frac{L}{2} \le \ell \le \frac{L}{2}-1 \tag{4.246}$$

with

$$h(\theta_j,\theta_{j'})_{\ell+mL} = h(\theta_j,\theta_{j'})_{\ell} \quad \text{for any integer } m. \tag{4.247}$$

Now, from equations (4.98) and (4.104), the discrete convolution (4.245) becomes

$$c_{j,\ell} = \frac{\Delta\theta}{4\pi} \sum_{j'=0}^{J-1} \sin\theta_{j'} \, \mathrm{DFT}^{-1}\Big(\mathrm{DFT}\big(g_{j',n}\big)_k \, \mathrm{DFT}\big(h(\theta_j,\theta_{j'})_n\big)_k \Big)_{\ell}, \tag{4.248}$$

where the DFTs are one-dimensional with respect to the longitude. The integration with respect to latitude is formulated here simply as the rectangle rule for numerical integration. Thus, one achieves at least some computational efficiency without introducing additional errors, while also keeping true to the spherical curvature of the integration domain. This procedure of applying the DFT (or, FFT) to the spherical convolution has acquired the name of the 1-D spherical FFT method (Haagmans et al. 1993).

It should be noted that the DFT formulation above still contains truncation and cyclic convolution errors, where the latter can be eliminated as discussed in Section 4.4, and the former occurs with any other numerical integration of the convolution over a limited area.

4.6 Discrete Filters and Windows

The discrete versions of the filters and windows introduced in Chapter 3, with some exceptions, follow directly from a sampling of the corresponding continuous functions. In electrical and communications engineering filters have a much greater importance as signals are massaged and manipulated to have various desirable characteristics. They are expounded invariably in terms of the time and temporal frequency domains. The discrete filters are called digital filters and the methods of their design in the engineering applications occupy many textbooks. Therefore, it would be remiss not to mention them here, at least to introduce common terminology, but for most geodetic and geophysical applications it suffices to adopt discretized versions of the filters and windows already discussed.

In general, a *digital system* is any operation that takes a digital input (i.e., a sequence of samples or data) and produces a digital output (cf. equation (3.66)):

$$y_\ell = h(g_\ell), \tag{4.249}$$

where h is the system function and the sequences, g_ℓ and y_ℓ, from a general perspective are infinite ($-\infty < \ell < \infty$), bounded, and absolutely summable. As usual, it is assumed that the sampling interval is constant and the same for the input and the output.

A linear, time-invariant digital system is called a *digital filter*; and, analogous to the case of continuous functions a digital filter is a discrete convolution (cf. equation (3.67)),

$$y_\ell = h_\ell \# g_\ell. \tag{4.250}$$

The filter is characterized by the (possibly infinite) sequence, h_ℓ. *Shift-invariance* (called *time-invariance* in temporal domain applications), also known as *translation-invariance*, implies that shifting the index of the input sequence by a certain number also shifts the index of the output sequence by that same number; in other words, the filter is independent of the index origin.

The output sequence, y_ℓ, is equal to the filter sequence, h_ℓ, if the input is a digital impulse, that is, the Kronecker delta, equation (4.22). From equation (4.35), there is

$$y_\ell = h_\ell \# \delta_\ell = \Delta x \sum_{n=-\infty}^{\infty} h_n \delta_{\ell-n}^{(\Delta x)} = h_\ell; \tag{4.251}$$

and, therefore, the filter sequence, h_ℓ, is also called the *impulse response* of the filter.

If the sequence, h_ℓ, consists of a *finite* number of non-zero elements, then the filter is known as a *finite-length impulse response* (FIR) filter; and, if h_ℓ is an infinite sequence of non-zero elements then the filter is an *infinite-length impulse response* (IIR) filter. For the IIR filter, it is required, of course, that the convolution

(4.250) exists. The filter is said to be *stable* if the output exists (is bounded) for every bounded input. It can be shown that a filter is stable if and only if the impulse response sequence, h_ℓ, is absolutely summable, equation (4.2). The discussion here is restricted to a few examples of the FIR filter, also called a moving-average filter, that are low-pass filters adapted from the filters discussed in Chapter 3.

Typical filters applied to real-valued data are also real and symmetric (see Chapter 3). In order to preserve the average of the original data sequence the zero-frequency spectral component of the filter should be unity, and this then dictates the scale of the filter sequence. By equation (4.20), the spectrum of the filter is also real and symmetric. The rectangle filter, already encountered in Section 4.2.1, as the rectangle sequence, is repeated here with slightly different definition (for even n, cf. equation (4.26)),

$$h_\ell^{(n)} = \begin{cases} \dfrac{1}{n\Delta x}, & -\dfrac{n}{2} \le \ell \le \dfrac{n}{2} - 1 \\ 0, & \text{otherwise} \end{cases} \tag{4.252}$$

Recalling that a filter is a convolution, here given by equation (4.35), the factor, Δx, is included in the denominator. Its frequency response, using equation (4.29), is given by

$$\tilde{H}^{(n)}(f) = \frac{1}{n} \frac{\sin(n\pi\Delta x f)}{\sin(\pi\Delta x f)} e^{i\pi\Delta x f}, \tag{4.253}$$

which equals unity at zero frequency, as obtained, for example, by applying l'Hôpital's rule to the limit, $f \to 0$.

Similarly, the discrete triangle (Bartlett or Fejér) filter may be defined by

$$\left(h_{\text{Bartlett}}^{(n)}\right)_\ell = h_\ell^{(n)} \# h_{\ell-1}^{(n)} = \begin{cases} \dfrac{1}{n\Delta x}\left(1 - \dfrac{|\ell|}{n}\right), & -n \le \ell \le n-1 \\ 0, & \text{otherwise} \end{cases} \tag{4.254}$$

With the translation property (4.19) and the convolution theorem (4.36), the frequency response is given by

$$\tilde{H}_{\text{Bartlett}}^{(n)}(f) = \frac{1}{n} F_n(2\pi\Delta x f), \tag{4.255}$$

where the Fejér kernel, $F_n(x)$, also periodic, is defined by (Exercise 4.14)

$$F_n(x) = \frac{1}{n} \frac{\sin^2(nx/2)}{\sin^2(x/2)}. \tag{4.256}$$

The response at the origin is unity, $\tilde{H}_{\text{Bartlett}}^{(n)}(0) = 1$.

The discrete rectangle and Bartlett filters are FIR filters. Other examples include the cosine window functions, equation (3.141), such as the Hann window, treated as filters. The discretization of these functions is straightforward and follows the discretization of the window sequences (below) with appropriate modification to accommodate the desideratum that the spectral component at the origin equals 1. On the other hand, the discretized Gaussian function, strictly speaking is an IIR filter, although practically its non-zero values are sufficiently small beyond some finite ℓ_0 so as to be negligible. Taken as in equation (4.30),

$$\left(h_{\text{Gauss}}^{(\beta)}\right)_\ell = \frac{1}{\beta} e^{-\pi(\ell\Delta x/\beta)^2}, \quad \beta > 0, \tag{4.257}$$

where the parameter for the filter function satisfies $\beta \gg \Delta x$ (see Figure 4.1), it does preserve the average of the filtered data, since the frequency response, given by equation (4.31), or its approximation,

$$\tilde{H}_{\text{Gauss}}^{(\beta)}(f) \simeq e^{-\pi(\beta f)^2}, \tag{4.258}$$

equals unity at $f = 0$.

Higher-dimensioned filters in Cartesian coordinates follow directly as for the continuous case, formed by the product of one-dimensional filters according to the separation of variables, as in equation (3.153).

The window (or taper) functions are discretized in effect along with the data function to which they are applied. For purposes of analysis, one may then also consider their spectra from this perspective. Unlike the filter sequences whose frequency responses are unity at the frequency origin, the window sequences as defined here in the space domain are equal to 1 at the origin. In that case, for example, the discrete rectangle window, $b_\ell^{(N)}$, is given by equation (4.26), the discrete Bartlett window is given by (compare with equation (4.254))

$$\left(u_{\text{Bartlett}}^{(N)}\right)_\ell = \begin{cases} 1 - \dfrac{2|\ell|}{N}, & -\dfrac{N}{2} \le \ell \le \dfrac{N}{2} - 1 \\ 0, & \text{otherwise} \end{cases} \tag{4.259}$$

and, the discrete Hann window is

$$\left(u_{\text{Hann}}^{(N)}\right)_\ell = \begin{cases} \dfrac{1}{2} + \dfrac{1}{2}\cos\left(\dfrac{2\pi\ell}{N}\right), & -\dfrac{N}{2} \le \ell \le \dfrac{N}{2} - 1 \\ 0, & \text{otherwise} \end{cases} \tag{4.260}$$

Their Fourier transforms for $|f| \le f_{\mathcal{N}} = 1/(2\Delta x)$ are, respectively,

$$\tilde{U}_{\text{Bartlett}}^{(N)}(f) = \Delta x F_{N/2}(2\pi\Delta x f) = \frac{2}{N} \frac{\sin^2(N\pi\Delta x f/2)}{\sin^2(\pi\Delta x f)}, \tag{4.261}$$

$$\tilde{U}_{\text{Hann}}^{(N)}(f) = \frac{1}{2}\tilde{B}^{(N)}(f) + \frac{1}{4}\left(\tilde{B}^{(N)}\left(f - \frac{1}{N\Delta x}\right) + \tilde{B}^{(N)}\left(f + \frac{1}{N\Delta x}\right)\right), \tag{4.262}$$

where $F_{N/2}$ is the Fejér kernel, equation (4.256), and $\tilde{B}^{(N)}$ is the spectrum of $b_\ell^{(N)}$. Due to the discretization they are very slightly aliased versions of the transforms of their continuous cousins (Section 3.6). Section 5.7.2, with equation (5.366), also recommends an alternative normalization of the data taper sequence.

The discrete version of the prolate spheroidal wave functions, considered also as tapers, is derived by setting up a concentration problem for sequences, analogous to the case for continuous functions (Section 3.6.2). For space- (index-) limited sequences, $(g_N)_{\ell}$, where $(g_N)_\ell = 0$, $\ell < -N/2$ or $\ell \geq N/2$, it is desired to find the sequence with maximum energy concentrated in a particular bandwidth, $|f| \leq f_0 < f_N$, with f_N equal to the Nyquist frequency. In other words (cf. equation (3.164)), the ratio,

$$\lambda(f_0) = \frac{\int\limits_{-f_0}^{f_0} \left|\tilde{G}_N(f)\right|^2 df}{\int\limits_{-f_N}^{f_N} \left|\tilde{G}_N(f)\right|^2 df} = \frac{\Delta x^2 \sum\limits_{\ell=-N/2}^{N/2-1} \sum\limits_{\ell'=-N/2}^{N/2-1} (g_N)_\ell (g_N)_{\ell'} \int\limits_{-f_0}^{f_0} e^{i2\pi\Delta x(\ell'-\ell)f} df}{\Delta x \sum\limits_{\ell=-N/2}^{N/2-1} (g_N)_\ell^2}, \tag{4.263}$$

should be maximized. The numerator on the far right follows from equation (4.16) and the denominator reflects Parseval's theorem, equation (4.21). From equation (3.163),

$$\int\limits_{-f_0}^{f_0} e^{i2\pi\Delta x(\ell'-\ell)f} df = \frac{\sin\left(2\pi\Delta x f_0\left(\ell'-\ell\right)\right)}{\pi\Delta x\left(\ell'-\ell\right)}; \tag{4.264}$$

hence, equation (4.263) becomes

$$\lambda(f_0) = \frac{\sum\limits_{\ell=-N/2}^{N/2-1} \sum\limits_{\ell'=-N/2}^{N/2-1} (g_N)_\ell (g_N)_{\ell'} \dfrac{\sin\left(2\pi\Delta x f_0\left(\ell'-\ell\right)\right)}{\pi\left(\ell'-\ell\right)}}{\sum\limits_{\ell=-N/2}^{N/2-1} (g_N)_\ell^2} = \frac{\mathbf{g}^{\mathrm{T}}\mathbf{A}\mathbf{g}}{\mathbf{g}^{\mathrm{T}}\mathbf{g}}, \tag{4.265}$$

where $\mathbf{A}$ is an $N \times N$ matrix with elements, $\dfrac{\sin\left(2\pi\Delta x f_0\left(\ell'-\ell\right)\right)}{\pi\left(\ell'-\ell\right)}$, and $\mathbf{g}$ is the vector of sequence elements. Assuming for the moment that $\mathbf{g}$ is a continuous variable, taking the differentials of $\lambda(f_0)$ and $\mathbf{g}$, i.e., $\delta(\lambda(f_0)\mathbf{g}^{\mathrm{T}}\mathbf{g}) = \delta(\mathbf{g}^{\mathrm{T}}\mathbf{A}\mathbf{g})$, and applying the product rule for derivatives, one obtains

$$\delta\lambda(f_0)\mathbf{g}^{\mathrm{T}}\mathbf{g} = 2\delta\mathbf{g}^{\mathrm{T}}\left(\mathbf{A}\mathbf{g} - \lambda(f_0)\mathbf{g}\right). \tag{4.266}$$

Thus, if $\lambda(f_0)$ is maximum then $\delta\lambda$ vanishes for arbitrarily small $\delta\mathbf{g}$, or

$$\mathbf{A}\mathbf{g} - \lambda(f_0)\mathbf{g} = \mathbf{0}. \tag{4.267}$$

This is now a standard eigenvector/eigenvalue problem. Because **A** is symmetric and invertible, the eigenvalues (concentration ratios) as in the continuous case (Section 3.6.2) are real, distinct, and positive by equation (4.263), but finite in number. Thus, they can be ordered,

$$1 > \lambda_0(f_0) > \ldots > \lambda_{N-1}(f_0) > 0. \tag{4.268}$$

The eigenvector, $\boldsymbol{g} = \psi_0^{(N,f_0)}$, associated with the maximum eigenvalue, $\lambda_0(f_0)$, solves the concentration problem. However, the matrix, **A**, is poorly conditioned, where the ratio of largest to smallest eigenvalues typically is many orders of magnitude. Fortunately, the solution to equation (4.267) is also a solution to a finite difference equation analogous to the Sturm-Liouville differential equation, which offers a stable means to determine the eigenvectors. The algorithms to compute these are outside the present scope; see, e.g. (Percival and Walden 1993, p.386; Gruenbacher and Hummels 1994). The elements of the eigenvector form a *discrete prolate spheroidal sequence* (dpss). Figure 4.7 shows examples of the dpss, $(\psi_0^{(N,f_0)})_\ell$, and its spectrum, equation (4.12),

$$\tilde{\Psi}_0^{(N,f_0)}(f) = \Delta x \sum_{\ell=-N/2}^{N/2-1} \left(\psi_0^{(N,f_0)}\right)_\ell e^{-i2\pi\Delta x f \ell}, \tag{4.269}$$

for the cases, $N = 50$, $\Delta x = 1$, $f_0 = 10/N = 0.4 f_{\mathcal{N}}$ and $f_0 = 2/N = 0.08 f_{\mathcal{N}}$. Each of these sequences is essentially limited in energy to the band of frequencies, $[-f_0, f_0]$.

The dpss may be viewed as the discretization of the prolate spheroidal wave functions encountered in Section 3.6.2. Indeed, the elements of the matrix, **A**, equation (4.265), are discrete samples of the kernel function, $\Delta x \cdot d_1(x, x'; f_0)$, equation (3.163). It is noted in Section 3.6.2 that the one-dimensional case and generalizations to the Cartesian plane and to the sphere share similar properties, since the corresponding kernel functions are essentially identical in the sense of being the transform of either a rectangle or cap function (assuming the spectral concentration region is a rectangle or cap). The same holds for the discretization in higher-dimensioned domains, and

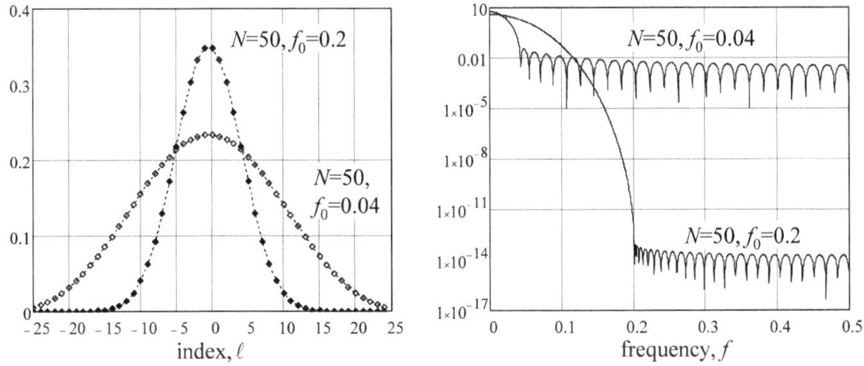

Figure 4.7: Discrete prolate spheroidal sequences, $(\psi_0^{(N,f_0)})^\ell$, (left) and their amplitude spectra, $\tilde{\Psi}_0^{(N,f_0)}(f)$, (right) for $Nf_0 = 10 = 50\cdot 0.2$ and $Nf_0 = 2 = 50\cdot 0.04$.

the details are only in the numerical computation of the sequences, which are left to other sources, notably Simons and Wang (2011).

As in the case of extent-limited functions on the continuous domain, the finite eigenvectors,

$$\boldsymbol{\psi}_j^{(N,f_0)} = \left(\left(\psi_j^{(N,f_0)}\right)_{-N/2} \quad \cdots \quad \left(\psi_j^{(N,f_0)}\right)_{N/2-1} \right)^{\mathrm{T}}, \tag{4.270}$$

corresponding to the eigenvalues, $\lambda_j(f_0)$, form an orthogonal basis for the index-limited sequences. These basis sequences are also known as *Slepian sequences*. Again, the eigenvalues are mostly either close to unity or close to zero, where the number of significant ("near-unity") values is the Shannon number, $2N\Delta xf_0$. The space of index-limited sequences that are almost band-limited is approximately spanned by the corresponding eigenvectors. Each eigenvector, $\boldsymbol{\psi}_j^{(N,f_0)}$, $j = 0,1,\ldots, 2N\Delta xf_0-1$, is maximally concentrated in succession according to the ratios, $\lambda_j(f_0)$, in the spectral band, $[-f_0, f_0]$. Thus, each one could also serve as a taper sequence that has minimal spectral leakage, which has important application in power spectral density estimation (Section 5.7.2). Figure 4.8 shows the amplitude spectra of several low-order dpss's, $\boldsymbol{\psi}_j^{(N,f_0)}$, for $\Delta x = 1$, $N = 60$, $f_0 = 0.05$, and corresponding eigenvalues, $\lambda_j(f_0)$; the first $2N\Delta xf_0 = 6$ ($j = 0,\ldots, 5$) are close to 1 in value.

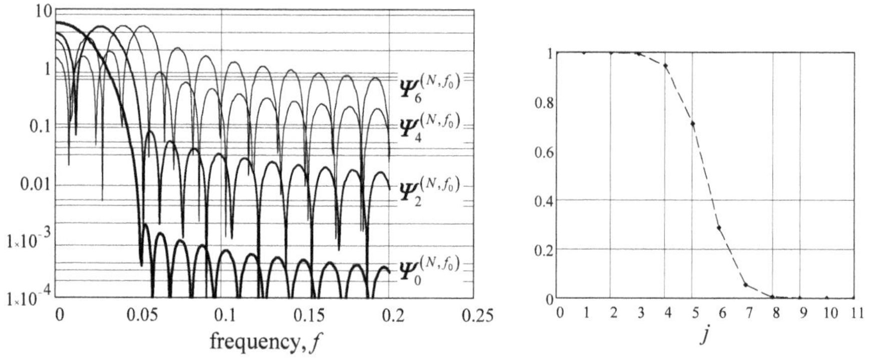

Figure 4.8: DPSS amplitude spectra, $|\tilde{\Psi}_j^{(N,f_0)}(f)|$, $j = 0, 2, 4, 6$, for $N = 60$ and $f_0 = 0.05$ (left); and eigenvalues, $\lambda_j(f_0)$, $j = 0, 1, \ldots, 11$ (right). The Shannon number is $2N\Delta xf_0 = 6$. Note that on the left only part of the frequency domain, $|f| \le f_N = 0.5$, is shown.

***

EXERCISES

Exercise 4.1

Derive equation (4.27) with $r = e^{ix}$ using the identity,

$$\sum_{k=0}^{n-1} r^k = \frac{1-r^n}{1-r}, \quad r \neq 1, \tag{E4.1}$$

by showing that this implies

$$\sum_{k=-n}^{n} r^k = \frac{r^{-(n+1/2)} - r^{n+1/2}}{r^{-1/2} - r^{1/2}}. \tag{E4.2}$$

Exercise 4.2

Prove the Fourier transform of the rectangle sequence, equation (4.29). Hint: Use the Dirichlet kernel, equation (4.27), and note that it is symmetric, $D_n(-x) = D_n(x)$, to show first that $\tilde{B}^{(n)}(f) = \Delta x\, D_{n/2}(2\pi\Delta x f) - \Delta x e^{-i2\pi\Delta x n f}$ (n even).

Exercise 4.3

Prove the Fourier transform pair (4.36) using the indicated procedure.

Exercise 4.4

Prove the Fourier transform pair (4.37) using the indicated procedure, making use of the periodic delta function, equation (4.25).

Exercise 4.5

For a single sinusoid, $g(x)$, with frequency, $\pm f_g$, prove that the aliasing frequency is given by equation (4.58). Hint: Assume $f_N < |f_g|$ and make use of equation (4.57) to show that

$$|f_a| = \left(\left|f_g\right| + f_N - 2nf_N\right) - f_N. \tag{E4.3}$$

Show also that the formula holds if $|f_g| \leq f_N$, verifying that there is no aliasing.

Exercise 4.6

Prove equation (4.68) by first showing that it is N if $\mathrm{mod}_N(\ell - j) = 0$. Then write

$$\sum_{k=k_0}^{k_0+N-1} e^{i\frac{2\pi}{N}k(\ell-j)} = e^{i\frac{2\pi}{N}k_0(\ell-j)} \sum_{k=0}^{N-1} e^{i\frac{2\pi}{N}k(\ell-j)}, \tag{E4.4}$$

and, using (E4.1) with $r = e^{ix}$ and $x = 2\pi(\ell - j)/N$, show that (E4.4) is zero if $\mathrm{mod}_N(\ell - j) \neq 0$.

Exercise 4.7

Show that $\tilde{G}'_k = \Delta x \sum_{\ell=-N/2}^{N/2-1} \tilde{g}_\ell e^{-i\frac{2\pi}{N}k\ell} = \Delta x \sum_{\ell=0}^{N-1} \tilde{g}_\ell e^{-i\frac{2\pi}{N}k\ell} = \tilde{G}_k$ for a discrete periodic sequence, $\tilde{g}_\ell$. Hint: separate the negative and positive wave numbers in the left sum and make use of the periodicity of $\tilde{g}_\ell$.

Exercise 4.8

Prove discrete Fourier transform pairs (4.78) through (4.83) in a manner similar to the proofs for corresponding pairs (2.20) through (2.24) and (2.77).

Exercise 4.9

Prove the equivalence (4.84) in a manner similar to the proof of the corresponding equivalence (2.26).

Exercise 4.10

Derive equation (4.136) from equation (4.135) using the defined periodicity of $\tilde{h}_{N'}$, equation (4.131).

Exercise 4.11

Prove the expression for the two-dimensional discrete convolution, equation (4.149), in terms of the corresponding cyclic convolution, equation (4.148), using the same procedure as for the one-dimensional case.

Exercise 4.12

Derive the minimum-norm estimate of the real-valued Fourier-Legendre transform, equation (4.209), from the corresponding complex-valued estimate, equation (4.208), using the relationship (2.228).

Exercise 4.13

Verify equations (4.236) and (4.237) using the indicated procedure.

Exercise 4.14

Show that the Fejér kernel, equation (4.256), is also given by

$$F_n(x) = \frac{1}{n}\sum_{k=0}^{n-1} D_k(x), \tag{E4.5}$$

where $D_k(x)$ is the Dirichlet kernel, equation (4.27), as well as

$$F_{n+1}(x) = \frac{1}{n+1} D_{n/2}^2(x), \quad n \text{ even.} \tag{E4.6}$$

Then show that the frequency response of the discrete Bartlett filter, equation (4.254), is given by equation (4.255). Hint: the Bartlett filter is the convolution of discrete rectangle filters.

Chapter 5

Correlation and Power Spectrum

5.1 Introduction

Correlation is generally associated with statistics and the degree to which two random quantities are related. The concept may be extended to non-random signals that have definite structure, although in many cases there is a physical reason for the correlation that is better explained by an analytic model based on well founded theory. The subsurface geologic configuration of the Earth and the external gravitational field are evidently correlated because the physical model of the mass-density integral, equation (6.17), gives a causal relationship. The gravitational and magnetic fields may also be correlated on account of a physical model, Poisson's relationship, equation (6.230), although the rather stringent assumptions associated with the model often are not realized. The correlation between the external gravitational field and the geometry of the topography, either on land or on the ocean bottom is also well established, but ostensibly strongest at the shorter wavelengths. Similarly, if there is correlation between the gravitational and magnetic fields it tends to be at the medium to shorter wavelengths.

These examples refer to the correlation between different geophysical signals, but just as important is the correlation of a signal with itself at different spatial (or temporal) scales. This auto-correlation is just another way of describing the energy or power of a signal at different wavelengths. For example, the height of the geoid (approximately sea level) above some geometric reference surface, such as an ellipsoid, has most of its energy at long wavelengths so that between points near one another it is highly correlated. The gradients of the gravitational vector, on the other hand, are dominated by short-wavelength energy and are virtually uncorrelated between points that are far apart. Understanding the distribution of signal strength in its spectral domain is crucial when designing mobile systems that measure the signal. The accuracy and spatial resolution of the measurements essentially dictate what parts of the spectrum are observable; conversely, the desired or interesting part of the spectrum prescribes the performance characteristics of the measurement system. For example, in order to model and interpret the deep lithospheric structure at tectonic plate boundaries, it is a wasted effort to make extensive point measurements of gravity at high spatial resolution, when a gravity model derived much more efficiently from an airborne survey, or even a satellite mission, yields sufficient resolution to help develop a particular hypothesis on the evolutionary tectonic processes. Conversely, measurements of the gravitational or magnetic field by a satellite system that typically requires an averaging time of several seconds have little hope of discerning very local features in the field (say, wavelengths shorter than a few kilometers) given that the speed of a low-orbiting satellite (several hundred kilometers in altitude) is about 7 km/s.

Correlation functions and their cousins, the covariance functions, play a significant role in the statistical analysis of random signals and associated optimal estimation theory and geostatistics (Section 6.6). In particular, the central feature of least-squares collocation in physical geodesy is knowledge of the covariance function of the disturbing potential or the gravity anomaly. Analogously, the variogram is essential to the method of kriging. Knowing or modeling the mutual correlation between different types of geophysical signals allows an operational approach to estimating one from the other that relies less on theoretical models than on statistical, correlative relationships.

This chapter develops the concepts of correlation with emphasis on the spectral domain. Even for applications strictly in the space domain, determining the correlative structure of a geophysical signal or between such signals often can be done more efficiently numerically using spectral domain techniques. This mirrors exactly the computation of convolutions by using the Fourier transforms of the functions being convolved. The spectral techniques take on greater importance in correlative analysis when proper modeling in specific spectral bands is required. Clearly, geostatistical estimation of high-frequency signals necessitates a correlation model that is reasonably accurate at high frequencies.

Before invoking a statistical interpretation of geophysical signals, one may first define correlation from a strictly deterministic view for functions given in one or more spatial dimensions. Then, only a slight practical modification (although a more involved theoretical consideration) is necessary with the introduction of random, or stochastic, processes whose correlation and covariance functions have many of the same properties as their counterparts for deterministic functions. Whether by the statistical or deterministic approach, the functions that represent geophysical signals, as always, are assumed to be real and bounded (although the presentation could include complex functions; indeed, many of the properties of correlation hold in this more general approach).

Correlation in statistics is defined usually as the normalized covariance between two random variables, that is, the covariance divided by the square root of the product of the variances of the two variables. This may be extended to covariance functions for stochastic, or random, processes (e.g., Priestley 1981) by defining the correlation function as the covariance function normalized by the variance. In this case, one is simply a scaled version of the other. Here, an alternative definition is chosen, also used by Maybeck (1979), Papoulis (1991), and Bendat and Piersol (1986). The correlation and covariance functions are related instead on account of a bias in the process. That is, the correlation function operates on random variables like the covariance, but without their means removed. Thus, the covariance function is the correlation function for centralized processes (i.e., whose means are removed). For deterministic (non-stochastic) functions, the correlation function is then also similar to the statistical correlation function; while the covariance function, having a rather more statistical connotation, is not specifically used in the deterministic realm.

As usual, functions on the continuous one-dimensional domain are considered first, with natural extensions to the plane, including finite-energy and periodic functions, the latter emphasized for the spherical domain. A new set of functions is also introduced, known as finite-power functions, which then prepares the stage for stochastic processes in Cartesian space, leading also to stochastic processes on the sphere. Finally practical applications demand also correlation and covariance functions for sequences of any of the above mentioned types. The Chapter concludes with a discussion on the estimation of covariance and power spectral density.

5.2 Correlation of Finite-Energy Functions

Let g and h represent real, absolutely integrable, finite-energy, functions on the continuous domain, $-\infty < x < \infty$, that is, functions satisfying equations (2.49) and (2.50). Then the *(cross-) correlation function* of g and h is defined by

$$\overset{\circ}{\phi}_{g,h}(x) = \int_{-\infty}^{\infty} g(x')h(x'+x)\,dx'. \tag{5.1}$$

It is bounded for all x according to Schwarz's inequality (1.22). If $g = h$, then $\overset{\circ}{\phi}_{g,g}(x)$ is known as the *(auto-) correlation function* of g. The "cross-" and "auto-" qualifiers are included only when emphasis is needed or confusion might arise. The argument of the correlation function is also known as the shift, or lag, distance between the arguments of the two functions, g and h. Physically, the correlation may be interpreted roughly as the amount of similarity between the functions as the shift distance increases, where zero lag for the auto-correlation function clearly yields the greatest similarity (see also equation (5.12)). The "o" notation is used for finite-energy functions in order to distinguish this correlation from that of the finite-power functions (Sections 5.3). As introduced above, some authors prefer to normalize the correlation function so that $\overset{\circ}{\phi}_{g,g}(0) = 1$; except where noted they are left un-normalized in this text. As an aside, the correlation function for complex functions is given by

$$\overset{\circ}{\phi}_{g,h}(x) = \int_{-\infty}^{\infty} g^*(x')h(x'+x)\,dx', \tag{5.2}$$

which is useful in some derivations even for real functions since their Fourier spectra generally are complex. Indeed, as already noted, the entire development of correlation functions could be based strictly on complex functions, but following previous chapters it is mostly eschewed here as an unnecessary generalization since all applications are for real signals in geodesy and geophysics.

Figure 5.1 illustrates typical auto-correlation functions. Salient features include the value at the origin, which is always positive and greater than the magnitude of the

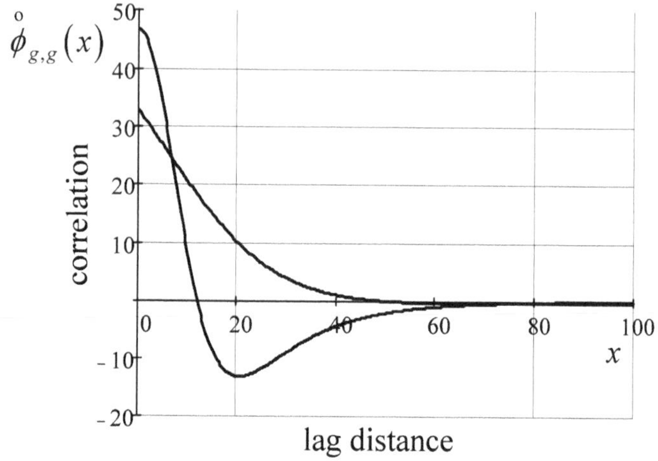

Figure 5.1: Sample auto-correlation functions obtained from an analytic model.

correlation at any other x, and that (for $x \neq 0$) it can be both positive and negative. The "width" of the initial attenuation defines the point of "decorrelation" of the function as the lag distance increases; it is called the *correlation distance* and has various formal definitions. For example, it is the point where the correlation is half the value at the origin. Also, it is noted that the function eventually (as $x \to \pm\infty$) decorrelates completely (zero correlation). These and other properties (such as symmetry with respect to the origin) are proved below.

The correlation function, $\overset{\circ}{\phi}_{g,h}(x)$, is similar (but not identical) to the convolution; in fact, the correlation function for real functions can be formulated as the convolution,

$$\overset{\circ}{\phi}_{g,h}(x) = \int_{-\infty}^{\infty} g(x')h(x+x')\,dx' = \int_{-\infty}^{\infty} g(-x')h(x-x')\,dx' = g(-x)*h(x), \tag{5.3}$$

obtained by a change in integration variable, $x' \to -x'$, and making use of equation (3.4). If g is an even function ($g(-x) = g(x)$) and real, then, clearly, $\overset{\circ}{\phi}_{g,h}(x) = g(x)*h(x)$. In any case, the units of the correlation function are the same as those for the convolution—the product of the units of g and h multiplied by the units of x.

Because it is a kind of convolution the correlation for finite-energy functions is continuous and has a Fourier transform,

$$\mathcal{F}\left(\overset{\circ}{\phi}_{g,h}(x)\right) \equiv \overset{\circ}{\Phi}_{g,h}(f) = \int_{-\infty}^{\infty} \phi_{g,h}(x)e^{-i2\pi xf}\,dx, \tag{5.4}$$

which, with the convolution theorem (3.17), is given by

$$\begin{aligned}\overset{\circ}{\Phi}_{g,h}(f) &= \mathcal{F}\left(g(-x)\right)\mathcal{F}\left(h(x)\right) = G(-f)H(f)\\ &= G^*(f)H(f)\end{aligned} \tag{5.5}$$

in view of the symmetry property (2.72) of Fourier transforms and the equivalence relation (2.84). In fact, this also holds for complex-valued functions, since in this case,

$$\overset{\circ}{\phi}_{g,h}(x) = g^*(-x) * h(x), \tag{5.6}$$

and $\mathcal{F}(g^*(-x)) = G^*(f)$; and, hence, the result follows by the convolution theorem. The inverse transform of $\overset{\circ}{\Phi}_{g,h}$, therefore, is

$$\overset{\circ}{\phi}_{g,h}(x) = \int_{-\infty}^{\infty} G^*(f) H(f) e^{i2\pi fx} df. \tag{5.7}$$

The correlation function, like the convolution, can be obtained directly, via the inverse Fourier transform, from the Fourier transforms of the correlated functions. If either of the functions, g and h, is band-limited, then clearly, from equation (5.5), so is the correlation function, $\overset{\circ}{\phi}_{g,h}(x)$.

Setting $x = 0$ in equations (5.2) and (5.7) yields Parseval's theorem (2.87),

$$\int_{-\infty}^{\infty} g^*(x') h(x') dx' = \int_{-\infty}^{\infty} G^*(f) H(f) df. \tag{5.8}$$

As already noted with equation (2.89), the spectrum of the correlation function, $\overset{\circ}{\phi}_{g,g}(x)$, given by

$$\overset{\circ}{\Phi}_{g,g}(f) = |G(f)|^2, \tag{5.9}$$

indicates the distribution of the total energy of g per frequency; and, is called the *(auto-) energy spectral density* of g. And, $\overset{\circ}{\Phi}_{g,h}(f)$ is the *(cross-) energy spectral density* of g and h. The units of the energy spectral density are the units of the product of the Fourier transforms of g and h.

The correlation function and its spectrum have a number of important and useful properties, listed and proved below. They hold also for complex functions, unless otherwise noted for the particular given formulation. These properties must be ensured when modeling a correlation function from empirical values; and, they facilitate the propagation of correlation to linear operations of functions, such as derivatives and convolutions.

1. $\overset{\circ}{\phi}_{g,g}(0) > 0$; (5.10)
2. $\overset{\circ}{\phi}_{g,h}(x)$ is real if g and h are real; (5.11)
3. $\left|\overset{\circ}{\phi}_{g,h}(x)\right|^2 \le \overset{\circ}{\phi}_{g,g}(0)\overset{\circ}{\phi}_{h,h}(0)$, $-\infty < x < \infty$; (5.12)
4. $\lim_{x\to\pm\infty} \overset{\circ}{\phi}_{g,h}(x) = 0$; (5.13)

5. $\overset{\circ}{\phi}_{g,h}(-x) = \overset{\circ}{\phi}_{h,g}(x)$, if g and h are real; (5.14)

6. $\frac{d}{dx}\overset{\circ}{\phi}_{g,h}(x) = \overset{\circ}{\phi}_{g,\frac{dh}{dx}}(x) = -\overset{\circ}{\phi}_{\frac{dg}{dx},h}(x)$, if g and h are real; (5.15)

7. $\overset{\circ}{\phi}_{g*h,g*h}(x) = \overset{\circ}{\phi}_{g,g}(x) * \overset{\circ}{\phi}_{h,h}(x)$; (5.16)

8. $\overset{\circ}{\Phi}_{g,g}(f)$ is real and $\overset{\circ}{\Phi}_{g,g}(f) \geq 0$; (5.17)

9. $\overset{\circ}{\Phi}_{g,h}(-f) = \overset{\circ}{\Phi}_{h,g}(f)$ if g and h are real; (5.18)

10. $\overset{\circ}{\Phi}_{\frac{dg}{dx},h}(f) = -i2\pi f G^*(f)H(f) = -\overset{\circ}{\Phi}_{g,\frac{dh}{dx}}(f)$; (5.19)

11. $\overset{\circ}{\Phi}_{\frac{d^pg}{dx^p},\frac{d^qh}{dx^q}}(f) = i^{q-p}(2\pi f)^{p+q} G^*(f)H(f) = i^{q-p}(2\pi f)^{p+q}\overset{\circ}{\Phi}_{g,h}(f)$; (5.20)

12. $\overset{\circ}{\Phi}_{g*h,g*h}(f) = \overset{\circ}{\Phi}_{g,g}(f)\overset{\circ}{\Phi}_{h,h}(f) = |G(f)|^2\,|H(f)|^2$. (5.21)

Inequality (5.10) is obvious from equation (5.1) for functions that are not identically zero, since g is a finite-energy function (equation (2.50)). It means that the total energy of the function is positive (only the zero-function has zero energy). Property (5.11) is self-evident from the correlation definition, equations (5.1) or (5.2). The proof of inequality (5.12) is based on Schwarz's inequality (1.22) and is left as Exercise 5.1. Equation (5.13) says that as the lag increases the two functions eventually decorrelate completely. This follows mathematically from the condition that the functions are integrable and themselves go to zero at infinity. For a proof of equation (5.14) consider a change in the variable of integration, $u' = x' - x$, and apply equation (5.1),

$$\overset{\circ}{\phi}_{g,h}(-x) = \int_{-\infty}^{\infty} g(x')h(x'-x)\,dx' = \int_{-\infty}^{\infty} g(u'+x)h(u')\,du' \\ = \overset{\circ}{\phi}_{h,g}(x) \tag{5.22}$$

In the case, $g = h$, the auto-correlation function, $\overset{\circ}{\phi}_{g,g}(x)$, is an even function, i.e., symmetric with respect to the origin. With $h = g$, hence $H(f) = G(f)$, in equation (5.5), equation (5.17) must also hold—the auto-energy spectral density of a function is real and non-negative. This also means that the auto-correlation, $\overset{\circ}{\phi}_{g,g}(x)$, by definition is a *positive semidefinite function* (Papoulis 1977, p.315). With equivalence relation (2.84) applied to equation (5.5) $\overset{\circ}{\Phi}_{g,g}(f)$ is also an even function for real functions. More generally, equation (5.18) follows from equation (5.14) and the symmetry property (2.72) of Fourier transforms.

Provided the derivatives exist, the first equality of equation (5.15) is a direct consequence of the correlation definition, equation (5.1); and, the second equality is proved, again, with a change in variable, $u = x' + x$,

$$\frac{d}{dx}\overset{\circ}{\phi}_{g,h}(x)=\frac{d}{dx}\left(\int_{-\infty}^{\infty}g(u-x)h(u)\,du\right)=-\int_{-\infty}^{\infty}h(u)g'(u-x)\,du$$

$$=-\overset{\circ}{\phi}_{h,\frac{dg}{dx}}(-x) \tag{5.23}$$

$$=-\overset{\circ}{\phi}_{\frac{dg}{dx},h}(x)$$

where g' denotes the derivative of g with respect to its argument, and where the last equality comes from equation (5.14). Equations (5.15) exemplify the *law of propagation of correlations* from a function to its derivatives. Equations (5.19) and (5.20) thus follow directly from property (2.75). Finally, the spectrum of the correlation function of a convolution, $c(x) = (g*h)(x)$, given by equation (5.9), has the equivalent expression,

$$\overset{\circ}{\Phi}_{c,c}(f)=\left|C(f)\right|^2=\left|G(f)H(f)\right|^2=\left|G(f)\right|^2\left|H(f)\right|^2=\overset{\circ}{\Phi}_{g,g}(f)\overset{\circ}{\Phi}_{h,h}(f), \tag{5.24}$$

using the convolution theorem (3.17) in the second equality. This proves equation (5.21) and also equation (5.16), again by the convolution theorem. That is, the correlation function of a convolution is the convolution of correlation functions.

Extension to higher Cartesian dimensions is straightforward, as in Chapters 2 and 3. The two-dimensional correlation function for real finite-energy functions, $g(x_1, x_2)$ and $h(x_1, x_2)$, is given by

$$\overset{\circ}{\phi}_{g,h}(x_1,x_2)=\int_{-\infty}^{\infty}\int_{-\infty}^{\infty}g(x_1',x_2')h(x_1'+x_1,x_2'+x_2)\,dx_1'dx_2'; \tag{5.25}$$

and, the energy spectral density is, analogous to equation (5.5),

$$\overset{\circ}{\Phi}_{g,h}(f_1,f_2)=G^*(f_1,f_2)H(f_1,f_2). \tag{5.26}$$

The following properties, similar to corresponding properties in the one-dimensional case, are easily proved as above (Exercise 5.2).

1. $\overset{\circ}{\phi}_{g,g}(0,0) > 0;$ (5.27)
2. $\left|\overset{\circ}{\phi}_{g,h}(x_1, x_2)\right|^2 \le \overset{\circ}{\phi}_{g,g}(0,0)\overset{\circ}{\phi}_{h,h}(0,0);$ (5.28)
3. $\overset{\circ}{\phi}_{g,h}(-x_1, x_2) = \overset{\circ}{\phi}_{h,g}(x_1, -x_2)$, if g and h are real; (5.29)
4. $\dfrac{\partial}{\partial x_k}\overset{\circ}{\phi}_{g,h}(x_1,x_2)=\overset{\circ}{\phi}_{g,\frac{\partial h}{\partial x_k}}(x_1,x_2)=-\overset{\circ}{\phi}_{\frac{\partial g}{\partial x_k},h}(x_1,x_2),\ k=1,2;$ (5.30)
5. $\overset{\circ}{\Phi}_{g,g}(f_1, f_2)$ is real and $\overset{\circ}{\Phi}_{g,g}(f_1, f_2) \ge 0;$ (5.31)

6. $\overset{\circ}{\Phi}_{g,h}(-f_1, f_2) = \overset{\circ}{\Phi}_{h,g}(f_1, -f_2)$, if g and h are real; (5.32)

7. $\overset{\circ}{\Phi}_{\frac{\partial^{p_1+p_2} g}{\partial x_1^{p_1} \partial x_2^{p_2}}, \frac{\partial^{q_1+q_2} h}{\partial x_1^{q_1} \partial x_2^{q_2}}}(f_1, f_2) = i^{q_1 - p_1 + q_2 - p_2} (2\pi f_1)^{p_1+q_1} (2\pi f_2)^{p_2+q_2} \overset{\circ}{\Phi}_{g,h}(f_1, f_2)$; (5.33)

where $p_1, p_2, q_1, q_2 \geq 0$.

It may be advantageous in some applications to assume that the two-dimensional correlation function is *isotropic*, i.e., that the correlation between two functions depends only on the magnitude of the lag, not on its orientation. Formally, the Cartesian coordinates in equation (5.25) are replaced by polar coordinates (Figure 2.12),

$$x_1 = s \cos \omega, \quad x_2 = s \sin \omega, \tag{5.34}$$

and subsequently the correlation function is averaged over the angle, ω,

$$\overset{\circ}{\phi}_{g,h}(s) = \frac{1}{2\pi} \int_0^{2\pi} \int_{-\infty}^{\infty} \int_{-\infty}^{\infty} g(x_1', x_2') h(x_1' + x_1, x_2' + x_2) \, dx_1' dx_2' d\omega. \tag{5.35}$$

In this case, the energy spectrum is the Hankel transform of the correlation function; see equation (2.180). It is also the average of the two-dimensional Fourier transform over all possible rotations of the coordinate system with respect to the angle, ω; which, in turn, is the average of the Fourier transform over all possible rotations of the corresponding frequency domain (see equation (2.188)),

$$\overset{\circ}{\Phi}_{g,h}(\bar{f}) = \frac{1}{2\pi} \int_0^{2\pi} G^*(f_1, f_2) H(f_1, f_2) \, dv, \tag{5.36}$$

where, as in equations (2.177) and (2.178), $\bar{f} = \sqrt{f_1^2 + f_2^2}$, with $f_1 = \bar{f} \cos v, f_2 = \bar{f} \sin v$.

Finally, consider the correlation between real sequences, g_ℓ and h_ℓ, sampled uniformly at intervals of Δx from real finite-energy functions and assumed summable. The correlation sequence is defined by

$$\left(\overset{\circ}{\phi}_{g,h}\right)_\ell = \Delta x \sum_{n=-\infty}^{\infty} g_n h_{n+\ell}, \quad -\infty < \ell < \infty, \tag{5.37}$$

where the notation on the left side should *not* be interpreted as a sample of the continuous correlation function (however, see also equation (5.51)). The Fourier transform of the correlation sequence is the (cross-) energy spectral density of the sequences. It is periodic (period, $1/\Delta x$; see equation (4.16)) and, analogous to equation (5.5), is (Exercise 5.3)

$$\overset{\circ}{\tilde{\Phi}}_{g,h}(f) = \tilde{G}^*(f) \tilde{H}(f). \tag{5.38}$$

Clearly, the correlation sequence and its Fourier transform have properties similar to equations (5.10) through (5.18). For example,

1. $\left(\overset{\circ}{\phi}_{g,g}\right)_0 > 0;$ (5.39)

2. $\left|\left(\overset{\circ}{\phi}_{g,h}\right)_\ell\right|^2 \le \left(\overset{\circ}{\phi}_{g,g}\right)_0 \left(\overset{\circ}{\phi}_{h,h}\right)_0, \; -\infty < \ell < \infty;$ (5.40)

3. $\lim_{\ell \to \pm\infty} \left(\overset{\circ}{\phi}_{g,h}\right)_\ell = 0;$ (5.41)

4. $\left(\overset{\circ}{\phi}_{g,h}\right)_{-\ell} = \left(\overset{\circ}{\phi}_{h,g}\right)_\ell$, if g_l and h_l are real; (5.42)

5. $\overset{\circ}{\tilde{\Phi}}_{g,g}(f)$ is real and $\overset{\circ}{\tilde{\Phi}}_{g,g}(f) \ge 0;$ (5.43)

6. $\overset{\circ}{\tilde{\Phi}}_{g,h}(-f) = \overset{\circ}{\tilde{\Phi}}_{h,g}(f)$ if g_l and h_l are real; (5.44)

The proofs of these adhere closely to those given for the correlation function, $\overset{\circ}{\phi}_{g,h}(x)$.

Furthermore, the extension to higher Cartesian dimensions follows naturally. For two dimensions, one has

$$\left(\overset{\circ}{\phi}_{g,h}\right)_{\ell_1,\ell_2} = \Delta x_1 \Delta x_2 \sum_{n_1=-\infty}^{\infty} \sum_{n_2=-\infty}^{\infty} g_{n_1,n_2} h_{n_1+\ell_1,n_2+\ell_2}, \tag{5.45}$$

which also satisfies, analogous to equation (5.29),

$$\left(\overset{\circ}{\phi}_{g,h}\right)_{-\ell_1,\ell_2} = \left(\overset{\circ}{\phi}_{h,g}\right)_{\ell_1,-\ell_2}. \tag{5.46}$$

The corresponding energy spectrum is

$$\overset{\circ}{\tilde{\Phi}}_{g,h}\left(f_1, f_2\right) = \tilde{G}^*\left(f_1, f_2\right) \tilde{H}\left(f_1, f_2\right), \tag{5.47}$$

with

$$\overset{\circ}{\tilde{\Phi}}_{g,h}\left(-f_1, f_2\right) = \overset{\circ}{\tilde{\Phi}}_{h,g}\left(f_1, -f_2\right). \tag{5.48}$$

It is periodic in f_1 and f_2 with respective periods, $1/\Delta x_1$ and $1/\Delta x_2$. These properties all hold for real g_ℓ and h_ℓ, and require appropriate modification if they are complex sequences.

As noted above, the correlation sequence, $\left(\overset{\circ}{\phi}_{g,h}\right)_\ell$, in general, is not a sample of the correlation function, $\overset{\circ}{\phi}_{g,h}(x)$, although it is a reasonable approximation. In fact, it is a sample if the parent functions, g and h, are band-limited by the Nyquist frequency, $f_N = 1/(2\Delta x)$. Indeed, with equation (4.8) for the Fourier spectra of the samples, g_ℓ and h_ℓ, equation (5.38) becomes for $-f_N \le f < f_N$,

$$\overset{\circ}{\tilde{\Phi}}_{g,h}(f) = \overset{\circ}{\Phi}_{g,h}(f) + H(f)\sum_{\substack{k=-\infty \\ k\neq 0}}^{\infty} G^*\left(f + \frac{k}{\Delta x}\right) + G^*(f)\sum_{\substack{k'=-\infty \\ k'\neq 0}}^{\infty} H\left(f + \frac{k'}{\Delta x}\right) + \sum_{\substack{k=-\infty \\ k\neq 0}}^{\infty} G^*\left(f + \frac{k}{\Delta x}\right)\sum_{\substack{k'=-\infty \\ k'\neq 0}}^{\infty} H\left(f + \frac{k'}{\Delta x}\right) \tag{5.49}$$

If the parent functions are band-limited by the Nyquist frequency then the sums all vanish, and the spectrum of the correlation sequence equals the spectrum of the correlation function over the principal part of the spectral domain,

$$\overset{\circ}{\tilde{\Phi}}_{g,h}(f) = \overset{\circ}{\Phi}_{g,h}(f), \quad -f_{\mathcal{N}} \le f < f_{\mathcal{N}}. \tag{5.50}$$

Since the correlation function, $\overset{\circ}{\phi}_{g,h}(x)$, is also band-limited, the Fourier transform of its samples, $\overset{\circ}{\phi}_{g,h}(\ell\Delta x)$, also equals $\overset{\circ}{\Phi}_{g,h}(f)$ in the interval, $-f_{\mathcal{N}} \le f < f_{\mathcal{N}}$ (see equation (4.49)); and thus, from equation (5.50),

$$\overset{\circ}{\phi}_{g,h}(\ell\Delta x) = \int_{-\infty}^{\infty} \overset{\circ}{\Phi}_{g,h}(f) e^{i2\pi f\ell\Delta x} df = \int_{-f_{\mathcal{N}}}^{f_{\mathcal{N}}} \overset{\circ}{\tilde{\Phi}}_{g,h}(f) e^{i2\pi f\ell\Delta x} df = \left(\overset{\circ}{\phi}_{g,h}\right)_{\ell}, \tag{5.51}$$

in this case. It is the result that follows also directly from the Whittaker-Shannon sampling theorem (Section 4.2.3). That is, the samples of the correlation function can fully reconstruct it.

5.3 Correlation of Finite-Power Functions

One may wish to consider functions continuing in the space domain that do not necessarily attenuate to zero; that is, $\lim_{|x|\to\infty} g(x) \neq 0$. Then, equation (2.49) is not satisfied and the Fourier transform does not exist in the sense of equation (2.56). Moreover, such functions are not square-integrable and thus possess infinite energy. Of course, infinitely extended geophysical signals are an idealization; and, in practice, they are defined either over a finite Cartesian region or on the sphere. The former have finite energy, and spherical signals, like periodic signals, are treated separately, where, even though they are formally extended over $(-\infty, \infty)$, they can be restricted to the fundamental period without loss of information. Nevertheless, a geophysical signal in Cartesian space often is viewed as continuing significantly beyond its observable part. Then, there are two options to analyze such a signal from the available data. Either, it is assumed to continue periodically, with the observable part forming one period; or, it is assumed to equal zero outside the window of observability, defining it as a truncated function, for example, by equation (3.132). Neither option may be particularly satisfactory as a physical interpretation. If one is interested primarily in analyzing a signal in terms of its correlation and corresponding correlation spectrum, then a third option is possible, whereby the signal is assumed to be a member of a new class of functions, whose Fourier transforms in the usual sense do not exist, but whose correlation functions still do. These are called finite-power functions.

Consider a function whose energy over a finite interval in time, T, is finite, and recall from physics that *power* is work, or energy, per unit time. That is, the energy of a system dissipated over some time interval divided by that interval is the total power, also called average power. A *finite-power function*, then, is a function whose total power exists in the limit as the interval becomes infinite, considered now for functions on the spatial domain,

$$\lim_{T\to\infty}\frac{1}{T}\int_{-T/2}^{T/2}\left|g(x)\right|^2 dx < \infty. \tag{5.52}$$

The usual regularity conditions are assumed, that the function is bounded and piecewise continuous with at most a finite number of jump-discontinuities. Clearly, finite-energy functions (those satisfying equation (2.50)) have zero total power in the limit. For example, the rectangle function, $b(x)$, equation (2.90), is a finite-energy function, but $1 - b(x)$ is only a finite-power function (Exercise 5.4).

Periodic functions, $\tilde{g}(x)$ (period, P), also are not finite-energy functions over the entire domain, $-\infty < x < \infty$, but are finite-power functions. Indeed, consider the following integrals, for all positive integers, n,

$$I_n = \frac{1}{nP}\int_{-nP/2}^{nP/2}\left|\tilde{g}(x)\right|^2 dx. \tag{5.53}$$

I_n is the power over n periods of the function. Now, $|\tilde{g}(x)|^2$ is also periodic with period P and n copies of $|\tilde{g}(x)|^2$ are contained in the interval, $(-nP/2, nP/2)$. Hence, I_n is equal to I_1, since

$$I_n = \frac{1}{nP}n\int_{-P/2}^{P/2}\left|\tilde{g}(x)\right|^2 dx = I_1. \tag{5.54}$$

That is, the total (or, average) power of a periodic function is the same over one period as over many periods. Thus,

$$\lim_{T\to\infty}\frac{1}{T}\int_{-T/2}^{T/2}\left|\tilde{g}(x)\right|^2 dx = \lim_{n\to\infty}\frac{1}{nP}\int_{-nP/2}^{nP/2}\left|\tilde{g}(x)\right|^2 dx = \lim_{n\to\infty} I_1 = I_1 < \infty, \tag{5.55}$$

showing that $\tilde{g}(x)$ is a finite-power function. Periodic functions and their correlations are treated separately and in greater detail in Section 5.4.

Here, we consider real, *non-periodic*, finite-power functions, g and h. Their (cross-) correlation function is defined by

$$\phi_{g,h}(x) = \lim_{T\to\infty} \frac{1}{T} \int_{-T/2}^{T/2} g(x')h(x'+x)\,dx'. \tag{5.56}$$

By Schwarz's inequality (1.22),

$$\left| \frac{1}{T} \int_{-T/2}^{T/2} g(x')h(x'+x)\,dx' \right|^2 \le \frac{1}{T} \int_{-T/2}^{T/2} \left|g(x')\right|^2 dx' \frac{1}{T} \int_{-T/2+x}^{T/2+x} \left|h(u)\right|^2 du, \tag{5.57}$$

where the second integral (with $u = x' + x$) in the limit with respect to T does not depend on x. Therefore, by inequality (5.52), $|\phi_{g,h}(x)|$, hence $\phi_{g,h}(x)$, is bounded. Moreover, only those functions, g and h, are considered for which their correlation function has a Fourier transform. As before, $\phi_{g,g}(x)$ is the *auto*-correlation function of g. The correlation function may be defined generally for complex-valued functions, but present applications are restricted to real geophysical signals; and, as before, the correlation function for real finite-power functions is also real. The symbol for correlation now omits the over-script, "o", indicating that the correlation is for finite-power functions. The units of $\phi_{g,h}$ are simply the product of the units of g and h, unlike the units of $\overset{\circ}{\phi}_{g,h}$.

Because it is assumed that the correlation of finite-power functions has a Fourier transform, it vanishes as the lag distance, x, increases, and thus the functions, g and h, decorrelate completely. As a limit of integrals that are all continuous, the correlation function is continuous; and, where needed we assume that it is differentiable. However, instances may be considered where the Fourier transform includes Dirac delta functions.

For the usual cases, the correlation and its Fourier spectrum have properties identical to those of the correlation function of real, finite-energy, functions, properties (5.10) through (5.15). Aside from those already mentioned,

1. $\phi_{g,g}(0) > 0;$ (5.58)

2. $\left|\phi_{g,h}(x)\right|^2 \le \phi_{g,g}(0)\phi_{h,h}(0), \quad -\infty < x < \infty;$ (5.59)

3. $\phi_{g,h}(-x) = \phi_{h,g}(x);$ (5.60)

4. $\dfrac{d}{dx}\phi_{g,h}(x) = \phi_{g,\frac{dh}{dx}}(x) = -\phi_{\frac{dg}{dx},h}(x).$ (5.61)

Inequality (5.58) follows directly from the definition of correlation function, equation (5.56); and, inequality (5.59) is a consequence of inequality (5.57), where each integral on the right side in the limit is the auto-correlation at the origin ($x = 0$). The proof of equation (5.60) is obtained with a change in integration variables, $u = x' - x$,

$$\phi_{g,h}(-x) = \lim_{T\to\infty}\frac{1}{T}\int_{-T/2-x}^{T/2-x} g(u+x)h(u)\,du$$
$$= \lim_{T\to\infty}\frac{1}{T}\left(\int_{-T/2-x}^{-T/2} h(u)g(u+x)\,du + \int_{-T/2}^{T/2} h(u)g(u+x)\,du - \int_{T/2-x}^{T/2} h(u)g(u+x)\,du\right)$$
$$= 0 + \phi_{h,g}(x) - 0 \tag{5.62}$$

where the first and third integrals, divided by T, vanish in the limit because g and h are bounded. The first of equations (5.61), which assume differentiability of the indicated functions, follows directly from equation (5.56) and the second is left as Exercise 5.5.

The Fourier transform of the correlation function for finite-power signals, g and h, is

$$\mathcal{F}\left(\phi_{g,h}\right) \equiv \Phi_{g,h}(f) = \int_{-\infty}^{\infty} \phi_{g,h}(x)e^{-i2\pi fx}\,dx, \tag{5.63}$$

and the inverse transform reverts to the correlation function,

$$\phi_{g,h}(x) = \mathcal{F}^{-1}\left(\Phi_{g,h}\right) = \int_{-\infty}^{\infty} \Phi_{g,h}(f)e^{i2\pi fx}\,df. \tag{5.64}$$

For $x = 0$, equation (5.64) yields

$$\phi_{g,h}(0) = \int_{-\infty}^{\infty} \Phi_{g,h}(f)\,df, \tag{5.65}$$

so that with equation (5.56),

$$\lim_{T\to\infty}\frac{1}{T}\int_{-T/2}^{T/2} g(x')h(x')\,dx' = \int_{-\infty}^{\infty} \Phi_{g,h}(f)\,df, \tag{5.66}$$

which is the form of *Parseval's theorem* for finite-power functions.

For $g = h$,

$$\lim_{T\to\infty}\frac{1}{T}\int_{-T/2}^{T/2} \left|g(x)\right|^2 dx = \int_{-\infty}^{\infty} \Phi_{g,g}(f)\,df, \tag{5.67}$$

showing that the total power of g equals the integral of $\Phi_{g,g}(f)$, and whence its name as the *power spectral density* (PSD) of g. Similar to the energy spectral density, it represents the power distribution of the function over the spectral domain, even though the function itself has no Fourier spectrum in the usual sense. For different functions, g and h, one may use the less ambiguous term, *cross*-power spectral density, but the context in which $\Phi_{g,g}(f)$ is used often makes this added qualifier unnecessary. From equation (5.64), the units of the PSD are the units of the correlation function (the product of the units of g and h) divided by the units of frequency.

The PSD is the principal means to describe the frequency content of finite-power functions, which, as suggested above, are the most likely to be encountered in many geophysical and geodetic applications, at least in an approximate way. But, unlike the energy spectral density, the PSD is not related to the Fourier transforms, as in equation (5.5), since these do not exist. Instead, consider the truncation of the finite-power function to a finite interval, equation (3.132),

$$g_T(x) = b\left(\frac{x}{T}\right)g(x). \tag{5.68}$$

Similarly, let $h_T(x)$ be the windowed part of h. The functions, g_T and h_T, are finite-energy functions that have Fourier transforms, respectively, $G_T(f)$ and $H_T(f)$. Their correlation function is given by equation (5.1); and, the energy spectral density is, from equation (5.5),

$$\overset{\circ}{\Phi}_{g_T,h_T}(f) = G_T^*(f)H_T(f). \tag{5.69}$$

Referring to Figure 5.2, it is readily verified that

$$\int_{-T/2}^{T/2} g(x')h(x'+x)\,dx' = \int_{-T/2}^{T/2} g_T(x')h_T(x'+x)\,dx' - \begin{cases} \displaystyle\int_{T/2}^{T/2-x} g_T(x')h(x'+x)\,dx', & x \ge 0 \\ \displaystyle\int_{-T/2-x}^{-T/2} g_T(x')h(x'+x)\,dx', & x < 0 \end{cases} \tag{5.70}$$

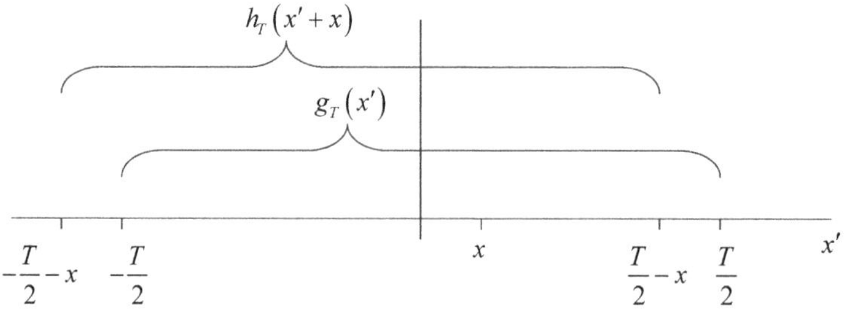

Figure 5.2: Domains for truncated functions, $g_T(x)$ and $h_T(x'+x)$. For example, if $x' = T/2 - x$, then $h_T(x'+x) = h_T(T/2)$.

Dividing by T and taking limits as $T \to \infty$, the second integrals on the right side vanish since g and h are bounded and the integration intervals are finite for any x. Therefore,

$$\lim_{T\to\infty}\frac{1}{T}\int_{-T/2}^{T/2} g(x')h(x'+x)\,dx' = \lim_{T\to\infty}\frac{1}{T}\int_{-T/2}^{T/2} g_T(x')h_T(x'+x)\,dx'. \tag{5.71}$$

According to equation (5.3), the integral on the right side is the correlation function, $\mathring{\phi}_{g_T,h_T}(x)$; hence, in view of equation (5.56), one obtains

$$\phi_{g,h}(x) = \lim_{T\to\infty}\frac{1}{T}\mathring{\phi}_{g_T,h_T}(x). \tag{5.72}$$

Applying the Fourier transform, since it exists by assumption, and using equation (5.69), there is finally a relationship between the PSD and the Fourier transforms of the (truncated) functions,

$$\Phi_{g,h}(f) = \lim_{T\to\infty}\frac{1}{T}G_T^*(f)H_T(f). \tag{5.73}$$

A comparison of equations (5.73) and (5.5) shows that the energy spectral density of the truncated finite-power functions (divided by T) is an approximation of the PSD,

$$\Phi_{g,h}(f) \approx \frac{1}{T}\mathring{\Phi}_{g_T,h_T}(f), \tag{5.74}$$

that improves as the truncation domain, T, increases. Note that the units of the PSD differ from those of the energy spectral density.

Properties of the PSD are analogous to those of the energy spectral density, equations (5.17), (5.18), and (5.20), and follow directly from equation (5.73).

1. $\Phi_{g,g}(f)$ is real and $\Phi_{g,g}(f) \ge 0$; (5.75)
2. $\Phi_{g,h}(-f) = \Phi_{h,g}(f)$ if g and h are real; (5.76)
3. $\Phi_{\frac{d^p g}{dx^p}\ \frac{d^q h}{dx^q}}(f) \quad i^{q\ \ p}(2\pi f)^{p\ \ q}\,\Phi_{g\ h}(f),\ p, q \ge 0;$ (5.77)
4. $\Phi_{c,c}(f) = \lim_{T\to\infty}\frac{1}{2T}\left|G_T(f)\right|^2\left|H_T(f)\right|^2 = \lim_{T\to\infty}\frac{1}{2T}\mathring{\Phi}_{g_T,g_T}(f)\mathring{\Phi}_{h_T,h_T}(f)$ (5.78)

where $c(x)$ is the convolution,

$$c(x) = \int_{-\infty}^{\infty} g(x')h(x-x')\,dx', \tag{5.79}$$

between a finite-power function, $g(x)$, and a *finite-energy* function, $h(x)$. The proof of equation (5.78) begins with the definition of the finite-energy convolution for the truncated functions, $g_T(x)$ and $h_T(x)$,

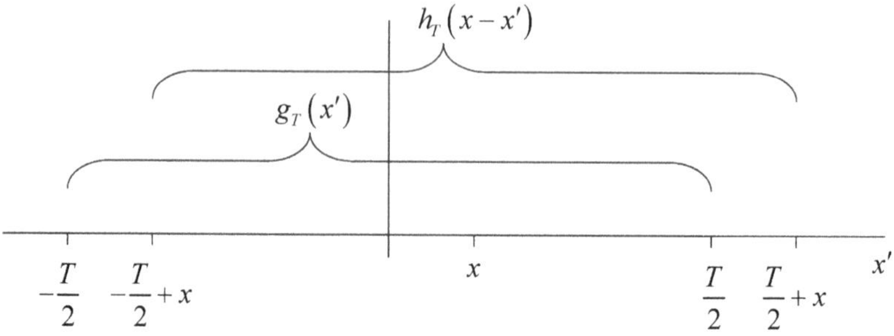

Figure 5.3: Domains for truncated functions, $g_T(x)$ and $h_T(x-x')$.

$$c_T(x) = (g_T * h_T)(x) = \int_{-T/2}^{T/2} g_T(x') h_T(x-x')\,dx', \tag{5.80}$$

which is also a truncated function, $c_T(x) = 0$ if $|x| > T$, as seen with the help of the Figure 5.3. By the convolution theorem, its Fourier transform is

$$C_T(f) = G_T(f) H_T(f). \tag{5.81}$$

Now, by equation (5.9) the Fourier transform of the correlation function of $c_T(x)$ is

$$\overset{\circ}{\Phi}_{c_T,c_T}(f) = |C_T(f)|^2 = |G_T(f)|^2 |H_T(f)|^2. \tag{5.82}$$

From equation (5.72),

$$\phi_{c,c}(x) = \lim_{T\to\infty} \frac{1}{2T} \overset{\circ}{\phi}_{c_T,c_T}(x), \tag{5.83}$$

since the domain on which $c_T(x)$ is non-zero has length, $2T$; and, taking Fourier transforms,

$$\Phi_{c,c}(f) = \lim_{T\to\infty} \frac{1}{2T} \overset{\circ}{\Phi}_{c_T,c_T}(f). \tag{5.84}$$

Together with equation (5.82) and the convolution theorem this leads to equation (5.78).

Extension of real, non-periodic, finite-power functions and their correlations to higher Cartesian dimensions follows the usual generalization. In two dimensions, finite-power functions are defined to be piecewise continuous and bounded on the plane such that they are square-integrable in the sense of equation (5.52),

$$\lim_{T_1\to\infty} \lim_{T_2\to\infty} \frac{1}{T_1 T_2} \int_{-T_1/2}^{T_1/2} \int_{-T_2/2}^{T_2/2} |g(x_1', x_2')|^2\, dx_1' dx_2' < \infty. \tag{5.85}$$

Their corresponding correlation function, given by

$$\phi_{g,h}(x_1,x_2) = \lim_{T_1\to\infty}\lim_{T_2\to\infty}\frac{1}{T_1T_2}\int_{-T_2/2}^{T_2/2}\int_{-T_1/2}^{T_1/2} g(x_1',x_2')h(x_1'+x_1,x_2'+x_2)\,dx_1'dx_2', \tag{5.86}$$

is assumed to have a Fourier transform. Properties of this correlation function and their proofs are analogous to the one-dimensional case, equations (5.58) through (5.61).

1. $\phi_{g,g}(0,0) > 0;$ (5.87)
2. $\left|\phi_{g,h}(x_1,x_2)\right|^2 \le \phi_{g,g}(0,0)\phi_{h,h}(0,0), \quad -\infty < x_1 < \infty, \quad -\infty < x_2 < \infty;$ (5.88)
3. $\phi_{g,h}(-x_1,x_2) = \phi_{h,g}(x_1,-x_2);$ (5.89)
4. $\frac{\partial}{\partial x_k}\phi_{g,h}(x_1,x_2) = \phi_{g,\frac{\partial h}{\partial x_k}}(x_1,x_2) = -\phi_{\frac{\partial g}{\partial x_k},h}(x_1,x_2), \quad k=1,2;$ (5.90)

where, for the latter, differentiability is assumed where indicated.

The two-dimensional PSD is the Fourier transform of the two-dimensional correlation function,

$$\Phi_{g,h}(f_1,f_2) = \int_{-\infty}^{\infty}\int_{-\infty}^{\infty}\phi_{g,h}(x_1,x_2)e^{-i2\pi(f_1x_1+f_2x_2)}dx_1dx_2, \tag{5.91}$$

with units equal to the units of the cross-correlation function divided by the product of the frequency units in each dimension. It is related to the Fourier transforms of the truncated finite-power functions,

$$g_{T_1,T_2}(x_1,x_2) = b\left(\frac{x_1}{T_1},\frac{x_2}{T_2}\right)g(x_1,x_2), \quad h_{T_1,T_2}(x_1,x_2) = b\left(\frac{x_1}{T_1},\frac{x_2}{T_2}\right)h(x_1,x_2), \tag{5.92}$$

where $b(x_1, x_2)$ is the two-dimensional rectangle function, equation (2.161), according to

$$\Phi_{g,h}(f_1,f_2) = \lim_{T_1\to\infty}\frac{1}{T_1}\lim_{T_2\to\infty}\frac{1}{T_2}G^*_{T_1,T_2}(f_1,f_2)H_{T_1,T_2}(f_1,f_2). \tag{5.93}$$

Properties of the two-dimensional PSD and their proofs are similar to those of the two-dimensional energy spectral density, properties (5.31) through (5.33), specialized to real, finite-power, functions,

1. $\Phi_{g,g}(f_1,f_2)$ is real and $\Phi_{g,g}(f_1,f_2) \ge 0;$ (5.94)
2. $\Phi_{g,h}(-f_1,f_2) = \Phi_{h,g}(f_1,-f_2);$ (5.95)
3. $\Phi_{\frac{\partial^{p_1+p_2}g}{\partial x_1^{p_1}\partial x_2^{p_2}},\frac{\partial^{q_1+q_2}h}{\partial x_1^{q_1}\partial x_2^{q_2}}}(f_1,f_2) = i^{q_1-p_1+q_2-p_2}(2\pi f_1)^{p_1+q_1}(2\pi f_2)^{p_2+q_2}\Phi_{g,h}(f_1,f_2),$ (5.96)

where $p_1, p_2, q_1, q_2 \geq 0$.

Isotropic correlation functions of finite-power functions are defined just as for finite-energy functions, equation (5.35), by averaging over all directions,

$$\phi_{g,h}(s) = \lim_{T_1\to\infty}\lim_{T_2\to\infty}\frac{1}{T_1T_2}\frac{1}{2\pi}\int_0^{2\pi}\int_{-T_2/2}^{T_2/2}\int_{-T_1/2}^{T_1/2} g(x_1',x_2')\,h(x_1'+x_1,x_2'+x_2)\,dx_1'dx_2'd\omega. \tag{5.97}$$

For the truncated finite-power functions, $g_{T_1,T_2}(x_1, x_2)$ and $h_{T_1,T_2}(x_1, x_2)$, the integrals become the finite-energy isotropic correlation function, equation (5.35); then,

$$\phi_{g,h}(s) = \lim_{T_1\to\infty}\lim_{T_2\to\infty}\frac{1}{T_1T_2}\overset{\circ}{\phi}_{g_{T_1,T_2},h_{T_1,T_2}}(s). \tag{5.98}$$

Taking the Hankel transform on both sides yields, with equation (5.36),

$$\Phi_{g,h}(\bar{f}) = \lim_{T_1\to\infty}\lim_{T_2\to\infty}\frac{1}{T_1T_2}\frac{1}{2\pi}\int_0^{2\pi} G^*_{T_1,T_2}(f_1,f_2)\,H_{T_1,T_2}(f_1,f_2)\,dv, \tag{5.99}$$

If a finite-power geophysical signal is sampled at regular intervals then it must satisfy the square-summability condition,

$$\lim_{N\to\infty}\frac{1}{N}\sum_{\ell=-N/2}^{N/2-1}|g_\ell|^2 < \infty, \tag{5.100}$$

a condition that is also assumed if the signal intrinsically is defined on a discrete domain. In two dimensions, this condition is

$$\lim_{N_1\to\infty}\lim_{N_2\to\infty}\frac{1}{N_1N_2}\sum_{\ell_1=-N_1/2}^{N_1/2-1}\sum_{\ell_2=-N_2/2}^{N_2/2-1}|g_{\ell_1,\ell_2}|^2 < \infty. \tag{5.101}$$

As before, only those finite-power sequences are considered that are real and non-periodic and whose correlation sequence, defined by

$$(\phi_{g,h})_\ell = \lim_{N\to\infty}\frac{1}{N}\sum_{n=-N/2}^{N/2-1} g_n h_{n+\ell}, \tag{5.102}$$

has a corresponding Fourier transform,

$$\tilde{\Phi}_{g,h}(f) = \Delta x\sum_{\ell=-\infty}^{\infty}(\phi_{g,h})_\ell\, e^{-i2\pi\Delta x f\ell}, \quad -f_N \leq f < f_N, \tag{5.103}$$

which has period equal to the reciprocal of the sampling interval, $1/\Delta x$ (see equation (4.16)).

Defining truncated finite-power sequences, $(g_N)_\ell$ and $(h_N)_\ell$, for example,

$$\left(g_N\right)_\ell = \begin{cases} g_\ell, & -\dfrac{N}{2} \le \ell \le \dfrac{N}{2} - 1 \\ 0, & \text{otherwise} \end{cases} \tag{5.104}$$

where $T = N\Delta x$, and replicating the proof for continuous finite-power functions, it is readily shown (Exercise 5.6) that the correlation sequences of the truncated and the complete finite-power sequences are related according to (analogous to equation (5.72))

$$\left(\phi_{g,h}\right)_\ell = \lim_{N\to\infty} \frac{1}{N\Delta x}\left(\overset{\circ}{\phi}_{g_N,h_N}\right)_\ell. \tag{5.105}$$

Therefore, the power spectrum of the sequence is given in terms of the Fourier transforms of the truncated sequences by (compare with equation (5.73))

$$\tilde{\Phi}_{g,h}(f) = \lim_{N\to\infty} \frac{1}{N\Delta x} \tilde{G}_N^*(f)\tilde{H}_N(f). \tag{5.106}$$

As in the case of the energy correlation function, the sequence, $(\phi_{g,h})_\ell$, is not a sample of $\phi_{g,h}(x)$, unless the sequences, $(g_N)_\ell$ and $(h_N)_\ell$, are band-limited by the corresponding Nyquist frequency. Indeed, suppose that the truncated parent functions are band-limited,

$$G_T(f) = 0, \quad H_T(f) = 0, \quad |f| > f_N, \quad \text{for all } T. \tag{5.107}$$

Then, since the truncated functions have finite energy, from equation (5.51), the correlation sequence of the truncated samples is the sample of the correlation function of the truncated functions,

$$\left(\overset{\circ}{\phi}_{g_N,h_N}\right)_\ell = \overset{\circ}{\phi}_{g_T,h_T}(\ell\Delta x), \quad \text{for all } T. \tag{5.108}$$

Substituting this and equation (5.105) into equation (5.72), one finds,

$$\phi_{g,h}(\ell\Delta x) = \lim_{T\to\infty}\frac{1}{T}\overset{\circ}{\phi}_{g_T,h_T}(\ell\Delta x) = \lim_{N\to\infty}\frac{1}{N\Delta x}\left(\overset{\circ}{\phi}_{g_N,h_N}\right)_\ell = \left(\phi_{g,h}\right)_\ell, \tag{5.109}$$

thus proving that the correlation sequence is a sample of the correlation function if the truncated parent functions are band-limited by the Nyquist frequency.

For two dimensions, the correlation array is

$$\left(\phi_{g,h}\right)_{\ell_1,\ell_2} = \lim_{N_1\to\infty}\lim_{N_2\to\infty}\frac{1}{N_1N_2}\sum_{n_1=-N_1/2}^{N_1/2-1}\sum_{n_2=-N_2/2}^{N_2/2-1} g_{n_1,n_2} h_{n_1+\ell_1,n_2+\ell_2}, \tag{5.110}$$

with corresponding Fourier transform, $\tilde{\Phi}_{g,h}(f_1,f_2)$, which is expressed in terms of the Fourier transforms of the truncated finite-power arrays as

$$\tilde{\Phi}_{g,h}(f_1,f_2) = \lim_{N_1\to\infty}\lim_{N_2\to\infty}\frac{1}{N_1\Delta x_1 N_2 \Delta x_2}\tilde{G}^*_{N_1,N_2}(f_1,f_2)\tilde{H}_{N_1,N_2}(f_1,f_2). \tag{5.111}$$

Properties of the correlation and its Fourier transform for one- and higher-dimensioned, real, finite-power sequences in Cartesian space follow immediately in analogy to the correlation function in the finite-energy case. They may be stated without proof.

1. $(\phi_{g,g})_0 > 0;$ (5.112)
2. $\left|(\phi_{g,h})_\ell\right|^2 \le (\phi_{g,g})_0 (\phi_{h,h})_0, \quad -\infty < \ell < \infty;$ (5.113)
3. $(\phi_{g,h})_{-\ell} = (\phi_{h,g})_\ell;$ (5.114)
4. $\tilde{\Phi}_{g,g}(f)$ is real and $\tilde{\Phi}_{g,g}(f) \ge 0;$ (5.115)
5. $\tilde{\Phi}_{g,h}(-f) = \tilde{\Phi}_{h,g}(f).$ (5.116)

The main developments in this Section show that much of the correlation structure for finite-power functions is virtually identical to that of finite-energy functions. That is, with a modest, yet significant, mathematical modification involving limits the concept of correlation can be extended to a larger class of functions that are merely bounded over the entire line or plane. On the one hand this conforms better to a view of geophysical signals that are not limited to a local domain; but, on the other hand, it also shows that implementations in this view entail approximations since the theoretical limits cannot be attained in practice. This becomes more apparent in Section 5.7 on the empirical determination of the correlation (covariance) function and the PSD. The finite-power formalism sets the stage for the most common interpretation for such signals as realizations of stochastic processes. This adds another mathematical modification, as well as another wrinkle, to the practical (numerical) determination and analysis of correlation. Before embarking on this extension, the special cases of periodic and spherical functions are elaborated, where even though they may be interpreted as having finite power, limiting their domain to a single period yields less demanding formulations for the correlation and its transform.

5.4 Correlation of Periodic Functions

The previous section shows that periodic functions, viewed over the infinite domain, $-\infty < x < \infty$, are finite-power functions, but over a single period, P, they clearly are of the finite-energy variety,

$$\int_0^P \left|\bar{g}(x)\right|^2 dx < \infty, \tag{5.117}$$

where the usual conditions on $\tilde{g}(x)$ of boundedness and piecewise continuity are assumed. Since one period sufficiently specifies such functions, obviating the infinite limit in equation (5.56), facilitates the correlations defined in this case by

$$\tilde{\phi}_{\tilde{g},\tilde{h}}(x) = \frac{1}{P}\int_0^P \tilde{g}(x')\tilde{h}(x'+x)\,dx', \tag{5.118}$$

where $\tilde{g}$ and $\tilde{h}$ are real and have the same period, P. The over-script, "o", is omitted on the correlation function because of its close relationship to finite-power functions, and the tilde, "~", should suffice to avoid confusion. Indeed, the correlation function itself is periodic with period, P, which is easily verified; and, it has the same units as the correlation for finite-power functions. The periodic extension is always implicitly assumed. For example, if $\tilde{h}$ is known over a single period, evaluating the correlation at $x \neq 0$ requires $\tilde{h}$ in the extended domain.

The correlation function for periodic functions may also be expressed as a cyclic convolution, equation (3.33),

$$\frac{1}{P}\tilde{g}(-x)\tilde{*}\tilde{h}(x) = \frac{1}{P}\int_0^P \tilde{g}(-x')\tilde{h}(x-x')\,dx' = \frac{1}{P}\int_{-P}^0 \tilde{g}(x')\tilde{h}(x+x')\,dx' = \tilde{\phi}_{\tilde{g},\tilde{h}}(x). \tag{5.119}$$

Hence, its Fourier series transform is given by the convolution theorem (3.35) for periodic functions,

$$\left(\Phi_{\tilde{g},\tilde{h}}\right)_k = \frac{1}{P}G_k^* H_k, \quad -\infty < k < \infty, \tag{5.120}$$

and

$$\tilde{\phi}_{\tilde{g},\tilde{h}}(x) = \frac{1}{P}\sum_{k=-\infty}^{\infty}\left(\Phi_{\tilde{g},\tilde{h}}\right)_k e^{i\frac{2\pi}{P}kx}. \tag{5.121}$$

The correlation function, $\tilde{\phi}_{\tilde{g},\tilde{h}}(x)$, enjoys the same properties as the correlation function for finite-energy functions, properties (5.10)–(5.12) and (5.14)–(5.16), as does its Fourier spectrum, $(\Phi_{\tilde{g},\tilde{h}})_k$, properties (5.17)–(5.21), with appropriate modifications (the reader is encouraged to verify these properties). The Fourier transform of the periodic correlation function is also called the *discrete power spectrum*.

Extension to higher dimensions follows immediately. For example, the correlation between two real periodic functions on the plane is

$$\tilde{\phi}_{\tilde{g},\tilde{h}}(x_1,x_2) = \frac{1}{P_1P_2}\int_0^{P_1}\int_0^{P_2} \tilde{g}(x_1',x_2')\tilde{h}(x_1'+x_1,x_2'+x_2)\,dx_1'dx_2'; \tag{5.122}$$

and, the power spectrum is related to the Fourier series transforms of the functions,

$$\left(\Phi_{\tilde{g},\tilde{h}}\right)_{k_1,k_2} = \frac{1}{P_1 P_2} G^*_{k_1,k_2} H_{k_1,k_2}. \tag{5.123}$$

If the real periodic functions are sampled at constant interval, $\Delta x = P/N$, the correlation of the sequences is defined by the periodic sequence,

$$\left(\tilde{\phi}_{\tilde{g},\tilde{h}}\right)_\ell = \frac{1}{N}\sum_{n=0}^{N-1} \tilde{g}_n \tilde{h}_{n+\ell}; \tag{5.124}$$

with power spectrum, also periodic and discrete,

$$\left(\tilde{\Phi}_{\tilde{g},\tilde{h}}\right)_k = \frac{1}{N\Delta x} \tilde{G}^*_k \tilde{H}_k, \tag{5.125}$$

where $\tilde{G}_k$, $\tilde{H}_k$ are the DFTs given by equation (4.70). This is known as a *periodogram* and is easily computed from actual data using the FFT. For $\tilde{h}_\ell = \tilde{g}_\ell$ the periodogram is also proportional to the amplitude spectrum and constitutes the simplest form of spectral analysis of a signal. It is only an approximation of the power spectral density of functions that are known only by their samples over a finite extent. Section (5.7.2) elaborates the practical estimation of the power spectral density from data.

If both parent functions, $\tilde{g}(x)$ and $\tilde{h}(x)$, are band-limited by the Nyquist wave number then $\tilde{G}_k = G_k$ and $\tilde{H}_k = H_k$ by equation (4.122); hence, in view of equation (5.120),

$$\left(\tilde{\Phi}_{\tilde{g},\tilde{h}}\right)_k = \frac{1}{N\Delta x} \tilde{G}^*_k \tilde{H}_k = \frac{1}{N\Delta x} G^*_k H_k = \left(\Phi_{\tilde{g},\tilde{h}}\right)_k. \tag{5.126}$$

Furthermore, equation (5.120) shows that $\tilde{\phi}_{\tilde{g},\tilde{h}}(x)$ is band-limited, which means that its samples also have the same Fourier transform. Therefore, its samples constitute the correlation sequence, $(\phi_{\tilde{g},\tilde{h}})_\ell$, in this case.

5.5 Correlation of Functions on the Sphere

Functions on the sphere are of principal importance in geophysics and geodesy for the obvious reason that many terrestrial phenomena, whether static or temporal, are associated with the nearly spherical domain of the Earth's surface. Thus, although in some cases a planar approximation suffices, correlations of signals on the sphere have a special interest. As before, spherical functions are viewed as being periodic, certainly in longitude, $(-\infty < \lambda < \infty)$, but in some sense also in co-latitude. However, as in previous chapters, the notation need not be overburdened with the "~" symbol. The general definition of correlation is completely analogous to the case of periodic functions on the line or plane, equations (5.118) or (5.122), and for real functions is given by

$$\phi_{g,h}(\psi,\zeta) = \frac{1}{4\pi}\iint_\Omega g(\theta,\lambda)\, h(\theta',\lambda')\, d\Omega, \tag{5.127}$$

where the points (θ, λ) and (θ', λ') (Figure 5.4) are related, as for the convolution on the sphere, by equations (2.196) and (2.197), repeated here for convenience,

$$\cos\psi = \cos\theta\cos\theta' + \sin\theta\sin\theta'\cos(\lambda' - \lambda), \tag{5.128}$$

$$\tan\zeta = -\frac{\sin\theta'\sin(\lambda' - \lambda)}{\sin\theta\cos\theta' - \cos\theta\sin\theta'\cos(\lambda' - \lambda)}. \tag{5.129}$$

Unlike the convolution, equation (3.44), however, the point, (ψ, ζ), is the argument of the correlation and is fixed within the integration. That is, the integration is performed over all pairs of points, (θ, λ) and (θ', λ'), separated by the fixed spherical distance, ψ, and oriented by the fixed angle, ζ. With this constraint, the differential element can be $d\Omega = \sin\theta d\theta d\lambda$ or $d\Omega = \sin\theta' d\theta' d\lambda'$. Note that the units of the spherical correlation function are simply the product of the units of g and h.

As with the spherical convolution, some restrictions on the correlation function are needed in order to find a practical relationship between the power spectrum and the Fourier-Legendre spectra of the functions g and h. Specifically, it is assumed that the correlation function is *isotropic*; that is, it is independent of the angle, ζ. This is accomplished by simply modifying the definition, equation (5.127), to include an average over ζ,

$$\phi_{g,h}(\psi) = \frac{1}{8\pi^2}\int_{\zeta=0}^{2\pi}\iint_{\Omega} g(\theta,\lambda)\,h(\theta',\lambda')\,d\Omega\,d\zeta. \tag{5.130}$$

Because of the imposed isotropy, these correlation functions are defined only for $0 \le \psi \le \pi$. The Legendre spectrum of $\phi_{g,h}(\psi)$ thus depends only on harmonic degree, not order; and, according to equations (2.205) and (2.206), the Legendre transform pair is

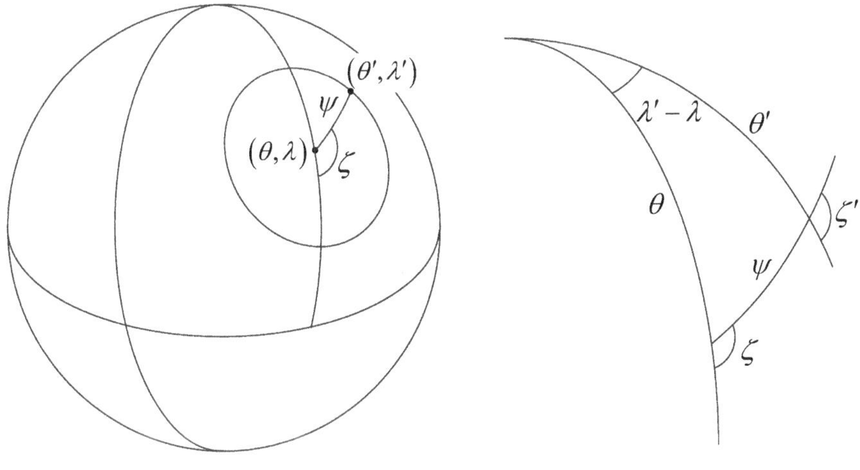

Figure 5.4: Geometry of points on the sphere for the definition of correlation function.

$$\left(\Phi_{g,h}\right)_n = \frac{1}{2}\int_0^{\pi} \phi_{g,h}(\psi) P_n(\cos\psi)\sin\psi\, d\psi, \tag{5.131}$$

$$\phi_{g,h}(\psi) = \sum_{n=0}^{\infty}(2n+1)\left(\Phi_{g,h}\right)_n P_n(\cos\psi). \tag{5.132}$$

The sequence, $(\Phi_{g,h})_n$, is called the (isotropic, cross-) power spectrum of g and h.

Substituting equation (5.130) into equation (5.131), one obtains

$$\left(\Phi_{g,h}\right)_n = \frac{1}{2}\int_{\psi=0}^{\pi}\left(\frac{1}{8\pi^2}\int_{\zeta=0}^{2\pi}\iint_{\Omega} g(\theta,\lambda)h(\theta',\lambda')\, d\Omega\, d\zeta\right) P_n(\cos\psi)\sin\psi\, d\psi, \tag{5.133}$$

which, with the addition theorem for the spherical harmonics, equation (2.254), and an exchange of summation and integration, becomes

$$\left(\Phi_{g,h}\right)_n = \frac{1}{2n+1}\sum_{m=-n}^{n}\frac{1}{4\pi}\iint_{\Omega} g(\theta,\lambda)\bar{Y}_{n,m}(\theta,\lambda)\left(\frac{1}{4\pi}\iint_{\Omega} h(\theta',\lambda')\bar{Y}_{n,m}(\theta',\lambda')\sin\psi\, d\psi\, d\zeta\right) d\Omega. \tag{5.134}$$

The coordinates θ, λ are fixed in the inner integral and serve as origin for coordinates, ψ, ζ. The integral, itself, is invariant with a change of coordinates from ψ, ζ to θ', λ' since it simply represents the average of $h(\theta', \lambda')\bar{Y}_{n,m}(\theta', \lambda')$ over the sphere. Therefore, with equation (2.220) for the Fourier-Legendre transform, the power spectrum is

$$\left(\Phi_{g,h}\right)_n = \frac{1}{2n+1}\sum_{m=-n}^{n} G_{n,m} H_{n,m}. \tag{5.135}$$

A similar expression holds if the Fourier-Legendre spectra are formulated in terms of complex coefficients, equation (2.222),

$$\left(\Phi_{g,h}\right)_n = \frac{1}{2n+1}\sum_{m=-n}^{n}\left(G^c_{n,m}\right)^* H^c_{n,m}, \tag{5.136}$$

obtained as in equation (5.135) by substituting equation (2.255) in equation (5.133); or, it can be verified using equation (2.228). It is thus shown that just like for periodic functions in Cartesian coordinates, the Legendre power spectrum of functions on the sphere, $(\Phi_{\tilde{g},\tilde{h}})_n$, is represented directly in terms of the Fourier-Legendre transforms, $G_{n,m}$ and $H_{n,m}$, of the correlated functions. In this case, it is an average over all wave numbers, m (compare with equation (5.36)).

If $g = h$, then

$$\left(\Phi_{g,g}\right)_n = \frac{1}{2n+1}\sum_{m=-n}^{n} G^2_{n,m} = \frac{1}{2n+1}c_n(g), \tag{5.137}$$

where

$$c_n(g) = \sum_{m=-n}^{n} G_{n,m}^2 \tag{5.138}$$

is called the *degree-variance* in Section 5.6.5, and differs by the factor, $2n+1$, from the definition of the power spectrum.

Properties of the isotropic spherical correlation function and its spectrum are equivalent to those of the Cartesian version, specifically properties (5.10)–(5.12) and (5.16)–(5.18).

1. $\phi_{g,g}(0) > 0;$ (5.139)

2. $\left|\phi_{g,h}(\psi)\right| \le \phi_{g,g}(0)\phi_{h,h}(0);$ (5.140)

3. $\left(\Phi_{g,g}\right)_n$ is real and $\left(\Phi_{g,g}\right)_n \ge 0;$ (5.141)

4. $\left(\Phi_{g,h}\right)_n = \left(\Phi_{h,g}\right)_n;$ (5.142)

5. $\left(\Phi_{g*h,g*h}\right)_n = \left(\Phi_{g,g}\right)_n H_n^2 = \dfrac{1}{2n+1}\left(\Phi_{g,g}\right)_n \left(\Phi_{h,h}\right)_n, \quad h = h(\theta).$ (5.143)

The convolution, $g*h$, in equation (5.143) is defined by equation (3.58) where $h(\theta)$ is interpreted as a function defined on the sphere. Its proof then follows readily from the convolution theorem (3.60), and equation (5.137). Indeed, by the former, the Fourier-Legendre spectrum of $g*h$ is $G_{n,m}H_n$; and, therefore, by the latter, the spectrum of the auto-correlation function of $g*h$ is

$$\left(\Phi_{g*h,g*h}\right)_n = \frac{1}{2n+1}\sum_{m=-n}^{n} G_{n,m}^2 H_n^2 = \left(\Phi_{g,g}\right)_n H_n^2. \tag{5.144}$$

Now, the auto-correlation of the spherical function, h, is defined as in equation (5.130) (that is, with $g = h$); and, therefore, its Legendre spectrum, from equation (5.137) and in view of equation (2.229), is

$$\left(\Phi_{h,h}\right)_n = \frac{1}{2n+1}\sum_{m=-n}^{n} H_{n,0}^2 = H_{n,0}^2 = (2n+1)H_n^2, \tag{5.145}$$

which proves the second part of equation (5.143). There is no analogue, however, to equation (5.16); i.e., the correlation function of a convolution on the sphere is not the same as the convolution (on the sphere) of corresponding correlation functions. Nevertheless, equation (5.143) provides an important relationship for the power spectrum of a convolution in terms of the power spectra of the convolved functions.

Analogous to equation (5.30), the law of propagation of correlations for horizontal (tangential) derivatives holds also for correlation functions on the sphere. That is, the averaging operator that defines the isotropic correlation, equation (5.130), and the derivative operators, $\partial/\partial\theta$ and $\partial/(\sin\theta\,\partial\lambda)$, are commutative. This non-trivial result was proved by Moritz (1972, Ch.8). Thus, for example, by the chain rule for differentiation,

$$\begin{aligned}\phi_{\frac{\partial g}{\partial \theta'}\frac{\partial h}{\sin\theta\partial\lambda}}(\psi) &= \frac{\partial^2 \phi_{g,h}(\psi)}{\partial\theta \sin\theta'\partial\lambda'} = \frac{1}{\sin\theta'}\frac{\partial}{\partial\lambda'}\left(\frac{d\phi_{g,h}}{d\psi}\frac{\partial\psi}{\partial\theta}\right) \\ &= \frac{1}{\sin\theta'}\left(\frac{d^2\phi_{g,h}}{d\psi^2}\frac{\partial\psi}{\partial\theta}\frac{\partial\psi}{\partial\lambda'} + \frac{d\phi_{g,h}}{d\psi}\frac{\partial^2\psi}{\partial\lambda'\partial\theta}\right)\end{aligned} \tag{5.146}$$

Similar to equations derived in Exercise 6.5, where ζ' is the angle at (θ', λ') analogous to ζ at (θ, λ) (Figure 5.4),

$$\frac{\partial\psi}{\partial\lambda'} = \sin\theta'\sin\zeta', \quad \frac{\partial\psi}{\partial\theta} = -\cos\zeta. \tag{5.147}$$

From the second equation and equation (5.129), it can be derived that

$$\frac{\partial^2\psi}{\partial\lambda'\partial\theta} = \sin\zeta\frac{\partial\zeta}{\partial\lambda'} = \sin\zeta\frac{\sin\theta'\cos\zeta'}{\sin\psi}. \tag{5.148}$$

Hence,

$$\phi_{\frac{\partial g}{\partial \theta'}\frac{\partial h}{\sin\theta\partial\lambda}}(\psi,\zeta) = -\sin\zeta'\cos\zeta\frac{d^2}{d\psi^2}\phi_{g,h}(\psi) + \sin\zeta\cos\zeta'\frac{1}{\sin\psi}\frac{d}{d\psi}\phi_{g,h}(\psi). \tag{5.149}$$

Similar equations hold for correlation functions among other derivatives of g and h.

These correlation functions of the derivatives, depending also on the orientation, ζ, of the arc, ψ (ζ' depends on ψ and ζ), are not isotropic, even if the correlation between *g* and *h* is isotropic. It also means that the Legendre spectra of correlation functions of horizontal derivatives do not readily propagate from the spectra of *g* and *h*, as in Cartesian space, equations (5.33) or (5.96). However, a plausible approximation exists for the magnitude of the horizontal gradient of g, equation (2.270),

$$\left|\nabla_{(\theta,\lambda)} g(\theta,\lambda)\right| \doteq d_H g(\theta,\lambda) = \sqrt{\left(\frac{\partial g}{\partial\theta}\right)^2 + \left(\frac{1}{\sin\theta}\frac{\partial g}{\partial\lambda}\right)^2}. \tag{5.150}$$

The auto-correlation function, equation (5.130), at the origin, $\psi = 0$, is

$$\phi_{d_H g, d_H g}(0) = \frac{1}{4\pi}\iint\limits_{\Omega}\left(\left(\frac{\partial g}{\partial\theta}\right)^2 + \left(\frac{1}{\sin\theta}\frac{\partial g}{\partial\lambda}\right)^2\right)d\Omega. \tag{5.151}$$

Substituting the Fourier-Legendre series, equation (2.221) yields

$$\phi_{d_H g,d_H g}(0)=\frac{1}{4\pi}\iint_{\Omega}\left(\left(\sum_{n=0}^{\infty}\sum_{m=-n}^{n}G_{n,m}\frac{\partial\bar{Y}_{n,m}}{\partial\theta}\right)^2+\left(\frac{1}{\sin\theta}\sum_{n=0}^{\infty}\sum_{m=-n}^{n}G_{n,m}\frac{\partial\bar{Y}_{n,m}}{\partial\lambda}\right)^2\right)d\Omega$$
$$=\sum_{n=0}^{\infty}\sum_{m=-n}^{n}G_{n,m}\sum_{n'=0}^{\infty}\sum_{m'=-n'}^{n'}G_{n',m'}\frac{1}{4\pi}\iint_{\Omega}\left(\frac{\partial\bar{Y}_{n,m}}{\partial\theta}\frac{\partial\bar{Y}_{n',m'}}{\partial\theta}+\frac{1}{\sin^2\theta}\frac{\partial\bar{Y}_{n,m}}{\partial\lambda}\frac{\partial\bar{Y}_{n',m'}}{\partial\lambda}\right)d\Omega \quad (5.152)$$

With equation (2.274), the quadruple sum collapses to a double sum,

$$\phi_{d_H g,d_H g}(0)=\sum_{n=0}^{\infty}n(n+1)\sum_{m=-n}^{n}G_{n,m}^2$$
$$=\sum_{n=0}^{\infty}n(n+1)(2n+1)\left(\Phi_{g,g}\right)_n \quad (5.153)$$

Comparing this to equation (5.132) with $\psi=0$,

$$\phi_{d_H g,d_H g}(0)=\sum_{n=0}^{\infty}(2n+1)\left(\Phi_{d_H g,d_H g}\right)_n, \quad (5.154)$$

it is thus tempting to set

$$\left(\Phi_{d_H g,d_H g}\right)_n=n(n+1)\left(\Phi_{g,g}\right)_n. \quad (5.155)$$

However, equations (5.153) and (5.154) only hold for $\psi=0$ and the formal definition of the power spectrum, equation (5.133), does not yield such a simple relationship. Nevertheless, it is a decomposition of the total power into spectral components by degree, and as such it is a reasonable and often-used approximation. In view of equation (2.301), one may even approximate, $(\Phi_{d_H g,h})_n=\sqrt{n(n+1)}(\Phi_{g,h})_n$.

A similar argument is made for the magnitude of the total gradient, equation (2.269), of a potential function, $v(\theta,\lambda,r)$, equation (6.164), evaluated on the sphere of radius, R,

$$\left|\nabla v(\theta,\lambda,r)_{r=R}\right|=\sqrt{\left(\frac{1}{R}\frac{\partial v}{\partial\theta}\bigg|_{r=R}\right)^2+\left(\frac{1}{R\sin\theta}\frac{\partial v}{\partial\lambda}\bigg|_{r=R}\right)^2+\left(\frac{\partial v}{\partial r}\bigg|_{r=R}\right)^2}. \quad (5.156)$$

As in equation (5.151), the correlation function at the origin is

$$\phi_{|\nabla v|,|\nabla v|}(0)=\frac{1}{4\pi}\iint_{\Omega}\left(\left(\frac{1}{R}\frac{\partial v}{\partial\theta}\bigg|_{r=R}\right)^2+\left(\frac{1}{R\sin\theta}\frac{\partial v}{\partial\lambda}\bigg|_{r=R}\right)^2+\left(\frac{\partial v}{\partial r}\bigg|_{r=R}\right)^2\right)d\Omega. \quad (5.157)$$

From equation (2.290), one readily obtains (Exercise 5.7)

$$\phi_{|\nabla v|,|\nabla v|}(0) = \frac{1}{R^2}\sum_{n=1}^{\infty}(2n+1)(n+1)\sum_{m=-n}^{n}V_{n,m}^2, \tag{5.158}$$

where $V_{n,m}$ is the Fourier-Legendre spectrum of $v(\theta, \lambda, R)$. One might then reason that the Legendre spectrum of the (isotropic) correlation function, $\phi_{|\nabla v|,|\nabla v|}(\psi)$, can be approximated by

$$\left(\Phi_{|\nabla v|,|\nabla v|}\right)_n = \frac{1}{R^2}(n+1)\sum_{m=-n}^{n}V_{n,m}^2. \tag{5.159}$$

As noted for the power spectrum in equation (5.137), $(\Phi_{d_H g, d_H g})_n$ and $(\Phi_{|\nabla v|,|\nabla v|})_n$ differ by the factor, $2n + 1$, from the respective degree-variances of the magnitudes of the horizontal and total gradients.

Using tensor spherical harmonics, Rummel and Van Gelderen (1992) show that a similar interpretation of spectral decomposition by degree holds for the magnitude of second-order horizontal derivatives,

$$\begin{aligned} d_C g(\theta,\lambda) &= \sqrt{\left(\frac{\partial^2 g}{\partial x_1^2} - \frac{\partial^2 g}{\partial x_2^2}\right)^2 + \left(2\frac{\partial^2 g}{\partial x_1 \partial x_2}\right)^2} \\ &= \sqrt{\left(\cot\theta\frac{\partial g}{\partial\theta} + \frac{1}{\sin^2\theta}\frac{\partial^2 g}{\partial\lambda^2} - \frac{\partial^2 g}{\partial\theta^2}\right)^2 + 4\left(\frac{\cos\theta}{\sin^2\theta}\frac{\partial g}{\partial\lambda} - \frac{1}{\sin\theta}\frac{\partial^2 g}{\partial\theta\partial\lambda}\right)^2} \end{aligned} \tag{5.160}$$

where x_1, x_2 are local Cartesian coordinates (Figure 1.1). In this case,

$$\left(\Phi_{d_C g, d_C g}\right)_n = n(n+1)(n-1)(n+2)\left(\Phi_{g,g}\right)_n. \tag{5.161}$$

It is cautioned, again, that this and equations (5.155) and (5.159) are approximations of the power spectrum of the magnitudes of horizontal derivatives. The correlation functions of the derivatives themselves are derivable in the spatial domain, but are complicated as illustrated by equation (5.149). Like any bounded function on the sphere the horizontal derivatives have Fourier-Legendre transforms, but no exact spectral relationships exist among them as indicated at the end of Section 2.6.3. Even when averaging to obtain isotropic correlation functions for the horizontal derivatives, the power spectral relationships are only approximate. The law of propagation of correlations of derivatives in the spherical spectral domain is thus hardly as straightforward as on the plane, e.g., equations (5.90) and (5.96).

The approximate relationship between planar and spherical isotropic correlation functions and, correspondingly, between their transforms follows from the asymptotic relationship between Legendre polynomials and Bessel functions, equation (2.329). With $2\pi\bar{f} = \sqrt{n(n+1)}/R$ and $s = R\psi$, this is

$$P_n(\cos\psi) \simeq J_0(2\pi\bar{f}s), \quad \text{for small } \psi. \tag{5.162}$$

The planar isotropic correlation function of finite-power functions, $\phi_{g,h}(s)$, equation (5.97), is the inverse Hankel transform of the corresponding PSD, equation (5.99),

$$\phi_{g,h}(s) = 2\pi \int_0^\infty \Phi_{g,h}(\bar{f}) J_0(2\pi \bar{f} s) \bar{f}\, d\bar{f}. \tag{5.163}$$

Discretizing this (with $d\bar{f} \simeq 1/(2\pi R)$) and substituting equation (5.162) yields

$$\phi_{g,h}(s) \simeq \sum_{n=0}^{\infty} \frac{n}{2\pi R^2} \Phi_{g,h}\left(\frac{n}{2\pi R}\right) P_n(\cos\psi), \tag{5.164}$$

where $2\pi \bar{f} = n/R$ is used. Comparing this with the spherical correlation function, equation (5.132), it is seen that

$$(2n+1)\left(\Phi_{g,h}\right)_n \simeq \frac{n}{2\pi R^2} \Phi_{g,h}(\bar{f}), \quad \text{for large } n, \quad \text{and where } \bar{f} = \frac{n}{2\pi R}. \tag{5.165}$$

This relationship between the planar PSD and the spherical power spectrum holds only for isotropic correlation functions. Note that for $g = h$ the left side is also the degree-variance, equation (5.138).

For arrays of data sampled regularly on the sphere, as in equation (4.156), the correlation function could be formulated as a discretized version of equation (5.130). This might then lead to approximating the correlation function empirically from the samples, as suggested by equations (5.51) or (5.109) in the Cartesian case. However, the samples are used more commonly to approximate the Legendre spectrum, that in turn approximates the correlation function according to equations (5.135) and (5.132).

The correlation function and sequences defined in these foregoing sections introduce basic concepts from a purely non-statistical perspective, that is, for functions that are deterministic (non-random), and whose correlation function has a Fourier transform. They offer one approach to establish, or at least hypothesize, a physical relationship (correlation) between functions if no other deterministic model is available that could describe their interdependence. In fact, this approach goes beyond a simple regression model for an empirical physical relationship since it yields a complete distribution of correlations over all scales of a function, i.e., over its spectral domain. Moreover, we are now in a position to transition from the strictly deterministic perspective to the development and implementation of spectral analysis of random processes. It becomes evident, however, that the consequent practical results are virtually identical to the deterministic ones. In effect, even if one assumes that certain geophysical signals have a stochastic underpinning, the fact that usually only a single realization exists means that with appropriate assumptions one reverts to the deterministic expressions to calculate correlations and power spectra. Nevertheless, the stochastic interpretation facilitates the assessment of the errors in these calculations and the determination of methods to reduce them.

5.6 Stochastic Processes

The concept of correlation up to now has been applied exclusively to deterministic functions; no notions of probability and statistics have entered the discussions, even though that is where correlation is perhaps more familiar. There are cases, however, when a geophysical signal is better characterized as a random quantity. For example, the gravity anomaly, topographic height, and magnetization depicted in Figures 5.6, 5.10, and 5.13 (Sections 5.6.2, 5.6.7, and 5.7.1) each could be construed as realization of a random process along a profile. Indeed, any attempt at a deterministic formulation of such a signal in one region, already requiring an innumerable set of parameters, likely does not hold for the signal in a neighboring region. Of course, only one realization of the gravitational or magnetic field is observable; it has been essentially fixed as a realization of a process and we don't have the luxury of observing repeated creations of the Earth. This is one argument against the interpretation of the field (or any similar signal) as being random. Yet the signal in the neighboring region may have very similar characteristics and this *stationarity*, or spatial invariance of similitude, again lends favor to the stochastic interpretation. Moreover, one may then apply the full power of estimation theory to the spectral analysis of these signals. Indeed, the widely implemented optimal estimation methods of kriging in geophysics data analysis (Cressie 1991) and least-squares collocation in physical geodesy (Moritz 1980) are predicated on this interpretation. The latter is elaborated in Section 6.6 with formulation in the frequency domain. In addition, with the stochastic interpretation one is able to evaluate estimators of the power spectral density using statistical concepts, such as expectation that predicts whether an estimate is biased, or consistency, which answers the question: is it worth the effort to obtain or analyze more data in order to obtain a better estimate?

Much of the groundwork for the spectral analysis of signals from the stochastic viewpoint was laid by Norbert Wiener during and after World War II for applications in communications and control theory. Because of the fundamental assumption of stationarity it is easily recognized that a random signal, even if only a single realization is the object of analysis, does not possess the usual Fourier transform because, in the first place, this assumption precludes the absolute integrability of the signal. In addition, a realization of a stochastic process is not continuous in the usual sense of a deterministic function. Wiener's developments are based on the more general approach to Fourier transforms as noted in Section 1.3 (see also Priestley 1981). The present exposition is based on the notion that the random signal is a finite-power function. The interest is less in its Fourier transform than in its power spectral density, where the only difference is that the correlation function now assumes the character of a covariance function in the usual statistical sense.

The theory that leads toward the determination of the PSD of a random process is the theory of stochastic processes, which the next section briefly reviews, beginning with some basic concepts in probability that prime the reader to think in terms of random quantities.

5.6.1 Probability, Random Variables, and Processes

Randomness is a state of uncertainty or non-determinism, and working with such a state requires a concept of *probability*. The subjective view of probability theory is ultimately adopted here, also known as *Bayesian probability*, wherein probability is interpreted as a "degree of belief" in an outcome (De Finetti 1970; see also Papoulis 1991, Ch.1, for a review), rather than an objective measure of relative frequency. The latter is based on experimentation, where outcomes come by chance and probability is assigned based on multiple realizations. For many geophysical signals in the space domain this not a feasible approach since one cannot perform experiments many times—there is only one Earth. In contrast, viewing the probability of earthquakes at any location lends itself more to the alternative *frequentist* approach. However, the basic rules and tools of probability derive from the same fundamental theory.

Probability theory was developed axiomatically by A.N. Kolmogorov as a deductive theory, based on a few fundamental assertions about probability measures for any particular outcome, or realization, of an experiment that has a random component (his work was published originally in a German monograph and translated into Russian and English; Kolmogorov 1956). Formally, the collection of all possible elementary outcomes of such an experiment is called the *sample space*, Ξ. The sample space can be discrete or continuous. The classic examples of discrete sample spaces include the numbers on a die or the numbers on lottery tickets. Rolling the die or drawing a ticket at random then is the experiment. Instrument errors are the random part of a measurement and could be any real number and thus are outcomes from a continuous sample space (even though very large errors are unlikely, in theory, there may be no particular limit and the sample space could be taken as the entire real line). The value of the topographic height relative to a mean reference at the same point (latitude and longitude) on all the rocky planets in our galaxy also comes from a continuous sample space of real numbers with no particular predefined bound.

In certain cases, one may restrict the sample space to be a particular subset of the real numbers, such as the set of integers when the random variable is the cycle ambiguity of the phase measurement of a radar signal (such as GPS). Or, the range of values may be a limited to an interval of real numbers; for example, the orientation of volcanic flows is an angle between zero and 2π. Some geophysical signals have only integer values, particularly random events in time, such as the number of magnetic reversals in Earth's history, or the number of earthquakes at a location. On the other hand, the random magnitude of an earthquake is a positive real number. With our interest focused on continuous signals in geodesy and geophysics, it is assumed, unless otherwise indicated, that the sample space is the set of real numbers, $\Xi = (-\infty, \infty)$, where residual geopotential fields, for example, may have either sign relative to a reference field.

Combinations of outcomes, such as all outcomes of topographic height less than 100 m, are called *events*; and, therefore, events are subsets of the sample space. The sample space, itself, is an event, as is "no outcome" (the empty set). It can be shown using elementary set operations (e.g., Maybeck 1979, Ch.3) that events, $A_a = \{\xi : \xi \le a, \xi \in \Xi\}$, for any real number, a, can generate all possible events including isolated points and open, closed, and half-open intervals on the real line. For example, the event, $\{\xi : a_1 < \xi \le a_2, \xi \in \Xi\}$ is the intersection of the events, $A^*_{a_1}$ and A_{a_2}, where $A^*_{a_1}$ is the complement of A_{a_1}, or the mutually exclusive event that joins with A_{a_1} to form the entire sample space.

Each event, A, thus generated, has a probability measure associated with it. This probability is a function that maps the field of all possible events to the interval, [0,1], with the following axiomatic properties. If P(A) denotes the probability of the event, A, then these axioms are

$$\mathrm{P}(A) \ge 0; \tag{5.166}$$

$$\mathrm{P}(\Xi) = 1; \tag{5.167}$$

$$\mathrm{P}(A \text{ or } B) = \mathrm{P}(A) + \mathrm{P}(B), \text{ provided } A \text{ and } B \text{ are mutually exclusive.} \tag{5.168}$$

The latter is the probability of either event being realized (under the mutual exclusivity assumption). It may be generalized to

$$\mathrm{P}(A_1 \text{ or } A_2 \text{ or } \dots \text{ or } A_n) = \sum_{j=1}^{n} \mathrm{P}(A_j), \tag{5.169}$$

provided A_j and A_k are mutually exclusive for all $j \ne k$, and where the integer, n, is finite or countably infinite. Denoting the *conditional* probability, P(A | B), as the probability of event A, given that event B has occurred, the probability of both events happening is

$$\mathrm{P}(A \text{ and } B) = \mathrm{P}(A \mid B)\mathrm{P}(B), \tag{5.170}$$

which can be derived from the axioms (Hoel 1971, p.13ff). Events A and B are said to be *independent* if and only if

$$\mathrm{P}(A \text{ and } B) = \mathrm{P}(A)\mathrm{P}(B). \tag{5.171}$$

Independence and mutual exclusivity are entirely different concepts, where independence implies that the occurrence of one event carries no information on the possibility of the other occurring. Independence allows both to occur, which means they are not mutually exclusive; and, in fact, mutual exclusivity implies *dependence* (if A occurs then B cannot, if mutually exclusive; hence, B depends on A).

Since the sample space is the set of all real numbers, the probability of the event of a single number is not defined in practical terms, since axioms (5.167) and (5.168) would imply that it could only have zero probability. Rather, probability on this sample space is defined for intervals of real numbers.

We do not need to delve further into axiomatic probability theory and leave that to excellent existing texts (e.g., Papoulis 1991, Parzen 1960). Indeed, the probability measure for some random signals in geophysics may not be known at all or well enough to perform computations of probability. It is enough, however, to know or assume that some probability measure exists in the background in order to talk about statistics and estimation, specifically in terms of *expectation* (Section 5.6.2).

Let us begin with the definition of a *random variable*—it is a variable associated with a putative experiment whose value is unknown until the conclusion of the experiment and depends on a probability measure. Once the random variable is *realized* as a result of the experiment then it is definite and no longer random. For example, it is known that a magnetometer is sensitive to the total magnetic field strength at a point, but also to errors associated with the sensing mechanism. The error depends on the particular condition of the sensor at the time of measurement and is likely to be different for each measurement—it is a random variable, making also the measurable quantity a random variable. However, the actual measurement value, which includes both the true magnetic field strength and a realization of the error, is no longer random. Of course, in this example the error is still not known unless one also knows the true magnetic field strength at the point of measurement. Knowing the probability measure associated with the randomness of the error would allow one to make some prediction about the error. Similar to the topography illustration above, one could also declare the magnetization, itself, a random quantity because the sources of this part of the field (magnetized material in the crust) are so complex that it cannot be characterized by a physical, deterministic model. The measurement, then, is a combination of the realizations of the random field and an instrument error.

Formally, a random variable is defined to be a function, or mapping, from the sample space to the *realization space* of real numbers, here denoted, Σ, that in our case of geodetic and geophysical applications is also the real interval, $(-\infty, \infty)$. Then, the random variable is also said to be continuous. Conceptually the two spaces, Ξ and Σ, are different, being, respectively, the domain and range spaces associated with the random variable defined as a function. Thus, a random variable, Z, formally is the function, $Z : \Xi \rightarrow \Sigma$, and one should write for its realization, $Z(\xi) = z$, where $\xi \in \Xi$, and $z \in \Sigma$. The less rigorous notation, $Z = z$, is adopted here, where the upper-case letter denotes the random variable and the lower-case its realization. Higher-dimensioned vector random variables are vectors of random variables from corresponding higher-dimensioned sample spaces to higher-dimensioned realization spaces. Similarly, complex random variables are complex-valued functions defined on a complex sample space; the real and imaginary parts are random variables.

The connection of probability to a random variable, Z, comes through events defined, as above, by

$$C_z = \left\{\xi : Z(\xi) \le z, \xi \in \Xi\right\}, \text{ for any real value, } z. \tag{5.172}$$

This event is the set of all elementary outcomes such that the realized random variable has values less than or equal to z. Since each event has an associated probability, one can define a function that maps the field of events to the interval, $[0,1]$. This is called the *cumulative distribution function* (also, *probability distribution function*),

$$F_Z(z) = \mathrm{P}(C_z) = \mathrm{P}(Z \le z), \tag{5.173}$$

where the second equality makes use of the short-hand notation. By axioms (5.166) and (5.167),

$$0 \le F_Z(z) \le 1, \tag{5.174}$$

$$\lim_{z\to\infty} F_Z(z) = 1, \tag{5.175}$$

$$\lim_{z\to-\infty} F_Z(z) = 0. \tag{5.176}$$

Also, by axiom (5.168), for $b > a$, $F_Z(b) = F_Z(a) + \mathrm{P}(a < Z \le b)$; which shows that F_Z is a monotonically increasing function.

For present applications it is assumed that the distribution function is piecewise continuous and differentiable (almost everywhere); in fact, differentiability, hence continuity, everywhere can be assumed without restricting our applications. Then, one has

$$p_Z(z) = \frac{d}{dz}F_Z(z) = \lim_{\Delta z\to 0}\frac{F_Z(z+\Delta z) - F_Z(z)}{\Delta z} = \lim_{\Delta z\to 0}\frac{\mathrm{P}(z \le Z \le z+\Delta z)}{\Delta z}, \tag{5.177}$$

where $p_Z(z)$ is known as the *probability density function* for the random variable, Z. Figure 5.5 shows a typical cumulative distribution function and the corresponding density function. It is easily seen from equation (5.177) that

$$F_Z(z) = \int_{-\infty}^{z} p_Z(z')\,dz', \tag{5.178}$$

since the realization space is $\Sigma_Z = (-\infty, \infty)$. Because F_Z is monotonically increasing, its derivative is non-negative,

$$p_Z(z) \ge 0, \quad \text{for all } z \in \Sigma_Z; \tag{5.179}$$

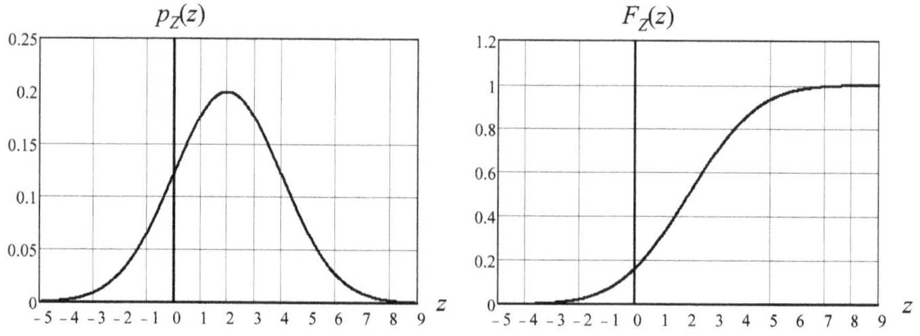

Figure 5.5: Example of a probability density function (left) and corresponding cumulative distribution function (right). Shown are the Gaussian density and distribution functions with parameters, $\mu = 2$ and $\sigma = 2$.

and

$$\int_{\Sigma_Z} p_Z(z)\,dz = \int_{-\infty}^{\infty} p_Z(z)\,dz = \lim_{z\to\infty} F_Z(z) = \mathrm{P}(Z < \infty) = \mathrm{P}(\Xi_Z) = 1. \tag{5.180}$$

The latter says that the density function is absolutely integrable, hence,

$$\lim_{z\to\pm\infty} p_Z(z) = 0. \tag{5.181}$$

The distribution function has no units, but the units of the density function are the inverse of the units of the random variable.

The probability density function depicted in Figure 5.5 is the *Gaussian*, or *normal*, density function, given by

$$p_Z(z) = \frac{1}{\sqrt{2\pi}\sigma} e^{-\frac{1}{2}\left(\frac{z-\mu}{\sigma}\right)^2}, \tag{5.182}$$

where μ and σ are two parameters that specify its location and its scale, or spread on the abscissa, and where the area under the function is fixed to unity. Its maximum is at $z = \mu$, and the inflection points, where the slopes of the function are maximum in absolute value, are at $z = \mu \pm \sigma$. The corresponding distribution function is

$$F_Z(z) = \frac{1}{\sqrt{2\pi}\sigma} \int_{-\infty}^{z} e^{-\frac{1}{2}\left(\frac{z'-\mu}{\sigma}\right)^2} dz', \tag{5.183}$$

which has no further analytic simplification. The importance of the Gaussian probability derives from the *Central Limit Theorem* which in its classical form states that the average of independent and identically distributed random variables, regardless of their distribution function, tends to behave like a Gaussian random variable as the number of variables increases. The requirements of independence and identical distributions can be relaxed in variants of the theorem (Hoel 1971, p.125; Parzen 1960, p.430ff). This means that many random phenomena in nature, generally interpretable as the accumulation or amalgamation of several random events, can be reasonably modeled as Gaussian. In particular, random errors in measurements using non-trivial instruments typically are Gaussian as they represent the combination of many random effects. There is one particularly interesting and relevant exception discussed in Section 5.6.6.

Another important random variable for our considerations is the uniformly distributed variable, with probability density function given by

$$p_Z(z) = \begin{cases} \dfrac{1}{b-a}, & a \le z \le b \\ 0, & \text{otherwise} \end{cases} \tag{5.184}$$

and corresponding distribution function,

$$F_Z(z) = \begin{cases} 0, & z < a \\ \dfrac{z-a}{b-a}, & a \le z \le b \\ 1, & z > b \end{cases} \tag{5.185}$$

The errors associated with the quantization or discretization of analog signals, also numerical round-off errors, are uniformly distributed.

Quantifying the probability measure of more than one random variable, say Z_1 and Z_2, requires a *joint probability* that defines the probability of both realizations occurring. In terms of distribution and density functions, equation (5.178) readily generalizes to

$$F_{Z_1,Z_2}(z_1,z_2) = \int_{-\infty}^{z_1}\int_{-\infty}^{z_2} p_{Z_1,Z_2}(z_1',z_2')\,dz_1'dz_2', \tag{5.186}$$

where the *marginal probability density functions* are given by

$$p_{Z_1}(z_1) = \int_{\Sigma_{Z_2}} p_{Z_1,Z_2}(z_1,z_2)\,dz_2, \quad p_{Z_2}(z_2) = \int_{\Sigma_{Z_1}} p_{Z_1,Z_2}(z_1,z_2)\,dz_1. \tag{5.187}$$

In agreement with equation (5.171), two random variables are said to be independent if (and only if)

$$p_{Z_1,Z_2}(z_1,z_2) = p_{Z_1}(z_1)\,p_{Z_2}(z_2). \tag{5.188}$$

We are now ready to define a stochastic process. In addition to elementary outcomes constituting a sample space, consider a deterministic parameter, such as time or the coordinate(s) of a point in a subset of the one- or higher-dimensioned space, $\mathbb{R}$, of real numbers. A *stochastic* (or, *random*) *process* is the function of two arguments that maps the product space, $\mathbb{R} \times \Xi$, to the realization space, Σ. Thus, the process is simply a collection of random variables, each defined on the same sample space and associated with a deterministic parameter. We may continue to use the notation, $g(x)$, to denote also a random process, but now for each fixed x, say $x = x_1$, it is implied that $g(x_1) \equiv g_{x_1}$ is a random variable with an assumed probability density function, $p_{g_{x_1}}(g_{x_1})$.

If the domain of the deterministic parameter is continuous, the process is called a *continuous (-parameter) stochastic process*; if it is discrete, it is called a *discrete (-parameter) stochastic process*. The result of a single realization of the stochastic process, which is also denoted, $g(x)$, unless confusion arises, is then a deterministic function of the parameter, x. In many cases, only a single realization of a presumed stochastic process is available; for example, the residual gravitational field may be viewed as a realization of a stochastic process locally on the plane or globally on the sphere. Under appropriate circumstances, statistical theory still allows that the stochastic properties of the process can be estimated from this single realization (*ergodicity*, Section 5.6.4).

Calling a process continuous, as above, refers generally to the continuity (or, *completeness*) of the domain of the deterministic parameter, not the continuity of the realized function, $g(x)$. It is easily imagined that the usual definition of continuity for $g(x)$ does not hold. As two values of the deterministic parameter approach each other, $x_1 \to x_2$, there is no reason to assume that the realizations converge, $g(x_1) \overset{?}{\to} g(x_2)$. For example, a stochastic process may be a random, noise-like propagation of some signal through time or space whose realization has no analytic formulation. Mathematically, this difficulty is circumvented by defining a special kind of continuity in the sense of an average over possible realizations (*stochastic continuity*, Section 5.6.4).

A particular type of stochastic process that does realize a continuous function in the usual sense is the *harmonic* (or, *periodic*) *process*, for example,

$$\tilde{g}(x) = a\cos(2\pi fx + Z), \tag{5.189}$$

where a and f are constants and Z is a random variable. For each realization of Z, $\tilde{g}(x)$ is a continuous (also, in this case, differentiable) deterministic function; and at each fixed x, $\tilde{g}_x$ is a random variable. The process may also be written as

$$\tilde{g}(x) = a(Z)\cos 2\pi fx + b(Z)\sin 2\pi fx, \tag{5.190}$$

where now the coefficients of the sinusoidal functions are random variables. In general, then, a Fourier series with random coefficients is a stochastic harmonic process. Similarly, a Fourier-Legendre series with random coefficients is a stochastic harmonic process on the sphere. These are studied in greater detail in Sections 5.6.5 and 5.6.6.

The probabilistic treatment of the stochastic process naturally involves the consideration of jointly distributed random variables; that is, the joint probabilities of g_{x_1}, g_{x_2}, g_{x_3}, etc. It is then equally clear that the description or formulation of the complete multivariate density function for the process, in the most general case, is difficult. The analysis simplifies considerably if the marginal probability density function for each random variable is the same. This leads to the concept of *stationarity* of the process. Moreover, one often obtains a fairly good description of the process from just the first- and second-order statistics of the process, that is, the covariance function. These are developed in the next section.

5.6.2 The Statistics of Stochastic Processes

While the cumulative distribution function completely characterizes a random variable, it is often sufficient to know only some essential features of its possible realizations in the long run or in some average sense. This is achieved through the *expectation* operator, defined by the sum of all possible realizations, each weighted by its probability of occurring. For a random variable, Z, with continuous sample space this sum turns to an integral; and, the *expected value* is given by

$$\mathcal{E}(Z) = \int_{\Sigma_Z} z\, p_Z(z)\, dz, \tag{5.191}$$

where, as always, it is assumed here that the probability density function exists. The expected value is also called the *ensemble average*, as it describes a weighted average over the total random ensemble of realizations. Another name is simply the *mean value*, denoted by

$$\mu_z = \mathcal{E}(Z). \tag{5.192}$$

If the probability density function is the uniform density function, equation (5.184), then the expected value is the same as the simple average of all possible realizations. The units of the expected value naturally are the units of the random variable. The expected value of any random variable is a deterministic, non-random quantity; therefore, it is amenable to deterministic analysis.

The expectation of a function of the random variable, say $h(Z)$, which is again a random variable, is given by

$$\mathcal{E}(h(Z)) = \int_{\Sigma_Z} h(z)\, p_Z(z)\, dz. \tag{5.193}$$

Thus, for example, it is easy to see that the expectation is a linear operator, since for constants, a and b, and with equation (5.180),

$$\mathcal{E}(aZ + b) = a\mathcal{E}(Z) + b. \tag{5.194}$$

A special case ($a = 0$) is the expected value of a constant, which equals that same constant. For two independent random variables equation (5.188) holds, and, therefore, the expectation of their product is

$$\begin{aligned}\mathcal{E}(Z_1 Z_2) &= \int_{\Sigma_{Z_2}} \int_{\Sigma_{Z_1}} z_1 z_2\, p_{Z_1,Z_2}(z_1 z_2)\, dz_1 dz_2 \\ &= \int_{\Sigma_{Z_1}} z_1 p_{Z_1}(z_1)\, dz_1 \int_{\Sigma_{Z_2}} z_2 p_{Z_2}(z_2)\, dz_2 \\ &= \mathcal{E}(Z_1)\,\mathcal{E}(Z_2) = \mu_{Z_1}\mu_{Z_2}\end{aligned} \tag{5.195}$$

More generally, if Z_1 and Z_2 are independent random variables, a similar derivation shows that

$$\mathcal{E}(h_1(Z_1)h_2(Z_2)) = \mathcal{E}(h_1(Z_1))\ \mathcal{E}(h_2(Z_2)). \tag{5.196}$$

For jointly distributed random variables, Z_1 and Z_2, there is also the analogue to Schwarz's inequality,

$$(\mathcal{E}(Z_1 Z_2))^2 \le \mathcal{E}(Z_1^2)\ \mathcal{E}(Z_2^2). \tag{5.197}$$

The proof can be found, e.g., in (Parzen 1960, p.363).

The *covariance* between two random variables, Z_1 and Z_2, is defined in terms of their *centralized* values as

$$\operatorname{cov}(Z_1,Z_2)=\mathcal{E}\left(\left(Z_1-\mu_{Z_1}\right)\left(Z_2-\mu_{Z_2}\right)\right)=\mathcal{E}\left(Z_1Z_2\right)-\mu_{Z_1}\mu_{Z_2}. \tag{5.198}$$

where the second equality follows from equation (5.194). If $Z_1 = Z_2$, the covariance is called the *variance*,

$$\sigma_Z^2\equiv\operatorname{var}(Z)=\mathcal{E}\left(\left(Z-\mu_Z\right)^2\right)=\mathcal{E}\left(Z^2\right)-\mu_Z^2. \tag{5.199}$$

If the covariance is zero, then Z_1 and Z_2 are said to be *uncorrelated.* If two random variables, Z_1 and Z_2, are independent then they are uncorrelated by equations (5.195) and (5.198),

$$\operatorname{cov}(Z_1, Z_2) = 0. \tag{5.200}$$

However, the converse is not necessarily true; being uncorrelated does not necessarily imply independence (e.g., Hoel 1971, p.150). On the other hand, jointly distributed Gaussian random variables that are uncorrelated are also independent.

The expected value is also know as the first *moment* of Z, and $\mathcal{E}(Z^2)$ is the second moment. Consider, then, the kth moment of Z, according to equation (5.193),

$$m_Z^{(k)}\equiv\mathcal{E}\left(Z^k\right)=\int_{\Sigma_Z}z^k p_Z(z)\,dz. \tag{5.201}$$

The moments of the random variable may be collected in a series, $M_z(t)$, known as the *moment-generating function* of Z,

$$M_Z(t)=\mathcal{E}\left(e^{tZ}\right)=\sum_{k=0}^{\infty}\frac{t^k}{k!}\mathcal{E}\left(Z^k\right)=\sum_{k=0}^{\infty}\frac{t^k}{k!}m_Z^{(k)}, \tag{5.202}$$

where the second equality follows from equations (1.7) and (5.194). Another important function in probability is the *characteristic function*, $\chi_z(\omega) = \mathcal{E}(e^{i\omega Z})$, that, like the cumulative distribution function, $F_z(z)$, completely specifies the probability of the random variable (Parzen 1960, p.395). Assuming all moments of Z exist, equation (5.201) and (5.202) then imply

$$\begin{aligned}\chi_Z(\omega)&=\mathcal{E}\left(e^{i\omega Z}\right)=M_Z(i\omega)\\&=\sum_{k=0}^{\infty}\frac{(i\omega)^k}{k!}\int_{\Sigma_Z}z^k p_Z(z)\,dz\\&=\int_{\Sigma_Z}\sum_{k=0}^{\infty}\frac{(i\omega)^k}{k!}z^k p_Z(z)\,dz\\&=\int_{\Sigma_Z}e^{i\omega z}p_Z(z)\,dz\end{aligned} \tag{5.203}$$

With $\omega = 2\pi f$ and $\Sigma_z = (-\infty, \infty)$, the complex conjugate of the characteristic function of Z is the Fourier transform of its probability density function,

$$\chi_Z^*(2\pi f) = \int_{-\infty}^{\infty} p_Z(z) e^{-i2\pi f z} dz. \tag{5.204}$$

The characteristic function is useful in deriving many probability properties of the random variable. For example, from $\chi_Z(\omega) = M_Z(i\omega)$ and equation (5.202), viewed as a Taylor series, its derivatives are the moments multiplied by i^k: $d^k\chi^*_Z(\omega)/d\omega^k|_{\omega=0} = m_Z^{(k)} i^k$. By equation (5.204) and appropriate application of equations (2.107) and (2.108), one has for the Gaussian probability density function, equation (5.182),

$$p_Z(z) = \gamma^{(\sqrt{2\pi}\sigma)}(z-\mu) \quad \leftrightarrow \quad \chi_Z^*(2\pi f) = e^{-2\pi^2(\sigma f)^2} e^{-i2\pi f \mu}, \tag{5.205}$$

where also the translation property (2.73) is used. Taking derivatives, it is readily shown that

$$\left.\frac{d}{d\omega}\chi_Z(\omega)\right|_{\omega=0} = i\mu, \quad \left.\frac{d^2}{d\omega^2}\chi_Z(\omega)\right|_{\omega=0} = i^2(\sigma^2 + \mu^2); \tag{5.206}$$

and, from these, it is easy to see with equation (5.199) that the mean and variance of the Gaussian random variable are μ and σ^2, respectively. The characteristic function can also be used to show that a linear combination of independent Gaussian random variables is again Gaussian (Exercise 5.8).

Although the Gaussian random variable is ubiquitous (Central Limit Theorem) and its probability function is completely determined by its mean and variance, geophysical random processes on the circle and on the sphere should not tacitly be assumed to be Gaussian, as illustrated in Sections 5.6.5 and 5.6.6. But, even in these cases, the first and second moments are usually sufficient for the analysis of a random process and the optimal estimation of its realization. Thus, the following developments concentrate on these moments without the necessarily specifying any particular (Gaussian or otherwise) probability function.

The expectation of a stochastic process, $g(x)$, at each point, x, is non-random; thus, it is a deterministic function of x, given by

$$\mu_g(x) = \mathcal{E}(g(x)) = \int_{\Sigma_{g_x}} g_x p_{g_x}(g_x) dg_x, \tag{5.207}$$

where p_{g_x} is the probability density function of the random variable, $g_x \equiv g(x)$, with realization space, Σ_{g_x}. With interest focused on the second moments among two stochastic processes, $g(u)$ and $h(v)$, consider the function,

$$\rho_{g,h}(u,v) \equiv \mathcal{E}(g_u h_v) = \int_{\Sigma_{h_v}} \int_{\Sigma_{g_u}} g_u h_v \, p_{g_u,h_v}(g_u, h_v) dg_u dh_v, \tag{5.208}$$

where p_{g_u, h_v} is the joint probability density function of the two random variables, $g_u \equiv g(u)$ and $h_v \equiv h(v)$. As for deterministic functions, $\rho_{g,h}$ is known as the *cross-correlation* function for the two stochastic processes; if $g = h$, then it is the *auto-correlation function* of g. In either case, $\rho_{g,h}$ is a deterministic function of the points, u and v, on the x-axis. Its units are those of the product, gh. The definition of correlation function is given under the assumption that the stochastic processes realize real-valued functions in accord with our principal applications. However, since we also express functions in terms of complex spectra, it is noted that a more general definition for complex-valued stochastic processes, analogous to equation (5.2), is

$$\rho_{g,h}(u,v) = \mathcal{E}\left(g_u^* h_v\right). \tag{5.209}$$

The *cross-* and *auto-covariance* functions are the corresponding correlation functions of the stochastic processes that at each point are centralized random variables, $g(u) - \mu_g(u)$ and $h(v) - \mu_h(v)$. The cross-covariance function is defined for complex-valued processes by

$$c_{g,h}(u,v) \equiv \mathcal{E}\left(\left(g_u - \mu_{g_u}\right)\left(h_v - \mu_{h_v}\right)\right) = \int\limits_{\Sigma_{h_v}} \int\limits_{\Sigma_{g_u}} \left(g_u - \mu_{g_u}\right)^* \left(h_v - \mu_{h_v}\right) p_{g_u, h_v}\left(g_u, h_v\right) dg_u dh_v; \tag{5.210}$$

and, the auto-covariance function is $c_{g,g}(u, v)$. Again, using the linearity of the expectation operator it is easy to show that

$$c_{g,h}(u,v) = \rho_{g,h}(u,v) - \mu_{g_u}\mu_{h_v}. \tag{5.211}$$

The value of the covariance at $u = v$ is called the (cross-) *variance*, denoted, if $g = h$, by

$$c_{g,g}(u,u) = \sigma_g^2(u). \tag{5.212}$$

For the case that $\Sigma_g = (-\infty, \infty)$ the correlation function does not necessarily attenuate to zero as the coordinate, u or v, goes to infinity. This means that, in general, the correlation function does not have a Fourier transform, since it may not be absolutely integrable. This limits the analysis of random signals in the spectral domain, and some additional constraints are imposed on the type of processes for our purposes. In particular, equation (5.211) shows that unless the mean values, μ_{g_u}, μ_{h_u}, also attenuate to zero, the correlation function does not attenuate to zero, even if the covariance function does. As indicated subsequently there is no particular loss in generality if we assume henceforth that for any stochastic process, $g(x)$,

$$\mu_g(x) = 0, \text{ for all } x. \tag{5.213}$$

It implies that correlation and covariance functions are the same. In order to distinguish between correlations of stochastic processes and deterministic functions, we refer only to the auto- and cross-*covariance* functions for the random processes (or, even, *ensemble* covariance function to refer specifically to the definition in terms of expectation). While equation (5.213) may appear overly restrictive, one can always create a new process by removing the mean (or an estimate of it) without changing its essential stochastic nature. This universal property already portends a further simplification that enables a spectral analysis of these processes, which is *invariance* of probability functions with respect to the origin of x.

In fact, many random processes in practice can be approximated well using random variables whose moments do not depend on the origin of the deterministic parameter. That is, the probabilities of the stochastic process are independent of the location on the x-axis. Specifically, the marginal probability density at every point, x, is the same; and all the joint probability density functions depend only on the differences between point coordinates. Such a process is called a *stationary* (or *homogeneous*) stochastic process. This is also equivalent to saying that the process is invariant with respect to a translation in x. Thus,

$$p_{g_x}\left(g_x\right)=p_g\left(g\right),\quad \text{for any } x; \tag{5.214}$$

and the mean and the variance of $g(x)$ are invariant with respect to x. The covariance function, equation (5.210), of two stationary processes, $g(x)$ and $h(x)$, depends only on the interval, $\xi = x_h - x_g$, of the two random variables of the processes,

$$\begin{aligned} c_{g,h}\left(\xi\right) &= \mathcal{E}\left(g_x h_{x+\xi}\right) \\ &= \int\limits_{\Sigma_h}\int\limits_{\Sigma_g} g_x h_{x+\xi}\, p_{g,h}\left(g_x, h_{x+\xi}\right) dg_x\, dh_{x+\xi}, \quad \text{for any } x \end{aligned} \tag{5.215}$$

which incorporates the stipulation that the means are zero, and where the notations for the joint density and the realization spaces are simplified, being independent of x. As for deterministic correlations, ξ is also called the lag distance, or simply, lag.

Stationarity for periodic processes (period, P), such as the harmonic process, equation (5.190), implies that the process is *rotation*-invariant. That is, if the deterministic parameter, x, is identified with the angle, $2\pi x/P$, of a circle, any rotation of the circle does not change the probabilities of the process. For stationary processes on the sphere, their covariance function depends only on the relative position of the random variables, as defined by equations (5.128) and (5.129).

Strict stationarity of the process demands that the joint probabilities of all orders are invariant under space or time translation (or, rotation for periodic processes). *Wide-sense* stationarity relaxes this to just the first- and second-order probabilities, which is assumed here; and, in fact, wide-sense Gaussian processes are necessarily also strictly stationary. If $c_{g,g}(\xi)$ decreases rapidly, i.e., g decorrelates quickly, with increasing $|\xi|$, then the random process becomes less predictable at a point given its realization at lag distance, ξ. If $c_{g,g}(\xi)$ decreases slowly, then the process differs less as $|\xi|$ increases and is more predictable from nearby realizations. Similar statements can, of course, be made for the general cross-covariance function, $c_{g,h}(\xi)$. Figure 5.6 shows auto-covariance functions for rough and smooth topography (determined empirically as in Section 5.7.1). The point of significant decorrelation is often characterized by the *correlation distance*, defined as the lag distance, ξ_c, at which

$$c_{g,g}\left(\xi_c\right)=c_{g,g}\left(0\right)/2 \tag{5.216}$$

(in electrical engineering for signals in time, $c_{g,g}(\xi_c) = e^{-1}c_{g,g}(0)$ is an alternative definition). In Figure 5.6, the correlation distance for the smoother topographic

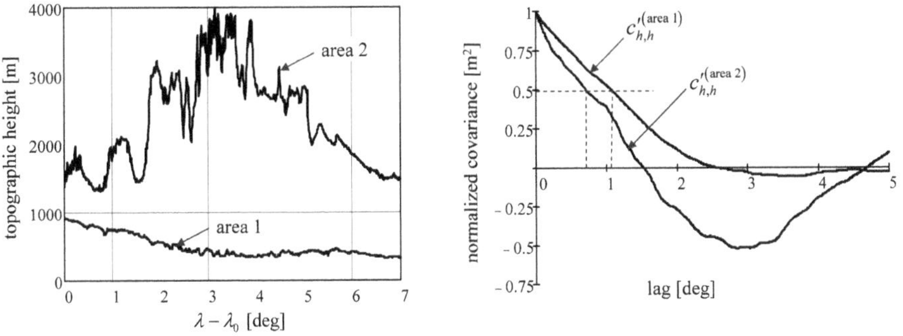

Figure 5.6: Longitude profiles of topography in the U.S. Midwest (latitude = 42.8°, λ_0 = 259°, "area 1") and the Rocky Mountains (latitude = 39°, λ_0 = 250°, "area 2") (left), and corresponding empirical normalized covariance functions, $c'_{h,h}(\xi) = c_{h,h}(\xi)/c_{h,h}(0)$.

profile of the U.S. Midwest is $\xi_c = 1.05°$; and, for the rougher Rocky Mountain profile it is $\xi_c = 0.70°$.

Stationarity on the line or plane necessarily implies that any realization of the process is not absolutely integrable (or square integrable). Indeed, stationarity implies that the process does not attenuate to zero as $x \to \pm\infty$, thus violating the condition of integrability (equation (2.49)). Hence, a stationary stochastic process on $(-\infty, \infty)$ behaves more like a finite-power signal, for which we also did not define a Fourier transform (Section 5.3). In addition, as noted in Section 5.6.1, realizations are not continuous in the usual sense since each realization is based on a probability. For these reasons a discussion of spectral analysis for stationary stochastic processes requires special care. The situation is less serious for processes on the circle and on the sphere, where one may assume that the process is a harmonic process. In any case, many of the properties of the correlation function for deterministic finite-power functions hold as well for the covariance functions of stochastic processes on any domain.

5.6.3 The Variogram

A cousin of the covariance function is the *variogram*, which plays a critical role in the optimal prediction and interpolation of random geophysical processes using the method of kriging. In this case, no information on the mean of the process is required, and although wide-sense stationarity may be assumed, a more general "intrinsic stationarity" (Cressie 1991) of the process is all that is needed. That is, one still assumes a kind of stationarity in the "variability" of the process, but the variance need not be constant. Instead, intrinsic stationarity means that the randomness *between* signals at different locations is stationary. It is described by the variogram, or semi-variogram (which is half of the variogram and more convenient in applications), instead of the covariance. The *semi-variogram* for a random process, $g(x)$, is defined as

$$\gamma_{g,g}(\xi) = \frac{1}{2}\operatorname{var}\left(g(x+\xi) - g(x)\right). \tag{5.217}$$

By assumption, the semi-variogram depends only on the difference of the deterministic parameter of the process and thus is also translation invariant. If the unknown mean of the process is a constant, independent of x, then from equation (5.199) with $Z = g(x + \xi) - g(x)$,

$$\begin{aligned}
\operatorname{var}\left(g(x+\xi) - g(x)\right) &= \mathcal{E}\left(\left(g(x+\xi) - g(x)\right)^2\right) - \left(\mathcal{E}\left(g(x+\xi) - g(x)\right)\right)^2 \\
&= \mathcal{E}\left(\left(g(x+\xi) - g(x)\right)^2\right) \\
&= \mathcal{E}\left(g^2(x+\xi) + g^2(x) - 2g(x+\xi)g(x)\right) \\
&= \operatorname{var}\left(g(x+\xi)\right) + \operatorname{var}\left(g(x)\right) - 2\operatorname{cov}\left(g(x+\xi), g(x)\right)
\end{aligned} \tag{5.218}$$

If the random process is also (wide-sense) stationary then the variances are equal; and, only in this case,

$$\gamma_{g,g}(\xi) = c_{g,g}(0) - c_{g,g}(\xi). \tag{5.219}$$

As the lag distance, ξ, between points increases the covariance function generally attenuates to zero, while the variogram reaches a maximum, also called the *sill*, i.e., $c_{g,g}(0)$.

The semi-variogram for a continuous stochastic process is zero at the origin,

$$\gamma_{g,g}(0) = 0, \tag{5.220}$$

which implies, since the semi-variogram is never negative, that

$$\gamma_{g,g}(\xi) \geq \gamma_{g,g}(0), \quad \text{for all } \xi. \tag{5.221}$$

Because of the translation invariance, the semi-variogram is also symmetric,

$$\gamma_{g,g}(-\xi) = \frac{1}{2}\operatorname{var}\left(g(x-\xi) - g(x)\right) = \frac{1}{2}\operatorname{var}\left(g(x) - g(x+\xi)\right) = \gamma_{g,g}(\xi). \tag{5.222}$$

Although it is rarely introduced in the geostatistical literature, one may also define the cross-semi-variogram for two stochastic processes, g and h,

$$\gamma_{g,h}(\xi) = \frac{1}{2}\operatorname{var}\left(g(x+\xi) - h(x)\right). \tag{5.223}$$

Other properties of the semi-variogram may be found in (Cressie 1991).

It is assumed here that the stochastic processes are wide-sense stationary, in which case the semi-variogram and the covariance function are simply mirror images as noted with equation (5.219). The only advantage of the semi-variogram over the covariance function then rests in not worrying about the presence of a (constant) mean in the process, as this is often not known in practice. However, there are ways to estimate it (e.g., Moritz 1980, Ch.16).

An important difference, however, between the covariance function and the semi-variogram, even in the case of stationary processes, is that the latter does not possess a standard Fourier transform since it does not attenuate to zero for large lag distances. Instead, one must invoke the Dirac delta function; and, from equations (2.121) and (5.219),

$$\Gamma_{g,g}(f) \equiv \mathcal{F}\left(\gamma_{g,g}(\xi)\right) = c_{g,g}(0)\delta(f) - C_{g,g}(f). \tag{5.224}$$

From a practical viewpoint, however, the variogram of a stationary process, g, can be determined from a spectral representation of g using

$$\gamma_{g,g}(\xi) = c_{g,g}(0) - \mathcal{F}^{-1}\left(C_{g,g}(f)\right), \tag{5.225}$$

(see also Cressie 1991, Section 2.5). As such the focus in Section 5.7 is on the determination of covariance functions and power spectral densities.

5.6.4 Non-Periodic Stationary Stochastic Processes

For local applications, the deterministic domain for a spectral analysis of stochastic processes that represent geophysical signals, may be approximated in one dimension by a line, or in two dimensions by a plane. If stationarity is to be retained, then in principle the process extends over the entire line or plane. It is assumed in this case that the deterministic structure of the process is neither periodic nor includes a constant, or zero-frequency, part. If there is a constant, then as noted in the previous section, one may re-define the process by removing it, or treat the constant separately as a parameter to be estimated. For the sake of simplicity in the development, processes on the line are considered first and this then easily generalizes to higher dimensions.

The added complication that $\lim_{x\to x_0} g(x)$ is not well defined for the realization of the stochastic process (i.e., it is not continuous anywhere) require a different limiting procedure and a notion of continuity called *stochastic continuity*. A stochastic process, $g(x)$, is stochastically continuous at x_0 if

$$\lim_{x\to x_0} \mathcal{E}\left(\left(g(x) - g(x_0)\right)^2\right) = 0; \tag{5.226}$$

that is, it is *continuous in the mean*. As before, denote the covariance function of $g(x)$ by $c_{g,g}(\xi)$. Stationarity of the process implies that

$$\begin{aligned}\mathcal{E}\left(\left(g(x) - g(x_0)\right)^2\right) &= \mathcal{E}\left(g^2(x)\right) + \mathcal{E}\left(g^2(x_0)\right) - 2\mathcal{E}\left(g(x_0)g(x)\right) \\ &= 2c_{g,g}(0) - 2c_{g,g}(\xi)\end{aligned} \tag{5.227}$$

where $\xi = x - x_0$. Then,

$$\lim_{x\to x_0} \mathcal{E}\left(\left(g(x) - g(x_0)\right)^2\right) = 2\lim_{\xi\to 0}\left(c_{g,g}(0) - c_{g,g}(\xi)\right) = 0; \tag{5.228}$$

and, hence, stochastic continuity for a stationary process is the same as the usual continuity of the covariance function at its origin. Moreover, continuity of the covariance function at the origin implies its continuity everywhere. Indeed, consider the difference in covariances for any two neighboring points, $c_{g,g}(\xi) - c_{g,g}(\xi + \delta\xi)$. Its squared magnitude is

$$\begin{aligned}
\left|c_{g,g}(\xi) - c_{g,g}(\xi+\delta\xi)\right|^2 &= \left|\mathcal{E}\left(g(0)g(\xi)\right) - \mathcal{E}\left(g(-\delta\xi)g(\xi)\right)\right|^2 \\
&= \left|\mathcal{E}\left(g(\xi)\left(g(0)-g(-\delta\xi)\right)\right)\right|^2 \\
&\le \mathcal{E}\left(g^2(\xi)\right)\mathcal{E}\left(\left(g(0)-g(-\delta\xi)\right)^2\right) \\
&= 2c_{g,g}(0)\left(c_{g,g}(0) - c_{g,g}(\delta\xi)\right)
\end{aligned} \tag{5.229}$$

where the first equality comes from equation (5.215), while stationarity and equation (5.227) imply the last equality. The inequality is Schwarz's inequality (5.197). If the covariance function is continuous at the origin, the last difference on the right side is arbitrarily small for all $\delta\xi$ within a given neighborhood of 0. Hence, the left is arbitrarily small for all $\xi + \delta\xi$ within a given neighborhood of ξ, thus demonstrating continuity at ξ.

To proceed with a spectral analysis in the spirit of finite-power signals, only the additional assumption is made that the covariance function of stationary stochastic processes, g and h, on the deterministic domain, $(-\infty, \infty)$, is absolutely integrable. The covariance function thus attenuates to zero as the lag distance, ξ, between the random variables, g_x and $h_{x+\xi}$, increases, as already implied in Section 5.6.2. Under these conditions it has a Fourier integral and its Fourier transform is given by

$$C_{g,h}(f) = \int_{-\infty}^{\infty} c_{g,h}(\xi) e^{-i2\pi f\xi} d\xi. \tag{5.230}$$

Analogous to the correlation function for finite-power functions, the Fourier transform of the covariance function, equation (5.230), for stationary stochastic processes on $(-\infty, \infty)$ is a power spectral density. Consider the truncated processes, $g_T(x)$ and $h_T(x)$, as in equation (5.68). Even though any particular realization is only stochastically continuous, it is possible to define them as integrable in a mean square sense (Priestley 1981). This modified interpretation is assumed wherever it is needed, in particular for the Fourier transforms,

$$G_T(f) = \int_{-T/2}^{T/2} g(x) e^{-i2\pi fx} dx, \qquad H_T(f) = \int_{-T/2}^{T/2} h(x) e^{-i2\pi fx} dx, \tag{5.231}$$

without a change in notation. The Fourier transforms are then also realizable random processes. The frequency, f, or the wave number, k, for periodic stochastic processes, in this case is the deterministic variable.

It is now proved that the Fourier transform of the covariance function is given by

$$C_{g,h}(f) = \lim_{T\to\infty}\left(\mathcal{E}\left(\frac{1}{T}G_T^*(f)H_T(f)\right)\right). \tag{5.232}$$

Substituting equations (5.231) for the Fourier transforms on the right side and taking the expectation inside the integrals,

$$\begin{aligned}\lim_{T\to\infty}\left(\mathcal{E}\left(\frac{1}{T}G_T^*(f)H_T(f)\right)\right) &= \lim_{T\to\infty}\frac{1}{T}\int_{-T/2}^{T/2}\int_{-T/2}^{T/2}\mathcal{E}\big(g(x)h(x')\big)e^{-i2\pi f(x'-x)}dxdx' \\ &= \lim_{T\to\infty}\frac{1}{T}\int_{-T/2}^{T/2}\int_{-T/2-x}^{T/2-x}c_{g,h}(\xi)e^{-i2\pi f\xi}d\xi dx\end{aligned} \tag{5.233}$$

where, as usual, g and h are assumed to be real processes, and the second equality follows with a change in integration variable, $\xi = x' - x$, and the fact that g and h are stationary. The region in the ξ, x-plane over which the integration takes place is shown in Figure 5.7.

Performing the integration with respect to x separately over the two triangles on either side of the x-axis, it is readily verified that

$$\begin{aligned}&\frac{1}{T}\int_{\xi=-T}^{0}\int_{-T/2-\xi}^{T/2}c_{g,h}(\xi)e^{-i2\pi f\xi}dxd\xi + \frac{1}{T}\int_{\xi=0}^{T}\int_{-T/2}^{T/2-\xi}c_{g,h}(\xi)e^{-i2\pi f\xi}dxd\xi \\ &= \int_{-T}^{T}\left(1-\frac{|\xi|}{T}\right)c_{g,h}(\xi)e^{-i2\pi f\xi}d\xi\end{aligned} \tag{5.234}$$

The parenthetical term is the Bartlett window or triangle function, $u_{\text{Bartlett}}^{(2T)}(\xi)$, equation (3.146). Since $u_{\text{Bartlett}}^{(2T)}(\xi) \le 1$ and the covariance function, assumed to have a Fourier transform, is absolutely integrable, the last integral in equation (5.234) is also

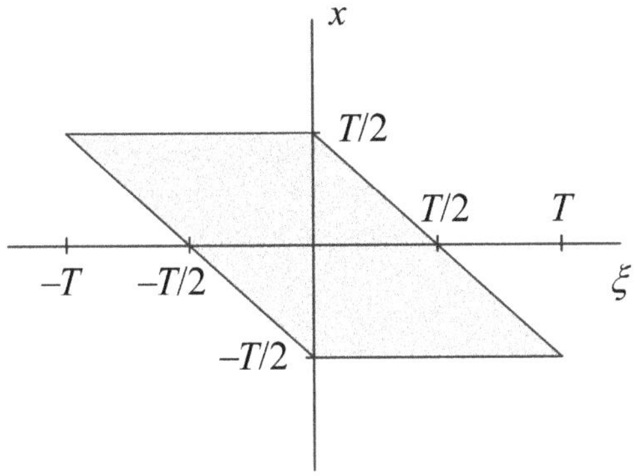

Figure 5.7: Region of integration for the variables, ξ, x, in equation (5.233).

absolutely integrable and the limit as $T \to \infty$ can be taken inside. Equation (5.233) thus becomes

$$\lim_{T\to\infty}\left(\mathcal{E}\left(\frac{1}{T}G_T^*(f)H_T(f)\right)\right) = \int_{-\infty}^{\infty} c_{g,h}(\xi)e^{-i2\pi f\xi}d\xi, \tag{5.235}$$

which proves equation (5.232) in view of equation (5.230).

The Fourier transform of the covariance function of stationary stochastic processes on the line thus has the same form, except for the expectation operator, as that of finite-power signals, equation (5.73), and is also called the (cross-) *power spectral density* of the processes. It is a consequence of an important result known as the *Wiener-Khintchine theorem*, which states that under rather general conditions, the covariance function of a stationary stochastic process can be written as a Fourier integral (Priestley 1981, p.218).

In general, the properties of the covariance function and its Fourier transform for a stationary stochastic process mirror those of the correlation of deterministic functions, equations (5.10) through (5.14), and (5.17), (5.18). Applying Schwarz's inequality (5.197) for the expectation operator to the covariance function yields

$$\left(c_{g,h}(\xi)\right)^2 = \left(\mathcal{E}\left(g_x h_{x+\xi}\right)\right)^2 \le \mathcal{E}\left(g_x^2\right)\mathcal{E}\left(h_{x+\xi}^2\right). \tag{5.236}$$

Since the processes are stationary, $\mathcal{E}(g_x^2) = c_{g,g}(0)$ and $\mathcal{E}(h_{x+\xi}^2) = c_{h,h}(0)$; hence,

$$\left(c_{g,h}(\xi)\right)^2 \le c_{g,g}(0)c_{h,h}(0). \tag{5.237}$$

The auto-covariance function thus reaches its maximum at the origin, $\xi = 0$,

$$c_{g,g}(\xi) \le c_{g,g}(0), \quad \text{for all } \xi. \tag{5.238}$$

This is the only location for the maximum if the process is not periodic. Note that the maximum value of the *cross*-covariance, $c_{g,h}$, is not necessarily at the origin. For a process that is a random constant, $g(x) = Z$, for all x, the covariance function is the variance of Z, $c_{g,g}(\xi) = \sigma_z^2$, for all ξ. The random constant is one of two processes considered here whose covariance function does not have the usual Fourier transform. The other process is white noise (see below). The Fourier transform of the covariance for the random constant formally is σ_z^2 times the Dirac delta function, equation (2.121).

The covariance function for real stationary processes is symmetric in the sense,

$$c_{g,h}(-\xi) = c_{h,g}(\xi); \tag{5.239}$$

which is easily proved from the stationarity,

$$\mathcal{E}\left(g_x h_{x-\xi}\right) = \mathcal{E}\left(g_{x+\xi}h_{x-\xi+\xi}\right) = \mathcal{E}\left(h_x g_{x+\xi}\right). \tag{5.240}$$

Finally, from equation (5.235), the Fourier transform of the *auto*-covariance function is non-negative, which means that it is a positive-semidefinite function. We assume, in fact, that the PSD is always strictly positive,

$$C_{g,g}(f) > 0, \tag{5.241}$$

and, therefore, that the auto-covariance function is positive definite.

Let $g(x)$ be a stationary stochastic process and let $h(x)$ be a *finite-energy* impulse response function for a convolution (Section 3.5). If $g(x)$ is the input to that system, then the output is also a stochastic process, where the convolution, equation (3.1), is re-written in its equivalent form as

$$\bar{g}(x) = (g * h)(x) = \int_{-\infty}^{\infty} g(x - x')h(x')\,dx'. \tag{5.242}$$

In view of equation (5.213), the expectation of this process is

$$\mathcal{E}(\bar{g}(x)) = \int_{-\infty}^{\infty} \mathcal{E}(g(x - x'))h(x')\,dx' = \mu_g \int_{-\infty}^{\infty} h(x')\,dx' = 0; \tag{5.243}$$

and, the covariance function is

$$\begin{aligned} \mathcal{E}(\bar{g}_x \bar{g}_{x+\xi}) &= \int_{-\infty}^{\infty}\int_{-\infty}^{\infty} \mathcal{E}(g(x-u)g(x+\xi-v))h(u)h(v)\,du dv \\ &= \int_{-\infty}^{\infty}\int_{-\infty}^{\infty} c_{g,g}(\xi+u-v)h(u)h(v)\,du dv \\ &= \int_{-\infty}^{\infty}\left(\int_{-\infty}^{\infty} c_{g,g}(v-u-\xi)h(u)\,du\right)h(v)\,dv \end{aligned} \tag{5.244}$$

where the last equality follows from the symmetry of the auto-covariance function, equation (5.239). With a change in integration variable, $w = v - \xi$, this becomes

$$\begin{aligned} \mathcal{E}(\bar{g}_x \bar{g}_{x+\xi}) &= \int_{-\infty}^{\infty}\left(\int_{-\infty}^{\infty} c_{g,g}(w-u)h(u)\,du\right)h(w+\xi)\,dw \\ &= \int_{-\infty}^{\infty}\left(c_{g,g}(w) * h(w)\right)h(w+\xi)\,dw \\ &= c_{g,g}(\xi) * h(\xi) * h(-\xi) \end{aligned} \tag{5.245}$$

which depends only on the difference in coordinates, ξ. Therefore, the convolution of a (wide-sense) stationary stochastic process with a deterministic function is also a (wide-sense) stationary stochastic process. Furthermore, since the correlation function of the impulse response, $h(x)$, is given by the convolution, equation (5.3),

$$\overset{\circ}{\phi}_{h,h}(x) = h(x) * h(-x), \tag{5.246}$$

equation (5.245) then shows that

$$c_{\bar{g},\bar{g}}(\xi) = c_{g,g}(\xi) * \overset{\circ}{\phi}_{h,h}(\xi). \tag{5.247}$$

That is, the covariance function of the convolution, equation (5.242), is the convolution of the ensemble covariance and deterministic correlation functions.

Being itself a deterministic function, $c_{g,h}(\xi)$ also enjoys the additional properties already reviewed for the correlation function of finite-power functions, equations (5.58) through (5.60) and (5.75), (5.76). The propagation laws for derivatives of stationary stochastic processes, equations (5.61) and (5.77), however, formally, must be cast in the context of stochastic differentiability, analogous to stochastic continuity,

$$\frac{d}{dx} g(x) = \lim_{\delta x \to 0} \mathcal{E}\left(\left(\frac{g(x+\delta x) - g(x)}{\delta x}\right)^2\right), \tag{5.248}$$

which exists if and only if the covariance function of g is twice differentiable at x (Priestley 1981, p.153). The corresponding details for the propagation laws are left to the interested reader.

A special stationary stochastic process, $w(x)$, that, like white light, has equal power at all frequencies is called continuous *white noise*. Its PSD is a constant, say,

$$C_{w,w}(f) = \sigma_w^2, \quad -\infty < f < \infty. \tag{5.249}$$

Without loss in generality, its mean value is defined to be zero. The covariance function is the inverse Fourier transform of the PSD, that is, proportional to the Dirac delta function, equation (2.112),

$$c_{w,w}(\xi) = \sigma_w^2 \delta(\xi). \tag{5.250}$$

Thus, the covariance for all non-zero lag distances is zero and the variance, technically, is infinite. The continuous white noise process has mostly theoretical value particularly in modeling hypothetical random processes (Marple 1987, Percival and Walden 1993), but strictly it is not a physically realizable process.

For stationary stochastic processes on the x_1,x_2-plane, the covariance function depends on the differences, $\xi_1 = x_1^{(h)} - x_1^{(g)}$ and $\xi_2 = x_2^{(h)} - x_2^{(g)}$,

$$\begin{aligned} c_{g,h}(\xi_1,\xi_2) &= \mathcal{E}\left(g_{x_1,x_2} h_{x_1+\xi_1,x_2+\xi_2}\right) \\ &= \int_{\Sigma_h}\int_{\Sigma_g} g_{x_1,x_2} h_{x_1+\xi_1,x_2+\xi_2}\, p_{g,h}\left(g_{x_1,x_2}, h_{x_1+\xi_1,x_2+\xi_2}\right) dg_{x_1,x_2}\, dh_{x_1+\xi_1,x_2+\xi_2} \end{aligned} \tag{5.251}$$

for any x_1, x_2. If the covariance function further does not depend on orientation; that is, the covariance only depends on the relative distance, $\sqrt{\xi_1^2 + \xi_2^2}$, between the points on the plane, then the covariance function is also *isotropic*. This is analogous to the case for the correlation of deterministic functions; however, an expression in terms of expectation requires a correspondingly isotropic joint probability density function, which is not pursued here.

The Fourier transform of the two-dimensional covariance function is

$$C_{g,h}(f_1, f_2) = \int_{-\infty}^{\infty}\int_{-\infty}^{\infty} c_{g,h}(\xi_1, \xi_2)\, e^{-i2\pi(f_1\xi_1 + f_2\xi_2)}\, d\xi_1\, d\xi_2; \tag{5.252}$$

and, the generalization of equation (5.235) for two dimensions is

$$\lim_{\substack{T_1 \to \infty \\ T_2 \to \infty}} \left(\mathcal{E}\left(\frac{1}{T_1 T_2} G^*_{T_1,T_2}(f_1, f_2)\, H_{T_1,T_2}(f_1, f_2) \right) \right) = \int_{-\infty}^{\infty}\int_{-\infty}^{\infty} c_{g,h}(\xi_1, \xi_2)\, e^{-i2\pi(f_1\xi_1 + f_2\xi_2)}\, d\xi_1\, d\xi_2 \tag{5.253}$$

Other properties of the covariance function, equivalent to equations (5.87) through (5.90) and (5.94) through (5.96), follow in a straightforward manner.

Instead of higher dimensions for the deterministic parameter, one may also consider higher dimensions of the random variable. Thus, let $\boldsymbol{g}(x) = (g_1(x) \cdots g_m(x))^{\mathrm{T}}$ be a stationary random *vector* process. The expectation of the process is then also a vector,

$$\boldsymbol{\mu}_{\boldsymbol{g}} = \mathcal{E}(\boldsymbol{g}(x)), \quad \text{with } \mu_{g_j} = \mathcal{E}(g_j) = \int_{\Sigma_{g_j}} g_j p_{g_j}(g_j)\, dg_j, \tag{5.254}$$

assumed here, as usual, to be zero. For vector processes, $\boldsymbol{g}(x)$ and $\boldsymbol{h}(x)$, the covariance now becomes an $m \times n$ covariance matrix function,

$$\mathbf{c}_{\boldsymbol{g},\boldsymbol{h}}(\xi) = \begin{pmatrix} c_{g_1,h_1}(\xi) & \cdots & c_{g_1,h_n}(\xi) \\ \vdots & \ddots & \vdots \\ c_{g_m,h_1}(\xi) & \cdots & c_{g_m,h_n}(\xi) \end{pmatrix}, \tag{5.255}$$

with elements of the form of equation (5.215). The matrix function, $\mathbf{c}_{\boldsymbol{g},\boldsymbol{h}}(\xi)$, should not be confused with the covariance matrix for a one-dimensional process where the elements are covariances of g and h for different values of ξ (as in equation (6.326)). Thus, depending on the elements, $g_j(x)$, of the vector process, $\mathbf{c}_{\boldsymbol{g},\boldsymbol{g}}(\xi)$ at a single point may be neither symmetric nor positive definite.

In a heuristic way, one may speak of the predictability of a stationary process, that is, its likely realization, on the basis of its covariance function. It seems reasonable that in many cases the converse should hold. That is, one *realization* of the process, or the set of realized values of the random variables at all points in the domain of the deterministic parameter, should in some way reflect its moments, even though the probability densities are not known. In fact, it is highly desirable if one could infer the moments of the process without the luxury of repeated realizations needed to estimate the probabilities. This is precisely the idea behind the property of *ergodicity*, which is also a natural extension within the framework of the Bayesian interpretation of probability (cf. the introduction to Section 5.6). A stationary process is ergodic if the moments of the process associated with the underlying probabilities can be estimated in some sense from an average over the domain of the deterministic parameter, that is, from a single realization of the process. Stationarity does not necessarily imply ergodicity. For example, a randomly constant process, $g(x) = Z$, for all x, while stationary, yields no information on the mean from a single realization. However, an ergodic process is always stationary by definition, and, conversely, a non-stationary process cannot be ergodic. Strict and wide-sense ergodicity associate directly with strict and wide-sense stationarity.

An *estimator* is defined as a function of random quantities that yields an *estimate* of a parameter if the actual realizations of the random quantities are substituted. Ergodicity is established if the estimators of the moments of the process, using one of its realizations, have certain desirable properties. Let $\hat{\theta}$ be an estimator of a moment, θ. One desirable property is that it should be *unbiased*, which is to say that

$$\mathcal{E}(\hat{\theta}) = \theta. \tag{5.256}$$

Another is that it should be *consistent*, meaning that as the number, n, of observables increases, the *mean square error* of the estimator, $\hat{\theta}(n)$, approaches zero,

$$\lim_{n\to\infty} \mathcal{E}\left(\left(\hat{\theta}(n)-\theta\right)^2\right) = 0. \tag{5.257}$$

For an unbiased estimator, consistency means that its variance decreases to zero with more and more observations.

A stationary process is ergodic with respect to a particular moment if the corresponding average-based estimator of that moment, using realizations over the deterministic domain of the process, is unbiased and consistent. For example, suppose that the deterministic domain for the stationary stochastic process, $g(x)$, is $\mathbb{R} = (-\infty, \infty)$, and that an estimator of the mean, $\mu_g = \mathcal{E}(g) = 0$, for some finite domain, $T > 0$, is given by

$$\hat{\mu}_g^{(T)} = \frac{1}{T}\int_{x-T/2}^{x+T/2} g(x')\,dx', \tag{5.258}$$

which, as a function of x, is also a stochastic process. Clearly, $\mathcal{E}(\hat{\mu}_g^{(T)}) = \mu_g$; i.e., $\hat{\mu}_g^{(T)}$ is an unbiased estimator of μ_g. Then, $g(x)$ is ergodic with respect to the mean (or, *mean-ergodic*) if

$$\lim_{T\to\infty} \operatorname{var}\left(\hat{\mu}_g^{(T)}\right) = \lim_{T\to\infty} \mathcal{E}\left(\left(\hat{\mu}_g^{(T)} - \mu_g\right)^2\right) = 0. \tag{5.259}$$

Substituting equation (5.258) into $\operatorname{var}(\hat{\mu}_g^{(T)})$, according to equation (5.199), and evaluating the resulting double integral, as in equations (5.233) and (5.234), it it is readily shown that equation (5.259) holds if the covariance function of $g(x)$ is integrable (*Slutsky's theorem*; Papoulis 1991, p.430).

Similarly a stationary process is *covariance-ergodic* if the estimator (compare with equation (5.56)),

$$\hat{c}_{g,g}^{(T)}(\xi)=\frac{1}{T}\int_{-T/2}^{T/2}\left(g(x)-\mu_g\right)\left(g(x+\xi)-\mu_g\right)dx, \tag{5.260}$$

is unbiased and has vanishing variance as $T\rightarrow\infty$. Unbiasedness is easy to prove,

$$\mathcal{E}\left(\hat{c}_{g,g}^{(T)}(\xi)\right)=\frac{1}{T}\int_{-T/2}^{T/2}\mathcal{E}\left(\left(g(x)-\mu_g\right)\left(g(x+\xi)-\mu_g\right)\right)dx=\frac{1}{T}\int_{-T/2}^{T/2}c_{g,g}(\xi)\,dx=c_{g,g}(\xi). \tag{5.261}$$

If the process is Gaussian, then it can be shown (Bendat and Piersol 1986, p.272) that the variance of the estimator, $\hat{c}_{g,g}(\xi)$, in the limit as $T\rightarrow\infty$, vanishes, thus proving that the process is covariance-ergodic (i.e., that the covariance estimator is consistent). These statements may be generalized to more than one stochastic process and an estimator for their cross-covariance function, $\hat{c}_{g,h}^{(T)}$. Therefore, under the premise of ergodicity the estimates based on the space averages, equations (5.258) and (5.260), are reasonable approximations of the moments, mean and covariance, respectively. In practice ergodicity is often assumed without much concern since one is presented de facto with a single realization of a geophysical signal, and by the Central Limit Theorem Gaussianity is a reasonable assumption.

It is noted that while the estimate of the covariance function thus derived converges in the mean to $c_{g,h}(\xi)$ as $T\rightarrow\infty$, the periodogram, similarly derived from a realization of processes, does not converge to the power spectral density, $C_{g,h}(f)$, equation (5.232). This is revisited in Section 5.7.2, where also modifications to the periodogram are proposed that make it a consistent estimator of the PSD.

5.6.5 Periodic Stochastic Processes

The process, illustrated by equation (5.190), is one example of a class of stochastic processes, called harmonic processes, whose realizations satisfy the conditions of being expressible as a Fourier series, where each element of the Fourier series spectrum is a random variable (compare with equation (2.8)),

$$\tilde{g}(x)=\frac{1}{P}\sum_{k=-\infty}^{\infty}G_k(Z)\,e^{i\frac{2\pi}{P}kx}. \tag{5.262}$$

The period of the realized process is P; and, the randomness of the complex Fourier coefficients is denoted by its dependence on a real random variable, Z, as

$$G_k(Z)=A_k(Z)+iB_k(Z). \tag{5.263}$$

It is assumed that for each A_k and B_k and for all wave numbers, k, the random variable, Z, is independently realized from the same distribution. The wave number is the deterministic variable; and, therefore, the Fourier spectrum is a (discrete) stochastic process in the spectral domain. Again, the dependence of the process, $\tilde{g}(x)$, on the random variables is implicit.

If the process is stationary then its mean,

$$\mathcal{E}\left(\tilde{g}(x)\right)=\frac{1}{P}\sum_{k=-\infty}^{\infty}\mathcal{E}\left(G_k(Z)\right)e^{i\frac{2\pi}{P}kx}, \tag{5.264}$$

must be invariant with respect to x, which can only be true if

$$\mathcal{E}\left(G_k(Z)\right)=0, \quad \text{for all } k\neq 0. \tag{5.265}$$

Setting $G_0 \equiv 0$ then also yields $\mathcal{E}(\tilde{g}_x)=0$ for all x, conforming to the simplifying assumption, equation (5.213). Thus, the covariance function, given by equation (5.210), is

$$\begin{aligned}\tilde{c}_{\tilde{g},\tilde{g}}(x,x')&=\frac{1}{P^2}\mathcal{E}\left(\sum_{k=-\infty}^{\infty}G_k^*(Z)e^{-i\frac{2\pi}{P}kx}\sum_{k'=-\infty}^{\infty}G_{k'}(Z)e^{i\frac{2\pi}{P}k'x'}\right)\\&=\frac{1}{P^2}\sum_{k=-\infty}^{\infty}\sum_{k'=-\infty}^{\infty}\mathcal{E}\left(G_k^*(Z)G_{k'}(Z)\right)e^{i\frac{2\pi}{P}(k'x'-kx)}\end{aligned} \tag{5.266}$$

This depends on the difference, $\xi = x' - x$, in agreement with a stationary process, only if

$$\mathcal{E}\left(G_k^*(Z)G_{k'}(Z)\right)=0, \quad \text{for } k\neq k'; \tag{5.267}$$

that is, the random Fourier coefficients are mutually uncorrelated. The converse is also evident (that is, equation (5.267) implies that the process is stationary). In summary, the Fourier spectrum of a periodic, stationary process comprises zero-mean and mutually uncorrelated components. Its covariance function, equation (5.266), may be written as

$$\tilde{c}_{\tilde{g},\tilde{g}}(\xi)=\frac{1}{P}\sum_{k=-\infty}^{\infty}\left(C_{\tilde{g},\tilde{g}}\right)_k e^{i\frac{2\pi}{P}k\xi}, \tag{5.268}$$

where the Fourier spectrum (power spectrum of $\tilde{g}$) is

$$\left(C_{\tilde{g},\tilde{g}}\right)_k=\frac{1}{P}\mathcal{E}\left(\left|G_k(Z)\right|^2\right). \tag{5.269}$$

The covariance function is also periodic with period, P.

For two periodic, stationary, zero-mean, stochastic processes, $\tilde{g}(x)$ and $\tilde{h}(x)$, the cross-covariance function is

$$\tilde{c}_{\tilde{g},\tilde{h}}(\xi)=\frac{1}{P}\sum_{k=-\infty}^{\infty}\left(C_{\tilde{g},\tilde{h}}\right)_k e^{i\frac{2\pi}{P}k\xi}, \tag{5.70}$$

where its Fourier-series spectrum is given by

$$\left(C_{\tilde{g},\tilde{h}}\right)_k=\frac{1}{P}\mathcal{E}\left(G_k^*(Z)H_k(Z)\right), \tag{5.271}$$

and where, as before, $\mathcal{E}(G_k(Z)) = 0 = \mathcal{E}(H_k(Z))$ and $\mathcal{E}(G_k^*(Z)H_{k'}(Z)) = 0$ for $k \neq k'$.

In addition to the properties already established for covariance functions of stationary stochastic processes (Sections 5.6.2 and 5.6.4), the covariance functions for the periodic stationary stochastic processes, equation (5.262), also follow the rules for propagation of correlation functions of deterministic derivatives, equations (5.15). Noting that $\xi = x^{(h)} - x^{(g)}$, it is easily verified that

$$\tilde{c}_{\tilde{g},\frac{d\tilde{h}}{dx}}(\xi)=-\tilde{c}_{\frac{d\tilde{g}}{dx},\tilde{h}}(\xi)=\frac{d}{d\xi}\tilde{c}_{\tilde{g},\tilde{h}}(\xi). \tag{5.272}$$

Similarly, the discrete power spectrum of the derivatives of periodic processes satisfy properties analogous to equations (5.19) and (5.20); for example, from equations (5.270) and (5.272),

$$\left(C_{\frac{d\tilde{g}}{dx},\tilde{h}}\right)_k=-i\frac{2\pi k}{P}\left(C_{\tilde{g},\tilde{h}}\right)_k=-\left(C_{\tilde{g},\frac{d\tilde{h}}{dx}}\right)_k. \tag{5.273}$$

Generalization to higher dimensions is straightforward. In two dimensions, the covariance function is

$$\tilde{c}_{\tilde{g},\tilde{h}}(\xi_1,\xi_2)=\frac{1}{P_1P_2}\sum_{k_1=-\infty}^{\infty}\sum_{k_2=-\infty}^{\infty}\left(C_{\tilde{g},\tilde{h}}\right)_{k_1,k_2} e^{i2\pi\left(\frac{k_1\xi_1}{P_1}+\frac{k_2\xi_2}{P_2}\right)}. \tag{5.274}$$

where

$$\left(C_{\tilde{g},\tilde{h}}\right)_{k_1,k_2}=\frac{1}{P_1P_2}\mathcal{E}\left(G_{k_1,k_2}^*(Z)H_{k_1,k_2}(Z)\right), \tag{5.275}$$

and the Fourier spectral components have zero mean and are mutually uncorrelated for different wave numbers. Properties generalize as for correlation functions; e.g., the law of propagation of covariances for derivatives, equation (5.30), is

$$\tilde{c}_{\tilde{g},\frac{\partial\tilde{h}}{\partial x_j}}(\xi_1,\xi_2)=-\tilde{c}_{\frac{\partial\tilde{g}}{\partial x_j},\tilde{h}}(\xi_1,\xi_2)=\frac{\partial}{\partial\xi_j}\tilde{c}_{\tilde{g},\tilde{h}}(\xi_1,\xi_2), \quad j=1,2. \tag{5.276}$$

To determine the conditions under which the stationary periodic process is ergodic, consider estimators of the mean and of the covariance function,

$$\hat{\mu}_{\tilde{g}}=\frac{\Delta x}{P}\sum_{j=0}^{N-1}\tilde{g}(x_j), \tag{5.277}$$

$$\hat{\tilde{c}}_{\tilde{g},\tilde{g}}(\xi) = \frac{\Delta x}{P}\sum_{j=0}^{N-1}\tilde{g}^*(x_j)\tilde{g}(x_j+\xi), \tag{5.278}$$

where $x_j = j\Delta x$, $P = N\Delta x$, and the complex conjugate on the real process, $\tilde{g}(x)$, is included in order to derive the ergodicity conditions for the process. Both $\hat{\mu}_{\tilde{g}}$ and $\hat{\tilde{c}}_{\tilde{g},\tilde{g}}(\xi)$ are random quantities. Substituting equation (5.262) the expression for the estimator, $\hat{\mu}_{\tilde{g}}$, becomes with equation (4.68),

$$\hat{\mu}_{\tilde{g}} = \frac{\Delta x}{P^2}\sum_{k=-\infty}^{\infty}G_k(Z)\sum_{j=0}^{N-1}e^{i\frac{2\pi}{N}kj} = \frac{1}{P}\sum_{k=-\infty}^{\infty}G_{kN}(Z). \tag{5.279}$$

Mean-ergodicity requires that as N increases, the variance of $\hat{\mu}_{\tilde{g}}$ vanishes. In fact, as $N \to \infty$ (and $\Delta x \to 0$ such that $P = N\Delta x$ is constant), the only non-zero spectral component is $G_0(Z)$ and $\hat{\mu}_{\tilde{g}} \to G_0(Z)/P$. This has zero variance only if $G_0(Z) \equiv G_0$ is a *non-random* constant. Hence, only then is the process mean-ergodic; and, as already assumed, $G_0 = 0$.

Similarly, the covariance estimator in terms of the random spectrum is obtained by substituting equation (5.262) in equation (5.278) and using equation (4.68),

$$\begin{aligned}\hat{\tilde{c}}_{\tilde{g},\tilde{g}}(\xi) &= \frac{\Delta x}{P^3}\sum_{k=-\infty}^{\infty}\sum_{k'=-\infty}^{\infty}G_k^*(Z)G_{k'}(Z)e^{i\frac{2\pi}{P}k'\xi}\sum_{j=0}^{N-1}e^{i\frac{2\pi}{N}(k'-k)j}\\ &= \frac{1}{P^2}\sum_{k=-\infty}^{\infty}\sum_{\substack{k'=-\infty\\ \mathrm{mod}_N(k'-k)=0}}^{\infty}G_k^*(Z)G_{k'}(Z)e^{i\frac{2\pi}{P}k'\xi}\end{aligned} \tag{5.280}$$

With equations (5.267) through (5.269), the mean of $\hat{\tilde{c}}_{\tilde{g},\tilde{g}}(\xi)$ is

$$\mathcal{E}\left(\hat{\tilde{c}}_{\tilde{g},\tilde{g}}(\xi)\right) = \frac{1}{P^2}\sum_{k=-\infty}^{\infty}\mathcal{E}\left(\left|G_k(Z)\right|^2\right)e^{i\frac{2\pi}{P}k\xi} = \tilde{c}_{\tilde{g},\tilde{g}}(\xi); \tag{5.281}$$

and, therefore, it is an unbiased estimator. In the limit as $N \to \infty$ the double sum in the second of equations (5.280) becomes a single sum. Then, the variance of the estimator is

$$\lim_{N\to\infty}\mathrm{var}\left(\hat{\tilde{c}}_{\tilde{g},\tilde{g}}(\xi)\right) = \mathcal{E}\left(\left(\frac{1}{P^2}\sum_{k=-\infty}^{\infty}\left|G_k(Z)\right|^2 e^{i\frac{2\pi}{P}k\xi} - \tilde{c}_{\tilde{g},\tilde{g}}(\xi)\right)^2\right); \tag{5.282}$$

and, it is zero only if

$$\tilde{c}_{\tilde{g},\tilde{g}}(\xi) = \frac{1}{P^2}\sum_{k=-\infty}^{\infty}\left|G_k(Z)\right|^2 e^{i\frac{2\pi}{P}k\xi}, \tag{5.283}$$

for any realizations of the random variables, Z. That is, by comparison to equation (5.268), it vanishes only if

$$\left(C_{\tilde{g},\tilde{g}}\right)_k = \frac{1}{P}\left|G_k\left(z_k\right)\right|^2, \tag{5.284}$$

or, in view of equation (5.269), only if

$$\mathcal{E}\left(\left|G_k\left(Z\right)\right|^2\right) = \left|G_k\left(z_k\right)\right|^2, \tag{5.285}$$

for any realizations, z_k, of the random variables, Z. If this holds, then the process is covariance-ergodic. However, since the expectation on the left side is a constant (for any particular k), independent of the realization (all realizations refer to the same distribution), this condition cannot hold if the amplitudes on the right side are different for different realizations.

Indeed, covariance-ergodicity thus puts rather strong conditions on the random variables, $G_k(Z)$. For example, if the Fourier spectral components, $G_k(Z)$, are independent, zero-mean Gaussian random variables, with variances, σ_k^2, then the periodic stochastic process is a Gaussian process (at each point, x, the random variable, $\tilde{g}(x)$, is Gaussian; see Exercise 5.8). Since the spectrum comprises independent, hence mutually uncorrelated, components, by equation (5.267) the process is also stationary. However, the condition (5.285) for ergodicity implies that

$$\sigma_k^2 = \left|G_k\left(z_k\right)\right|^2, \tag{5.286}$$

for any set of realizations, z_k. This is violated in almost all cases, and a Gaussian periodic stationary process cannot be ergodic.

On the other hand, suppose that the Fourier spectrum of the stochastic process, $\tilde{g}$, has the form

$$G_k\left(Z\right) = \sqrt{P}Q_k e^{iZ}, \quad k \neq 0, \quad \text{and} \quad G_0 = 0, \tag{5.287}$$

where the Q_k are arbitrary constants and the random variable, Z, is uniformly distributed on the interval, $\Sigma_z = [0, 2\pi]$, and independently realized. By equations (5.193) and (5.184) it is easy to see that $\mathcal{E}(G_k(Z)) = 0$ for all k. The random variables, $G_k(Z)$ and $G_{k'}(Z)$, for $k \neq k'$ are uncorrelated (equation (5.200)) because the random variable, Z, is independently realized; and, therefore, according to equation (5.267) the process is stationary. Moreover, for any realization, z_k, of the random variable, Z,

$$\mathcal{E}\left(\left|G_k\left(Z\right)\right|^2\right) = \mathcal{E}\left(P \cdot Q_k^2 e^{iZ} e^{-iZ}\right) = P \cdot Q_k^2 = \left|G_k\left(z_k\right)\right|^2, \tag{5.288}$$

which satisfies equation (5.285), and thus the process is covariance-ergodic in this case.

In general, if periodic, stationary, stochastic processes, $\tilde{g}$ and $\tilde{h}$, are mean- and covariance-ergodic, and if $N \to \infty$ ($\Delta x \to 0$, such that $P = N\Delta x$ is constant), then the sums in equations (5.277) and (5.278) become integrals. In particular, the variance of the covariance estimator, $\hat{\tilde{c}}_{\tilde{g},\tilde{h}}(\xi)$, in this case is zero. Therefore, in the limit for any realization of the processes, the estimate, $\hat{\tilde{c}}_{\tilde{g},\tilde{h}}(\xi)$, is identical to the ensemble covariance function, $\tilde{c}_{\tilde{g},\tilde{h}}(\xi)$. The estimate is also the space-averaged correlation function, given by equation (5.118); hence, for covariance-ergodic processes,

$$\hat{\tilde{c}}_{\tilde{g},\tilde{h}}(\xi) = \frac{1}{P}\int_0^P \tilde{g}(x)\tilde{h}(x+\xi)\,dx = \tilde{\phi}_{\tilde{g},\tilde{h}}(\xi) = \tilde{c}_{\tilde{g},\tilde{h}}(\xi). \tag{5.289}$$

Moreover, its Fourier spectrum, equation (5.120), is then

$$\left(C_{\tilde{g},\tilde{h}}\right)_k = \left(\Phi_{\tilde{g},\tilde{h}}\right)_k = \frac{1}{P}G_k^*(z_k)H_k(z_k), \tag{5.290}$$

where the z_k are realizations of the random variable, Z; or, $G_k(z_k)$ and $H_k(z_k)$ are the Fourier-series spectra of the realizations of $\tilde{g}(x)$ and $\tilde{h}(x)$. It is noted that the collection of random variables, $\{G_k(Z)\}$, constitutes a discrete stochastic process that is *not* stationary in the frequency domain since the variance of each coefficient depends on the deterministic scaling constants, Q_k (equation (5.287)). Also, the covariance function and its spectrum are never determined exactly from any particular *finite* set of realizations, $\tilde{g}(x_j)$, $j = 0,\ldots, N-1$ (unless it is band-limited according to the Whittaker-Shannon sampling theorem).

5.6.6 Stochastic Processes on the Sphere

The harmonic stochastic process extends to two dimensions also with the spherical harmonics as basis functions. Referring to equation (2.221), at each point, (θ, λ), on the sphere,

$$g(\theta,\lambda) = \sum_{n=0}^{\infty}\sum_{m=-n}^{n} G_{n,m}(Z)\bar{Y}_{n,m}(\theta,\lambda), \tag{5.291}$$

is a random variable, governed by the random variable, Z, independently realized for each n and m from the same distribution. As in the one-dimensional case, the process, $g(\theta, \lambda)$, is stationary only if

$$\mathcal{E}\left(G_{n,m}(Z)\right) = 0, \quad \text{for all } (n,m) \neq (0,0); \tag{5.292}$$

otherwise, the expectation of the process would depend on (θ,λ) and violate stationarity. As before, it is assumed that $G_{0,0} = 0$.

The auto-covariance function of a stationary process, $g(\theta, \lambda)$, depends only on the relative coordinates, ψ, ζ, given by equations (5.128) and (5.129), between any two corresponding random variables, $g_{\theta,\lambda}$ and $g_{\theta',\lambda'}$. And, again, the simplifying assumption is made that the covariance function should depend only on the spherical distance, ψ, i.e., that it is isotropic. Then, the covariance function, given by

$$\begin{aligned} c_{g,g}(\theta,\lambda,\theta',\lambda') &= \mathcal{E}\left(g(\theta,\lambda)g(\theta',\lambda')\right) \\ &= \sum_{n=0}^{\infty}\sum_{m=-n}^{n}\sum_{n'=0}^{\infty}\sum_{m'=-n'}^{n'} \mathcal{E}\left(G_{n,m}(Z)G_{n',m'}(Z)\right)\bar{Y}_{n,m}(\theta,\lambda)\bar{Y}_{n',m'}(\theta',\lambda') \end{aligned} \tag{5.293}$$

is isotropic only if the coefficients, $G_{n,m}(Z)$, of unlike degree and order are uncorrelated,

$$\mathcal{E}\left(G_{n,m}(Z)G_{n',m'}(Z)\right) = 0, \quad \text{for } (n,m) \neq (n',m'), \tag{5.294}$$

and if their variances do not depend on the order,

$$\mathcal{E}\left(G_{n,m}^2(Z)\right) = \left(C_{g,g}\right)_n, \quad \text{for all } |m| \leq n. \tag{5.295}$$

That is, with the addition formula (2.254), these conditions imply that

$$c_{g,g}(\psi) = \sum_{n=0}^{\infty}(2n+1)\left(C_{g,g}\right)_n P_n(\cos\psi), \tag{5.296}$$

and that $G_{n,m}^2(Z)$ is an unbiased estimator of $(C_{g,g})_n$, the Legendre spectrum of $c_{g,g}(\psi)$. Similar conditions on the random Fourier-Legendre coefficients, $G_{n,m}(Z)$ and $H_{n,m}(Z)$, of two processes, $g(\theta, \lambda)$ and $h(\theta, \lambda)$, yield stationarity and an isotropic cross-covariance function, $c_{g,h}(\psi)$. The discrete cross-power spectrum of g and h in this case is given by

$$\left(C_{g,h}\right)_n = \mathcal{E}\left(G_{n,m}(Z)H_{n,m}(Z)\right), \quad \text{for all } |m| \leq n. \tag{5.297}$$

Analogous to the special case of white noise on the line (equation (5.249)) the Legendre spectrum of white noise on the sphere is unity multiplied by a "variance", σ_w^2, and the covariance function is then

$$c_{w,w}(\psi) = \sigma_w^2 \sum_{n=1}^{\infty}(2n+1)P_n(\cos\psi), \tag{5.298}$$

where the assumption of zero mean implies no zero-degree harmonic.

Essential properties of the covariance function on the sphere follow those already confirmed for the periodic and non-periodic processes (Sections 5.6.2, 5.6.4, and 5.6.5). As in one dimension, the covariance functions of the derivatives of processes g and h on the sphere can be determined by the usual propagation laws established for the deterministic correlation functions, for example, illustrated by equation (5.146). Indeed, the proof that the expectation and derivative operators are commutative is much easier to prove. If $\mathcal{D}_\theta$ and $\mathcal{D}_\lambda$ represent these derivatives then

$$c_{\mathcal{D}_\theta g, \mathcal{D}_\lambda h} = \mathcal{E}((\mathcal{D}_\theta g)(\mathcal{D}_\lambda h) = \mathcal{D}_\theta \mathcal{D}_\lambda \mathcal{E}(gh) = \mathcal{D}_\theta \mathcal{D}_\lambda c_{g,h}, \tag{5.299}$$

since the expectation operator integrates over probability space (see equation (5.215)), while the derivatives operate on the space of the deterministic parameters, θ, λ. However, as in the case of the deterministic correlation functions, the Legendre spectrum of the covariance function is not derivable directly from the realized Fourier-Legendre transforms of corresponding ergodic processes, although one may consider approximations as in equations (5.155) or (5.161).

To determine the conditions for ergodicity of a stationary process, g, on the sphere, consider estimators of the mean and of the covariance function. For the one-dimensional covariance-ergodic periodic process, equation (5.289) shows that the *integral* space-average correlation function for any realization equals the covariance function of the process. Therefore, instead of starting with finite sums as in equations (5.277) and (5.278), it is sufficient to go directly to the limiting case of integrals and find the conditions under which the variances of these estimators vanish. Thus, let

$$\hat{\mu}_g = \frac{1}{4\pi}\iint_{\Omega} g(\theta,\lambda)\,d\Omega = G_{0,0}(Z), \tag{5.300}$$

$$\hat{c}_{g,g}(\psi) = \frac{1}{8\pi^2}\int_{\zeta=0}^{2\pi}\iint_{\Omega} g(\theta,\lambda)\,g(\theta',\lambda')\,d\Omega\,d\zeta. \tag{5.301}$$

Following the same derivation as in equations (5.131) through (5.136), the covariance estimator is also

$$\hat{c}_{g,g}(\psi) = \sum_{n=0}^{\infty}\sum_{m=-n}^{n} G_{n,m}^2(Z)\,P_n(\cos\psi). \tag{5.302}$$

In principle, the estimator, $\hat{\mu}_g$, is a random constant, but its variance vanishes only if that constant is not random; in fact, it is already set to zero and, therefore, the process is mean-ergodic. The expected value of the covariance function estimator, in view of equations (5.295) and (5.296), is given by

$$\mathcal{E}\left(\hat{c}_{g,g}(\psi)\right) = \sum_{n=1}^{\infty}\sum_{m=-n}^{n} \mathcal{E}\left(G_{n,m}^2(Z)\right)P_n(\cos\psi) = \sum_{n=1}^{\infty}(2n+1)\left(C_{g,g}\right)_n P_n(\cos\psi) = c_{g,g}(\psi). \tag{5.303}$$

The estimator is thus unbiased and its variance is

$$\operatorname{var}\left(\hat{c}_{g,g}(\psi)\right) = \mathcal{E}\left(\left(\sum_{n=0}^{\infty}\sum_{m=-n}^{n} G_{n,m}^2(Z)\,P_n(\cos\psi) - c_{g,g}(\psi)\right)^2\right), \tag{5.304}$$

which, when equation (5.296) is substituted, vanishes only if

$$\left(C_{g,g}\right)_n = \frac{1}{2n+1}\sum_{m=-n}^{n} G_{n,m}^2(Z), \tag{5.305}$$

for all independent realizations of Z (note that the left side is a non-random quantity). This is the condition for (covariance-) ergodicity on the sphere.

A stationary random process on the sphere with Gaussian random variables, $G_{n,m}(Z)$, cannot be ergodic because the right side of equation (5.305) would vary with each realization of the process. On the other hand, consider the set of $2n + 1$ random coefficients, $G_{n,m}(Z)$, of degree, n, that consists of random points on a unit hypersphere, $\Omega^{(2n)}$, embedded in $(2n + 1)$-dimensional Cartesian space, $\mathbb{R}^{(2n+1)}$, and scaled by the "radius", Q_n,

$$G_{n,m}(Z) = Q_n u_m^{(2n+1)}(Z), \quad -n \le m \le n, \tag{5.306}$$

where $u_m^{(2n+1)}(Z)$ is the mth component of a unit vector, $\boldsymbol{u}^{(2n+1)}$, in $\mathbb{R}^{(2n+1)}$ having uniformly random direction in this space, and Q_n is a constant. The endpoint of each unit vector thus represents a random point from the uniform distribution on $\Omega^{(2n)}$. The probability density function of this distribution is $p = 1/A_{2n}$, where $A_{2n} = 2\pi^{n+1/2}/\Gamma(n+1/2)$ is the area of $\Omega^{(2n)}$ and Γ is the gamma function. (This probability distribution is the limiting case of the von Mises-Fisher distribution (Sra 2012) as the concentration (inverse scale) parameter vanishes.) It can be assumed without loss in generality that the origin point for the density function on this hypersphere is such that $\mathcal{E}(G_{n,m}(Z)) = 0$. For a particular degree, n, the $2n + 1$ coefficients are uncorrelated. To see this, let $u_m^{(2n+1)} = \boldsymbol{u}^{(2n+1)} \cdot \boldsymbol{e}_m = \cos\psi_m$, where $\boldsymbol{e}_m$ is the unit vector along the mth Cartesian axis and ψ_m is the angle between $\boldsymbol{u}^{(2n+1)}$ and $\boldsymbol{e}_m$. Then, since $\cos(\psi_m + \pi) = -\cos(\psi_m)$ and using equation (5.193), one obtains

$$\begin{aligned}\mathcal{E}\left(G_{n,m}(Z)G_{n,m'}(Z)\right) &= \frac{Q_n^2}{A_{2n}} \int\limits_{\Omega^{(2n)}} u_m^{(2n+1)}(z)\, u_{m'}^{(2n+1)}(z)\, d\Omega^{(2n)} \\ &= -\frac{Q_n^2}{A_{2n}} \int\limits_{\Omega^{(2n)}} u_m^{(2n+1)}(z)\, u_{m'}^{(2n+1)}(z)\, d\Omega^{(2n)}\end{aligned} \tag{5.307}$$

noting that the integration is independent of the handedness of the axes (right- or left-handed) (Moritz 1980, p.287). The left side being equal to both a quantity and its negative shows that the only possibility is $\mathcal{E}(G_{n,m}(Z)G_{n,m'}(Z)) = 0$. Therefore, the random coefficients, $G_{n,m}(Z)$, are uncorrelated for $|m| \le n$; but they are not independent in view of equation (5.305). On the other hand, it is assumed that they are independent (hence uncorrelated) for $n \ne n'$.

The variances,

$$\mathcal{E}\left(G_{n,m}^2(Z)\right) = \frac{Q_n^2}{A_{2n}} \int\limits_{\Omega^{(2n)}} \left(u_m^{(2n+1)}(z)\right)^2 d\Omega^{(2n)} = \frac{Q_n^2}{A_{2n}} \int\limits_{\Omega^{(2n)}} \cos^2\psi_m\, d\Omega^{(2n)}, \tag{5.308}$$

do not depend on the order, m. Indeed, because $d\Omega^{(2n)} = \sin^{2n-1}\psi_m\, d\psi_m\, d\Omega^{(2n-1)}$ (Lovisolo and da Silva 2001), equation (5.308) becomes

$$\mathcal{E}\left(G_{n,m}^2(Z)\right) = \frac{Q_n^2}{A_{2n}} \int_0^{\pi} \cos^2\psi_m \sin^{2n-1}\psi_m\, d\psi_m \int\limits_{\Omega^{(2n-1)}} d\Omega^{(2n-1)} = Q_n^2 \frac{A_{2n-1}}{A_{2n}} \int_{-1}^{1} x^2\left(1-x^2\right)^{n-1} dx, \tag{5.309}$$

with the last equality due to a change in integration variable, $x = \cos\psi_m$. The integral and its antecedent factor reduce to $1/(2n+1)$ (Exercise 5.9), so that

$$\mathcal{E}\left(G_{n,m}^2(Z)\right) = \frac{Q_n^2}{2n+1}. \tag{5.310}$$

By equations (5.294) and (5.295), the covariance function is isotropic and the left side of equation (5.310) may be identified with $(C_{g,g})_n$. Therefore, because $\boldsymbol{u}^{(2n+1)}$ is a unit vector in $\mathbb{R}^{(2n+1)}$, one has

$$\sum_{m=-n}^{n} G_{n,m}^2(Z) = Q_n^2 \sum_{m=-n}^{n} \left(u_m^{(2n+1)}(Z)\right)^2 = Q_n^2 = (2n+1)\left(C_{g,g}\right)_n, \tag{5.311}$$

for any realization of Z, and equation (5.305) is satisfied—the process is covariance-ergodic. Moritz (1980, p.288ff) also gives an alternative representation of an ergodic process on the sphere in terms of rotation group space.

Since the variance, equation (5.304), vanishes, the isotropic covariance function of an ergodic process, g, on the sphere is identical to the "space-average" defined by the integral in equation (5.301) for any particular realization. Of course, any *practical* realization of g constitutes only a finite number of discrete data on the sphere and either the integral is approximated or the equivalent series, equation (5.302), is truncated, resulting in an approximate covariance function.

The variance of the process, g, is $c_{g,g}(0)$, or, from equation (5.296),

$$\sigma_g^2 = \sum_{n=0}^{\infty} (2n+1)\left(C_{g,g}\right)_n; \tag{5.312}$$

and, the *degree-variances* (or, variances per degree) are defined by

$$c_n(g) = (2n+1)\left(C_{g,g}\right)_n = \sum_{m=-n}^{n} G_{n,m}^2\left(z_{n,m}\right). \tag{5.313}$$

One might also say that the power spectrum, $(C_{g,g})_n$, is the *degree-and-order-variance*, although the order is hidden due to the averaging of the covariance function over azimuths. For this reason, the degree-variances, $C_n(g)$, rather than $(C_{g,g})_n$, are often used to characterize the power spectrum of a geophysical signal on the sphere.

5.6.7 Coherency

The previous sections defined the similarity between various types of functions and processes through the use of the correlation (and covariance) function. The Fourier transform of the correlation or covariance function, the power spectral density, describes this similarity in the frequency domain. In particular, a measure of the linearity relating two stochastic geophysical processes, g and h, is the *coherency*, defined by the magnitude of the cross-power spectral density, $C_{g,h}(f)$, normalized at each frequency,

$$W_{g,h}(f) = \frac{\left|C_{g,h}(f)\right|}{\sqrt{C_{g,g}(f)C_{h,h}(f)}}, \tag{5.314}$$

where $C_{g,g}(f)$ and $C_{h,h}(f)$ are the corresponding auto-PSDs. It is shown below that $0 \le W_{g,h}(f) \le 1$; and $W_{g,h}(f) = 0$ means complete incoherency, while $W_{g,h}(f) = 1$ implies a linear relationship at the frequency, f. Coherency was introduced by Norbert Wiener (1930, pp.182-195) for stochastic processes whose power spectral densities are given formally by equation (5.232) (Foster and Guinzy 1967). And, indeed, it is principally for random processes that coherency makes sense. The coherency for deterministic functions, whose correlation is not based on some kind of indeterminacy that is eliminated by averaging (or, expectation, in the case of stochastic processes), is identically equal to unity, as also shown below. This has important consequences in the calculation of coherency from data because they are (deterministic) realizations of a process, even if that underlying process is presumed random.

For example, define the coherency for finite-energy functions, g and h, analogously by

$$\overset{\circ}{W}_{g,h}(f) = \frac{\left|\overset{\circ}{\Phi}_{g,h}(f)\right|}{\sqrt{\overset{\circ}{\Phi}_{g,g}(f)\overset{\circ}{\Phi}_{h,h}(f)}}, \tag{5.315}$$

where the energy spectral density is given by equation (5.4). Substituting the latter, it is easily seen that

$$\overset{\circ}{W}_{g,h}(f) = \frac{\left|G^*(f)H(f)\right|}{\sqrt{G^*(f)G(f)H^*(f)H(f)}} = \frac{|G(f)||H(f)|}{|G(f)||H(f)|} = 1, \quad \text{for all } f, \tag{5.316}$$

and regardless of a linear relationship between g and h. Thus, coherency, as defined, is not a particularly useful concept for these types of functions; and, the cross-energy spectral density must suffice to measure their relationship in the frequency domain. The same conclusion holds for deterministic finite power functions, including periodic functions.

On the other hand, substituting the power spectral density for stochastic processes, equation (5.232), into equation (5.314), one has

$$\begin{aligned} W_{g,h}(f) &= \frac{\left|\lim\limits_{T\to\infty}\left(\frac{1}{T}\mathcal{E}\left(G_T^*(f)H_T(f)\right)\right)\right|}{\sqrt{\lim\limits_{T\to\infty}\left(\frac{1}{T}\mathcal{E}\left(\left|G_T(f)\right|^2\right)\right)\lim\limits_{T\to\infty}\left(\frac{1}{T}\mathcal{E}\left(\left|H_T(f)\right|^2\right)\right)}} \\ &= \lim_{T\to\infty}\frac{\left|\mathcal{E}\left(G_T^*(f)H_T(f)\right)\right|}{\sqrt{\mathcal{E}\left(\left|G_T(f)\right|^2\right)\mathcal{E}\left(\left|H_T(f)\right|^2\right)}} \end{aligned} \tag{5.317}$$

if all the limits exist. The expectation cannot be taken outside the non-linear operations of division, multiplication, and square root, and there is no further simplification. Applying Schwarz's inequality for the expectation operator, equation (5.197),

$$\left(\mathcal{E}\left(\left|G_T^*(f)H_T(f)\right|\right)\right)^2 \le \mathcal{E}\left(\left|G_T(f)\right|^2\right)\mathcal{E}\left(\left|H_T(f)\right|^2\right), \tag{5.318}$$

then, dividing by T, and taking the limit as $T \to \infty$, one obtains

$$0 \le \left|C_{g,h}(f)\right|^2 \le \left|C_{g,g}(f)\right|^2 \left|C_{h,h}(f)\right|^2, \tag{5.319}$$

and, hence,

$$0 \le W_{g,h}(f) \le 1. \tag{5.320}$$

Suppose that two stochastic processes are related by a *linear* system, for example, where g is the input and $\bar{g}$ is the output,

$$\bar{g}(x) = \int_{-\infty}^{\infty} g(x')h(x - x')\,dx', \tag{5.321}$$

and where h is a deterministic function with a Fourier transform, $H(f)$, and where the integral, interpreted for the process in the mean-square sense, exists. The Fourier transform of the corresponding truncated process, equation (5.231), is related to the transform of the output according to the convolution theorem (3.17) by

$$\bar{G}_T(f) = G_T(f)H(f). \tag{5.322}$$

Hence, the Fourier transform of their cross-correlation function is given by equation (5.232),

$$\begin{aligned} C_{g,\bar{g}}(f) &= \lim_{T\to\infty}\left(\mathcal{E}\left(\frac{1}{T}G_T^*(f)G_T(f)H(f)\right)\right) \\ &= \lim_{T\to\infty}\left(\mathcal{E}\left(\frac{1}{T}G_T^*(f)G_T(f)\right)\right)H(f) \\ &= C_{g,g}(f)H(f) \end{aligned} \tag{5.323}$$

Also, from equation (5.247) and the convolution theorem (3.17),

$$C_{\bar{g},\bar{g}}(f) = C_{g,g}(f)\left|H(f)\right|^2 \tag{5.324}$$

In this case, the coherency is unity for all frequencies,

$$W_{g,\bar{g}}(f) = \frac{\left|C_{g,g}(f)H(f)\right|}{\sqrt{C_{g,g}(f)C_{g,g}(f)H^*(f)H(f)}} = 1, \tag{5.325}$$

as expected for a linear relationship.

As shown in Section 5.7.2 (equation (5.363)), the cross-power spectral density of two processes, g and h, is estimated at discrete frequencies by the periodogram, equation (5.125), based on data in the interval, T,

$$\left(\hat{C}_{g,h}\right)_k = \frac{1}{N\Delta x}\left(\tilde{G}_T^*\right)_k\left(\tilde{H}_T\right)_k . \tag{5.326}$$

A corresponding estimate of the coherency is always unity even if the two processes are completely incoherent at all wave numbers, k,

$$\left(\hat{W}_{g,h}\right)_k = \frac{\left|\frac{1}{N\Delta x}\left(\tilde{G}_T^*\right)_k\left(\tilde{H}_T\right)_k\right|}{\sqrt{\frac{1}{N\Delta x}\left(\tilde{G}_T^*\right)_k\left(\tilde{G}_T\right)_k\frac{1}{N\Delta x}\left(\tilde{H}_T^*\right)_k\left(\tilde{H}_T\right)_k}} = \frac{\left|\left(\tilde{G}_T^*\right)_k\right|\left|\left(\tilde{H}_T\right)_k\right|}{\sqrt{\left|\left(\tilde{G}_T^*\right)_k\right|^2\left|\left(\tilde{H}_T^*\right)_k\right|^2}} = 1. \tag{5.327}$$

The (unmodified) periodogram is thus not a good estimate of the power spectral density for the purpose of calculating coherency. Instead, if coherency is to be determined from realizations of ergodic processes, then special care is needed to emulate the averaging that is associated with expectation, as discussed in Section 5.7.2.

In some cases, however, an averaging process is already built into the formulation of the power spectral density. For example, consider the azimuthal average of the correlation function for two-dimensional finite-power functions, equation (5.97), and its Hankel transform, equation (5.99). A corresponding coherency, defined by

$$W_{g,h}\left(\bar{f}\right) = \frac{\left|\Phi_{g,h}\left(\bar{f}\right)\right|}{\sqrt{\Phi_{g,g}\left(\bar{f}\right)\Phi_{h,h}\left(\bar{f}\right)}}, \tag{5.328}$$

leads to

$$W_{g,h}\left(\bar{f}\right) = \lim_{T_1\to\infty}\lim_{T_2\to\infty}\frac{\left|\int\limits_{\nu=0}^{2\pi} G_{T_1,T_2}^*\left(f_1,f_2\right)H_{T_1,T_2}\left(f_1,f_2\right)d\nu\right|}{\sqrt{\int\limits_{\nu=0}^{2\pi}\left|G_{T_1,T_2}\left(f_1,f_2\right)\right|^2 d\nu\int\limits_{\nu=0}^{2\pi}\left|H_{T_1,T_2}\left(f_1,f_2\right)\right|^2 d\nu}}. \tag{5.329}$$

Schwarz's inequality (1.22) guarantees that $0 \le W_{g,h}(\bar{f}) \le 1$. If g and $\bar{g}$ are related by a linear system with isotropic system function, h, where the spectral relationship between truncated functions is as in equation (3.31),

$$\bar{G}_{T_1,T_2}(f_1,f_2) = G_{T_1,T_2}(f_1,f_2)H(\bar{f}), \tag{5.330}$$

then the coherency, again, is unity for all frequencies,

$$\begin{aligned} W_{g,\bar{g}}(\bar{f}) &= \lim_{T_1\to\infty}\lim_{T_2\to\infty} \frac{\left|\int_{\nu=0}^{2\pi} \left|G_{T_1,T_2}(f_1,f_2)\right|^2 H(\bar{f})\,d\nu\right|}{\sqrt{\int_{\nu=0}^{2\pi} \left|G_{T_1,T_2}(f_1,f_2)\right|^2 d\nu \int_{\nu=0}^{2\pi} \left|G_{T_1,T_2}(f_1,f_2)H(\bar{f})\right|^2 d\nu}} \\ &= \lim_{T_1\to\infty}\lim_{T_2\to\infty} \frac{\left|H(\bar{f})\right| \int_{\nu=0}^{2\pi} \left|G_{T_1,T_2}(f_1,f_2)\right|^2 d\nu}{\left|H(\bar{f})\right| \sqrt{\int_{\nu=0}^{2\pi} \left|G_{T_1,T_2}(f_1,f_2)\right|^2 d\nu \int_{\nu=0}^{2\pi} \left|G_{T_1,T_2}(f_1,f_2)\right|^2 d\nu}} = 1 \end{aligned} \tag{5.331}$$

Similarly, the coherency for functions on the sphere with isotropic cross-correlation function, equation (5.130), may be defined as

$$\left(W_{g,h}\right)_n = \frac{\left|\left(\Phi_{g,h}\right)_n\right|}{\sqrt{\left(\Phi_{g,g}\right)_n \left(\Phi_{h,h}\right)_n}}. \tag{5.332}$$

Substituting the cross-power spectral density, equation (5.135), this becomes

$$\left(W_{g,h}\right)_n = \frac{\left|\frac{1}{2n+1}\sum_{m=-n}^{n} G_{n,m}H_{n,m}\right|}{\sqrt{\frac{1}{2n+1}\sum_{m=-n}^{n} G_{n,m}^2 \frac{1}{2n+1}\sum_{m=-n}^{n} H_{n,m}^2}} = \frac{\left|\sum_{m=-n}^{n} G_{n,m}H_{n,m}\right|}{\sqrt{\sum_{m=-n}^{n} G_{n,m}^2 \sum_{m=-n}^{n} H_{n,m}^2}}, \tag{5.333}$$

Again, Schwarz's inequality (1.23) implies that

$$0 \le \left(W_{g,h}\right)_n \le 1. \tag{5.334}$$

If the functions on the sphere are related by an isotropic convolution,

$$\bar{g}(\theta,\lambda) = \frac{1}{4\pi}\iint_{\Omega} g(\theta',\lambda')h(\psi)\,d\Omega, \tag{5.335}$$

then their Legendre spectra are related by

$$\bar{G}_{n,m} = G_{n,m}H_n. \tag{5.336}$$

In this case, the coherency is

$$\left(W_{g,\bar{g}}\right)_n = \frac{\left|\sum_{m=-n}^{n} G_{n,m} G_{n,m} H_n\right|}{\sqrt{\sum_{m=-n}^{n} G_{n,m}^2 \sum_{m=-n}^{n} G_{n,m}^2 H_n^2}} = \frac{\left|H_n\right| \sum_{m=-n}^{n} G_{n,m}^2}{\left|H_n\right| \sqrt{\sum_{m=-n}^{n} G_{n,m}^2 \sum_{m=-n}^{n} G_{n,m}^2}} = 1, \quad \text{for all } n, \quad (5.337)$$

again, as one expects for a linear system.

Figure 5.8 shows the coherency, equation (5.333), between independent global models of topography (including ocean depths) and the radial component of gravitation. A moving average over 16 harmonic degrees is applied to highlight the trend in increasing coherency from low to higher frequencies, reflecting the ability of the Earth's crust to support topography at the shorter wavelengths, whereas it is isostatically compensated at lower frequencies. At the very short wavelengths the coherency drops again, as the subsurface geologic inhomogeneities compete with the topographic geometry as the source for the gravitational signal. It is noted that all radial derivatives are linearly related to each other at each harmonic degree, so that the Figure 5.8 also represents the coherency between topography and either gravitational potential or vertical gravitational gradient. Several geophysical examples of coherency are given in (Watts 2001, Sect. 5.6). A particular application is the feasibility study on the use satellite-altimetry-derived gravity to infer sea floor topography (Smith and Sandwell 1994).

Another method to characterize coherency in the frequency domain is by the difference in phases, equation (2.60), between corresponding spectral components (von Frese et al. 1997). If the phase difference is zero then the spectral components are linearly related (at that frequency). Viewing the spectra of two functions, g and h, as two-dimensional vectors in the complex plane (Figure 5.9), their "inner product", $G^*(f)\bullet H(f) = \mathrm{Re}(G(f))\mathrm{Re}(H(f)) + \mathrm{Im}(G(f))\mathrm{Im}(H(f))$ is proportional to the cosine of the phase difference,

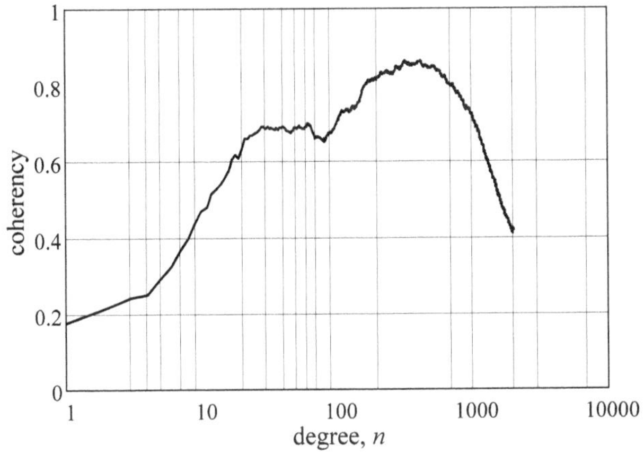

Figure 5.8: Coherency of global topography (including ocean depths) and the gravitational field of the Earth.

$$V(f) = \cos\left(\phi_G(f) - \theta_H(f)\right) = \frac{G^*(f) \cdot H(f)}{|G(f)||H(f)|}. \tag{5.338}$$

The *phase coherency*, $V(f)$, lies in the interval, $[-1,1]$, and $V(f) = 1$ implies $H(f) = aG(f)$ for some $a > 0$. Unlike the coherency based on PSDs, which obliterates phase information, the phase coherency may be applied directly to the DFTs of sampled data. Figure 5.10 contrasts two sets of regional profiles of topography and gravity anomaly. The phase coherency (Figure 5.11) is based on their DFTs and is smoothed to reveal the coherency trends. Using a cutoff of 0.5, it is clear that the topography and gravity anomaly are not coherent at any wavelengths in the north Midwest of the U.S., which is relatively flat in terrain but contains the gravitationally significant mid-continent rift. The coherency is generally positive in mountainous regions, and relatively strong in this case, as also evident in Figure 5.10 (right), at wavelengths greater than 20 km (frequency, 0.05 cy/km) and shorter than 4 km.

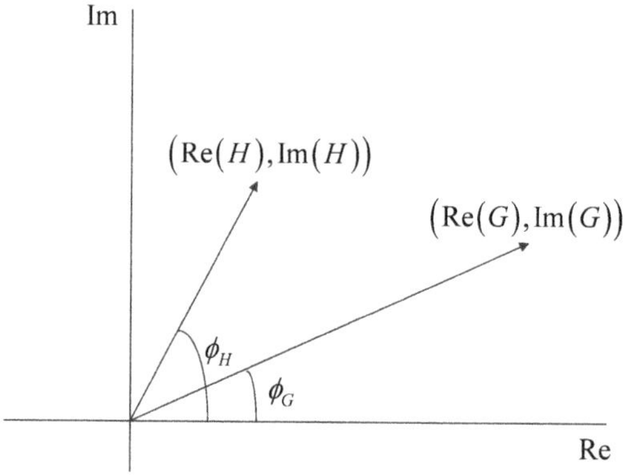

Figure 5.9: Spectral components, $G(f)$, $H(f)$, and their phases in the complex plane.

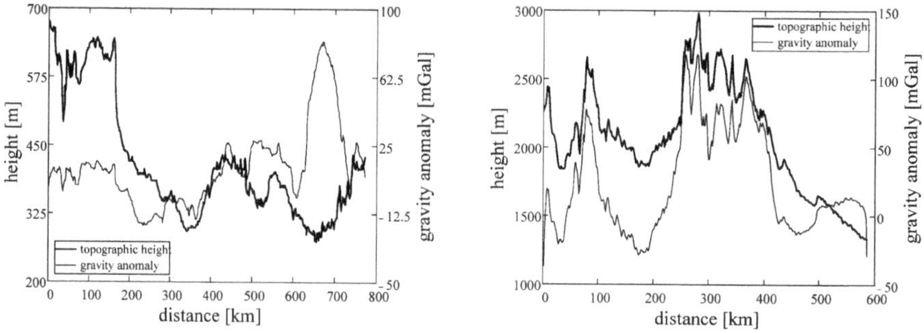

Figure 5.10: Profiles of topography and gravity anomaly in two contrasting regions of the U.S. (Dakotas/Minnesota on the left and Colorado on the right).

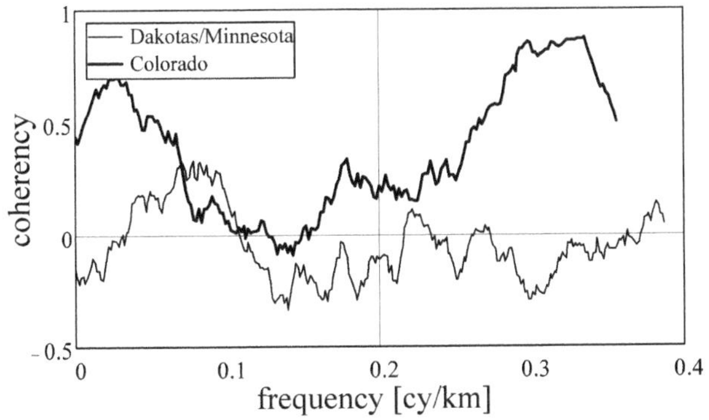

Figure 5.11: Phase coherency between topography and gravity anomaly in two regions of the U.S. represented by the profiles of Figure 5.10.

5.6.8 Covariances for Discrete Processes

The correlation between sequences is a sample of the correlation function of the parent functions only if the latter are band-limited by the Nyquist frequency defined by the sampling interval, Δx (e.g., equations (5.51) and (5.109)). In the case of discrete stochastic processes, however, it is easily shown that their covariance is a sample of the covariance function for the continuous parent processes, whether or not they are band-limited. Indeed, let $g_j = g(x_j)$, $h_j = h(x_j)$ be stochastic sequences (sequences of random variables) where the deterministic parameter, $x_j = j\Delta x$, is simply an element of a discrete (countably infinite) subset of the continuous parameter, x. We abbreviate the deterministic parameter by an integer. The covariance between any two random variables, g_j and h_k, is given by equation (5.210),

$$c_{g,h}(j,k) = \int_{\Sigma_{h_k}} \int_{\Sigma_{g_j}} \left(g_j - \mu_{g_j}\right)\left(h_k - \mu_{h_k}\right) p_{g_j,h_k}\left(g_j, h_k\right) dg_j\, dh_k. \tag{5.339}$$

As usual, the integration domains are the realization spaces of the random variables. It follows immediately that the discrete covariance is a sample of the continuous covariance function. Formula (5.339) holds also with straightforward modification in notation for discrete processes defined on higher-dimensioned domains of the deterministic parameters (e.g., g_{j_1,j_2}).

If the continuous processes are stationary (our only consideration), so are the corresponding discrete processes, and their covariances depend only on the difference between integers, $\ell = k - j$. The covariances, $(c_{g,h})_\ell$, then form a discrete sequence, with

$$\left(c_{g,h}\right)_\ell = c_{g,h}\left(\ell \Delta x\right). \tag{5.340}$$

The Fourier transform of the covariance sequence, according to equation (4.16), is given by the periodic function,

$$\tilde{C}_{g,h}(f) = \Delta x \sum_{\ell=-\infty}^{\infty} \left(c_{g,h}\right)_\ell e^{-i2\pi\Delta x \ell f}, \quad -f_\mathcal{N} \le f < f_\mathcal{N}, \tag{5.341}$$

where $f_\mathcal{N} = 1/(2\Delta x)$ is the Nyquist frequency. Also, the inverse Fourier transform is assumed to hold,

$$\left(c_{g,h}\right)_\ell = \int_{-f_\mathcal{N}}^{f_\mathcal{N}} \tilde{C}_{g,h}(f) e^{i2\pi\Delta x \ell f} df. \tag{5.342}$$

The Fourier transform is related to the PSD of the parent processes by equation (4.8),

$$\tilde{C}_{g,h}(f) = \sum_{k=-\infty}^{\infty} C_{g,h}(f + 2kf_\mathcal{N}). \tag{5.343}$$

That is, the PSD of the discrete processes is *aliased* by the PSD of the continuous processes at frequencies higher than the Nyquist frequency. Other properties of the covariances for discrete stochastic processes follow exactly those of the correlations for finite-power sequences, equations (5.112) through (5.116).

5.7 Estimation of the Covariance and PSD

The covariance function plays a critical role in physical geodesy in the optimal estimation of quantities related to the gravitational potential (least-squares collocation), as does the semi-variogram in optimal prediction and interpolation of random geophysical signals (kriging). In many cases, the statistics of the signals (gravitational or otherwise) are not known, or only poorly approximated, since only a single realization is available. Thus, a reasonable determination of the covariance function or the semi-variogram is often the first important step in a stochastically based operational approach to optimal estimation of geophysical quantities. It is assumed here that the corresponding random processes are stationary, so that the semi-variogram and covariance function are related according to equation (5.219); and, therefore, the determination of the latter is sufficient (see also Ver Hoef et al. 2004). Moreover, if the data are the only or primary information from which to determine the covariance function, then one must assume also that the data constitute a realization of a covariance-ergodic process. Finally, without significant loss of generality, it is supposed that the random processes have zero mean. If there is a constant (known) mean value, it can be removed without altering the correlation statistics. If it is not known then it may be estimated, but this leads to possible biases in further estimation of the covariance, which is not discussed here. It is rare, especially with the existence of well determined spherical harmonic models, that covariance functions are estimated directly from globally distributed data. Rather, a stochastic interpretation, when needed, is implied directly by the spherical power spectrum or the associated degree-variances, equation (5.313). Therefore, the emphasis in this section is on the estimation of *local* covariances and PSDs with the planar approximation. However, the last sub-section, Section 5.7.3, concludes with an introduction to the estimation of the spherical power spectrum from data on a spherical domain with truncated polar caps.

5.7.1 Covariance Function Estimation

Under these various introductory suppositions, an unbiased (if the means are known and removed) and consistent estimator of the covariance function is given by (see also equation (5.260))

$$\hat{c}_{g,h}(\xi) = \frac{1}{T} \int_{-T/2}^{T/2} g(x) h(x+\xi)\, dx. \tag{5.344}$$

For finite sequences of realizations, g_n and h_n, $n = -N/2,\ldots, N/2 - 1$ (presumably uniformly spaced at interval, Δx), sampled from continuous parent processes, a corresponding estimator of the covariance at discrete lag distances, $\ell\Delta x$, is

$$(\hat{c}_{g,h})_\ell = \frac{1}{N-\ell} \sum_{n=-N/2}^{N/2-\ell-1} g_n h_{n+\ell}, \quad \ell = 0,\ldots, N-1. \tag{5.345}$$

The sequence, $(\hat{c}_{g,h})_\ell$, is known as the *correlogram*. By the symmetry property of the covariance function, equation (5.239), auto-covariance estimates need only be calculated for $\ell \geq 0$; however, for cross-covariances, both $(\hat{c}_{g,h})_\ell$ and $(\hat{c}_{h,g})_\ell$, are required,

$$(\hat{c}_{g,h})_{-\ell} = (\hat{c}_{h,g})_\ell = \frac{1}{N-\ell} \sum_{n=-N/2}^{N/2-\ell-1} h_n g_{n+\ell}, \quad \ell = 0,\ldots, N-1. \tag{5.346}$$

Since the discrete covariance is a sample of the continuous covariance function (Section 5.6.8), the estimator is also unbiased and consistent. That is,

$$\mathcal{E}\left(\left(\hat{c}_{g,h}\right)_\ell\right) = \frac{1}{N-\ell} \sum_{n=-N/2}^{N/2-\ell-1} \mathcal{E}\left(g_n h_{n+\ell}\right) = \frac{1}{N-\ell} \sum_{n=-N/2}^{N/2-\ell-1} \left(c_{g,h}\right)_\ell = \left(c_{g,h}\right)_\ell; \tag{5.347}$$

and, as $\Delta x \to 0$ and $N \to \infty$ while $T = N\Delta x$ remains constant, $(\hat{c}_{g,h})_\ell$ approaches the consistent estimator, equation (5.345).

The estimator for a given lag distance, $\ell\Delta x$, combines only $N - \ell$ pairs of data values, which reduces to just a single product for $\ell = N - 1$. Clearly, then, as the lag increases so does the error in the estimate of $c_{g,h}(\ell\Delta x)$; and, it is recommended that the maximum effective lag, $\ell_{\max}$, should be much less than N. The deterioration of the estimate is completely analogous to the edge effect in the practical evaluation of convolutions of functions whose domains are truncated to some finite interval (Section 4.4). Thus, the estimator is reliable only for shorter wavelengths of the signals, certainly significantly shorter than $N\Delta x$, and due care must be exercised by the practitioner.

A more serious drawback is the fact that the estimator can give unrealistic auto-covariances, such as $|(\hat{c}_{g,g})_\ell| > (\hat{c}_{g,g})_0$, for $\ell > 0$, especially for larger lags distances. This violates a central property of auto-covariance functions, equation (5.238), and could lead to covariance matrices that are not positive definite (Marple 1987, p.148; Percival and Walden 1993, p.193) and variances of estimation errors that are negative. The remedy is to define an alternative estimator, which is, however, biased,

$$\left(\breve{c}_{g,h}\right)_\ell = \frac{1}{N}\sum_{n=-N/2}^{N/2-\ell-1} g_n h_{n+\ell}, \quad \ell = 0,\ldots,N-1, \tag{5.348}$$

where the bias is given by $-(\ell/N)(c_{g,h})_\ell$,

$$\mathcal{E}\left(\left(\breve{c}_{gh}\right)_\ell\right) = \frac{N-\ell}{N}\left(c_{g,h}\right)_\ell. \tag{5.349}$$

On the other hand, this estimator is still *asymptotically* unbiased, since the bias vanishes as $N \to \infty$. Because the estimator, $(\breve{c}_{g,h})_\ell$, always yields positive definite covariances, it is often preferred despite the bias.

For two-dimensional processes, the unbiased estimator of the covariance function is

$$(\hat{c}_{g,h})_{\ell_1,\ell_2} = \frac{1}{N_1-\ell_1}\frac{1}{N_2-\ell_2}\sum_{n_1=-N_1/2}^{N_1/2-\ell_1-1}\sum_{n_2=-N_2/2}^{N_2/2-\ell_2-1} g_{n_1,n_2} h_{n_1+\ell_1,n_2+\ell_2}, \\ \ell_1 = 0,\ldots,N_1-1, \quad \ell_2 = 0,\ldots,N_2-1 \tag{5.350}$$

and the biased version replaces the antecedent factors by $1/(N_1N_2)$. Estimates corresponding to negative lags, $\ell_1\Delta x_1$ and/or $\ell_2\Delta x_2$, $\ell_1 = -N_1+1,\ldots,0$, $\ell_2 = -N_2+1,\ldots,0$, are obtained as in equation (5.346). When it is more convenient (and also justifiable) to work with isotropic covariance functions (e.g., when comparing planar and spherical covariance functions; analogous to equation (5.165)), an azimuthal average of $(\hat{c}_{g,h})_{\ell_1,\ell_2}$ may be defined,

$$\hat{c}_{g,h}(\ell\Delta s) = \frac{1}{L_\ell}\sum_{\substack{\ell_1,\ell_2 \\ \ni \ell = \left\lfloor \sqrt{(\ell_1\Delta x_1)^2+(\ell_2\Delta x_2)^2}/\Delta s \right\rfloor}} \hat{c}_{g,h}(\ell_1\Delta x_1, \ell_2\Delta x_2) \tag{5.351}$$

where $\Delta s = \sqrt{\Delta x_1^2 + \Delta x_1^2}$, $\lfloor\cdot\rfloor$ denotes rounding down to an integer, and L_ℓ is the total number of estimates averaged per radial lag, $\ell\Delta s$.

Estimates of the auto- and cross-covariance functions of the topography and gravity anomaly along the profiles in Figure 5.10 are shown in Figure 5.12. For ease of comparison they are scaled according to

$$\left(\hat{c}'_{g,g}\right)_\ell = \left(\hat{c}_{g,g}\right)_\ell \big/ \left(\hat{c}_{g,g}\right)_0, \quad \left(\hat{c}'_{g,h}\right)_\ell = \left(\hat{c}_{g,h}\right)_\ell \Big/ \sqrt{\left(\hat{c}_{g,g}\right)_0\left(\hat{c}_{h,h}\right)_0}. \tag{5.352}$$

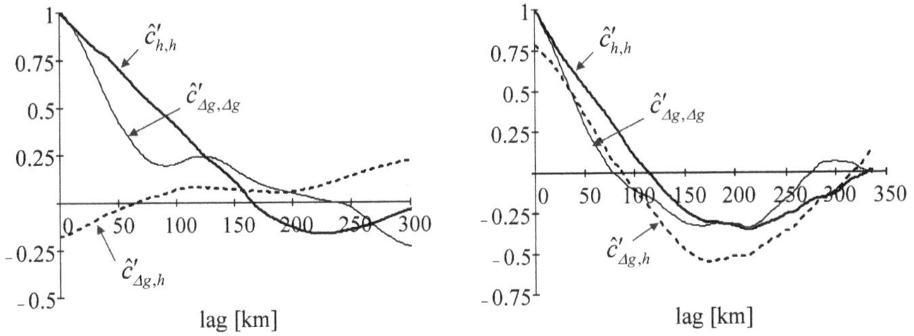

Figure 5.12: Normalized auto- and cross-covariance function estimates by the correlogram method for the gravity anomaly, Δg, and topographic height, h, left and right, respectively, for the profiles, left and right, of Figure (5.10). The covariance estimates are discrete and connected by lines for clarity.

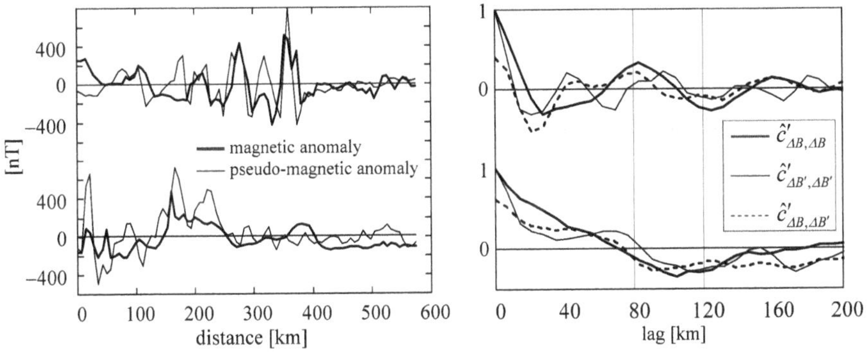

Figure 5.13: Left: The magnetic anomaly, ΔB, from data along two profiles in Colorado and the pseudo-magnetic anomaly, $\Delta B'$, implied by gravity gradients derived from EGM2008. Right: the normalized auto- and cross-covariance function estimates obtained by the correlogram method. The data are discrete, as are the covariance estimates, and connected by lines for clarity.

The more evident positive strong correlation between topography and gravity along the Colorado profile is reflected in the cross-covariance estimates; whereas, the incidental 180° phase shifts between parts of the gravity and topography profiles in the northern Midwest are responsible for the indicated negative cross-variance in the left panel of the figure.

Another example of auto- and cross-covariance estimation according to the correlogram method is shown in Figure 5.13 for both the magnetic anomaly, ΔB, and the pseudo-magnetic anomaly, $\Delta B'$, derived from gravitational gradients, equation (6.234), along two profiles in the Colorado region. The gradients are derived from the Earth Gravitational Model 2008 (EGM2008) (Section 6.3.1) and the mean mass density and magnetic susceptibility values are assumed to be 2670 kg/m^3 and 0.05, respectively. The top profile exhibits short-wavelength correlation, while the longer-wavelength correlations are more evident in the bottom profile. These characteristics are reflected in the corresponding covariance estimates (again, scaled according to equation (5.352) for easier visualization)—the correlation distance, equation (5.216), for the bottom profile is larger than for the top profile.

Evoking the relationship between correlation and convolution, e.g., equation (5.3), the covariance function estimate, equation (5.345), is also representable as a discrete convolution. This opens the possibility to make the computation of the estimate more efficient by using the FFT, as detailed in Section 4.4. However, in order to avoid the consequent cyclic convolution error, consider that equation (5.345) is the same as

$$\left(\hat{c}_{g,h}\right)_\ell = \frac{1}{N-\ell}\sum_{n=-N/2}^{N/2-1}\left(g_{2N}^0\right)_n\left(h_{2N}^0\right)_{n+\ell}, \quad \ell = 0,\ldots,N-1, \tag{5.353}$$

where the sequences of realized values, g_n and h_n, are augmented (padded) with zeros,

$$\left(g_{2N}^0\right)_n = \begin{cases} g_n, & -\dfrac{N}{2} \le n \le \dfrac{N}{2}-1 \\ 0, & \dfrac{N}{2} \le n \le \dfrac{3N}{2}-1 \end{cases} \tag{5.354}$$

$$\left(h_{2N}^0\right)_n = \begin{cases} h_n, & -\dfrac{N}{2} \le n \le \dfrac{N}{2}-1 \\ 0, & \dfrac{N}{2} \le n \le \dfrac{3N}{2}-1 \end{cases} \tag{5.355}$$

and, therefore, all terms in equation (5.353) with $n = N/2 - \ell,\ldots, N/2 - 1$ are zero. Assuming now that both sequences, $(g_{2N}^0)_n$ and $(h_{2N}^0)_n$, are periodic (period = $2N$) and denoting their periodic extensions by $(\tilde{g}_{2N}^0)_n$ and $(\tilde{h}_{2N}^0)_n$, respectively, equation (5.353) is equivalent to

$$\left(\hat{c}_{g,h}\right)_\ell = \frac{1}{N-\ell}\sum_{n=-N}^{N-1}\left(\tilde{g}_{2N}^0\right)_n\left(\tilde{h}_{2N}^0\right)_{n+\ell}, \quad \ell = 0,\ldots,N-1, \tag{5.356}$$

since all terms with $n = -N,\ldots,-N/2-1$ and $n = N/2,\ldots, N-1$ are zero. The right side is also a discrete cyclic *convolution*, which is evident by replacing the summation index, n, with $-n$, and then making use of the $2N$-periodicity of $(\tilde{g}_{2N}^0)_n$ and $(\tilde{h}_{2N}^0)_n$,

$$\begin{aligned}
\left(\hat{c}_{g,h}\right)_\ell &= \frac{1}{N-\ell}\sum_{n=-N+1}^{N}\left(\tilde{g}_{2N}^0\right)_{-n}\left(\tilde{h}_{2N}^0\right)_{-n+\ell} \\
&= \frac{1}{N-\ell}\left(-\left(\tilde{g}_{2N}^0\right)_N\left(\tilde{h}_{2N}^0\right)_{N+\ell} + \left(\sum_{n=-N}^{N-1}\left(\tilde{g}_{2N}^0\right)_{-n}\left(\tilde{h}_{2N}^0\right)_{\ell-n}\right) + \left(\tilde{g}_{2N}^0\right)_{-N}\left(\tilde{h}_{2N}^0\right)_{\ell-N}\right) \\
&= \frac{1}{N-\ell}\sum_{n=-N}^{N-1}\left(\tilde{g}_{2N}^0\right)_{-n}\left(\tilde{h}_{2N}^0\right)_{\ell-n} \\
&= \frac{1}{(N-\ell)\Delta x}\left(\tilde{g}_{2N}^0\right)_{-\ell} \# \left(\tilde{h}_{2N}^0\right)_\ell
\end{aligned} \tag{5.357}$$

The last equality follows from the definition of the cyclic convolution, equation (4.98), and the fact that the summand is periodic and any sequence of length, $2N$, may be summed. Therefore, by the convolution theorem (4.101),

$$\left(\hat{c}_{g,h}\right)_{\ell} = \frac{1}{(N-\ell)\Delta x} \mathrm{DFT}^{-1}\left(\mathrm{DFT}\left(\left(\tilde{g}_{2N}^{0}\right)_{-n}\right)_{k} \mathrm{DFT}\left(\left(\tilde{h}_{2N}^{0}\right)_{n}\right)_{k}\right)_{\ell}, \tag{5.358}$$

where only the values for $|\ell| = 0,\ldots, \ell_{\max} \le N - 1$ would be used. A similar formula holds for the corresponding biased estimator, equation (5.348).

Zero-padding effectively removes any cyclic convolution error; that is, equation (5.357) is identical to equation (5.345). Unlike the convolution of a signal with a known kernel, zero-padding is applied to both sequences in this case since neither one is known beyond the given data interval. Without zero padding, the FFT method, implemented to speed up the computation of the covariance estimate, introduces a cycle convolution error similar to equation (4.136), which biases the estimator.

5.7.2 PSD Estimation

Estimation of the power spectral density is one of the key elements of spectral analysis and falls into two general categories, parametric and non-parametric estimation. With the former a particular model of the PSD is assumed and corresponding parameters are estimated. One such approach is discussed more thoroughly in Section 6.4.1 in connection with covariance function modeling for the Earth's gravitational field. The non-parametric techniques are based strictly on the data and no attempt is made to create a particular model. The Fourier transform of the data plays a key role as shown, for example, by equations (5.125) and (5.232). Technically, the determination of the *power* spectral density applies only to finite-power functions, which includes periodic functions (e.g., functions on the sphere) and stationary stochastic processes (the focus in this section), although in practice these conditions can only be assumed and never quite fulfilled with a finite amount of data. A large volume of literature is devoted to ameliorating the effects of this truncation. The essential aspects of the associated non-parametric methods applied to stochastic processes are covered in this section.

Since the power spectral density is the Fourier transform of the covariance function, a natural estimator of the PSD is the Fourier transform of the correlogram, equation (5.345). As in the previous section, let g_n and h_n, be finite sequences of uniformly sampled stationary stochastic processes, given for $n = -N/2,\ldots, N/2 - 1$, and which are used to create estimates of the covariance functions, $(\hat{c}_{g,h})_\ell$ and $(\hat{c}_{h,g})_\ell$, $\ell = 0,\ldots, \ell_{\max}$, where $\ell_{\max} \le N - 1$ is the maximum lag number for which the estimates are deemed accurate. The Fourier series transform of $(\hat{c}_{g,h})_\ell$ is

$$\hat{C}_{g,h}(f) = \Delta x \sum_{\ell=-\ell_{\max}}^{\ell_{\max}-1} \left(\hat{c}_{g,h}\right)_{\ell} e^{-i2\pi\Delta x \ell f}, \quad |f| \le f_{\mathcal{N}} = \frac{1}{2\Delta x}, \tag{5.359}$$

where $(\hat{c}_{g,h})_{-\ell} = (\hat{c}_{h,g})_{\ell}$, as in equation (5.346). Now, with the rectangle sequence, equation (4.26), and equation (5.347), the expectation of this estimator (a random process) is

$$\mathcal{E}\left(\hat{C}_{g,h}(f)\right) = \Delta x \sum_{\ell=-\infty}^{\infty} b_{\ell}^{(2\ell_{\max})}\left(c_{g,h}\right)_{\ell} e^{-i2\pi\Delta x\ell f}, \tag{5.360}$$

since the estimator, $(\hat{c}_{g,h})_{\ell}$, is unbiased. The right side is the Fourier transform, equation (4.16), of a product of sequences. By the dual convolution theorem (4.37) for sequences, it is a convolution in the frequency domain,

$$\mathcal{E}\left(\hat{C}_{g,h}(f)\right) = \tilde{C}_{g,h}(f) \tilde{*} \tilde{B}^{(2\ell_{\max})}(f) = \int_{-f_{\mathcal{N}}}^{f_{\mathcal{N}}} \tilde{C}_{g,h}(f')\tilde{B}^{(2\ell_{\max})}(f-f')\,df', \tag{5.361}$$

where $\tilde{B}^{(2\ell_{\max})}(f)$ is the Fourier transform of $b_{\ell}^{(2\ell_{\max})}$, equation (4.29), and $\tilde{C}_{g,h}(f)$ is the Fourier transform of $(c_{g,h})_{\ell}$.

Although the correlogram, $(\hat{c}_{g,h})_{\ell}$, is unbiased, the corresponding estimator of the PSD *is* biased (in part) due to spectral leakage (Section 3.6), characterized by the frequency response of the discrete rectangle window. Moreover, because the covariance function is represented by a discrete sequence, the estimator only holds for frequencies less than the Nyquist frequency and, therefore, is aliased (additionally biased) particularly at higher frequencies (for broad-band geophysical signals).

Instead of the unbiased covariance estimate, consider using the biased version, equation (5.348), in equation (5.359). Also, suppose for ease of notation that $\ell_{\max} = N-1$ (but the inaccuracy for these large lags should be kept in mind). Then, with the aforementioned symmetry of the covariances, and again for $|f| \le f_{\mathcal{N}}$, one obtains

$$\begin{aligned} \breve{C}_{g,h}(f) &= \Delta x \sum_{\ell=-N+1}^{N-1} \left(\breve{c}_{g,h}\right)_{\ell} e^{-i2\pi\Delta x\ell f} \\ &= \frac{\Delta x}{N}\left(\sum_{n=-N/2}^{N/2-1} g_n h_n + \sum_{\ell=1}^{N-1}\sum_{n=-N/2}^{N/2-\ell-1}\left(h_n g_{n+\ell} + g_n h_{n+\ell}\right)e^{-i2\pi\Delta x\ell f}\right) \\ &= \frac{\Delta x}{N}\sum_{j=-N/2}^{N/2-1}\sum_{k=-N/2}^{N/2-1} h_j g_k e^{-i2\pi\Delta x(j-k)f} \end{aligned} \tag{5.362}$$

The third equality is easily checked (Exercise 5.10). Compared to equation (5.125), this is the *periodogram* for $f = f_k = k/(N\Delta x)$ under the assumption that the finite sequences are extended periodically, $g_n = \tilde{g}_n$ and $h_n = \tilde{h}_n$,

$$\breve{C}_{g,h}(f_k) = \frac{1}{N\Delta x}\tilde{G}_k^* \tilde{H}_k = \left(\Phi_{\tilde{g},\tilde{h}}\right)_k. \tag{5.363}$$

As before, this PSD estimator is biased due to spectral leakage and aliasing. Indeed, the expectation of the first equation (5.362) is given, according to equation (5.349), by

$$\begin{aligned}\mathcal{E}\left(\breve{C}_{g,h}(f)\right) &= \Delta x \sum_{\ell=-N+1}^{N-1} \mathcal{E}\left(\left(\breve{c}_{g,h}\right)_\ell\right) e^{-i2\pi\Delta x \ell f} \\ &= \Delta x \sum_{\ell=-N+1}^{N-1} \left(1-\frac{|\ell|}{N}\right)\left(c_{g,h}\right)_\ell e^{-i2\pi\Delta x \ell f}\end{aligned} \tag{5.364}$$

which is the Fourier transform of the product of the covariance sequence and the discrete Bartlett window, $(u^{(2N)}_{\text{Bartlett}})_\ell$, equation (4.259). The dual convolution theorem (4.37) then says it is also the convolution of the Fourier transforms of $(c_{g,h})_\ell$ and of $(u^{(2N)}_{\text{Bartlett}})_\ell$, equation (4.261),

$$\mathcal{E}\left(\breve{C}_{g,h}(f)\right) = \mathcal{F}\left(\left(c_{g,h}\right)_\ell \left(u^{(2N)}_{\text{Bartlett}}\right)_\ell\right) = \int_{-f_N}^{f_N} \tilde{C}_{g,h}(f')\tilde{U}^{(2N)}_{\text{Bartlett}}(f-f')\,df'. \tag{5.365}$$

Figure 5.14 compares the frequency responses of the discrete rectangle and Bartlett windows, $\tilde{B}^{(2N)}(f)$ and $\tilde{U}^{(2N)}_{\text{Bartlett}}(f)$, contrasting the character of the spectral leakage for each of the PSD estimators. Although the main lobe of the rectangle response is narrower, there is less spectral leakage in the estimate of the PSD at the higher frequencies when using the periodogram rather than the DFT of the unbiased correlogram (Figure 5.15). Since both $b_\ell^{(2\ell_{\max})}$ and $(u^{(2N)}_{\text{Bartlett}})_\ell$ are unity at the origin, so are the integrals of their Fourier transforms, independent of N. As the number of samples increases to infinity (i.e., $N \to \infty$) these transforms approach infinity at the spectral origin and their first zero approaches zero, meaning that their oscillations

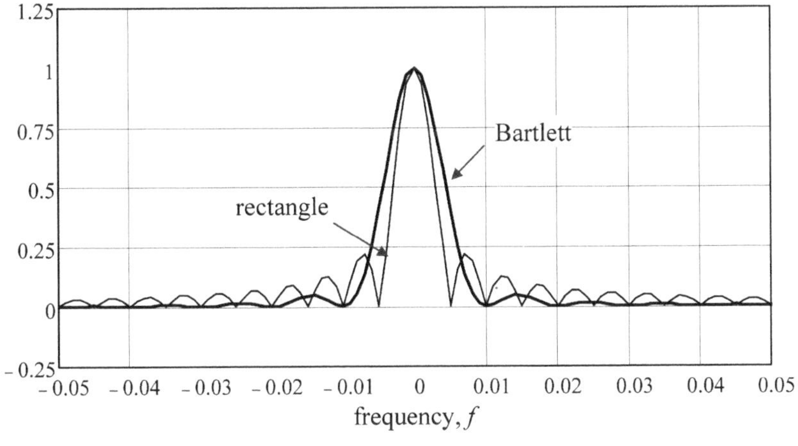

Figure 5.14: Fourier transforms (amplitude spectra) of the discrete rectangle and Bartlett windows, $b_\ell^{(2N)}$ and $(u^{(2N)}_{\text{Bartlett}})_\ell$, for $N = 100$ and $\Delta x = 1$. Both are normalized to unity at the origin to facilitate the comparison.

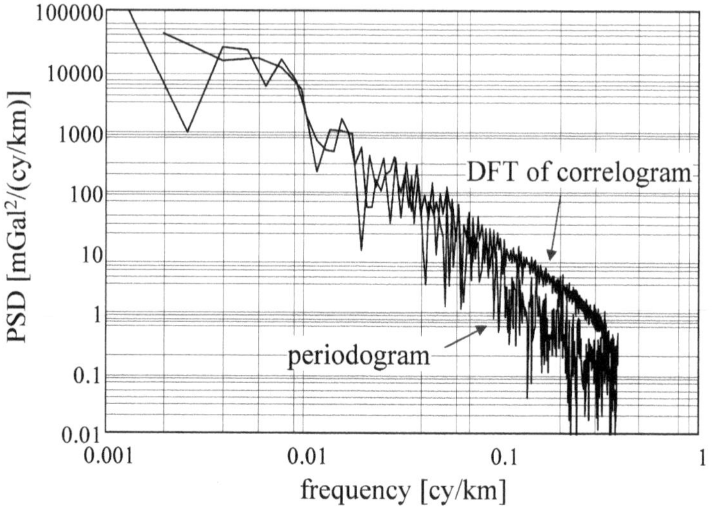

Figure 5.15: The periodogram, $\breve{C}_{\Delta g,\Delta g}(f_k)$, versus the DFT of the correlogram, $\hat{C}_{\Delta g,\Delta g}(f_k)$, for the gravity anomaly profile in Figure 5.10 (left). The discrete points of these estimates are connected by lines for clarity.

(Figure 5.14) become more and more dense. Intuitively, the transforms approach the Dirac delta function and the bias due to spectral leakage vanishes. Thus, the periodogram, $\breve{C}_{g,h}(f_k)$, is an asymptotically unbiased estimator of $\tilde{C}_{g,h}(f)$. However, the bias due to aliasing remains—the difference between $\tilde{C}_{g,h}(f)$ in equation (5.365) and the true PSD, $C_{g,h}(f)$.

Spectral leakage at higher frequencies and thus estimator bias can be reduced further by applying a taper or window function to the data (Section 3.6). Let u_n be a window sequence (data taper) such that $u_n = 0$ for $n < -N/2$ and $n \geq N/2$. For reasons that become apparent below, the window sequence is normalized,

$$\bar{u}_n = \bar{u}_n = \left(\frac{1}{N} \sum_{n=-N/2}^{N/2-1} u_n^2 \right)^{-1/2} u_n, \tag{5.366}$$

so that

$$\frac{1}{N} \sum_{n=-N/2}^{N/2-1} \bar{u}_n^2 = 1. \tag{5.367}$$

(The rectangle sequence, equation (4.26), already satisfies this normalization.) Applying the taper sequence one obtains modified data sequences, $g_n^{(u)} = \bar{u}_n g_n$ and $h_n^{(u)} = \bar{u}_n h_n$. As in equation (5.348), the corresponding biased covariance estimator is, for $\ell = 0,\ldots, N-1$,

$$\left(\breve{c}_{g,h}^{(u)}\right)_\ell = \frac{1}{N} \sum_{n=-N/2}^{N/2-\ell-1} g_n^{(u)} h_{n+\ell}^{(u)} = \frac{1}{N} \sum_{n=-N/2}^{N/2-\ell-1} \bar{u}_n \bar{u}_{n+\ell} g_n h_{n+\ell}. \tag{5.368}$$

We may disregard the correlogram approach based on the *unbiased* covariance estimator in view of the foregoing analysis. As in equation (5.362), the modified data sequences lead to

$$\breve{C}_{g,h}^{(u)}(f) = \Delta x \sum_{\ell=-N+1}^{N-1} \left(\breve{c}_{g,h}^{(u)}\right)_\ell e^{-i2\pi\Delta x\ell f} = \frac{\Delta x}{N} \sum_{j=-N/2}^{N/2-1} \sum_{k=-N/2}^{N/2-1} h_j^{(u)} g_k^{(u)} e^{-i2\pi\Delta x(j-k)f}, \tag{5.369}$$

and, for discrete frequencies, $f_k = k/(N\Delta x)$, to a PSD estimator according to the periodogram method,

$$\breve{C}_{g,h}^{(u)}(f_k) = \frac{1}{N\Delta x} \tilde{H}_k^{(u)} \left(\tilde{G}_k^{(u)}\right)^*, \tag{5.370}$$

where $\tilde{G}_k^{(u)} = \text{DFT}(g_n^{(u)})_k$, $\tilde{H}_k^{(u)} = \text{DFT}(h_n^{(u)})_k$, assuming both sequences are continued periodically. The estimator, $\breve{C}_{g,h}^{(u)}(f)$, of the PSD is called the *direct spectral estimator*, or also the *modified periodogram* (for $f = f_k$).

The expected value of the corresponding covariance estimator is (for $\ell = 0,\ldots, N-1$)

$$\begin{aligned} \mathcal{E}\left(\left(\breve{c}_{g,h}^{(u)}\right)_\ell\right) &= \frac{1}{N} \sum_{n=-N/2}^{N/2-\ell-1} \mathcal{E}\left(g_n^{(u)} h_{n+\ell}^{(u)}\right) = \frac{1}{N} \sum_{n=-N/2}^{N/2-\ell-1} \bar{u}_n \bar{u}_{n+\ell} \mathcal{E}\left(g_n h_{n+\ell}\right) \\ &= \left(c_{g,h}\right)_\ell \frac{1}{N} \sum_{n=-N/2}^{N/2-\ell-1} \bar{u}_n \bar{u}_{n+\ell} \end{aligned} \tag{5.371}$$

Similarly (again, for $\ell = 0,\ldots, N-1$) and in view of equation (5.346),

$$\mathcal{E}\left(\left(\breve{c}_{g,h}^{(u)}\right)_{-\ell}\right) = \frac{1}{N} \sum_{n=-N/2}^{N/2-\ell-1} \bar{u}_n \bar{u}_{n+\ell} \mathcal{E}\left(h_n g_{n+\ell}\right) = \left(c_{g,h}\right)_{-\ell} \frac{1}{N} \sum_{n=-N/2}^{N/2-\ell-1} \bar{u}_n \bar{u}_{n+\ell}. \tag{5.372}$$

Now, since $\bar{u}_n = 0$ for $n < -N/2$ and $n \geq N/2$, the last sum above is also

$$\Delta x \sum_{n=-N/2}^{N/2-\ell-1} \bar{u}_n \bar{u}_{n+\ell} = \Delta x \sum_{n=-\infty}^{\infty} \bar{u}_n \bar{u}_{n+\ell} = \left(\mathring{\phi}_{\bar{u},\bar{u}}\right)_\ell, \tag{5.373}$$

by the definition of the discrete correlation, equation (5.37), which, therefore, is also zero for $\ell \leq -N$ and $\ell \geq N$. By equation (5.38) the Fourier transform of this correlation sequence is

$$\mathring{\Phi}_{\bar{u},\bar{u}}(f) = \tilde{\bar{U}}^*(f) \tilde{\bar{U}}(f), \tag{5.374}$$

where $\tilde{\bar{U}}(f)$ is the Fourier transform, equation (4.16), of the normalized data taper (considered an infinite, finite-energy sequence for $-\infty < n < \infty$).

From the first of equations (5.369) and equations (5.371) through (5.373), the expectation of the PSD estimator, $\breve{C}_{g,h}^{(u)}(f)$, is

$$\begin{aligned}\mathcal{E}\left(\breve{C}_{g,h}^{(u)}(f)\right) &= \Delta x \sum_{\ell=-N+1}^{N-1} \mathcal{E}\left(\left(\breve{c}_{g,h}^{(u)}\right)_\ell\right) e^{-i2\pi\Delta x \ell f} \\ &= \Delta x \sum_{\ell=-\infty}^{\infty} \left(c_{g,h}\right)_\ell \frac{1}{N\Delta x} \left(\overset{\circ}{\phi}_{\bar{u},\bar{u}}\right)_\ell e^{-i2\pi\Delta x \ell f}\end{aligned} \tag{5.375}$$

This is the Fourier transform of a product of sequences. Hence, by the dual convolution theorem (4.37) it is the convolution, equation (3.33), of the Fourier transforms of the covariance and taper correlation sequences,

$$\begin{aligned}\mathcal{E}\left(\breve{C}_{g,h}^{(u)}(f)\right) &= \frac{1}{N\Delta x} \Delta x \sum_{\ell=-\infty}^{\infty} \left(c_{g,h}\right)_\ell \left(\overset{\circ}{\phi}_{\bar{u},\bar{u}}\right)_\ell e^{-i2\pi\Delta x \ell f} \\ &= \frac{1}{N\Delta x} \int_{-f_{\mathcal{N}}}^{f_{\mathcal{N}}} \tilde{C}_{g,h}(f') \overset{\circ}{\tilde{\Phi}}_{\bar{u},\bar{u}}(f - f')\, df'\end{aligned} \tag{5.376}$$

For the special case of the discrete rectangle window, $u_n = b_n^{(N)}$, equations (4.29) and (4.261) show that

$$\overset{\circ}{\tilde{\Phi}}_{\bar{u},\bar{u}}(f) = \left(\tilde{B}^{(N)}(f)\right)^* \tilde{B}^{(N)}(f) = N\Delta x \tilde{U}_{\text{Bartlett}}^{(2N)}(f), \tag{5.377}$$

in agreement with equation (5.365).

Consider the integral of the expected value of the modified periodogram, $\breve{C}_{g,h}^{(u)}(f)$, from the first of equations (5.376),

$$\begin{aligned}\int_{-f_{\mathcal{N}}}^{f_{\mathcal{N}}} \mathcal{E}\left(\breve{C}_{g,h}^{(u)}(f)\right) df &= \frac{1}{N} \sum_{\ell=-\infty}^{\infty} \left(c_{g,h}\right)_\ell \left(\overset{\circ}{\phi}_{\bar{u},\bar{u}}\right)_\ell \int_{-f_{\mathcal{N}}}^{f_{\mathcal{N}}} e^{-i2\pi\Delta x \ell f}\, df \\ &= \frac{1}{N\Delta x} \left(c_{g,h}\right)_0 \left(\overset{\circ}{\phi}_{\bar{u},\bar{u}}\right)_0 \\ &= \left(c_{g,h}\right)_0 \frac{1}{N} \sum_{n=-N/2}^{N/2-1} \bar{u}_n^2\end{aligned} \tag{5.378}$$

where the integral on the right, using equation (3.163) with $f_0 = f_{\mathcal{N}} = 1/(2\Delta x)$, is easily seen to vanish unless $\ell = 0$ and equation (5.373) yields the last equality. With the normalization, equation (5.367), and $\ell = 0$ in equation (5.343), the total energy of the expectation of the PSD estimator equals the true total spectral energy,

$$\int_{-f_N}^{f_N} \mathcal{E}\left(\breve{C}^{(u)}_{g,h}(f)\right) df = \left(c_{g,h}\right)_0 = \int_{-f_N}^{f_N} C^{(u)}_{g,h}(f)\, df. \tag{5.379}$$

When applying an arbitrary taper, such as the discrete Hann window, equation (4.260), a significant amount of the data is reduced in magnitude, which affects the magnitude of the computed periodogram. Equation (5.379) verifies that an appropriate normalization of the taper, while still changing the amplitude of the data, nevertheless ensures that the energy of periodogram, hence its overall magnitude, is not unduly biased.

Of particular interest in data tapers is the discrete prolate spheroidal sequence, $u_\ell = (\psi_0^{(N,f_0)})_\ell$, that has the special property of concentrating the maximum amount of energy in the spectral band, $|f| \le f_0$, for a given data span (Section 4.6). That is, the spectral leakage at the higher frequencies is minimized. The energy spectral densities, $\overset{\circ}{\tilde{\Phi}}_{\bar{u},\bar{u}}(f)$, of the discrete rectangle, discrete Hann, and discrete prolate spheroidal tapers are compared in Figure 5.16 for $N = 50$ and $\Delta x = 1$. The half bandwidth for the latter is chosen as $f_0 = 0.02$ in order to approximate a cutoff frequency that is equivalent to the first zero of the rectangle window spectrum at $f = 1/(N\Delta x)$. The formulas for $\overset{\circ}{\tilde{\Phi}}_{\bar{u},\bar{u}}(f)$ are equation (5.377) for the discrete rectangle window, and equations (4.262) and (4.269) (multiplied by corresponding normalizations, equation (5.367)) substituted into equation (5.374), respectively, for the discrete Hann and prolate spheroidal windows. The side-lobes of the discrete prolate spheroidal taper

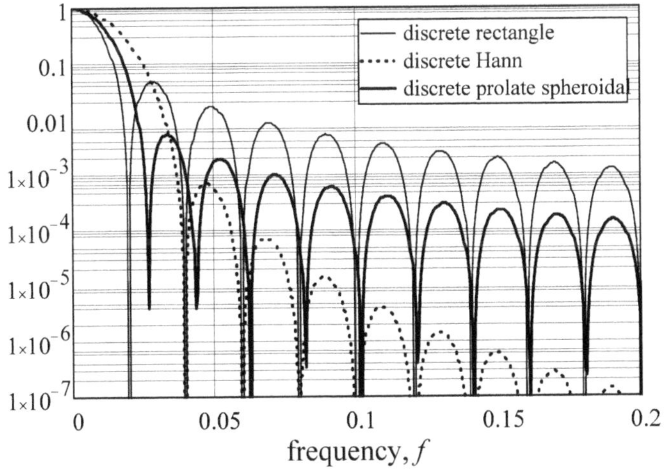

Figure 5.16: Amplitudes of energy spectral densities, $\overset{\circ}{\tilde{\Phi}}_{u,u}(f)$, for several window sequences with $N = 50$, $\Delta x = 1$, and normalized to unity at the origin for ease of comparison. The half bandwidth for the discrete prolate spheroidal sequence is chosen as $f_0 = 0.02$. Only part of the total frequency domain, $0 \le f \le f_N = 0.5$, is shown.

in this case ($f_0 = 0.02$) are an order of magnitude smaller than those of the rectangle window, when both densities are normalized to unity at the origin.

An example of spectral estimation of (local) geopotential fields is illustrated as follows for a profile of magnetic anomalies in the Dakotas/Minnesota region (Figure 5.17). The profile contains data at a spatial resolution of about $\Delta x = 1.33$ km. Together with the original profile this figure shows the tapered data (thicker line), as well as the corresponding normalized discrete prolate spheroidal (dps) taper for a half-bandwidth of $f_0 = 1/(N\Delta x)$, which corresponds to the first zero of the rectangle spectrum, as in Figure 5.16. Note that the taper is not zero at the end points of the data domain. The periodograms of the original data and the tapered data are shown in Figure 5.18. These have been smoothed for improved visualization. The PSD estimate from the tapered data is generally less in magnitude at the high frequencies, ostensibly reflecting a reduction in bias due to spectral leakage. Figure 5.18 includes the periodogram for the data tapered by the discrete normalized Hann window, equation (4.260); see also Figure 5.16. There is only a slight difference between these two modified periodograms.

Even though the modified periodogram reduces the bias due to spectral leakage, it is still a biased estimator (but an asymptotically unbiased estimator of $\tilde{C}_{g,h}(f)$, provided the taper function approaches unity as $T \to \infty$). It is also not a *consistent* estimator. That is, even if $N \to \infty$, it can be shown that the variance of the estimator, $\tilde{C}^{(u)}_{g,h}(f)$, does not reduce to zero (Brillinger 2001, p.125ff); in fact, the variance approaches the square of the true PSD of the sequence, $\tilde{C}^2_{g,h}(f)$ $(f \neq 0)$.

A number of methods might be considered to reduce the variance. Two classical methods appeal to the exact expression for the power spectral density, equation (5.232), where it is clear that the periodogram, equation (5.326), aside from the

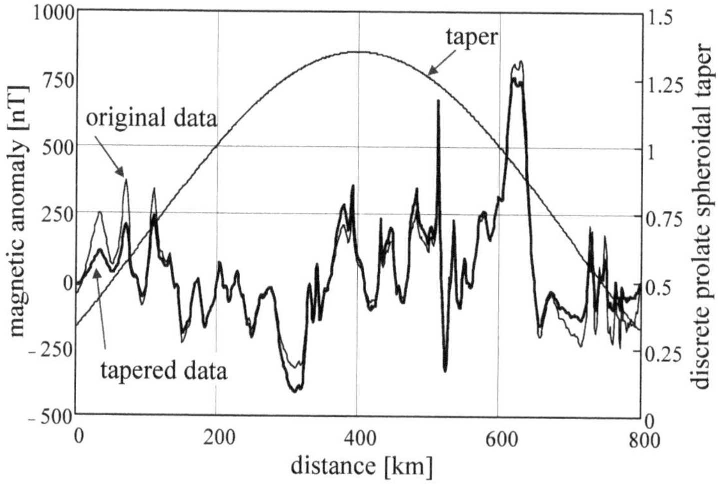

Figure 5.17: Magnetic anomaly data along a profile in longitude in the Dakotas/Minnesota area. The thinner curve represents the original data compared to the thicker curve that is the product of the data and the normalized discrete prolate spheroidal taper (also shown). The data and taper sequences are connected by lines for clarity.

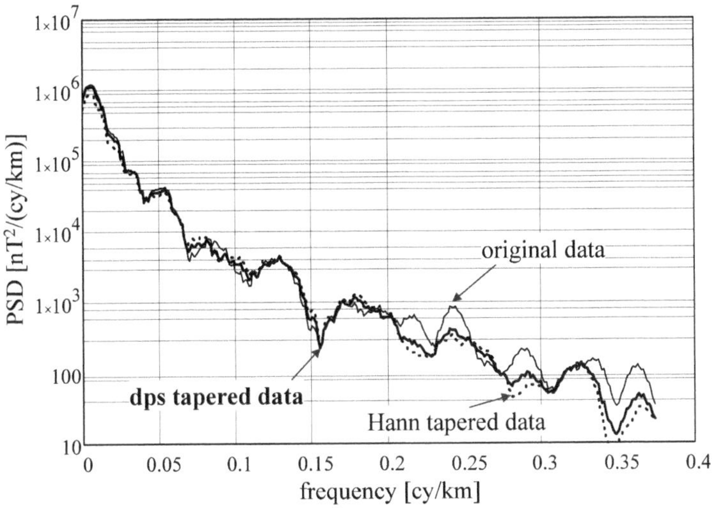

Figure 5.18: Periodograms of magnetic anomaly data (Figure 5.17) according to various normalized tapers. The thin solid line represents the un-tapered data, while the thick solid line corresponds to the discrete prolate spheroidal (dps) taper. The dotted line is the periodogram of the data tapered by the discrete Hann window.

sampling error, neglects the expectation, as well as the extension in the limit over an infinite data domain. Little can be done about the truncated domain, except to apply a taper to reduce the corresponding spectral leakage (or increase the extent of the space domain for the data). However, the expectation can be simulated by averaging different estimates of the PSD. This is the idea behind the *Bartlett Periodogram*, in which a data sequence is divided into a number of shorter sequences of equal length. The estimator is then the average of the periodograms of the shorter sequences. It is easily proved, using equation (5.199), that the variance of a sum of N identically and independently distributed random variables, Z_j, each having variance, σ_Z^2, is smaller than σ_Z^2 by the factor, $1/N$. Thus, assuming that the individual PSD estimators, corresponding to the disjoint data segments, are independent (and identically distributed by stationarity), the variance of the Bartlett Periodogram decreases with the number of such segments, making it a consistent estimator. Of course, an increase in the number of data segments (for a fixed original finite data sequence) diminishes the available *spectral* resolution of the lower frequencies (longer wavelengths); see Section 3.5.2. One remedy is the *Welsh Periodogram* that expands this concept by using overlapping segments, thus extending the domains of the segments for a given data sequence (or increasing the number of segments of the Bartlett size for lower estimator variance). On the other hand this creates the danger of averaging correlated periodograms (due to the overlap). For additional information on these modifications, the reader may consult (Marple 1987, p.153ff).

Instead of sacrificing spectral resolution, it may be more acceptable in some applications to lose spatial resolution in the PSD estimator. This is accomplished by averaging the PSD estimator over frequencies, which, analogous to a filter in the space domain, is a convolution in the frequency domain of $\breve{C}_{g,h}^{(u)}(f)$ with a smoothing function, $\tilde{V}^{(m)}(f)$,

$$\bar{\breve{C}}_{g,h}^{(u)}(f)=\int_{-f_N}^{f_N}\breve{C}_{g,h}^{(u)}(f')\tilde{V}^{(m)}(f-f')\,df', \tag{5.380}$$

where m is a parameter that controls the spectral band of the smoothing operation. By the dual cyclic convolution theorem (4.37) for sequences, $\bar{\breve{C}}_{g,h}^{(u)}(f)$ is the Fourier transform of the product of the covariance sequence and the inverse Fourier transform of the smoothing function,

$$\begin{aligned}\bar{\breve{C}}_{g,h}^{(u)}(f)&=\Delta x\sum_{\ell=-\infty}^{\infty}\left(\breve{c}_{g,h}^{(u)}\right)_\ell v_\ell^{(m)}e^{-i2\pi\ell\Delta xf}\\&=\Delta x\sum_{\ell=-N+1}^{N-1}\left(\breve{c}_{g,h}^{(u)}\right)_\ell v_\ell^{(m)}e^{-i2\pi\ell\Delta xf}\end{aligned} \tag{5.381}$$

where the second equality follows from the practical estimate of the covariance sequence, equation (5.348), being limited to a finite number of lags, and where for cross-covariances it is assumed that in addition one has $(\breve{c}_{g,h}^{(u)})_{-\ell}=(\breve{c}_{h,g}^{(u)})_\ell$. Thus, $\bar{\breve{C}}_{g,h}^{(u)}(f)$ is also the Fourier transform of a tapered covariance sequence—as one would expect, tapering the covariance sequence and (frequency-) filtering the power spectral density estimate are dual operations. The sequence, $v_\ell^{(m)}$, is the inverse Fourier transform of $\tilde{V}^{(m)}(f)$, and is called a *lag window* (Priestley 1981). The estimator, $\bar{\breve{C}}_{g,h}^{(u)}(f)$, may then also be called a *lag-window spectral estimator*, but many other appellations exist (Percival and Walden 1993, p.242). For discrete frequencies, $\bar{f}_k=k/(2N\Delta x)$, $-N\le k\le N-1$, and defining periodic sequences, $\tilde{v}_\ell^{(m)}$ and $(\tilde{\breve{c}}_{g,h}^{(u)})_\ell$, with $(\tilde{\breve{c}}_{g,h}^{(u)})_{-N}=0$ and period, $2N$, equation (5.381) becomes the DFT of the tapered covariance sequence,

$$\bar{\breve{C}}_{g,h}^{(u)}\left(\bar{f}_k\right)=\Delta x\sum_{\ell=-N}^{N-1}\left(\tilde{\breve{c}}_{g,h}^{(u)}\right)_\ell\tilde{v}_\ell^{(m)}e^{-i2\pi\ell\Delta x\bar{f}_k}. \tag{5.382}$$

It is customary to define the lag-window for the covariance sequence to be space-limited, such that

$$v_\ell^{(m)}=0\text{ for }|\ell|\ge N, \tag{5.383}$$

which does not change equations (5.381) and (5.380), but it does prevent $\tilde{V}^{(m)}(f)$ from being band-limited (Sections 2.3 and 3.5.2). For example, if equation (5.380) were to be a simple (unweighted) moving average, i.e., if $\tilde{V}^{(m)}(f)$ were the rectangle filter function, zero for all frequencies except $|f|\le f_m=1/(2m\Delta x)<f_N$, then $v_\ell^{(m)}$ would be non-zero for an infinitely many ℓ, as may be inferred from the inverse Fourier transform of $\tilde{V}^{(m)}(f)$, equation (4.264) with $f_0=f_m=1/(2m\Delta x)$ and $\ell'=0$. Thus,

imposing the additional condition (5.383) and normalizing to unity at the origin, $v_\ell^{(m)}$ becomes (Exercise 5.11)

$$v_\ell^{(m)} = \begin{cases} \text{sinc}\left(\dfrac{\ell}{m}\right), & |\ell| < N \\ 0, & |\ell| \geq N \end{cases} \tag{5.384}$$

An approximation of the simple moving average in the frequency domain is then described by the filter function,

$$\tilde{V}^{(m)}(f) = \Delta x \sum_{\ell=-N+1}^{N-1} \text{sinc}\left(\frac{\ell}{m}\right) e^{-i2\pi\Delta x \ell f}. \tag{5.385}$$

Figure 5.19 shows $\tilde{V}^{(m)}(f)$ for the simple moving average with $N = 100$, $\Delta x = 1$, and $m = 10$, implying a bandwidth, $2f_m = 1/(m\Delta x) = 0.1$. It is nearly, but not exactly, band-limited.

Clearly, other window sequences, $v_\ell^{(m)}$, can be used whose secondary lobes in the frequency domain attenuate more rapidly and where the parameter, m, specifies the bandwidth of the spectral smoothing (Percival and Walden 1993, Section 6.11, offer a detailed comparison). With the constraint, equation (5.383), one has

$$\tilde{V}^{(m)}(f) = \Delta x \sum_{\ell=-N+1}^{N-1} v_\ell^{(m)} e^{-i2\pi\Delta x \ell f}. \tag{5.386}$$

However, it should also be recognized that this spectral domain filter function acts in concert with the data taper to create the final spectral estimator. From equations (5.376) and (5.380),

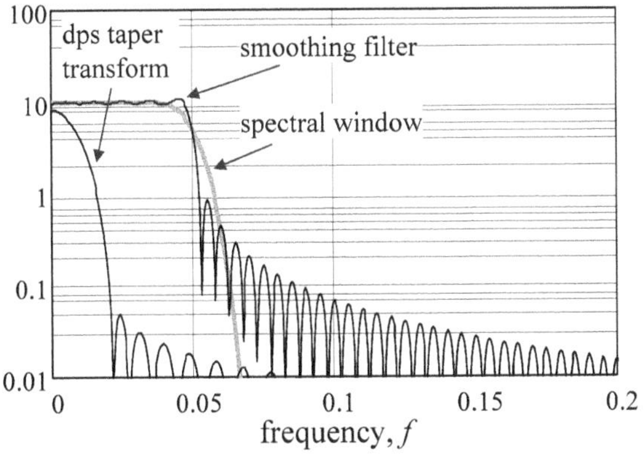

Figure 5.19: Fourier transforms of the discrete prolate spheroidal taper, $\tilde{\Psi}_0^{(N,f_0)}(f)$ ($N = 100, f_0 = 0.02$, $\Delta x = 1$); of the lag window, $\tilde{V}^{(m)}(f)$ ($m = 10$), for the (approximate) simple moving average in the spectral domain; and, their convolution, the spectral window, $\tilde{W}^{(u,m)}(f)$.

$$\mathcal{E}\left(\breve{\bar{C}}_{g,h}^{(u)}(f)\right) = \int_{-f_N}^{f_N} \mathcal{E}\left(\breve{C}_{g,h}^{(u)}(f')\right)\tilde{V}^{(m)}(f-f')\,df' \\ = \frac{1}{N\Delta x}\int_{-f_N}^{f_N} \tilde{W}^{(u,m)}(f-f'')\tilde{C}_{g,h}(f'')\,df'' \tag{5.387}$$

where (Exercise 5.12)

$$\tilde{W}^{(u,m)}(f) = \int_{-f_N}^{f_N} \tilde{\overset{\circ}{\Phi}}_{\bar{u},\bar{u}}(f')\tilde{V}^{(m)}(f-f')\,df'. \tag{5.388}$$

This scaled convolution of two Fourier transforms, one of the taper correlation and one of the lag-window, can be formulated also as a Fourier series of a sequence of products. According to the dual convolution theorem (4.37),

$$\tilde{W}^{(u,m)}(f) = \sum_{\ell=-N+1}^{N-1} \left(\overset{\circ}{\phi}_{\bar{u},\bar{u}}\right)_\ell v_\ell^{(m)} e^{-i2\pi\Delta x\ell f}, \tag{5.389}$$

where equations (5.373) or (5.383) imply the finite limits on the sum. $\tilde{W}^{(u,m)}(f)$ is called the "spectral window" (Percival and Walden 1993, p.244; see also Priestley 1981, p.436, who excludes the data taper, in which case, $\tilde{W}^{(u,m)}(f) = \tilde{V}^{(m)}(f)$); it is also displayed in Figure 5.19 for the case of the discrete prolate spheroidal taper, $(\psi_0^{(N,f_0)})_\ell$ ($N = 100, f_0 = 0.02$), and the simple moving average, equation (5.384).

It can be shown (Percival and Walden 1993, p.247) that the variance of the lag-window spectral estimator is given approximately by

$$\operatorname{var}\left(\breve{\bar{C}}_{g,h}^{(u)}(f)\right) \approx C_h \frac{\left(\tilde{C}_{g,h}^{(u)}(f)\right)^2}{N\Delta x} \int_{-f_N}^{f_N} \left(\tilde{V}^{(m)}(f)\right)^2 df, \tag{5.390}$$

where $C_h \geq 1$ is a *variance-inflation factor* that depends on the taper applied to the data (it is, however, a constant that does not depend on N—a formula is given by (ibid., p.251)). As expected, the variance of the lag-window spectral estimator decreases with an increase in the number of data, thus making it a consistent estimator. An analysis of the modifications to the periodogram should account for the effects of the data taper as well as the frequency filter. The taper, $\bar{u}_\ell$, is applied primarily to reduce the side lobes in the estimator of the PSD that are associated with the truncation of the data to a finite extent, thus reducing the bias in the PSD estimator due to spectral leakage. The filter smoothes the estimator in order to imitate the expectation operator in the theoretical formulation of the PSD, equation (5.232), and thereby also reduces its variance. Here, one must ensure that the spectral smoothing, as specified by the parameter, m, does not obliterate significant physical resonances in the PSD. Moreover, introducing the spectral window, or a finite bandwidth for the spectral estimator, causes additional bias as indicated in equations (5.387) and (5.388) (compared to equation (5.376)).

A relatively recent advance in spectral estimation makes use of multiple discrete prolate spheroidal tapers to reduce both the bias due to spectral leakage and the variance. This has come to be called *multi-taper spectral estimation* and was introduced by Thomson (1982). It is only sketched here and the interested reader is directed to (Percival and Walden 1993, Chapter 7) for a detailed exposition.

Recalling the concentration problem (Section 3.6.2) and its discrete version, starting with equation (4.263)) the solutions to the N-dimensional eigenvalue problem, equation (4.267),

$$\mathbf{Ag} - \lambda(f_0)\mathbf{g} = \mathbf{0}, \tag{5.391}$$

(matrix, $\mathbf{A}$, is populated by elements in equation (4.265)) are denoted $\boldsymbol{g} = \psi_n^{(N,f_0)}$, $n = 0,\ldots, N-1$, and are orthogonal N-dimensional eigenvectors with corresponding distinct and real eigenvalues, $\lambda_n(f_0)$. The eigenvectors form a basis for the space of index-limited sequences, where the ordered N eigenvalues (concentration ratios), equation (4.268), are close to unity for $k = 0,\ldots, 2N\Delta x f_0 - 1$ and attenuate rapidly to near zero for k greater than the Shannon number, $2N\Delta x f_0$ (Figure 4.8). Therefore, the eigenvectors, $\psi_k^{(N,f_0)}$, $k = 0,\ldots, 2N\Delta x f_0 - 1$, approximately span the space of index-limited sequences and their spectral energies are maximally concentrated in the band, $[-f_0, f_0]$. Used as data tapers they minimize the spectral leakage in the estimation of the PSD. Because they are orthogonal, the average of periodograms modified by such tapers creates a multi-taper estimator of the PSD that besides minimizing spectral leakage also has reduced variance (asymptotically it is a consistent estimator). This estimator may be formulated as

$$\breve{C}_{g,h}^{(mt)}(f) = \frac{1}{K}\sum_{k=1}^{K} \breve{C}_{g,h}^{(u_k)}(f), \quad |f| \le f_{\mathcal{N}}, \tag{5.392}$$

with $\breve{C}_{g,h}^{(u_k)}(f)$ given by equation (5.370) for the data taper equal to the kth-order dpss,

$$u_n^{(k)} = \left(\psi_k^{(N,f_0)}\right)_n, \quad n = -N/2,\ldots,N/2-1, \tag{5.393}$$

where the eigenvectors explicitly are given by equation (4.270).

All these PSD estimation techniques can be extended to two dimensions in Cartesian coordinates in the usual natural way. For example, one has overlapping *areas* in the case of the two-dimensional Welch periodogram, and the two-dimensional taper functions are designed as products of one-dimensional functions, one for each variable, as in equation (3.153).

5.7.3 Spherical Power Spectrum Estimation

The determination of the Fourier-Legendre spectrum of a spherical function requires a global, preferably uniform, distribution of samples. As noted in Section 6.3.1, areas lacking data for a global analysis of a geophysical signal, such as the

Earth's gravitational field, can be supplemented, though inexactly, with proxy data (topographic data in the case of the gravitational field) or with predicted data from previous models. This situation is much less common today compared to decades ago as satellite systems are able to collect data world wide and with high spatial resolution. However, global coverage depends on the inclination of the satellite orbit, that is, the angle of the orbit with respect to the equator. If the orbital inclination differs significantly from 90°, the data distribution from a satellite excludes two polar caps (Sneeuw and Van Gelderen 1997). For example, the GRACE and GOCE gravity mapping satellites (Section 6.3.1) have inclinations of 89° and 96.7°, respectively, the latter also for the magnetic field mapping satellites, Magsat and Ørsted (Section 6.3.2), thus leaving polar caps with diameters, about 220 km and 1500 km, that are effectively devoid of data. Estimating the Fourier-Legendre spectrum from data in these truncated, or windowed, global regions incurs corresponding spectral leakage error.

The analysis of this error follows a procedure similar to the Cartesian case based on a stochastic interpretation of the data functions and an estimation of the power spectrum. For the sake of generality the analysis is derived for the cross-power spectrum, but it readily specializes to the auto-power spectrum, and, in fact, the illustrated effect of spectral leakage is common to both cases. Thus, let g and h be stationary harmonic stochastic processes expressed as spherical harmonic series with random coefficients, $G_{n,m}(Z)$ and $H_{n,m}(Z)$, respectively, and represented by an isotropic cross-covariance function. The usual estimator of the power spectrum, $(C_{g,h})_n$, equation (5.297), is

$$\left(\hat{C}_{g,h}\right)_n = \frac{1}{2n+1}\sum_{m=-n}^{n} G_{n,m}(Z)H_{n,m}(Z), \tag{5.394}$$

where Z is a random variable that characterizes the processes, and

$$G_{n,m}(Z) = \frac{1}{4\pi}\iint_{\Omega} g(\theta,\lambda)\bar{Y}_{n,m}(\theta,\lambda)\,d\Omega, \quad H_{n,m}(Z) = \frac{1}{4\pi}\iint_{\Omega} h(\theta,\lambda)\bar{Y}_{n,m}(\theta,\lambda)\,d\Omega. \tag{5.395}$$

The estimator, $(\hat{C}_{g,h})_n$, is unbiased. Indeed, substituting equation (5.395) into equation (5.394) and using the addition theorem, equation (2.254), one has

$$\begin{aligned}\mathcal{E}\left(\left(\hat{C}_{g,h}\right)_n\right) &= \frac{1}{(4\pi)^2}\iint_{\Omega}\iint_{\Omega'} \mathcal{E}\left(g(\theta,\lambda)h(\theta',\lambda')\right)P_n(\cos\psi)\,d\Omega d\Omega' \\ &= \frac{1}{(4\pi)^2}\iint_{\Omega}\iint_{\Omega'} c_{g,h}(\psi)P_n(\cos\psi)\,d\Omega d\Omega'\end{aligned}. \tag{5.396}$$

With the series expansion for the isotropic covariance function, equation (5.296), most of the integrals collapse because one can choose the origin of coordinates arbitrarily, and the expectation becomes

$$\begin{aligned}\mathcal{E}\left(\left(\hat{C}_{g,h}\right)_n\right) &= \frac{1}{2}\int_{\psi=0}^{\pi}\sum_{n'=0}^{\infty}(2n'+1)\left(C_{g,h}\right)_{n'} P_{n'}(\cos\psi)P_n(\cos\psi)\sin\psi\,d\psi \\ &= \sum_{n'=0}^{\infty}(2n'+1)\left(C_{g,h}\right)_{n'}\frac{1}{2}\int_{\psi=0}^{\pi}P_{n'}(\cos\psi)P_n(\cos\psi)\sin\psi\,d\psi \\ &= \sum_{n'=0}^{\infty}(2n'+1)\left(C_{g,h}\right)_{n'}\frac{1}{2n+1}\delta_{n-n'} \\ &= \left(C_{g,h}\right)_n \end{aligned} \tag{5.397}$$

in view of equation (2.204). Hence, the estimator is unbiased and not subject to spectral leakage errors. It is assumed here and henceforth that the samples are sufficiently dense in distribution to avoid an aliasing error, which would be an issue for a coarsely, even if uniformly sampled function.

For the satellite-relevant case mentioned above, suppose that the data domain is truncated to a "belt", Ω_B, symmetric with respect to the equator,

$$\Omega_B = \left\{(\theta,\lambda) \mid \theta_s \le \theta \le \pi-\theta_s, 0 \le \lambda \le 2\pi\right\}; \tag{5.398}$$

and, consider the estimator,

$$\left(\breve{C}_{g,h}\right)_n = \frac{4\pi}{A_B}\frac{1}{2n+1}\sum_{m=-n}^{n}\frac{1}{4\pi}\iint_{\Omega_B} g(\theta,\lambda)\bar{Y}_{n,m}(\theta,\lambda)\,d\Omega\frac{1}{4\pi}\iint_{\Omega_B'} h(\theta',\lambda')\bar{Y}_{n,m}(\theta',\lambda')\,d\Omega', \tag{5.399}$$

where A_B denotes the area of the data region. Define the function,

$$P_n(\theta,\lambda,\theta',\lambda') = \sum_{m=-n}^{n}\bar{Y}_{n,m}(\theta,\lambda)\bar{Y}_{n,m}(\theta',\lambda') = (2n+1)P_n(\cos\psi), \tag{5.400}$$

which is a more explicit form of the Legendre polynomial when expressed according to the addition theorem, equation (2.254). Then, the expected value of the estimator is

$$\begin{aligned}\mathcal{E}\left(\left(\breve{C}_{g,h}\right)_n\right) &= \frac{1}{A_B 4\pi}\frac{1}{2n+1}\iint_{\Omega_B}\iint_{\Omega_B'} c_{g,h}(\psi)P_n(\theta,\lambda,\theta',\lambda')\,d\Omega' d\Omega \\ &= \frac{1}{A_B 4\pi}\frac{1}{2n+1}\iint_{\Omega_B}\iint_{\Omega_B'}\sum_{n'=0}^{\infty}(2n'+1)\left(C_{g,h}\right)_{n'}P_{n'}(\cos\psi)P_n(\theta,\lambda,\theta',\lambda')\,d\Omega' d\Omega\end{aligned} \tag{5.401}$$

The integrals do not collapse in this case because one is not free to rotate the coordinate system on account of the truncated limits in co-latitude. Substituting equation (5.400) for $P_{n'}(\cos\psi)$, and denoting

$$d_{n,n'} = \frac{1}{A_B 4\pi} \frac{1}{2n+1} \iint\limits_{\Omega_B} \iint\limits_{\Omega'_B} P_{n'}(\theta,\lambda,\theta',\lambda') P_n(\theta,\lambda,\theta',\lambda') \, d\Omega' d\Omega, \tag{5.402}$$

the expectation is

$$\mathcal{E}\left(\left(\breve{C}_{g,h}\right)_n\right) = \sum_{n'=0}^{\infty} d_{n,n'} \left(C_{g,h}\right)_{n'}, \tag{5.403}$$

which shows that the estimator is biased.

Because of the symmetry in the data region, Ω_B, about the polar axis, the integrals in $d_{n,n'}$ with respect to the longitudes, λ and λ', reduce to the constant, $4\pi^2$. And, the equatorial symmetry implies that the integrals with respect to the colatitudes, θ and θ', vanish if $n + n'$ is odd. It can be shown that these *coupling coefficients* thus become

$$d_{n,n'} = \frac{\pi}{A_B} \frac{\left(1 + (-1)^{n+n'}\right)^2}{2n+1} \sum_{m=0}^{\min(n,n')} \varepsilon_m \left(\int_{y=0}^{\cos\theta_s} \bar{P}_{n,m}(y) \bar{P}_{n',m}(y) \, dy \right)^2, \tag{5.404}$$

where $\varepsilon_0 = 1$ and $\varepsilon_m = 1/2$, $m > 0$, equation (2.212). The sum of all coefficients, $d_{n,n'}$, with respect to n' is equal to one, irrespective of n. Indeed, with the spherical Dirac delta function, equation (2.321), one obtains from equations (5.400) and (5.402),

$$\begin{aligned} \sum_{n'=0}^{\infty} d_{n,n'} &= \frac{1}{A_B 4\pi} \frac{1}{2n+1} \iint\limits_{\Omega_B} \iint\limits_{\Omega'_B} P_n(\theta,\lambda,\theta',\lambda') \sum_{n'=0}^{\infty} P_{n'}(\theta,\lambda,\theta',\lambda') \, d\Omega' d\Omega \\ &= \frac{1}{A_B} \frac{1}{2n+1} \iint\limits_{\Omega_B} \left(\frac{1}{4\pi} \iint\limits_{\Omega'_B} P_n(\theta,\lambda,\theta',\lambda') \delta_s(\theta,\lambda,\theta',\lambda') \, d\Omega' \right) d\Omega \\ &= \frac{1}{A_B} \frac{1}{2n+1} \iint\limits_{\Omega_B} P_n(\theta,\lambda,\theta,\lambda) \, d\Omega \end{aligned} \tag{5.405}$$

in view of equation (2.323). By equation (5.400), the last integrand is just $(2n + 1) \cdot 1$ and the result follows,

$$\sum_{n'=0}^{\infty} d_{n,n'} = 1. \tag{5.406}$$

Therefore, since by equation (5.404), $d_{n,n'} \geq 0$, spectral leakage at any degree, n, is characterized by the value of $1 - d_{n,n}$, where no leakage occurs if $d_{n,n} = 1$.

The left panel of Figure 5.20 indicates that the amount of spectral leakage in the estimated power spectrum for $\theta_s = 5°, 10°, 20°$ is less than about 5% if n is large (only about 0.36% for $\theta_s = 5°$). The right panel shows how that spectral leakage is distributed to neighboring harmonic degrees according to the coefficients, $d_{n,n'}$, for

$n = 20$ and $|n' - n| \leq 20$. Again, the spectral leakage is less significant for smaller polar caps. When necessary taper and multitaper functions, and, in particular, the solutions to the spectral concentration problem (Section 3.6.2), can be applied to reduce the spectral leakage (Dahlen and Simons 2008); see also (Albertella et al. 1999, Simons and Dahlen 2007) for methods of spherical analysis using the Slepian functions. Further details on the use of these functions for local analysis of spherical data in limited regions may be found, e.g., in (Wieczorek and Simons 2005).

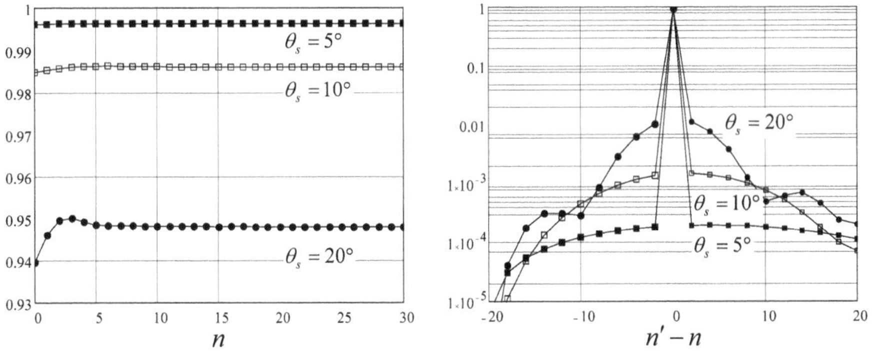

Figure 5.20: Coefficients, $d_{n,n}$ (left) and $d_{n,n'}$ (right), equation (5.404), that define the spectral leakage expected in the estimated power spectrum using data in an equatorial spherical belt, $\Omega_B = \{(\theta, \lambda)|\ \theta_s \leq \theta \leq \pi - \theta_s, 0 \leq \lambda \leq 2\pi\}$ for $\theta_s = 5°, 10°, 20°$.

* * *

EXERCISES

Exercise 5.1

Show that $\left|\overset{\circ}{\phi}_{g,h}(x)\right|^2 \leq \overset{\circ}{\phi}_{g,g}(0)\overset{\circ}{\phi}_{h,h}(0)$, $-\infty < x < \infty$, using a form of Schwarz's inequality.

Exercise 5.2

a) Show that for complex-valued functions, g and h,

$$\overset{\circ}{\phi}_{g,h}(-x_1, x_2) = \overset{\circ}{\phi}{}^{*}_{h,g}(x_1, -x_2), \tag{E5.1}$$

using the inverse Fourier transform of $\overset{\circ}{\Phi}_{g,h}(f_1, f_2)$. Then verify equation (5.29).

b) Show that $\frac{\partial}{\partial x_k}\overset{\circ}{\phi}_{g,h}(x_1, x_2) = \overset{\circ}{\phi}_{g,\frac{\partial h}{\partial x_k}}(x_1, x_2) = -\overset{\circ}{\phi}_{\frac{\partial g}{\partial x_k},h}(x_1, x_2)$, $k = 1,2$, by changing integration variables appropriately and using the result in a), above.

Exercise 5.3

Show that the energy spectral density of finite-energy sequences, g_ℓ and h_ℓ, is given by $\mathring{\tilde{\Phi}}_{g,h}(f) = \tilde{G}^*(f)\,\tilde{H}(f)$, making use of the fact that $\left(\mathring{\phi}_{g,h}\right)_\ell = g_{-\ell} \# h_\ell$ (prove this!) and the convolution theorem for sequences.

Exercise 5.4

Use the definition (5.52) to show that $1 - b(x)$ is a finite-power function, where $b(x)$ is the rectangle function.

Exercise 5.5

For finite-power functions, g and h, show that $\frac{d}{dx}\phi_{g,h}(x) = -\phi_{\frac{dg}{dx},h}(x)$. Hint: First show that $\phi_{g,h}(-x) = \phi^*_{h,g}(x)$; then use an appropriate change of integration variables in the derivative expression and Leibnitz's rule for differentiating an integral with variable limits.

Exercise 5.6

Following the same procedure that leads to equations (5.72) and (5.73), prove equations (5.105) and (5.106) for finite-power sequences. To prove (5.105), first show that

$$\lim_{N\to\infty}\frac{1}{N}\sum_{n=-N/2}^{N/2-1} g_n h_{n+\ell} = \lim_{N\to\infty}\frac{1}{N}\sum_{n=-N/2}^{N/2-1}\left(g_N\right)_n\left(h_N\right)_{n+\ell}; \tag{E5.2}$$

then apply this to the correlation sequence, $\left(\mathring{\phi}_{g_N,h_N}\right)_\ell$.

Exercise 5.7

Prove equation (5.158) by substituting equation (2.290) into equation (5.157).

Exercise 5.8

Let $Z = \sum_j a_j Z_j$ be a combination of independent Gaussian random variables, Z_j, where the a_j are constants. Using the characteristic function of Z, show that Z is a Gaussian random variable. Hint: using the property of independence, equation (5.195), first show that

$$\chi_Z(\omega) = \mathcal{E}\left(\prod_j e^{i\omega a_j Z_j}\right) = \prod_j \mathcal{E}\left(e^{i\omega a_j Z_j}\right). \tag{E5.3}$$

Exercise 5.9

Show by mathematical induction that the integral in equation (5.309) is

$$I_n = \frac{A_{2n-1}}{A_{2n}} \int_{-1}^{1} x^2 \left(1-x^2\right)^{n-1} dx = \frac{1}{2n+1}. \tag{E5.4}$$

Hint: Show that $I_n = \frac{2n(2n-1)}{2^2 n(n-1)} I_{n-1} - \frac{A_{2n-1}}{A_{2n}} \int_{-1}^{1} x^4 \left(1-x^2\right)^{n-2} dx$; then, integrating by parts show that $I_n = \frac{2}{3}\frac{A_{2n-1}}{A_{2n}}(n-1) \int_{-1}^{1} x^4 \left(1-x^2\right)^{n-2} dx$. Combine and apply the inductive step.

Exercise 5.10

Show that

$$\sum_{n=-N/2}^{N/2-1} g_n h_n + \sum_{\ell=1}^{N-1} \sum_{n=-N/2}^{N/2-\ell-1} \left(h_n g_{n+\ell} + g_n h_{n+\ell}\right) e^{-i2\pi\Delta x \ell f} = \sum_{j=-N/2}^{N/2-1} \sum_{k=-N/2}^{N/2-1} h_j g_k e^{-i2\pi\Delta x (j-k) f} \tag{E5.5}$$

by letting $\rho = e^{-i2\pi\Delta x f}$, writing out the sums on both sides explicitly, and showing that they are equivalent.

Exercise 5.11

Using equation (4.264) show that the lag window sequence for the approximate simple moving-average spectral estimator is given by the sinc function, equation (5.384), and that the filter function of the moving average, equation (5.380), is given by equation (5.385), where

$$v_\ell^{(m)} = \int_{-f_N}^{f_N} \tilde{V}^{(m)}(f) e^{i2\pi\ell\Delta x f} df. \tag{E5.6}$$

Exercise 5.12

With an appropriate change in integration variables and noting that the integrand is periodic, show that

$$\int_{-f_N}^{f_N} \overset{\circ}{\tilde{\Phi}}_{\bar{u},\bar{u}}(f'-f'')\tilde{V}^{(m)}(f-f')\,df' = \int_{-f_N}^{f_N} \overset{\circ}{\tilde{\Phi}}_{\bar{u},\bar{u}}(f')\tilde{V}^{(m)}(f-f''-f')\,df'. \tag{E5.7}$$

Then prove equations (5.387) and (5.388) using equation (5.376).

Chapter 6

Applications in Geodesy and Geophysics

6.1 Introduction

This chapter offers a sampling of applications in geodesy and geophysics where spectral methods find utility, either in the interpretation of signals, including simply constructing useful physical models in the spectral domain, or in facilitating numerical computation associated with convolutions or correlations. The illustrative focus is on geopotential fields (mostly gravitational, but also magnetic) and topography, viewed as a geospatial function on the Earth that is correlated to some extent with the fields due to its geometry. While the topography is a function of height on a two-dimensional surface and thus directly amenable to the spectral methods developed in previous chapters, the geopotential is a function in three-dimensional space, where its variation in the third dimension (radial distance from Earth's center) is of essentially different character than its horizontal variations. From a strictly numerical viewpoint, one might simply extend the spectral methods to three dimensions, but in free space the geopotential also is subject to a physical constraint that in most applications must be enforced. This is Laplace's equation, which is shown to dictate very precisely the relationship between the spectra of the geopotential on different geocentric spheres, or different horizontal planes in the local approximation. A mere generalization of spectral methods to the third dimension does not necessarily incorporate this essential constraint. Moreover, it is unusual to obtain geopotential data specifically in the vertical dimension; it is much more common to collect such data on roughly horizontal or spherical surfaces. Also, a source body of limited vertical extent, much smaller in scale than its horizontal extent (such as the Earth's lithosphere on continental scales), may be regarded approximately as a collapsed, or "condensed", infinitesimal layer on a spherical or horizontal surface. Thus, the applications of spectral methods considered here are restricted to the two dimensions elaborated throughout this text—the plane and sphere, with due mention of the relationships of spectra at different altitudes.

6.2 Spectrum of the Potential Function

Of the many physical attributes of the Earth that geophysicists use to study the Earth's interior, the gravitational and magnetostatic fields offer a global view of the internal structure at many different scales. The gravitational field also has important application in physical geodesy where the surface of constant potential (gravitation plus centrifugal acceleration) that closely approximated mean sea level, the *geoid*, defines the principal *shape* of the Earth (Torge and Müller 2012, Section 6.2.4). This surface is used as a reference for physical heights that determine the flow of water. Because these fields extend into the external space of the Earth, both geodetic and geophysical measurements can be made not only on the Earth's surface, but also on aircraft and satellites, thus affording efficiency, data uniformity, and global accessibility for modeling the fields and the sources that generate them.

6.2.1 Potential Function

A fundamental physical law, discovered by Isaac Newton, says that a point mass, a *monopole*, with mass, M, generates a force on another point mass, m, at a distance, d, given by,

$$\boldsymbol{f} = G\frac{Mm}{d^2}\boldsymbol{e}_d. \tag{6.1}$$

The force is a vector along the line joining the point masses where $\boldsymbol{e}_d$ is a unit vector along this line; and, G is a proportionality constant, called Newton's gravitational constant. The gravitational force thus is proportional in magnitude to the product of the masses and inversely proportional to the squared distance between them. An analogous inverse-square-distance law due to Charles Augustin de Coulomb holds for electric charges (Blakely 1995).

The gravitational field generated by M is defined as the force per mass element, m, thus, $\boldsymbol{f}/m$, or as the gravitational acceleration, $\boldsymbol{g} = -\boldsymbol{f}/m$ (negative, since it is an attraction force, and $\boldsymbol{e}_d$ is assumed to originate at the source). This force field is *conservative*, meaning that the work (energy) needed to move a test mass, m, in this field from one point to another does not depend on the path taken. No energy is lost (or gained) by choosing any particular path; that is, the force is not dissipative, as in the case of friction. The corresponding *potential* (potential energy per mass, m) is a scalar function, v_g, defined as the work per unit mass needed to bring it from infinity to a point in the force field (by definition, the potential is zero at infinity). For a differential displacement vector, $\boldsymbol{dr} = (dx\ dy\ dz)^{\mathrm{T}}$ (in Cartesian coordinates), a differential element of potential is then

$$dv_g = -\frac{1}{m}\boldsymbol{f}\cdot\boldsymbol{dr}, \tag{6.2}$$

where the sign is matter of convention (Kellogg 1953, p.53). This implies that

$$\nabla v_{\text{g}} = \boldsymbol{g}, \tag{6.3}$$

where the gradient operator, applied to v_g at its point of evaluation, is

$$\nabla = \left(\frac{\partial}{\partial x} \quad \frac{\partial}{\partial y} \quad \frac{\partial}{\partial z} \right)^{\text{T}}. \tag{6.4}$$

For the field generated by a point mass at (x', y', z'), it is easy to verify with distance in terms of Cartesian coordinate differences,

$$d = \sqrt{(x-x')^2 + (y-y')^2 + (z-z')^2}, \tag{6.5}$$

that the gravitational potential is

$$v_{\text{g}} = G\frac{M}{d}. \tag{6.6}$$

In the International System (SI) of Units, $G \simeq 6.67408 \times 10^{-11}\,\text{m}^3\text{kg}^{-1}\text{s}^{-2}$ (Mohr et al. 2016); and, with mass in kilograms, distance in meters, the units of the gravitational potential are m^2s^{-2}.

Magnetic monopoles do not exist in the theory of magnetostatics; and, the most elemental quantity is a magnetic dipole, conceptually generated by an infinitesimal current loop and having magnetic *dipole moment*, $\boldsymbol{p}$ (a vector with both direction and magnitude). It can be shown that the *scalar* potential of a magnetic dipole depends on the gradient of the inverse distance (Telford et al. 1990, p.66; Blakely 1995, p.75)

$$v_{\text{m}} = -\frac{\mu_0}{4\pi}\boldsymbol{p}\cdot\nabla\left(\frac{1}{d}\right) = \frac{\mu_0}{4\pi}\boldsymbol{p}\cdot\nabla'\left(\frac{1}{d}\right), \tag{6.7}$$

where the gradient operator, ∇, is applied at the evaluation point of v_{m}, and ∇' is the gradient applied to the source point, and where it is easy to show from equation (6.5) that

$$\nabla\left(\frac{1}{d}\right) = -\nabla'\left(\frac{1}{d}\right). \tag{6.8}$$

Conceptually, $\boldsymbol{p}$ is the magnetic moment generated by a current, I, around a loop of infinitesimal area, dA,

$$\boldsymbol{p} = I\boldsymbol{n}_A dA, \tag{6.9}$$

where $\boldsymbol{n}_A$ is a unit vector normal to the area. Equation (6.7) is written in terms of SI units, where the constant, μ_0, in equation (6.7) is the permeability of free space, defined by $\mu_0 = 4\pi \times 10^{-7}\,\text{N} \cdot \text{A}^{-2}$ (Newton per Ampere-squared, using the derived unit, $1\ \text{N} = 1\ \text{kg} \cdot \text{m/s}^2$). The units of the magnetic dipole moment are $\text{A} \cdot \text{m}^2$; hence, the units of the magnetic potential are $\text{N} \cdot \text{A}^{-1}$. Again by convention (Kellogg 1953, p.53), the magnetostatic field is the negative gradient of the potential,

$$\boldsymbol{B} = -\nabla v_{\mathrm{m}}. \tag{6.10}$$

Formally, $\boldsymbol{B}$ is called the magnetic induction due to the magnetic field, $\boldsymbol{H}$, and given at points in free space above the Earth's surface by $\boldsymbol{B} = \mu_0 \boldsymbol{H}$. We may follow the common practice to denote the geomagnetic field with $\boldsymbol{B}$.

In spherical polar coordinates, the distance between the evaluation point, (θ, λ, r), and the source point, (θ', λ', r'), is (Figure 6.1)

$$d = \sqrt{r^2 + r'^2 - 2rr'\cos\psi}, \tag{6.11}$$

where $\cos\psi$ is given by equation (2.196). For $r > r'$ the reciprocal distance can be expanded in a series of Legendre polynomials, also called the *Coulomb expansion* (Cushing 1975, p.155), or a binomial series in which the Legendre polynomials are the coefficients (Arfken 1970),

$$\frac{1}{d} = \frac{1}{r}\sum_{n=0}^{\infty}\left(\frac{r'}{r}\right)^n P_n(\cos\psi), \tag{6.12}$$

which converges uniformly since $|P_n(\cos\psi)| \le 1$. If $r = r'$, then the Coulomb expansion also converges to the inverse distance, unless $\psi = 0$, when $1/d \to \infty$. Substituting the addition formula for spherical harmonics, equation (2.254), this becomes

$$\frac{1}{d} = \frac{1}{r}\sum_{n=0}^{\infty}\left(\frac{r'}{r}\right)^n \frac{1}{2n+1}\sum_{m=-n}^{n}\bar{Y}_{n,m}(\theta',\lambda')\bar{Y}_{n,m}(\theta,\lambda). \tag{6.13}$$

Thus, for fixed r and r', the Fourier-Legendre transform of the potential, equation (6.6), on the sphere of radius, r, due to a monopole source on the sphere of radius, r', is proportional to $r'^n\bar{Y}_{n,m}(\theta',\lambda')/\left(r^{n+1}(2n+1)\right)$, in view of equation (2.221). If the monopole is at the origin ($r' = 0$), then the Fourier-Legendre transform is zero except at degree, $n = 0$.

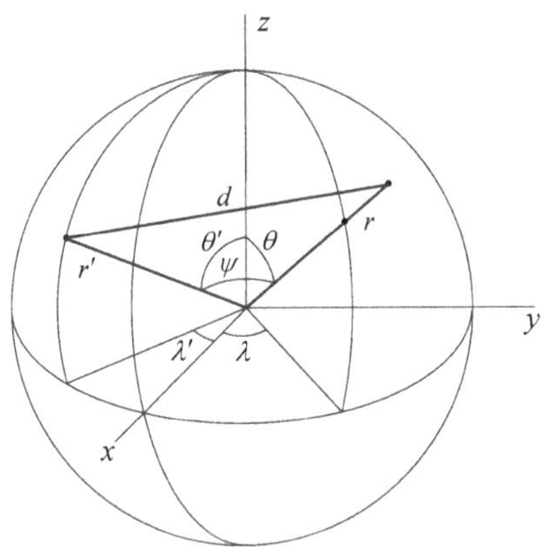

Figure 6.1: Spherical geometry for the distance, d, between points (θ', λ', r') and (θ, λ, r).

For a dipole at (θ', λ', r') and aligned in the radial direction, i.e., $\boldsymbol{p} = 0 \cdot \boldsymbol{e}_{\theta'} + 0 \cdot \boldsymbol{e}_{\lambda'} + p_{r'} \boldsymbol{e}_{r'}$ using unit vectors, $\boldsymbol{e}_{\theta'}$, $\boldsymbol{e}_{\lambda'}$, $\boldsymbol{e}_{r'}$ (Section 2.6.4), there is $\boldsymbol{p} \cdot \boldsymbol{\nabla}' = p_{r'} \partial / \partial r'$, where $p_{r'}$ is the radial component of the dipole moment. The radial derivative of the reciprocal distance with respect to r' is

$$\frac{\partial}{\partial r'}\left(\frac{1}{d}\right) = \frac{1}{r^2} \sum_{n=1}^{\infty} n \left(\frac{r'}{r}\right)^{n-1} P_n\left(\cos\psi\right). \tag{6.14}$$

Therefore, the Fourier-Legendre spectrum of this dipole, again with the addition theorem, is proportional to $nr'^{n-1}\bar{Y}_{n,m}(\theta', \lambda')/r^{n+1}$, $n \geq 1$. For a dipole at the origin ($r' = 0$) the spectrum is zero except for degree, $n = 1$, and thus the potential consists only of first-degree harmonics, $\bar{Y}_{1,m}(\theta, \lambda)$. A similar result holds for any other alignment of the dipole at the origin since by equation (2.247) any rotation of coordinates does not introduce harmonics of any other degree.

Instead of a single source point, consider many points, or, more generally, a continuous body of sources. In this case, it is convenient to define a density function. For example, a differential mass element, dM, is the mass-density times a differential spatial (e.g., volume) element of the body, db,

$$\rho = \frac{dM}{db}. \tag{6.15}$$

Similarly, a differential dipole element, $d\boldsymbol{p}$, is a density times a differential spatial element. In the case of the magnetostatic field, the density is called the dipole moment density, or dipole magnetization intensity, or simply the magnetization (Telford et al. 1990, p.64),

$$\eta = \frac{d\boldsymbol{p}}{db}. \tag{6.16}$$

By the superposition principle, the potential generated by many sources, including a continuum of sources, is the sum or integral of the potentials of the point sources. For a volumetric mass body, b, the gravitational potential is

$$v_{\mathrm{g}}\left(\theta, \lambda, r\right) = G \iiint_b \frac{\rho\left(\theta', \lambda', r'\right)}{d} db, \tag{6.17}$$

where $db = r'^2 dr' d\Omega' = r'^2 dr' \sin\theta' d\theta' d\lambda'$. Similarly, the magnetic potential for a magnetized body is

$$v_{\mathrm{m}}\left(\theta, \lambda, r\right) = -\frac{\mu_0}{4\pi} \iiint_b \eta\left(\theta', \lambda', r'\right) \cdot \nabla\left(\frac{1}{d}\right) db, \tag{6.18}$$

where, again, the gradient operator applies to the evaluation point, (r, θ, λ).

It is noted that the magnetic scalar potential is defined here under the assumption that there are no source currents and the electric displacement flux density is neglected. According to one of Maxwell's differential equation (Jackson 1999), one has,

$$\boldsymbol{\nabla} \times \boldsymbol{B} = \mu_0 \boldsymbol{J} + \mu_0 \varepsilon_0 \frac{\partial \boldsymbol{E}}{\partial t}, \tag{6.19}$$

where $\boldsymbol{B}$ is the magnetic field (also magnetic induction, or magnetic flux density), $\boldsymbol{J}$ is the current density, $\boldsymbol{E}$ is the electric field (also electric flux density), the gradient operator is with respect to space, t is time, and the constant, ε_0, is the permittivity of free space. If $\boldsymbol{J} = 0$ and the electric displacement flux density is neglected $(\partial(\varepsilon_0 \boldsymbol{E})/\partial t \approx 0)$, then

$$\boldsymbol{\nabla} \times \boldsymbol{B} = 0, \tag{6.20}$$

which implies that there exists a scalar function, v_{m}, such that its gradient is the magnetic field, as in equation (6.10).

It is straightforward to show (Exercise 6.1) that for points, (x, y, z), in free space,

$$\nabla^2 \left(\frac{1}{d} \right) = 0, \tag{6.21}$$

where $\nabla^2 = \boldsymbol{\nabla} \cdot \boldsymbol{\nabla}$ is the *Laplace differential operator* applied at the point, (x, y, z) and d is given by equation (6.5). Therefore, both the gravitational and the magnetic potential satisfy *Laplace's equation* in free space,

$$\nabla^2 v = 0. \tag{6.22}$$

Solutions to this differential equation are called harmonic functions. It is assumed here that all sources are either on or within a closed boundary (e.g., a sphere) that also contains the coordinate origin. If the sources are piecewise continuous and bounded, then v is bounded and continuous everywhere (Kellogg 1953, pp.151, 160), including the boundary; and, therefore, it is absolutely integrable on any closed surface that contains the sources. Moreover, the potential approaches zero at infinity, e.g., in spherical polar coordinates,

$$\lim_{r \to \infty} v(\theta, \lambda, r) = 0. \tag{6.23}$$

6.2.2 Fourier Spectrum of the Potential

In the local Cartesian domain (Figure 1.1), it is assumed that all sources are on or below the x_1, x_2-plane, that is, $x'_3 \leq 0$ (here, with some exceptional cases, a piecewise continuous three-dimensional distribution of sources is assumed, although Section 6.4.2 also considers infinitesimal layers of continuous density). Formally, the gravitational potential is not always absolutely integrable on the x_1, x_2 plane, in which case its transform does not exist. For example, with (in local coordinates)

$$d = \sqrt{\left(x_1 - x_1'\right)^2 + \left(x_2 - x_2'\right)^2 + \left(x_3 - x_3'\right)^2} \tag{6.24}$$

and equation (6.6), the potential on the x_1, x_2-plane at $x_3 = 0$, due to a point source with mass, M, located at $(0,0,-|x_3'|)$, $x'_3 < 0$, is

$$v_g\left(x_1, x_2, 0\right) = \frac{GM}{\sqrt{x_1^2 + x_2^2 + x_3'^2}}, \tag{6.25}$$

By changing to polar coordinates as in Figure 2.12, it is easy to show that

$$\int_{-\infty}^{\infty}\int_{-\infty}^{\infty} \left|v_g\left(x_1, x_2, 0\right)\right| dx_1 dx_2 = GM \int_{s=0}^{\infty}\int_{\omega=0}^{2\pi} \frac{s ds d\omega}{\sqrt{s^2 + x_3'^2}} = 2\pi GM \left.\sqrt{s^2 + x_3'^2}\right|_{s=0}^{\infty} \to \infty, \tag{6.26}$$

where s is defined by equation (2.175). Nonetheless, one may calculate the Fourier transform, in this case the Hankel transform, equation (2.180) (Gradshteyn and Ryzhik 1980, p.682; see also equation (2.337)),

$$2\pi GM \int_0^{\infty} \frac{1}{\sqrt{s^2 + x_3'^2}} s J_0\left(2\pi s \bar{f}\right) ds = \frac{GM}{\bar{f}} e^{-2\pi|x_3'|\bar{f}}, \quad \bar{f} \neq 0, \tag{6.27}$$

where $\bar{f}$ is given by equation (2.178). Thus, the Hankel transform exists for all frequencies except at the origin, $(f_1, f_2) = (0.0)$. A similar result holds for $x_3 > 0$. The inverse Hankel transform, equation (2.181),

$$2\pi GM \int_0^{\infty} e^{-2\pi|x_3'|\bar{f}} J_0\left(2\pi \bar{f} s\right) d\bar{f} = \frac{GM}{\sqrt{x_3'^2 + s^2}}, \tag{6.28}$$

however, converges to v_g everywhere on or above the plane, $x_3 = 0$ (Gradshteyn and Ryzhik 1980, p.707). For an infinite horizontal plate of thickness, c, constant mass-density, ρ_0, the potential on or above the plate does not exist, since from equation (6.17), transformed to polar coordinates in the x_1, x_2-plane (equation (2.174)),

$$\begin{aligned} v_g\left(x_1, x_2, x_3\right) &= G\rho_0 \int_{\omega=0}^{2\pi}\int_{x_3'=-c}^{0}\int_{s=0}^{\infty} \frac{s ds}{\sqrt{s^2 + \left(x_3 - x_3'\right)^2}} dx_3' d\omega \\ &= \pi G\rho_0 \lim_{s\to\infty} \left(\begin{aligned} & s^2 \ln \frac{x_3 + c + \sqrt{s^2 + \left(x_3 + c\right)^2}}{x_3 + \sqrt{s^2 + x_3^2}} + x_3^2 - \left(x_3 + c\right)^2 \\ & + \left(x_3 + c\right)\sqrt{s^2 + \left(x_3 + c\right)^2} - x_3\sqrt{s^2 + x_3^2} \end{aligned} \right) \end{aligned} \tag{6.29}$$

as shown in (Hofmann-Wellenhof and Moritz 2005, p.132); the limit is infinite for any $x_3 \geq 0$. Thus, to consider the Fourier transform of the gravitational potential in Cartesian space, one must at least remove its mean value.

The magnetic potential, on the other hand, due to a single, vertically oriented dipole at $(0,0,-|x_3'|)$, $x_3' < 0$, with magnetic moment, p, from equation (6.7), is

$$v_{\mathrm{m}}\left(x_1,x_2,x_3\right) = \frac{\mu_0}{4\pi} p \frac{x_3 - x_3'}{\left(x_1^2 + x_2^2 + \left(x_3 - x_3'\right)^2\right)^{3/2}}, \tag{6.30}$$

which is absolutely integrable on any x_1, x_2-plane ($x_3 \geq 0$); and, its Fourier (Hankel) transform, equation (2.334), is

$$\mathcal{F}\left(v_{\mathrm{m}}\left(x_1,x_2,x_3\right)\right) = \frac{\mu_0}{2} p e^{-2\pi\left(x_3 - x_3'\right)\bar{f}}, \tag{6.31}$$

which holds for all frequencies. However, for an infinite plate of constant (vertical) dipole moment density, η_0, the potential is a constant. Indeed, the corresponding scalar magnetic potential, in view of equations (6.24), (6.17), and (6.18), is proportional to the vertical gravitational attraction of the constant-mass-density plate,

$$v_{\mathrm{m}}\left(x_1,x_2,x_3\right) = \frac{\mu_0\eta_0}{4\pi}\frac{\partial}{\partial x_3}\iiint_b \frac{1}{d}\,db \sim \frac{\partial}{\partial x_3} v_{\mathrm{g}}\left(x_1,x_2,x_3\right). \tag{6.32}$$

And, from equation (6.29), and $x_3 \geq 0$,

$$\begin{aligned} \frac{\partial}{\partial x_3} v_{\mathrm{g}} &= -2\pi G\rho_0 \lim_{s\to\infty}\left(c + x_3\sqrt{s^2 + x_3^2} - \left(x_3 + c\right)\sqrt{s^2 + \left(x_3 + c\right)^2}\right) \\ &= -2\pi G\rho_0 c \end{aligned} \tag{6.33}$$

Thus, also in this case the magnetic potential does not have the usual Fourier transform (it is proportional to the Dirac delta function, Section 2.4.3); and, in general, its average should first be removed.

For the following it is assumed that the Fourier integrals of potentials on any plane, $x_3 \geq 0$, exist with suitable adjustment or approximation (e.g., windowing, Section 3.6), except possibly at the origin of the spectral domain. As in the examples above, the variable, x_3, is viewed as a parameter of the Fourier transform that is restricted to operate in the first two dimensions. A three-dimensional Fourier transform would imply independent basis functions in three dimensions, but solutions to Laplace's equation (6.22) for the potential establish a dependence of the potential in the third dimension on its values in the x_1, x_2-plane, as shown below.

Thus, referring to equations (2.149) and (2.150), let V be the two-dimensional Fourier transform of v with respect to the coordinates, x_1, x_2, for some fixed x_3 in free space,

$$V(f_1, f_2; x_3) = \int_{-\infty}^{\infty}\int_{-\infty}^{\infty} v(x_1, x_2, x_3) e^{-i2\pi(f_1x_1 + f_2x_2)} dx_1 dx_2; \tag{6.34}$$

and, the inverse transform is given by

$$v(x_1, x_2, x_3) = \int_{-\infty}^{\infty}\int_{-\infty}^{\infty} V(f_1, f_2; x_3) e^{i2\pi(f_1x_1 + f_2x_2)} df_1 df_2. \tag{6.35}$$

Applying Laplace's equation (6.22) in Cartesian coordinates,

$$\nabla^2 v(x_1, x_2, x_3) = \left(\frac{\partial^2}{\partial x_1^2} + \frac{\partial^2}{\partial x_2^2} + \frac{\partial^2}{\partial x_3^2}\right) v(x_1, x_2, x_3) = 0, \tag{6.36}$$

yields

$$0 = \int_{-\infty}^{\infty}\int_{-\infty}^{\infty} \left(\frac{\partial^2}{\partial x_3^2} V(f_1, f_2; x_3) - (2\pi)^2 (f_1^2 + f_2^2) V(f_1, f_2; x_3)\right) e^{i2\pi(f_1x_1 + f_2x_2)} df_1 df_2 \tag{6.37}$$

for all $x_3 > 0$. This holds for arbitrary points, (x_1, x_2), which implies that the integrand must equal zero for all frequencies, giving a differential equation for V in terms of x_3, now interpreted as an independent variable,

$$\frac{\partial^2}{\partial x_3^2} V(f_1, f_2; x_3) - (2\pi)^2 \bar{f}^2 V(f_1, f_2; x_3) = 0. \tag{6.38}$$

The solution to equation (6.38) for any (f_1, f_2) is given by (as easily verified using back-substitution),

$$V(f_1, f_2; x_3) = C_1 e^{-2\pi\bar{f}x_3} + C_2 e^{2\pi\bar{f}x_3}, \tag{6.39}$$

where C_1 and C_2 are constants. The only possibility for the latter is $C_2 = 0$; otherwise $v \rightarrow \infty$ as $x_3 \rightarrow \infty$, contradicting condition (6.23). Since v is continuous everywhere and has a Fourier transform (with the stipulation of a restricted spectral domain for the gravitational potential) on the plane, $x_3 = 0$, the transform as a function of x_3 must also be continuous, and equation (6.39) holds for $x_3 = 0$, as well. Thus, let

$$V(f_1, f_2; 0) = V_0(f_1, f_2); \tag{6.40}$$

and, the solution (6.39) is given by

$$V(f_1, f_2; x_3) = V_0(f_1, f_2) e^{-2\pi\bar{f}x_3}. \tag{6.41}$$

In general, the two-dimensional spectra of the potential, v, at two levels, $x_3^{(1)}$ and $x_3^{(2)}$, are related according to

$$V\left(f_1,f_2;x_3^{(2)}\right)=V\left(f_1,f_2;x_3^{(1)}\right)e^{-2\pi\bar{f}\left(x_3^{(2)}-x_3^{(1)}\right)}. \tag{6.42}$$

The inverse Fourier transform (6.35) now becomes

$$v(x_1,x_2,x_3)=\int_{-\infty}^{\infty}\int_{-\infty}^{\infty}V_0(f_1,f_2)e^{-2\pi\bar{f}x_3}e^{i2\pi(f_1x_1+f_2x_2)}df_1df_2\,. \tag{6.43}$$

Then, for any integer $q\geq 0$,

$$\frac{\partial^q}{\partial x_3^q}v(x_1,x_2,x_3)=\int_{-\infty}^{\infty}\int_{-\infty}^{\infty}\left(-2\pi\bar{f}\right)^q V_0(f_1,f_2)e^{-2\pi\bar{f}x_3}e^{i2\pi(f_1x_1+f_2x_2)}df_1df_2, \tag{6.44}$$

which implies that the Fourier transforms of the vertical derivatives of v are given by

$$\mathcal{F}\left(\frac{\partial^q}{\partial x_3^q}v(x_1,x_2,x_3)\right)=\left(-2\pi\bar{f}\right)^q V(f_1,f_2;x_3). \tag{6.45}$$

With the additional Fourier transforms of horizontal derivatives, equations (2.152), one has the following general Fourier transform pairs of the potential for $p_1, p_2, q\geq 0$,

$$\frac{\partial^{p_1}}{\partial x_1^{p_1}}\frac{\partial^{p_2}}{\partial x_2^{p_2}}\frac{\partial^q}{\partial x_3^q}v(x_1,x_2,x_3)\leftrightarrow(i2\pi f_1)^{p_1}(i2\pi f_2)^{p_2}\left(-2\pi\bar{f}\right)^q V(f_1,f_2;x_3), \tag{6.46}$$

where, again, it is assumed that the Fourier transform, as well as all pertinent derivatives, of the potential exist. Because derivatives of the potential remove its constant part, their Fourier transforms exist even if the transform of the potential does not exist at zero frequency. For example, applying equation (6.45) with $q = 1$ to the Fourier transform of the potential of a mass monopole, equation (6.27), yields the Fourier transform of the gravitational attraction that exists for all frequencies.

Knowing the spectrum of any derivative of the potential at one level (constant x_3) easily generates its spectrum at any other level, $x_3 > 0$, according to equations (6.42) and (6.46). Thus, the analytic *continuation* (downward or upward continuation) of the potential or its derivatives is readily accomplished in the spectral domain. However, *downward* continuation of actual data, for example from aircraft altitude ($x_3^{(1)} > 0$) to ground level ($x_3^{(2)} = 0$), amplifies also measurement errors, especially at high frequencies according to the exponential in equation (6.42), thus causing instabilities in the continued spectrum. These are handled with special regularization (smoothing) techniques; see, e.g., Tikhonov and Arsenin (1977), and modern texts on least-squares estimation (e.g., Aster et al. 2005).

Formally the law of propagation of correlations, equation (5.90), extends to the derivatives of the potential with respect to the third dimension, x_3, simply by applying partial derivatives to equation (5.86),

$$\phi_{\frac{\partial^{q_1} v}{\partial x_3^{q_1}}, \frac{\partial^{q_2} v}{\partial x_3'^{q_2}}}\left(x_1, x_2; x_3, x_3'\right) = \frac{\partial^{q_1+q_2}}{\partial x_3^{q_1} \partial x_3'^{q_2}} \phi_{v,v}\left(x_1, x_2; x_3, x_3'\right), \tag{6.47}$$

assuming that such derivatives of the potential exist on the corresponding x_1, x_2-planes at x_3 and x_3'. With equation (6.46), the propagation law in the spectral domain, equation (5.96), is extended to

$$\Phi_{\frac{\partial^{p_1+p_2+q_1} v}{\partial x_1^{p_1} \partial x_2^{p_2} \partial x_3^{q_1}}, \frac{\partial^{p_1'+p_2'+q_2} v}{\partial x_1^{p_1'} \partial x_2^{p_2'} \partial x_3'^{q_2}}}\left(f_1, f_2; x_3, x_3'\right)$$
$$= \left(-1\right)^{q_1+q_2} i^{p_1'-p_1+p_2'-p_2} \left(2\pi f_1\right)^{p_1+p_1'} \left(2\pi f_2\right)^{p_2+p_2'} \left(2\pi \bar{f}\right)^{q_1+q_2} \Phi_{v,v}\left(f_1, f_2; x_3, x_3'\right) \tag{6.48}$$

6.2.3 Fourier-Legendre Spectrum of the Potential

The Laplace operator in spherical polar coordinates is given by

$$\nabla^2 = \frac{1}{r^2}\frac{\partial}{\partial r}\left(r^2 \frac{\partial}{\partial r}\right) + \frac{1}{r^2}\frac{1}{\sin\theta}\frac{\partial}{\partial\theta}\left(\sin\theta \frac{\partial}{\partial\theta}\right) + \frac{1}{r^2 \sin^2\theta}\frac{\partial^2}{\partial\lambda^2}, \tag{6.49}$$

and it is readily verified (Exercise 6.2) from equations (2.240) and (2.241) that

$$\nabla^2\left(r^n \bar{Y}_{n,m}\left(\theta,\lambda\right)\right) = \left(\frac{1}{r^2}\frac{\partial}{\partial r}\left(r^2\frac{\partial}{\partial r}\right) - \frac{n(n+1)}{r^2}\right)\left(r^n \bar{Y}_{n,m}\left(\theta,\lambda\right)\right) = 0, \tag{6.50}$$

as well as,

$$\nabla^2\left(\frac{1}{r^{n+1}}\bar{Y}_{n,m}\left(\theta,\lambda\right)\right) = \left(\frac{1}{r^2}\frac{\partial}{\partial r}\left(r^2\frac{\partial}{\partial r}\right) - \frac{n(n+1)}{r^2}\right)\left(\frac{1}{r^{n+1}}\bar{Y}_{n,m}\left(\theta,\lambda\right)\right) = 0. \tag{6.51}$$

The solutions to Laplace's equations in spherical coordinates are the *solid spherical harmonic functions*, $r^n\bar{Y}_{n,m}(\theta, \lambda)$ and $r^{-(n+1)}\bar{Y}_{n,m}(\theta, \lambda)$. For points in the free space outside a bounding sphere of radius, R (the *Brillouin* sphere), only the latter satisfy the condition (6.23). The complete solution in the exterior space is then a linear combination of all these solid spherical harmonics,

$$v\left(\theta,\lambda,r\right) = \sum_{n=0}^{\infty}\sum_{m=-n}^{n}\left(\frac{R}{r}\right)^{n+1} V_{n,m}\bar{Y}_{n,m}\left(\theta,\lambda\right), \quad r \ge R, \tag{6.52}$$

where, by introducing the ratio, R/r, the coefficients, $V_{n,m}$, have the same units as v and are given by the potential on the sphere of radius, R, according to equation (2.220),

$$V_{n,m} = \frac{1}{4\pi}\iint_{\Omega} v\left(\theta,\lambda,R\right)\bar{Y}_{n,m}\left(\theta,\lambda\right) d\Omega. \tag{6.53}$$

Thus, the set, $\{V_{n,m}\}$, is the Fourier-Legendre spectrum of $v(\theta, \lambda, R)$ on the bounding sphere, $r = R$; and, the spherical harmonic series converges to the potential on this sphere. Likewise, for $R_1 > R$, the set, $\{V_{n,m}^{(R_1)}\} = \{(R/R_1)^{n+1} V_{n,m}\}$, is the Fourier-Legendre spectrum of the function, $v(\theta, \lambda, R_1)$, on the sphere of radius, R_1. In general, the Fourier-Legendre spectra of harmonic functions on spheres with radii, R_1 and R_2, are related by

$$V_{n,m}^{(R_2)} = \left(\frac{R_1}{R_2}\right)^{n+1} V_{n,m}^{(R_1)}, \quad R_1 \geq R, \quad R_2 \geq R. \tag{6.54}$$

Unlike the Cartesian case, the potential exists, is continuous, and has a Fourier-Legendre spectrum defined at all wave numbers for bounded (spatially and functionally), piecewise continuous density distributions. The potential is analytic and hence all its derivatives exist if $r > R$. From equation (6.52) it is easily shown that the Fourier-Legendre transform pair of the qth radial derivative is

$$\left.\frac{\partial^q}{\partial r^q} v(\theta, \lambda, r)\right|_{r=R_1} \leftrightarrow (-1)^q \frac{(n+1)\cdots(n+q)}{R_1^q} V_{n,m}^{(R_1)}, \quad R_1 > R, \quad q \geq 1. \tag{6.55}$$

Equation (6.52) or (6.54) also shows that little can be gained by defining a spectral transform in the radial direction. Since the potential, constrained by Laplace's equation (6.22), is determined completely in free space by its values on the boundary, there is no independent radial structure.

Analytic continuation of the potential (or its derivatives) in the radial or vertical directions, according to equations (6.42) and (6.54) is equivalent to high-pass or low-pass filtering, depending on whether in the downward or upward directions, where the frequency response of the filter is $e^{-2\pi \bar{f}\left(x_3^{(2)} - x_3^{(1)}\right)}$, respectively, $(R_1/R_2)^{n+1}$. Figure 6.2 compares the upward continuation response for the potential to the Gaussian smoothing response.

The radial (vertical) derivatives, in contrast, are high-pass filters, whose frequency response, shown in (6.55), counteracts the upward continuation. For example, the vertical gradient of the gravitational attraction is particularly revealing of sharp contrasts in Earth's mass-density structure. And, although it also attenuates away from a source body, this is more than compensated by the increased sensitivity to the shorter wavelengths. Indeed, the geopotential field spectrum scales by $\sim n^q$ due to q radial derivatives of the field; while it scales by $(R_1/R_2)^{n+1+q}$ due to upward continuation. Specifically, suppose $R_1 = R$ and $R_2 = R + h$, where h is the altitude above Earth's mean radius, R. By combining both upward continuation and radial differentiation, there is no attenuation in the (scaled) potential spectrum, $V_{n,m}/R^q$ (the normalizing factor, R^q, accounts for units), if

$$(n+1)\cdots(n+q)\frac{R^{n+1+q}}{(R+h)^{n+1+q}} \geq 1, \tag{6.56}$$

or,

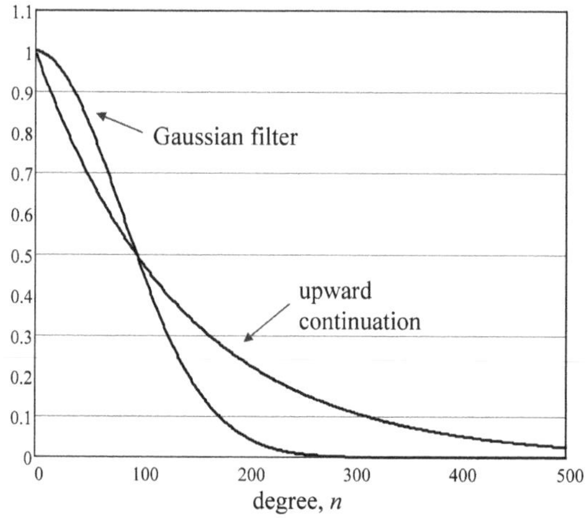

Figure 6.2: Comparison of the frequency responses of the Gaussian filter (equation (3.92), $\beta_s = 1/(1-\cos\psi_0)$, $\psi_0 = 1°$) and upward continuation (equation (6.54), $R_1/R_2 = 0.99267685$), normalized to unity at degree, $n = 0$.

$$h \le R\left(\left((n+1)\cdots(n+q)\right)^{\frac{1}{n+1+q}} - 1\right). \tag{6.57}$$

Similarly, in Cartesian space, by combining the upward continuation frequency response, $e^{-2\pi\bar{f}h}$ (equation (6.41)) and the (magnitude of the) differentiation frequency response, $(2\pi\bar{f})^q$ (equation (6.45)), there is no attenuation of the potential spectrum if

$$\left(2\pi\bar{f}\right)^q e^{-2\pi\bar{f}h} \ge 1. \tag{6.58}$$

In order to take care of units on the left side, define $\bar{f} = n/(2\pi R)$ (equation (2.331)), and, again, incorporate the factor, R^q, in the spectrum of the potential. Then, no attenuation occurs if

$$h \le R\frac{q}{n}\ln n. \tag{6.59}$$

As seen in Figure 6.3, for all altitudes, the spectrum generally is always greater with gravity gradiometry ($q = 2$) than with gravimetry ($q = 1$). Thus, in theory, *in situ* gravity gradiometry offers superior resolution compared to gravimetry on satellites and airborne systems. For example, at altitude, $h = 200$ km, a system that measures gravitation attenuates spectral components of the gravitational potential for degrees, $n \ge 163$, while a gradiometric system attenuates components only after degree, $n = 382$. Of course, many other factors determine the choice of one system over another, such as the precision and resolution of the instruments and the speed of the vehicle.

The spherical spectral connections among the potential and its upward continuation and differentiation (or, inversely, its integration) have been systematically developed

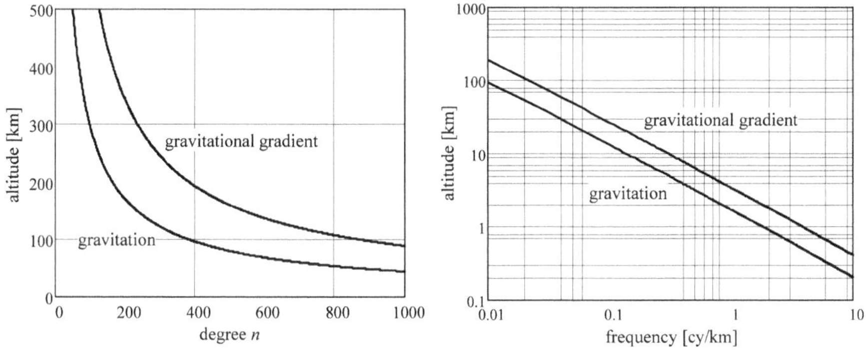

Figure 6.3: The area *under* each curve represents the combinations of altitude and maximum resolution for which there is no attenuation of the potential spectrum compared to its magnitude at ground level as recoverable either from the first or the second radial derivative of the gravitational potential. The left panel depicts the bounds for the spherical spectrum and the right panel for the equivalent Cartesian spectrum.

in physical geodesy by Rummel and Van Gelderen (1995), extending earlier work by Meissl (1971) who elucidated the spectral relationships among the first derivatives of the potential (Figure 6.4). Now also called the *Meissl Scheme*, further extensions were made to higher derivatives by Rummel (1997, p.377) and Van Gelderen and Rummel (2001). The latter are approximate for the horizontal derivatives and are based on the spectral relationships in Section (2.6.4); see also (Grafarend 2001). Each operation, upward continuation and radial differentiation, is represented by its frequency response according to equations (6.54) and (6.55). Corresponding transfer (i.e., filter) functions are derived in Section 6.2.4 using appropriate convolution theorems.

The law of propagation of isotropic correlations in the space and frequency domains, given by equation (5.149) and the approximate relationships (5.155) and (5.161), may be extended easily for the vertical derivatives using this scheme. For example, from the basic definition of the (isotropic) correlation function, equation (5.130), it is a straightforward derivation to propagate the correlation function of the potential to the cross-correlation between the first radial derivatives on the spheres of radii, $r = R_1 \geq R$ and $r = R_2 \geq R$,

$$\begin{aligned}
\phi_{\left.\frac{\partial v}{\partial r}\right|_{r=R_1},\left.\frac{\partial v}{\partial r}\right|_{r=R_2}}\left(\psi; R_1, R_2\right) &= \frac{1}{8\pi^2}\int_{\zeta=0}^{2\pi}\iint_{\Omega}\left.\frac{\partial}{\partial r}v(\theta,\lambda,r)\right|_{r=R_1}\left.\frac{\partial}{\partial r'}v(\theta',\lambda',r')\right|_{r'=R_2} d\Omega\, d\zeta \\
&= \left.\frac{1}{8\pi^2}\frac{\partial^2}{\partial r \partial r'}\int_{\zeta=0}^{2\pi}\iint_{\Omega} v(\theta,\lambda,r)\,v(\theta',\lambda',r')\, d\Omega\, d\zeta\right|_{r=R_1, r'=R_2} \qquad (6.60) \\
&= \left.\frac{\partial^2}{\partial r \partial r'}\phi_{v,v}\left(\psi; r, r'\right)\right|_{r=R_1, r'=R_2}
\end{aligned}$$

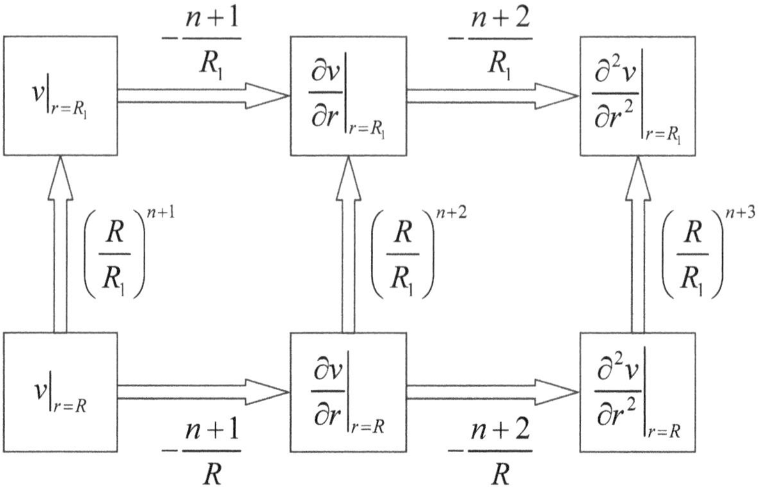

Figure 6.4: The Meissl Scheme connecting upward continuation and radial differentiation. The vertical arrows indicate the Legendre transforms of the isotropic transfer (filter) functions for upward continuation and the horizontal arrows indicate radial differentiation in the spectral domain. Reversing the arrows inverts the Legendre transforms yielding transforms of the inverse operations, downward continuation and the radial anti-derivative.

since the integral does not depend on the radial coordinate. Just as easy is the propagation in the spectral domain, where in this case,

$$\left(\Phi_{\frac{\partial v}{\partial r}\big|_{r=R_1},\frac{\partial v}{\partial r}\big|_{r=R_2}}\right)_n = \frac{(n+1)^2}{R_1 R_2}\left(\Phi_{v|_{r=R_1},v|_{r=R_2}}\right)_n, \tag{6.61}$$

and similarly for higher-order derivatives using equation (6.55).

6.2.4 Green's Functions and Their Inverses from Spectral Relationships

Many problems in physical geodesy involve integrals of the gravitational potential or its derivatives. The most famous are the Stokes and Pizzetti formulas that combine gravity anomalies on the geoid with Stokes's function to yield the geoid undulation and disturbing potential, respectively. There are numerous others, including the Vening-Meinesz integrals, the Molodensky/Moritz operators, the Poisson upward continuation integral, and corresponding inverse operators, as well as similar integrals that involve the topographic height. Each of these is recognized as or can be reduced to a convolution of two functions, g and h, on the sphere, given by equation (3.44). As such, they are filters that transform an input into an output with particular spectral properties. Adhering to previous terminology, the input, g, is the data function, and the filter function, h, is called the kernel of the convolution. In most cases, h depends only on the spherical distance between input and output points, and the convolution has the form given by equation (3.58). The kernel in the present case is also called a Green's function, deriving from the fact that the potential, and hence its derivatives, satisfy a partial differential equation with boundary conditions. Some well known and also not so well known Green's functions are derived from the spectral relationships between the input and output on the basis of the convolution theorem. For each particular convolution, however, the more common nomenclature in geodesy is used here, as well.

In order to put the various convolutions in physical geodesy in proper context, some preliminary definitions and concepts are required. The notation is also changed from previous chapters in order to conform to common convention. Details of the following may be found in (Hofmann-Wellenhof and Moritz 2005, Sansò and Sideris 2013). The gravitational potential, V, of the Earth is generated by its total mass-density distribution. It is customary to exclude the density of the atmosphere so that Laplace's equation (6.22) holds above the Earth's surface. For applications on the Earth's surface it is useful and convenient to include a (non-Newtonian) potential function, ϕ, whose gradient is the centrifugal acceleration due to Earth's rotation. The sum, $W = V + \phi$, is then known, in geodetic terminology, as the *gravity* potential, distinct from the gravitational potential due strictly to mass attraction. It is further advantageous to define a relatively simple reference potential, or *normal potential*, that accounts for the bulk of the gravity potential. The normal gravity potential, U, is defined as a gravity potential associated with a best-fitting ellipsoid, the *normal ellipsoid*, which rotates with the Earth and is also a surface of constant potential in the normal field. The difference between the actual and the normal gravity potentials is known as the *disturbing potential*, $T = W - U$; it thus excludes the centrifugal potential and satisfies Laplace's equation. The normal gravity potential accounts for approximately 99.9995% of the total gravity potential.

The gradient of the potential is an acceleration, equation (6.3), called simply gravity or gravitation, depending on whether the centrifugal acceleration is included. Normal gravity, $\boldsymbol{\gamma} = \nabla U$, comprises generally about 99.995% of the total gravity, although the difference in magnitudes, the *gravity disturbance*, $\delta g = |\mathbf{g}| - |\boldsymbol{\gamma}|$, can be as large as several parts in 10^4. A special kind of difference, called the *gravity anomaly* (also, *free-air gravity anomaly*), $\Delta g = |\mathbf{g}_P| - |\boldsymbol{\gamma}_Q|$, is defined as the difference between the magnitude of gravity at a point, P, and the magnitude of normal gravity at a corresponding point, Q, where $W_P = U_Q$, and P and Q are on the same perpendicular to the normal ellipsoid (Figure 6.5).

The surface of constant gravity potential, $W = W_0$, that closely approximates mean sea level is known as the *geoid*. The separation between the geoid and the ellipsoid is known as the *geoid undulation*, N, or also the *geoid height*. A simple Taylor expansion of the normal gravity potential along the ellipsoid perpendicular yields the following important relationship, known as *Bruns's formula* that relates the geometry of Earth's shape to a physical quantity, the disturbing potential,

$$N = \frac{T_{P_0}}{\gamma_{Q_0}}, \tag{6.62}$$

where $\gamma_{Q_0} = |\boldsymbol{\gamma}_{Q_0}|$ on the ellipsoid and the constant normal potential on the ellipsoid is defined to be $U_0 = W_0$.

The gravity anomaly is the gravity disturbance corrected for the evaluation of normal gravity at Q instead of P. This correction is $N\partial\gamma/\partial h = (\partial\gamma/\partial h)(T/\gamma)$, where h is height along the ellipsoid perpendicular. In spherical approximation and neglecting second-order terms, the gravity disturbance is given by (Hofmann-Wellenhof and Moritz 2005)

$$\delta g = -\frac{\partial T}{\partial r}, \tag{6.63}$$

where the sign is defined by convention (gravity considered as positive downward in opposition to the radial, or outward positive, derivative of the potential). The approximation, $\gamma \simeq GM/r^2$, where GM is the product of Newton's gravitational constant and Earth's total mass, yields $\partial\gamma/\partial h \simeq -2\gamma/r$; and, the gravity anomaly becomes

$$\Delta g = -\frac{\partial T}{\partial r} - \frac{2}{r}T, \tag{6.64}$$

which is an approximate form of the *Fundamental Equation of Physical Geodesy*.

The slope of the geoid with respect to the ellipsoid is equal to the angle between the corresponding perpendiculars to these surfaces (Figure 6.5). Also defined at arbitrary points, this angle is known as the *deflection of the vertical*, that is, the deflection of the plumb line (perpendicular to the surface, W = constant) relative to the perpendicular to the normal ellipsoid. The deflection angle has components, η, ξ, respectively, in the west and south directions, given with the same approximations as above by (ibid.)

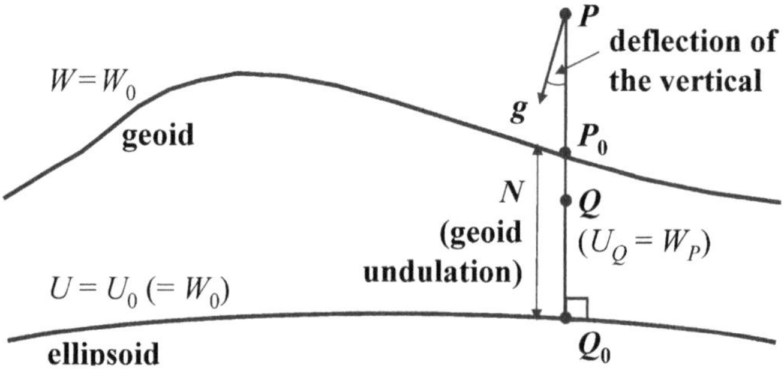

Figure 6.5: The geoid is an equipotential surface of the gravity field separated from the normal ellipsoid by the geoid undulation, N. The gravity anomaly at P is defined by $\Delta g_P = |\boldsymbol{g}_P| - |\boldsymbol{\gamma}_Q|$, and the gravity disturbance by $\delta g_P = |\boldsymbol{g}_P| - |\boldsymbol{\gamma}_P|$. The deflection of the vertical is the angle of the gravity vector relative to the normal to the ellipsoid at P.

$$\eta = -\frac{1}{\gamma}\frac{\partial T}{r\sin\theta\,\partial\lambda}, \quad \xi = \frac{1}{\gamma}\frac{\partial T}{r\partial\theta}, \tag{6.65}$$

or, with equation (2.270),

$$\frac{1}{\gamma r}\boldsymbol{\nabla}_{(\theta,\lambda)}T(\theta,\lambda,r) = \xi(\theta,\lambda,r)\boldsymbol{e}_\theta - \eta(\theta,\lambda,r)\boldsymbol{e}_\lambda, \tag{6.66}$$

where the signs on the derivatives, again, are a matter of convention.

Higher derivatives of the disturbing potential generally are formulated in local Cartesian coordinates since they represent primarily the structure of the local mass-density distribution. With the local coordinate system defined in Figure 1.1, the gradients of the gravity disturbance form a second-order tensor, $\Gamma = [\Gamma_{j,k}], j,k = 1, 2, 3,$

$$\Gamma = \boldsymbol{\nabla}\boldsymbol{\nabla}^{\mathrm{T}}T = \boldsymbol{\nabla}\delta\boldsymbol{g}^{\mathrm{T}} = \begin{pmatrix} \frac{\partial^2 T}{\partial x_1 \partial x_1} & \frac{\partial^2 T}{\partial x_1 \partial x_2} & \frac{\partial^2 T}{\partial x_1 \partial x_3} \\ \frac{\partial^2 T}{\partial x_2 \partial x_1} & \frac{\partial^2 T}{\partial x_2 \partial x_2} & \frac{\partial^2 T}{\partial x_2 \partial x_3} \\ \frac{\partial^2 T}{\partial x_3 \partial x_1} & \frac{\partial^2 T}{\partial x_3 \partial x_2} & \frac{\partial^2 T}{\partial x_3 \partial x_3} \end{pmatrix}. \tag{6.67}$$

Since all the derivatives of the potential are continuous is free space, the gradient tensor is symmetric and Laplace's equation shows that its trace is zero. Their relationships to second order derivatives with respect to spherical polar coordinates are (Reed 1973)

$$\Gamma_{11} = \frac{\cot\theta}{r^2}\frac{\partial T}{\partial\theta} + \frac{1}{r}\frac{\partial T}{\partial r} + \frac{1}{r^2\sin^2\theta}\frac{\partial^2 T}{\partial\lambda^2} \tag{6.68}$$

$$\Gamma_{12} = \frac{\cot\theta}{r^2\sin\theta}\frac{\partial T}{\partial\lambda} - \frac{1}{r^2\sin\theta}\frac{\partial^2 T}{\partial\theta\partial\lambda} \tag{6.69}$$

$$\Gamma_{13} = -\frac{1}{r^2\sin\theta}\frac{\partial T}{\partial\lambda} + \frac{1}{r\sin\theta}\frac{\partial^2 T}{\partial\lambda\partial r} \tag{6.70}$$

$$\Gamma_{22} = \frac{1}{r}\frac{\partial T}{\partial r} + \frac{1}{r^2}\frac{\partial^2 T}{\partial\theta^2} \tag{6.71}$$

$$\Gamma_{23} = \frac{1}{r^2}\frac{\partial T}{\partial\theta} - \frac{1}{r}\frac{\partial^2 T}{\partial\theta\partial r} \tag{6.72}$$

$$\Gamma_{33} = \frac{\partial^2 T}{\partial r^2} \tag{6.73}$$

The solution of Laplace's equation for the disturbing potential in terms of a spherical harmonic series is given as in equation (6.52) by

$$T(\theta,\lambda,r)=\frac{GM}{R}\sum_{n=2}^{\infty}\sum_{m=-n}^{n}\left(\frac{R}{r}\right)^{n+1}\delta C_{n,m}\bar{Y}_{n,m}(\theta,\lambda), \tag{6.74}$$

where the unit-less coefficients, $\delta C_{n,m}$, together with the adopted constants, GM, R, constitute the Fourier-Legendre transform, $\{GM\,\delta C_{n,m}/R\}$, of T on the sphere of radius, R. This spectrum is also the difference between the spectra of the total and reference gravity potentials. The unit-less coefficients of the total gravitational field are

$$C_{n,m}=C_{n,0}^{(\mathrm{ref})}+\delta C_{n,m}, \tag{6.75}$$

where by its symmetry the ellipsoidal reference field has only even-degree zonal harmonics. The coefficient, $\delta C_{0,0}$, is zero under the assumption that the reference field accounts completely for the monopole component of the total field. Also, by assigning the coordinate origin to the center of mass, the first-degree harmonic coefficients, $\delta C_{1,m}$, $m=-1,0,1$, vanish as shown later by equations (6.151). In further developments here, the reasonable assumption is made, consistent with a first-order approximation, that the Earth is a sphere of radius, R, and that this is a bounding sphere.

In general, if $V_{n,m}$ is the Fourier-Legendre spectrum of a potential, v (the input), on the sphere of radius, R, then by equation (6.54), its spectrum on the sphere of radius, $r > R$, is $V_{n,m}(R/r)^{n+1}$. Interpreting this as the spectrum of an output, the convolution theorem (3.60) for functions on the sphere yields immediately the corresponding convolution, equation (3.58), known as the *Poisson integral*,

$$v(\theta,\lambda,r)=\frac{1}{4\pi}\iint_{\Omega'}v(\theta',\lambda',R)\,K(\psi,r)\,d\Omega', \tag{6.76}$$

where, from equation (2.206),

$$K(\psi,r)=\sum_{n=0}^{\infty}(2n+1)\left(\frac{R}{r}\right)^{n+1}P_n(\cos\psi) \tag{6.77}$$

is known as *Poisson's kernel*. For $r > R$ it has the closed form (Exercise 6.3),

$$K(\psi,r)=\frac{R\left(r^2-R^2\right)}{d^3}, \tag{6.78}$$

where d is given by equation (6.11) with $r' = R$,

$$d=\sqrt{r^2+R^2-2rR\cos\psi}, \tag{6.79}$$

If $r = R$, then the Poisson kernel, $K(\psi, R)$, degenerates to the spherical Dirac delta function, equation (2.321), as expected (compare equations (6.76) for $r = R$ with equation (2.323); see also (Freeden et al. 1998, p.53)).

The Poisson integral is the *upward continuation* filter with frequency response given by $(R/r)^{n+1}$. Often it is written in the form,

$$v(\theta,\lambda,r)=\frac{R}{r}v(\theta,\lambda,R)+\frac{1}{4\pi}\iint\limits_{\Omega'}\big(v(\theta',\lambda',R)-v(\theta,\lambda,R)\big)K(\psi,r)\,d\Omega', \tag{6.80}$$

which is easily proved since the average of $K(\psi, r)$ over Ω' is its zero-degree spectral component, R/r. This form "localizes" the integrand to the evaluation point on the sphere and ameliorates the effect of the nearly singular kernel function for r close to R. As noted in Figure 6.2, upward continuation ($r > R$), as expressed by equations (6.76) or (6.80), is a *low-pass* filter since the high-frequency spectrum of v is attenuated by the factor, $(R/r)^{n+1}$.

Equation (6.76) holds for any function that satisfies Laplace's equation above the bounding sphere, that is, for any function that can be expanded in solid spherical harmonics, equations (6.52) or (6.74), where the radial attenuation per degree, n, has the form $(1/r)^{n+1}$. Applying the derivative approximations for the gravity disturbance and the gravity anomaly, equations (6.63) and (6.64), to the spherical harmonic series for the disturbing potential, equation (6.74), yields

$$\delta g(\theta,\lambda,r)=\frac{GM}{R^2}\sum_{n=2}^{\infty}\sum_{m=-n}^{n}(n+1)\left(\frac{R}{r}\right)^{n+2}\delta C_{n,m}\bar{Y}_{n,m}(\theta,\lambda), \tag{6.81}$$

and

$$\Delta g(\theta,\lambda,r)=\frac{GM}{R^2}\sum_{n=2}^{\infty}\sum_{m=-n}^{n}(n-1)\left(\frac{R}{r}\right)^{n+2}\delta C_{n,m}\bar{Y}_{n,m}(\theta,\lambda). \tag{6.82}$$

Thus, $r\delta g$ and $r\Delta g$ are harmonic in free space (depending on $r^{-(n+1)}$; see also equation (6.51)); and, for example, the upward continuation integral for the anomaly, from equation (6.80), is

$$\Delta g(\theta,\lambda,r)=\frac{R}{r}\Delta g(\theta,\lambda,R)+\frac{R}{4\pi r}\iint\limits_{\Omega'}\big(\Delta g(\theta',\lambda',R)-\Delta g(\theta,\lambda,R)\big)K(\psi,r)\,d\Omega'. \tag{6.83}$$

A similar result holds for the gravity disturbance, δg.

The cardinal problem in physical geodesy is the determination of the geoid undulation from measurements of gravity on the Earth's surface. This amounts to solving Laplace's equation for T using either equation (6.63) or (6.64) as boundary condition, corresponding to *Neumann* (second kind) and *Robin* (third kind) boundary-value problems, and subsequently applying Bruns's formula (6.62).

The boundary condition may be applied in the form of the spectrum of the gravity disturbance or the anomaly. From equations (6.81) or (6.82), the spectrum of δg or Δg on the sphere of radius, R, is $GM(n \pm 1)\,\delta C_{n,m}/R^2$. Therefore, the spectrum of T on the sphere of radius, $r > R$, is related to the spectrum of δg or Δg by the product, $\left(\frac{GM}{R^2}(n\pm1)\delta C_{n,m}\right)\left(\frac{R}{n\pm1}\left(\frac{R}{r}\right)^{n+1}\right)$. The spherical convolution theorem (3.60) then yields the convolutions for the disturbing potential for $r > R$. In terms of the gravity disturbance it is

$$T(\theta,\lambda,r)=\frac{R}{4\pi}\iint_{\Omega'}\delta g(\theta',\lambda',R)\,H(\psi,r)\,d\Omega', \tag{6.84}$$

which is known as *Hotine's integral*, where the *Hotine kernel* is

$$H(\psi,r)=\sum_{n=0}^{\infty}\frac{2n+1}{n+1}\left(\frac{R}{r}\right)^{n+1}P_n(\cos\psi); \tag{6.85}$$

and, in terms of the gravity anomaly, one obtains the *Pizzetti integral*,

$$T(\theta,\lambda,r)=\frac{R}{4\pi}\iint_{\Omega'}\Delta g(\theta',\lambda',R)\,S(\psi,r)\,d\Omega', \tag{6.86}$$

with the generalized Stokes kernel given by

$$S(\psi,r)=\sum_{n=2}^{\infty}\frac{2n+1}{n-1}\left(\frac{R}{r}\right)^{n+1}P_n(\cos\psi). \tag{6.87}$$

Stokes's integral for the geoid undulation (published in 1849 before Pizzetti's result in 1911) is obtained by setting $r = R$ in equation (6.86) and dividing by normal gravity, γ (Hofmann-Wellenhof and Moritz 2005). The inverse Fourier-Legendre transform for the gravity anomaly, equation (6.82), is missing first-degree harmonics, even if they exist for T. For this reason, also the Stokes kernel has no first-degree spectral components. If T does have these harmonics, the spectral coefficients must be obtained from other data. And, since the normal field contains the zero-degree harmonic, it is also absent from the Stokes kernel. No restriction on the first-degree harmonic is imposed on the Hotine kernel, but it could as well be defined by

$$H'(\psi,r)=\sum_{n=2}^{\infty}\frac{2n+1}{n+1}\left(\frac{R}{r}\right)^{n+1}P_n(\cos\psi), \tag{6.88}$$

if it is known that $\delta C_{0,0} = 0$ and $\delta C_{1,m} = 0$, or if they are determined otherwise.

The factors, $1/(n \pm 1)$, in the Legendre spectra of the Stokes and Hotine kernels represent the radial anti-derivative that attenuates the high frequencies of Δg, respectively δg, in determining T. That is, the disturbing potential is a *low-pass* filter of the gravity anomaly (or, the gravity disturbance) with the Stokes (respectively, the Hotine) kernel as filter function. The other factor, $(R/r)^{n+1}$, as with the Poisson integral, accounts for the attenuation of the field with upward continuation in free space. This also attenuates the high frequencies, but not nearly as severely as the anti-derivative response, as shown in Figure 6.6 (this figure excludes the Hotine kernel which is almost the same as the Stokes kernel).

The uniform convergence of the series in equations (6.85) and (6.87) is ensured if $r > R$, and closed formulas may be obtained with appropriate mathematical manipulations of the Coulomb expansion (6.12) (Hotine 1969, p.311; Hofmann-Wellenhof and Moritz 2005),

$$H(\psi, r) = \frac{2R}{d} - \ln\left(\frac{R + d - r\cos\psi}{r(1-\cos\psi)}\right), \tag{6.89}$$

$$S(\psi, r) = 2\frac{R}{d} + \frac{R}{r} - 3\frac{Rd}{r^2} - 5\frac{R^2}{r^2}\cos\psi - 3\frac{R^2}{r^2}\cos\psi \ln\frac{d + r - R\cos\psi}{2r}, \tag{6.90}$$

where d is given by equation (6.79). Note that if $\psi = 0$, then $d = r - R$, and $S(0, r)$ is well defined for $r > R$, as is $H(0, r)$ since,

$$\lim_{\psi \to 0} \frac{R + d - r\cos\psi}{r(1-\cos\psi)} = \lim_{y \to 1} \frac{R + d - ry}{r(1-y)} = \lim_{y \to 1}\left(\frac{R}{d} + 1\right) = \frac{r}{r - R}, \tag{6.91}$$

with $y = \cos\psi$ and where l'Hôpital's rule is applied to obtain the second equality.

In the limit as $r \to R$ both the Stokes and Hotine kernels have a singularity at $\psi = 0$, which is evident also from their Legendre spectra (recall that $P_n(\cos 0) = 1$ for all n). Thus, technically the convolution theorem cannot be applied in this case since these functions then are not bounded. On the other hand, because the potential and its first derivatives are continuous and bounded for *all* r, the Fourier-Legendre series for the gravity anomaly and disturbance, equations (6.81) and (6.82), both converge for $r = R$. In addition, the Legendre series for the Hotine and Stokes kernels, equations (6.85) and (6.87), converge for $r = R$ if $\psi > 0$, because $H(\psi, R)$ and $S(\psi, R)$ are well

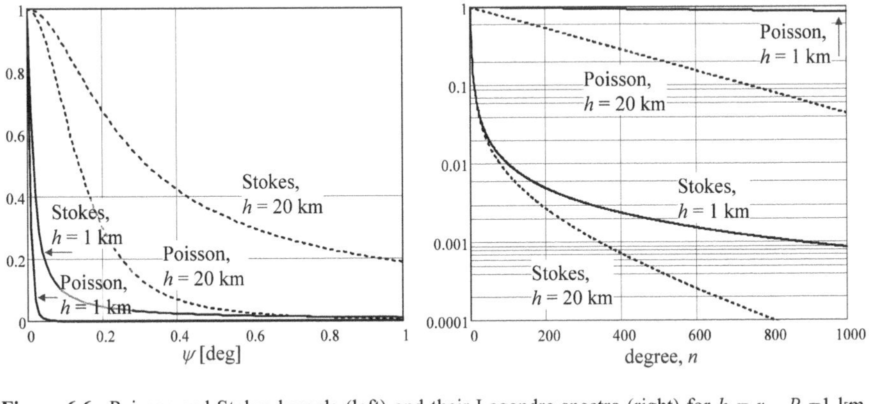

Figure 6.6: Poisson and Stokes kernels (left) and their Legendre spectra (right) for $h = r - R = 1$ km, 20 km. The kernels are normalized to unity at the origin.

defined for $\psi > 0$ by equations (6.89) and (6.90). Therefore, they have Legendre transforms, just like the spherical Dirac delta function.

However, the Stokes and Hotine kernels for $r = R$ are not as strongly singular as the Dirac delta function (or, the Poisson kernel) since they attenuate less rapidly near the origin, rather more like the reciprocal distance function, $1/d$ (Figure 6.6). This is another example of the inverse relationship between bandwidth and spatial extent (Section 3.5.2); that is, less attenuation in the space domain implies greater attenuation in the frequency domain, and vice versa. Substituting $r = R$ into equation (6.79), $d = 2R\sin(\psi/2)$ and these kernels become

$$H(\psi,R) \equiv H(\psi) = \frac{1}{\sin\frac{\psi}{2}} - \ln\left(\frac{1}{\sin\frac{\psi}{2}} + 1\right), \tag{6.92}$$

$$S(\psi,R) \equiv S(\psi) = \frac{1}{\sin\frac{\psi}{2}} + 1 - 6\sin\frac{\psi}{2} - 5\cos\psi - 3\cos\psi\ln\left(\sin\frac{\psi}{2} + \sin^2\frac{\psi}{2}\right). \tag{6.93}$$

With a change in the integration variables in equations (6.84) and (6.86) (with $r = R$) such that the pole of the spherical coordinate system passes through the computation point, (θ, λ), the differential area element on the unit sphere is $d\Omega' = \sin\psi\, d\psi\, d\zeta = 2\sin(\psi/2)\cos(\psi/2)\, d\psi\, d\zeta$, where ζ is the longitude of the integration point (Figure 2.13). The singularity of the integrand then disappears, whence these integrals are *weakly* singular (Section 6.5.1.1). The vanishing singularity is demonstrated by applying L'Hôpital's rule to the logarithmic terms in $H(\psi, R)$ and $S(\psi, R)$. For Hotine's kernel, with $x = \sin(\psi/2)$,

$$\begin{aligned}\lim_{\psi\to 0}\left(\sin\frac{\psi}{2}H(\psi,R)\right) &= \lim_{x\to 0}\left(x\ln\left(\frac{1}{x}+x\right)\right) = \lim_{x\to 0}\frac{\ln\left(\frac{1}{x}+x\right)}{1/x} = \lim_{x\to 0}\frac{\frac{x}{1+x^2}\left(\frac{1}{x^2}-1\right)}{1/x^2} \\ &= \lim_{x\to 0} x\frac{1-x^2}{1+x^2} = 0\end{aligned} \tag{6.94}$$

and, similarly, for Stokes's kernel,

$$\lim_{x\to 0} x\ln\left(x+x^2\right) = \lim_{x\to 0}\left(\frac{\ln\left(x+x^2\right)}{1/x}\right) = \lim_{x\to 0}\left(\frac{(1+2x)/\left(x+x^2\right)}{-1/x^2}\right) = \lim_{x\to 0}\left(-\frac{x+2x^2}{1+x}\right) = 0. \tag{6.95}$$

The practical benefit of integrals, such as the Poisson, Hotine, and Stokes-Pizzetti integrals lies in local applications, where because the kernels attenuate with distance the integrals may be truncated to a neighborhood of the evaluation point. This is advantageous if detailed data are available only locally and do not warrant a global spectral analysis. In particular, the geoid undulation is determined best from local gravity anomaly (or disturbance) data using *Stokes's integral*, derived from equation (6.86) and Bruns's formula (6.62),

$$N(\theta,\lambda)=\frac{R}{4\pi\gamma}\iint_{\Omega'}\Delta g(\theta',\lambda',R)S(\psi)\,d\Omega', \tag{6.96}$$

by truncating the integral to a neighborhood of (θ, λ) and supplementing this with a global Fourier-Legendre model as described in Section 3.6.3.

The spherical spectral relationships among the radial derivatives of the potential lead to other convolutions of a data signal. Using the Fourier-Legendre spectrum of the qth radial derivative of v on the sphere of radius, $r > R$, equation (6.55), the convolution theorem yields the following convolutions,

$$\frac{\partial^q}{\partial r^q}v(\theta,\lambda,r)=\frac{1}{4\pi R^q}\iint_{\Omega'}v(\theta',\lambda',R)L^{(q)}(\psi,r)\,d\Omega', \tag{6.97}$$

where, $q \geq 1$,

$$L^{(q)}(\psi,r)=(-1)^q\sum_{n=0}^{\infty}(2n+1)\prod_{j=1}^{q}(n+j)\left(\frac{R}{r}\right)^{n+q+1}P_n(\cos\psi). \tag{6.98}$$

For example, the kernel of the convolution for the first radial derivative ($q = 1$) is

$$\begin{aligned}L^{(1)}(\psi,r)&=-\sum_{n=0}^{\infty}(2n+1)(n+1)\left(\frac{R}{r}\right)^{n+2}P_n(\cos\psi)\\&=-Z(\psi,r)-\frac{R}{r}K(\psi,r)\end{aligned} \tag{6.99}$$

where

$$Z(\psi,r)=\sum_{n=1}^{\infty}(2n+1)n\left(\frac{R}{r}\right)^{n+2}P_n(\cos\psi), \tag{6.100}$$

which has the closed expression (Exercise 6.4),

$$Z(\psi,r)=\frac{R^3}{d^3}\left(5\cos\psi-\frac{r}{R}+\frac{d^2}{rR}-6\frac{rR}{d^2}\sin^2\psi\right). \tag{6.101}$$

Substituting equation (6.99) and using equation (6.76), the first derivative of the potential, equation (6.97), becomes

$$\frac{\partial}{\partial r}v(\theta,\lambda,r)=-\frac{1}{r}v(\theta,\lambda,r)-\frac{1}{4\pi R}\iint_{\Omega'}\left(v(\theta',\lambda',R)-v(\theta,\lambda,R)\right)Z(\psi,r)\,d\Omega', \tag{6.102}$$

where the fact that $Z(\psi, r)$ has no zero-degree spectral component (its integral over Ω' is zero) accounts for the second term in the integrand and thus allows a "localization" of the data signal. If $r = R$, then with $d_R^2 = 2R^2(1 - \cos\psi)$ and $Z(\psi, r) = -2R^3/d_R^3$, the radial derivative on this sphere is

$$\left.\frac{\partial}{\partial r}v(\theta,\lambda,r)\right|_{r=R} = -\frac{1}{R}v(\theta,\lambda,R)+\frac{R^2}{2\pi}\iint_{\Omega'}\frac{v(\theta',\lambda',R)-v(\theta,\lambda,R)}{d_R^3}d\Omega'. \tag{6.103}$$

The kernel now has a stronger singularity at its origin and the integral is defined as a Cauchy principal value (Klees and Lehmann 1998), which results from a specific limiting process and assumes that the data function is well behaved in the neighborhood of the evaluation point. The convolution for the derivative acts as a *high-pass* filter since it emphasizes the high degrees of the potential spectrum, as evident in its frequency response, $n(R/r)^{n+2}$, equation (6.100), if R/r is close to unity.

Equation (6.102) has a number of special applications in physical geodesy. For example, the inverse of the Hotine integral is obtained immediately from the definition of the gravity disturbance, equation (6.63),

$$\delta g(\theta,\lambda,r)=\frac{1}{r}T(\theta,\lambda,r)+\frac{1}{4\pi R}\iint_{\Omega'}\big(T(\theta',\lambda',R)-T(\theta,\lambda,R)\big)Z(\psi,r)\,d\Omega'. \tag{6.104}$$

And, comparing equations (6.63) and (6.64), the inverse to the Pizzetti integral is given by

$$\Delta g(\theta,\lambda,r)=-\frac{1}{r}T(\theta,\lambda,r)+\frac{1}{4\pi R}\iint_{\Omega'}\big(T(\theta',\lambda',R)-T(\theta,\lambda,R)\big)Z(\psi,r)\,d\Omega'. \tag{6.105}$$

These integrals may be used to obtain the gravity disturbance, respectively anomaly, if T is given on the sphere of radius, R, according to equation (6.62), in the form of the geoid undulation measured from satellite altimetry, equation (6.224). Typically, however, only the high-frequency spectrum of the gravity anomaly is derived from satellite altimetry using the corresponding spectral relationships in Cartesian coordinates (Section 6.4.2).

Since equation (6.102) holds for any function that is harmonic in free space, applied to $r\Delta g$ with due consideration of the radial derivative on the left side, and evaluated at $r = R$, it results in

$$\left.\frac{\partial}{\partial r}\Delta g(\theta,\lambda,r)\right|_{r=R} = -\frac{2}{R}\Delta g(\theta,\lambda,R)+\frac{R^2}{2\pi}\iint_{\Omega'}\frac{\Delta g(\theta',\lambda',R)-\Delta g(\theta,\lambda,R)}{d_R^3}d\Omega'. \tag{6.106}$$

This radial derivative is used by Moritz (1980) for upward and downward continuation of the gravity anomaly by Taylor series expansion in his solution to the Molodensky boundary-value problem (which determines the Earth's surface rather than the geoid directly from gravity measurements). It is usual to drop the first term since it is insignificant due to the relatively large value of R.

As a further example, apply equation (6.102) to the harmonic function, $r\delta g$, with $\delta g = -\partial T/\partial r$ on the left side, thus obtaining

$$\frac{\partial^2 T}{\partial r^2}(\theta,\lambda,r)=\frac{2}{r}\delta g(\theta,\lambda,r)+\frac{1}{4\pi r}\iint_{\Omega'}\big(\delta g(\theta',\lambda',R)-\delta g(\theta,\lambda,R)\big)Z(\psi,r)\,d\Omega'. \tag{6.107}$$

This equation may be used to validate airborne gravity gradiometry by upward continuing an existing ground data base of gravity disturbances.

Applying the horizontal derivatives of the disturbing potential, equations (6.65), to the Pizzetti integral, equation (6.86), yields the *Vening-Meinesz integrals*,

$$\begin{Bmatrix} \eta(\theta,\lambda,r) \\ \xi(\theta,\lambda,r) \end{Bmatrix} = \frac{1}{4\pi\gamma} \iint_{\Omega'} \Delta g(\theta',\lambda',R) \begin{Bmatrix} VM_\eta(\alpha,\psi,r) \\ VM_\xi(\alpha,\psi,r) \end{Bmatrix} d\Omega', \tag{6.108}$$

The kernels, $VM_{\eta,\xi}$ (α, ψ, r), are obtained by taking the derivatives of the Stokes kernel,

$$\begin{Bmatrix} VM_\eta(\alpha,\psi,r) \\ VM_\xi(\alpha,\psi,r) \end{Bmatrix} = \begin{Bmatrix} -\dfrac{R}{r\sin\theta}\dfrac{\partial S}{\partial\lambda} \\ \dfrac{R}{r}\dfrac{\partial S}{\partial\theta} \end{Bmatrix} = \frac{R}{r}\begin{Bmatrix} -\dfrac{1}{\sin\theta}\dfrac{\partial\psi}{\partial\lambda} \\ \dfrac{\partial\psi}{\partial\theta} \end{Bmatrix}\frac{\partial S}{\partial\psi} = \frac{R}{r}\begin{Bmatrix} \sin\zeta \\ -\cos\zeta \end{Bmatrix}\frac{\partial S}{\partial\psi}$$

$$= \frac{R}{r}\begin{Bmatrix} \sin\alpha \\ \cos\alpha \end{Bmatrix}\frac{\partial S}{d\psi}. \tag{6.109}$$

using the derivatives of equation (2.196) with equations (2.198) and (2.199) substituted (Exercise 6.5). The direction angle, $\zeta = \pi - \alpha$, is shown in Figure 2.13, where α is the azimuth of the integration point. With the derivative of Stokes's function, one obtains

$$\begin{Bmatrix} VM_\eta(\alpha,\psi,r) \\ VM_\xi(\alpha,\psi,r) \end{Bmatrix} = -\left(\frac{R}{r}\right)^3 \begin{Bmatrix} \sin\alpha \\ \cos\alpha \end{Bmatrix}.$$

$$\sin\psi\left(2\frac{r^3}{d^3} + 3\frac{r}{d} - 5 - 3\ln\frac{d+r-R\cos\psi}{2r} + 3\frac{R\cos\psi}{d+r-R\cos\psi}\left(\frac{r}{d}+1\right)\right) \tag{6.110}$$

Equation (6.108) is a general convolution, equation (3.44), for which no simple convolution theorem exists. That is, there is no exact Fourier-Legendre spectral relationship between the gravity anomaly and the deflection of vertical. On the other hand, as surface gradient of the potential, equation (6.66), the combined deflection components have the *vector* Fourier-Legendre transform according to equation (2.301), $\sqrt{n(n+1)}\delta C_{n,m}$, on the sphere of radius, R. Thus, we may characterize the frequency response of the Vening-Meinesz formulas approximately as $\sqrt{n(n+1)}/(\gamma(n-1))$, since $\gamma(n-1)\delta C_{n,m}$ is the Fourier-Legendre transform of the gravity anomaly.

The frequency response of the *inverse* Vening-Meinesz integral is then approximately $\gamma(n-1)/\sqrt{n(n+1)}$, but the integral, itself, is not a simple convolution on the sphere. It may be derived by starting with Green's first identity for differentiable functions, g and h (Kaplan 1973, p.325), on a patch of the sphere, $\Delta\Omega$, with boundary, B (a line),

$$\iint_{\Delta\Omega} g\nabla^2_{(\theta,\lambda)}h\, d\Omega + \iint_{\Delta\Omega} \nabla_{(\theta,\lambda)}g \cdot \nabla_{(\theta,\lambda)}h\, d\Omega = \int_B g\,\nabla_{(\theta,\lambda)}h \cdot \boldsymbol{n}\, dB, \tag{6.111}$$

where the Laplace-Beltrami operator, $\nabla^2_{(\theta,\lambda)}$, and surface gradient, $\boldsymbol{\nabla}_{(\theta,\lambda)}$, are given by equations (2.240) and (2.270), respectively; and, $\boldsymbol{n}$ is an outward unit vector on the boundary, B, and tangent to the sphere. Taking the limit as $\Delta\Omega \to \Omega$, the boundary vanishes as does the integral on the right side. Thus,

$$\iint_{\Omega} g\nabla^2_{(\theta,\lambda)} h\, d\Omega = -\iint_{\Omega} \boldsymbol{\nabla}_{(\theta,\lambda)} g \cdot \boldsymbol{\nabla}_{(\theta,\lambda)} h\, d\Omega. \tag{6.112}$$

Now, from equations (2.239), (6.74), and (6.81), the Fourier-Legendre spectra of $\nabla^2_{(\theta,\lambda)} T(\theta, \lambda, R)$ and $\Delta g(\theta, \lambda, r)$ are, respectively, $-(GM/R)n(n+1)\delta C_{n,m}$ and $(GM/R^2)(n-1)(R/r)^{n+2}\delta C_{n,m}$. By the convolution theorem (3.60), this implies the convolution,

$$\Delta g(\theta,\lambda,r) = -\frac{1}{4\pi R}\iint_{\Omega'} \nabla^2_{(\theta',\lambda')} T(\theta',\lambda',R) W(\psi,r)\, d\Omega', \tag{6.113}$$

where

$$W(\psi,r) = \sum_{n=2}^{\infty} \frac{(2n+1)(n-1)}{n(n+1)} \left(\frac{R}{r}\right)^{n+2} P_n(\cos\psi). \tag{6.114}$$

With the special case of Green's first identity, equation (6.112), the convolution is

$$\Delta g(\theta,\lambda,r) = \frac{1}{4\pi R}\iint_{\Omega'} \boldsymbol{\nabla}_{(\theta',\lambda')} T(\theta',\lambda',R) \cdot \boldsymbol{\nabla}_{(\theta',\lambda')} W(\psi,r)\, d\Omega'; \tag{6.115}$$

and, with equations (6.65) and derivatives with respect to θ', λ' analogous to equation (6.109), this becomes,

$$\Delta g(\theta,\lambda,r) = \frac{\gamma}{4\pi}\iint_{\Omega'} \left(\eta(\theta',\lambda',R) Y_\eta(\psi,\alpha',r) + \xi(\theta',\lambda',R) Y_\xi(\psi,\alpha',r)\right) d\Omega'. \tag{6.116}$$

where

$$\begin{Bmatrix} Y_\eta(\psi,\alpha',r) \\ Y_\xi(\psi,\alpha',r) \end{Bmatrix} = -\begin{Bmatrix} \sin\alpha' \\ \cos\alpha' \end{Bmatrix} \frac{\partial}{\partial\psi} W(\psi,r), \tag{6.117}$$

and α' is the forward azimuth at (θ', λ') of the great circle arc emanating from (θ, λ).

The closed expressions for $W(\psi,r)$ and $\partial W(\psi,r)/\partial\psi$, if $\psi > 0$, are given by (Hwang 1998),

$$W(\psi,r) = \frac{R}{r}\left(2\frac{R}{d} - \frac{R}{r}\ln\left(\frac{2r}{d + r - R\cos\psi}\right) - 2\ln\left(\frac{d + R - r\cos\psi}{r(1-\cos\psi)}\right)\right), \tag{6.118}$$

$$\frac{\partial W(\psi,r)}{\partial \psi}=\frac{R}{r}\sin\psi\left(-\frac{2R^2 r}{d^3}+\frac{R^2}{r}\frac{\frac{r}{d}+1}{d+r-R\cos\psi}-2r\frac{\frac{R}{d}+1}{d+R-r\cos\psi}+\frac{2}{1-\cos\psi}\right). \tag{6.119}$$

The function, $\partial W(\psi,r)/\partial\psi$, is not defined for any $r \geq R$ if $\psi=0$. For $r=R$, $d_R=2R\sin(\psi/2)$, hence,

$$\frac{\partial W(\psi,R)}{\partial \psi}=\frac{\cos(\psi/2)}{2\sin(\psi/2)}\left(-\frac{1}{\sin(\psi/2)}+\frac{3+2\sin(\psi/2)}{1+\sin(\psi/2)}\right). \tag{6.120}$$

Equation (6.116) is a kind of inverse Vening-Meinesz formula. It represents another way to compute the gravity anomaly from satellite altimetry, based on computed geoid slopes that are also the deflection components (substitute Bruns's equation (6.62) into equations (6.65)). Just like the Vening-Meinesz integral, its inverse, equation (6.116), is not a convolution that is amenable to a convolution theorem. Figure 6.7 compares the inverse Stokes kernel, $Z(\psi, R)$, and a combination of the inverse Vening-Meinesz kernels, $\sqrt{Y_\eta^2(\psi,\alpha',R)+Y_\xi^2(\psi,\alpha',R)}=\partial W(\psi,R)/\partial\psi$, as well as the corresponding frequency responses, $\{n\}$ and $\{(n-1)/\sqrt{n(n+1)}\}$. The latter are approximate as indicated above since neither is a formal Legendre transform. A practical application of these integrals, often performed in planar approximation (Section 6.5.1.2), requires smoothing to suppress the high-frequency response, more substantially with the inverse Stokes formula.

Planar approximations can be obtained for any of the convolutions by computing the limit as $R \to \infty$. An easier approach is based on the spectral relationships among the potential and its derivatives and upward continuation in free space as given by equations (6.41) and (6.46). The frequency response for upward continuation is $e^{-2\pi\bar{f}x_3}$; hence, the planar approximation of the Poisson kernel is the corresponding inverse Fourier transform (2.334),

$$\bar{K}(x_1,x_2,x_3)=\int_{-\infty}^{\infty}\int_{-\infty}^{\infty} e^{-2\pi\bar{f}x_3}e^{i2\pi(f_1x_1+f_2x_2)}df_1df_2=\begin{cases}\dfrac{1}{2\pi}\dfrac{x_3}{\left(x_1^2+x_2^2+x_3^2\right)^{3/2}}, & x_3>0\\ \delta(x_1,x_2), & x_3=0\end{cases} \tag{6.121}$$

where $\delta(x_1, x_2)$ is the two-dimensional Dirac delta function, equation (2.172).

In planar approximation the gravity anomaly and gravity disturbance are equivalent and equal to the negative vertical derivative of the disturbing potential,

$$\Delta g=\delta g=-\frac{\partial T}{\partial x_3}. \tag{6.122}$$

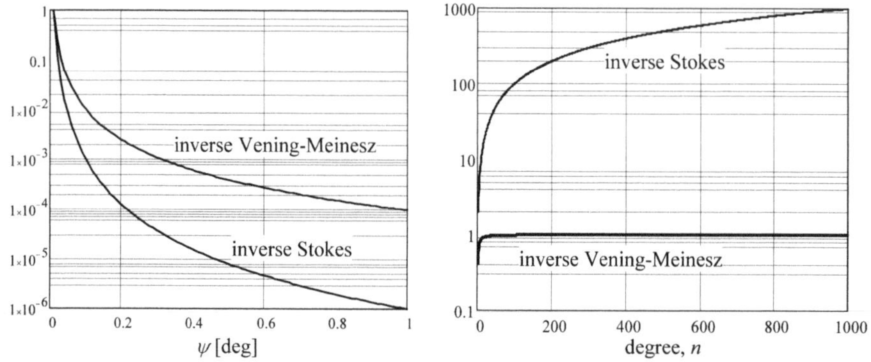

Figure 6.7: Inverse Stokes kernel, $Z(\psi, R)$, and inverse Vening-Meinesz kernel, $\partial W(\psi, R)/\partial\psi$, (left) and their Legendre spectra (right) on the sphere of radius, R. The kernels are "normalized" to unity at $\psi = 0.01°$ in order to show that the inverse Stokes kernel approaches infinity faster as $\psi \to 0$ than the inverse Vening-Meinesz kernel.

Thus from equations (6.45) and (6.41),

$$\mathcal{F}\left(T\left(x_1,x_2,x_3\right)\right) = \mathcal{F}\left(T\left(x_1,x_2,0\right)\right)e^{-2\pi x_3\bar{f}} = \frac{1}{2\pi\bar{f}}\mathcal{F}\left(\Delta g\left(x_1,x_2,0\right)\right)e^{-2\pi x_3\bar{f}}; \quad (6.123)$$

and, consequently, by the convolution theorem (3.26), the Stokes kernel is the inverse Fourier transform (2.337) of $e^{-2\pi\bar{f}x_3}/2\pi\bar{f}$,

$$\bar{S}\left(x_1,x_2,x_3\right) = \int_{-\infty}^{\infty}\int_{-\infty}^{\infty}\frac{e^{-2\pi\bar{f}x_3}}{2\pi\bar{f}}e^{i2\pi\left(f_1x_1+f_2x_2\right)}df_1df_2 = \frac{1}{2\pi}\frac{1}{\left(x_1^2+x_2^2+x_3^2\right)^{1/2}}. \quad (6.124)$$

Similarly, the spectral relationships in planar approximation between the gravity anomaly and the deflections of the vertical (equation (6.65)),

$$\eta = -\frac{1}{\gamma}\frac{\partial T}{\partial x_1}, \quad \xi = -\frac{1}{\gamma}\frac{\partial T}{\partial x_2}, \quad (6.125)$$

are given, in view of equations (6.46) and (6.123), by

$$\mathcal{F}\begin{Bmatrix}\eta\left(x_1,x_2,x_3\right)\\ \xi\left(x_1,x_2,x_3\right)\end{Bmatrix} = -\frac{1}{\gamma}\begin{Bmatrix}i2\pi f_1\\ i2\pi f_2\end{Bmatrix}\frac{1}{2\pi\bar{f}}\mathcal{F}\left(\Delta g\left(x_1,x_2,0\right)\right)e^{-2\pi x_3\bar{f}}. \quad (6.126)$$

Therefore, again by the convolution theorem, the Fourier transforms of the Vening-Meinesz kernels are $\left\{-if_1/\bar{f}, -if_2/\bar{f}\right\}e^{-2\pi\bar{f}x_3}$. Note that they are not well defined at the origin. Indeed, approaching (0,0) along the f_1-axis ($f_2 = 0$), the Fourier transform of the kernel for ξ is zero in the limit. But, if one approaches the origin along the line, $f_1 = f_2$, then $-if_2/\bar{f}$ in the limit ($f_1 = f_2 \to 0$) is $-i/\sqrt{2}$. Nevertheless, the inverse Fourier transforms (2.335) exist,

$$\begin{Bmatrix} \overline{VM}_\eta(x_1,x_2,x_3) \\ \overline{VM}_\xi(x_1,x_2,x_3) \end{Bmatrix} = \int_{-\infty}^{\infty}\int_{-\infty}^{\infty} \frac{e^{-2\pi \bar{f} x_3}}{\bar{f}} \begin{Bmatrix} -if_1 \\ -if_2 \end{Bmatrix} e^{i2\pi(f_1x_1+f_2x_2)} df_1 df_2 = \frac{1}{2\pi} \frac{1}{\left(x_1^2+x_2^2+x_3^2\right)^{3/2}} \begin{Bmatrix} x_1 \\ x_2 \end{Bmatrix}. \tag{6.127}$$

Substituting these kernels (and $1/\gamma$) into a convolution with the gravity anomaly then yields the Vening-Meinesz formulas in planar approximation,

$$\begin{Bmatrix} \eta(x_1,x_2,x_3) \\ \xi(x_1,x_2,x_3) \end{Bmatrix} = \frac{1}{2\pi\gamma} \int_{-\infty}^{\infty}\int_{-\infty}^{\infty} \frac{\Delta g(x_1',x_2',0)}{\left(s^2+x_3^2\right)^{3/2}} \begin{pmatrix} x_1 - x_1' \\ x_2 - x_2' \end{pmatrix} dx_1' dx_2'. \tag{6.128}$$

It is noted that these differ in sign from the standard formulas in (Hoffman-Wellenhof and Moritz 2005, p.126) where the coordinate differences in the integral are $x_1' - x_1$ and $x_2' - x_2$ (contrary to the definition of a convolution).

Again, from equations (6.45) and (6.41),

$$\mathcal{F}\left(\Delta g(x_1,x_2,x_3)\right) = 2\pi\bar{f}\; \mathcal{F}\left(T(x_1,x_2,0)\right) e^{-2\pi x_3 \bar{f}}, \tag{6.129}$$

thus leading to the frequency response in planar approximation of the inverse Pizzetti formula, $2\pi\bar{f}\, e^{-2\pi x_3 \bar{f}}$ (for $x_3 > 0$). This is a combination of responses, one from the vertical derivative of the potential and the other due to upward continuation of the gravity anomaly. The inverse Stokes kernel is then obtained from the Fourier transform pairs (2.334) and (2.338) by first adding and subtracting $1/(2\pi x_3 \bar{f})$,

$$\begin{aligned} \bar{Z}(x_1,x_2,x_3) &= \int_{-\infty}^{\infty}\int_{-\infty}^{\infty} 2\pi\bar{f} e^{-2\pi \bar{f} x_3} \left(1 + \frac{1}{2\pi x_3 \bar{f}} - \frac{1}{2\pi x_3 \bar{f}}\right) e^{i2\pi(f_1x_1+f_2x_2)} df_1 df_2 \\ &= \frac{1}{2\pi} \frac{1}{\left(x_1^2+x_2^2+x_3^2\right)^{3/2}} \left(\frac{3x_3^2}{x_1^2+x_2^2+x_3^2} - 1\right) \end{aligned} \tag{6.130}$$

In fact, this planar approximation also holds if $x_3 = 0$. However, as $x_3 \to 0$, the limiting frequency response, $2\pi\bar{f}$, attests to the practical instability at high frequencies when computing gravity anomalies from the potential by integration, which, therefore, requires adequate smoothing.

Laplace's equation (6.22) in Cartesian coordinates and equations (6.122) and (6.125) for the anomaly and the deflection components show that

$$\frac{1}{\gamma}\frac{\partial \Delta g}{\partial x_3} = -\frac{\partial \eta}{\partial x_1} - \frac{\partial \xi}{\partial x_2}. \tag{6.131}$$

Taking Fourier transforms and utilizing equations (6.46) and (2.153) yields

$$\mathcal{F}\left(\Delta g(x_1,x_2,x_3)\right) = i\frac{\gamma}{\bar{f}}\left(f_1\, \mathcal{F}\left(\eta(x_1,x_2,0)\right) + f_2\, \mathcal{F}\left(\xi(x_1,x_2,0)\right)\right) e^{-2\pi x_3 \bar{f}}. \tag{6.132}$$

Each term in the parentheses on the right side is a product of Fourier transforms; and, an inverse Fourier transform then leads to a sum of convolutions that is the planar approximation of the inverse Vening-Meinesz integral, equation (6.116),

$$\Delta g(x_1,x_2,x_3) = \gamma \int_{-\infty}^{\infty}\int_{-\infty}^{\infty} \left(\overline{Y}_\eta(x_1-x_1',x_2-x_2',x_3)\,\eta(x_1',x_2',0) + \overline{Y}_\xi(x_1-x_1',x_2-x_2',x_3)\,\xi(x_1',x_2',0) \right) dx_1'dx_2' \qquad (6.133)$$

where the kernels, $\overline{Y}_\eta$ and $\overline{Y}_\xi$, are the inverse Fourier transforms (2.335) of $if_1 e^{-2\pi x_3 \bar{f}}/\bar{f}$ and $if_2 e^{-2\pi x_3 \bar{f}}/\bar{f}$, respectively,

$$\begin{Bmatrix} \overline{Y}_\eta(x_1,x_2,x_3) \\ \overline{Y}_\xi(x_1,x_2,x_3) \end{Bmatrix} = -\frac{1}{2\pi}\frac{1}{\left(x_1^2+x_2^2+x_3^2\right)^{3/2}} \begin{Bmatrix} x_1 \\ x_2 \end{Bmatrix}. \qquad (6.134)$$

For $x_3 \to 0$ the kernels are well defined (except at the origin, $(x_1, x_2) = (0,0)$), but their Fourier transforms are not absolutely integrable. Yet, as also shown in Figure 6.7, the problem is not as severe as for the inverse Stokes kernel. Table 6.1 summarizes the planar approximations of the principal convolution kernels in physical geodesy and their Fourier transforms. The Stokes kernel is the most stable from the viewpoint of being a low-pass filter. Conversely, its inverse is least stable, amplifying the high-frequencies. Interestingly, the Vening-Meinesz kernel and its inverse are identical

Table 6.1: The principal kernels of convolutions in physical geodesy and their Fourier spectra, representing transformations among the potential and its first-order derivatives, as well as upward continuation. Notation: $s^2 = x_1^2 + x_2^2$, $\bar{f}^2 = f_1^2 + f_2^2$.

Name	Kernel (x_1,x_2,x_3)	Spher. form*	Fourier Spectrum	Legendre Spectrum†
Poisson, $\overline{K}$	$\frac{1}{2\pi}\frac{x_3}{\left(s^2+x_3^2\right)^{3/2}}$, $x_3>0$; $\delta(x_1,x_2)$, $x_3=0$	(6.78)	$e^{-2\pi\bar{f}x_3}$	$\left(\frac{R}{r}\right)^{n+1}$
Stokes, $\overline{S}$	$\frac{1}{2\pi}\frac{1}{\left(s^2+x_3^2\right)^{1/2}}$	(6.90)	$\frac{e^{-2\pi\bar{f}x_3}}{2\pi\bar{f}}$	$\frac{1}{n-1}\left(\frac{R}{r}\right)^{n+1}$
Inverse Stokes / radial derivative, $\overline{Z}$	$\frac{1}{2\pi}\frac{1}{\left(s^2+x_3^2\right)^{3/2}}\left(\frac{3x_3^2}{s^2+x_3^2}-1\right)$	(6.101)	$2\pi\bar{f}e^{-2\pi x_3\bar{f}}$	$n\left(\frac{R}{r}\right)^{n+2}$
Vening-Meinesz, $\begin{Bmatrix}\overline{VM}_\eta\\ \overline{VM}_\xi\end{Bmatrix}$	$\frac{1}{2\pi}\frac{1}{\left(s^2+x_3^2\right)^{3/2}}\begin{Bmatrix}x_1\\x_2\end{Bmatrix}$	(6.110)	$-i\frac{e^{-2\pi\bar{f}x_3}}{\bar{f}}\begin{Bmatrix}f_1\\f_2\end{Bmatrix}$	$\frac{\sqrt{n(n+1)}}{n-1}\left(\frac{R}{r}\right)^{n+1}$
Inverse Vening-Meinesz, $\begin{Bmatrix}\overline{Y}_\eta\\ \overline{Y}_\xi\end{Bmatrix}$	$-\frac{1}{2\pi}\frac{1}{\left(s^2+x_3^2\right)^{3/2}}\begin{Bmatrix}x_1\\x_2\end{Bmatrix}$	(6.117), (6.119)	$i\frac{e^{-2\pi x_3\bar{f}}}{\bar{f}}\begin{Bmatrix}f_1\\f_2\end{Bmatrix}$	$\frac{n-1}{\sqrt{n(n+1)}}\left(\frac{R}{r}\right)^{n+1}$

* equation numbers for the spherical forms of the kernels
† approximate for the Vening-Meinesz and inverse Vening-Meinesz kernels

except in sign. Transforming between radial and horizontal derivatives of the potential they behave somewhat similar to the Poisson kernel in that their spectra are approximately flat (for $x_3 = 0$); see also Figure 6.7.

6.3 Global Spectral Analysis

Any global geophysical signal may be analyzed in terms of its Fourier-Legendre spectrum. The classical signals are Earth's gravitational and magnetic fields. More recently the topography has received such analysis, though mostly as a supplement to gravitational analysis. Other signals such as terrestrial heat flow (Pollack et al. 1993), or spatial meteorological data on isobaric surfaces (Lorenz 1979) are developed specifically to reveal particular modes, as well as correlations with the scales of other geophysical signals. Some signals, such as the global hydrological variations in space and time (e.g., Wahr et al. 1998), can also be related to or are derived from a gravitational analysis. It is noted that even though the Earth's surface is not a sphere, it can be assumed that functions defined on the surface depend uniquely only on two coordinates, latitude and longitude, and as such are amenable to a Fourier-Legendre spectral analysis (the technical condition is that the Earth's surface is homeomorphic to a sphere). For potential fields, on the other hand, the sphere plays a special role as the basis for coordinates that permit a solution to Laplace's equation. In this case, the Fourier-Legendre spectrum refers to the field values exactly on a sphere of specific radius. Values at other points are determined by analytic continuation as constrained by Laplace's equation. Other choices of coordinates, such as ellipsoidal coordinates, yield other types of spectral analyses, which, however, are outside the present scope.

This section concentrates on the geopotential fields with an excursion here and there to the analysis of topographic heights. The global models for the potentials customarily are series expansions in terms of solid spherical harmonic functions, as given by equation (6.52), repeated here for convenience,

$$v(\theta,\lambda,r) = \sum_{n=0}^{\infty}\sum_{m=-n}^{n}\left(\frac{R}{r}\right)^{n+1} V_{n,m}\bar{Y}_{n,m}(\theta,\lambda), \quad r \geq R. \tag{6.135}$$

The coefficients, $V_{n,m}$, are known as *Stokes's constants* for the gravitational field and *Gauss's coefficients* for the magnetic field. In both cases they may also be called *multipoles* (from electrostatics) since they indicate the magnitudes of the essential geometry of the source distribution (Section 6.3.1.1). From a purely analytical perspective, the set, $\{V_{n,m}\}$, represents the Fourier-Legendre spectrum of the potential on the (bounding) sphere, $r = R$. On any other external sphere, $r = R_1 > R$, the spectrum is given with equation (6.54) by $\{V_{n,m}(R/R_1)^{n+1}\}$. The spherical harmonic expansion theoretically is valid only on the bounding sphere or in the exterior space that is assumed completely free of sources. For the gravitational field, this means one either neglects the masses of the atmosphere and extraterrestrial bodies, or one separately accounts for their effects (which also depend on time). Similarly, a relatively small component of the total magnetic field is generated by external sources (principally the interaction of solar radiation with the magnetosphere and ionosphere) that are also highly variable in time. The global part of this external-source field may be expressed in spherical harmonics, where the radial dependence in this case has the form, $(r/R)^n$ (see equation (6.50)).

6.3.1 Gravitational Potential Models

Analyses in recent years have emphasized the temporal changes in the gravitational field associated primarily with the hydrological mass fluxes on the Earth's surface, due to the seasonal variations in precipitation, runoff, and storage (the Amazon River basin being one of the largest signals) and the secular changes associated with melting of glaciers and polar land-bound ice sheets caused by climate change. The focus here is on the static field and the corresponding Fourier-Legendre spectrum.

One of the initial spherical harmonic analyses of the Earth's gravity field using global gravity data was done by Heiskanen (1938), although the objective was to investigate only the second-degree harmonics (see also Rapp 1998). A somewhat higher-degree expansion was constructed by Jeffreys (1943) from a global set of gravity values on a 10°-grid of latitudes and longitudes; the maximum degree and order was $n_{max} = 3$. This was followed by a model to maximum degree, $n_{max} = 8$, by Zhongolovich (1952) and a model to degree, $n_{max} = 4$, by Uotila (1962). A common method to determine the spectrum in those days and still in use today (with several modifications and mostly for the high end of the spectrum) is a numerical approximation of equation (2.220). Formulated with the transform of the transfer function from gravity anomaly to disturbing potential, that is, the Legendre transform, $1/(n - 1)$, of the Stokes kernel, equation (6.87), or simply the inverse of equation (6.82), this is

$$\delta C_{n,m} = \frac{R^2}{4\pi GM(n-1)} \iint_{\Omega} \Delta g(\theta,\lambda,R) \bar{Y}_{n,m}(\theta,\lambda)\, d\Omega, \quad n \geq 2. \tag{6.136}$$

Zero- and first-degree harmonics are established by other means (Section 6.3.1.1). Since gravity anomaly data are discretely distributed over the globe, the most straightforward spectral analysis approximates the integral by some form of numerical quadrature. For example, with data averaged on a regular grid, such as defined by equations (4.156), an approximation of the integral is

$$\delta C_{n,m} \simeq \frac{R^2}{4\pi GM(n-1)} \sum_{j=0}^{K-1} \sum_{\ell=0}^{M-1} \overline{\Delta g}(\theta_j,\lambda_\ell,R) \iint_{\Delta\Omega_{j,\ell}} \bar{Y}_{n,m}(\theta,\lambda)\, d\Omega, \tag{6.137}$$

where $\overline{\Delta g}$ is the average value over the (j, ℓ) grid cell, $\Delta\Omega_{j,\ell}$. The integrals of the spherical harmonic functions can be evaluated analytically and recursion formulas exist in terms of the degree and order (Paul 1978, Fukushima 2012b). Past practices reduced the gravity anomaly data to the geoid (Figure 6.5), thus making a further approximation of the geoid as a sphere. This creates an error on the order of the Earth's flattening, about 0.3%. Current methods account for these approximations with appropriate corrections to the data (e.g., Rapp and Pavlis 1990) and using instead an ellipsoidal harmonic transform that can be related back to the Fourier-Legendre transform (Jekeli 1988).

With increasingly high resolution in the data this form of spectral analysis was accompanied in the early days by the numerical challenge of solving for a quadratically growing number of harmonic coefficients (that number is $(n_{\max}+1)^2$). The first high-resolution model that adapted the fast Fourier transform (Section 4.3.2) to the analysis of spherical data, thus demonstrating high computational efficiency, was the OSU81 model ($n_{\max} = 180$) (Rapp 1981). The methodology, developed by Colombo (1980), uses equation (4.219) as an approximation to equation (6.136), reformulated using the complex spherical harmonics, $\bar{Y}^c_{n,m}$, as in equations (2.222) and (2.225). For the average data values one obtains (Exercise 6.6)

$$\begin{aligned}\delta C^c_{n,m} &= \frac{R^2}{4\pi GM(n-1)} \iint_{\Omega} \Delta g(\theta,\lambda,R)\bar{Y}^c_{n,m}(\theta,\lambda)\,d\Omega \\ &\simeq \frac{\sqrt{\varepsilon_m}\,\mathrm{sinc}\left(\dfrac{m}{M}\right)e^{-i\pi\frac{m}{M}}R^2}{4\pi GM(n-1)} \sum_{j=0}^{K-1} IP^{(j)}_{n,|m|}\,\mathrm{DFT}\left(\overline{\Delta g}_{j,\ell}\right)_m\end{aligned} \tag{6.138}$$

where $IP^{(j)}_{n,|m|} = \int\limits_{\theta=j\Delta\theta}^{(j+1)\Delta\theta} \bar{P}_{n,|m|}(\cos\theta)\sin\theta\,d\theta$, and the unit-less complex coefficients, $\delta C^c_{n,m}$, are related to $\delta C_{n,m}$ as in equation (2.228). Higher-degree expansions could now easily keep pace with the increase in global resolution of terrestrial gravity values, principally over the oceans as derived from satellite altimetry.

The method that avoids these approximations, in principle, is the series expansion of the gravity anomaly, or any other derivative of the disturbing potential, in terms of solid spherical harmonics, equation (6.82),

$$\Delta g(\theta,\lambda,r) = \frac{GM}{R^2}\sum_{n=2}^{n_{\max}}\sum_{m=-n}^{n}(n-1)\left(\frac{R}{r}\right)^{n+2}\delta C_{n,m}\bar{Y}_{n,m}(\theta,\lambda). \tag{6.139}$$

The only approximation in this expression is the truncation of the series to finite harmonic degree, $n_{\max}$. The data function is linearly related to the Fourier-Legendre spectrum and solving the latter is a matter of inverting a linear system of equations, one for each data point, (θ, λ, r). With more data than unknown parameters, $\delta C_{n,m}$, as well as random errors in the data, the solution is one of least-squares estimation. This method was proposed by Rapp (1969) who tested the analysis for $n_{\max} = 14$ with the spherical approximation, $r = R$, and simplified error covariances for the data. Modern computational capabilities can handle perhaps tens of thousands of unknown parameters without the spherical or error covariance approximations ($n_{\max} < 300$); however, longitudinal symmetry in the data grid (and the simplified error covariances) permits the use of fast Fourier transform techniques that lead to considerable efficiency and the feasibility of much higher expansion degree, $n_{\max}$. Corresponding least-squares methods of estimating the spherical spectrum are elaborated in Section 4.5.2. Both Earth Gravitational Models 1996 (EGM96,

$n_{max} = 360$, Lemoine et al. 1998) and EGM2008 ($n_{max} = 2190$, Pavlis et al. 2012, 2013) were constructed based on these methods. The analysis derived on the basis of equations (6.136) or (6.139) (or similar inverse Fourier-Legendre transforms of other data) is known as the *space-wise* approach, where the spectral estimation starts only after one has a complete set of observations in the space domain.

With the advent of Earth-orbiting satellites in the late 1950s and early 1960s came the opportunity to obtain accurate determinations of the gravitation spectrum from satellite tracking data gathered by stations around the world. The fundamental equation of motion of a body in space is derived from Newton's laws of motion and gravitation (Jekeli 2015),

$$\frac{d^2 \boldsymbol{x}}{dt^2} = \boldsymbol{a} + \boldsymbol{g}, \tag{6.140}$$

where $\boldsymbol{x}$ is a position vector, t is time, $\boldsymbol{a}$ is the specific force, or also the inertial acceleration, due to action forces, and $\boldsymbol{g}$ is the gravitational acceleration vector. This equation holds in a non-rotating, freely falling frame, i.e., an inertial frame.

A satellite in Earth's orbit, that is, in free-fall, experiences gravitational perturbations due to the global mass inhomogeneities of the Earth, as the Earth rotates under the orbit. Tracking the satellite as a function of time also yields the gravitational acceleration as a function of the time-varying position of the satellite. Since gravitation is the gradient of the gravitational potential, $\boldsymbol{g} = \boldsymbol{\nabla} V$, substituting equation (6.52) into equation (6.140) yields

$$\frac{d^2 \boldsymbol{x}}{dt^2} = \sum_{n=0}^{n_{max}} \sum_{m=-n}^{n} V_{n,m} \nabla \left(\left(\frac{R}{r} \right)^{n+1} \bar{Y}_{n,m} \left(\theta, \lambda + \omega_E t \right) \right) + \delta \boldsymbol{R}, \tag{6.141}$$

where ω_E is Earth's rate of rotation, t is time, and $\delta \boldsymbol{R}$ represents residual accelerations due to action forces (solar radiation pressure, atmospheric drag, Earth's albedo, *etc.*), gravitational tidal accelerations due to other bodies (moon, sun, planets), and all other consequent indirect effects (deformations of the Earth). The maximum degree, n_{max}, corresponds to the highest spatial resolution, equation (3.131), considered achievable from the spatial distribution of global data. The position vector on the left side of equation (6.141) is more explicitly, $\boldsymbol{x}(t) = \boldsymbol{x}(\theta(t), \lambda(t) + \omega_E t, r(t))$. It is numerically more convenient to transform the satellite position and velocity into Keplerian orbital elements (semi-major axis of the orbital ellipse, its eccentricity, its inclination to the equator, the angle of perigee, the right ascension of the node of the orbit, and the mean motion), all of which also change in time, but most much more slowly (Kaula 1966, Seeber 2003).

In the most general case ($n_{max} > 2$ and $\delta \boldsymbol{R} \neq \boldsymbol{0}$) there is no analytic solution to equation (6.141) or its transformations to other types of coordinates. The positions of the satellite are observed by ranging techniques (radar or laser) and the unknowns to be solved are the coefficients, $V_{n,m}$. Numerical integration algorithms have been specifically adapted to this problem and extremely sophisticated models for $\delta \boldsymbol{R}$ are employed with additional unknown parameters to be solved in order to estimate as accurately as possible the gravitational coefficients (e.g., Cappelari et al. 1976, Pavlis et al. 1999). The entire procedure falls under the broad category of dynamic orbit determination, and the corresponding gravitational spectrum modeling is classified as the *time-wise* approach, as opposed to the space-wise approach of equations (6.136) or (6.139). The partial derivatives of equation (6.141) with respect to the unknown parameters, $\boldsymbol{p} = \{\ldots, V_{n,m}, \ldots\}$, are integrated numerically in time, yielding estimates for the matrix, $\mathbf{A} = \partial \boldsymbol{x}/\partial \boldsymbol{p} = \{\ldots, \partial \boldsymbol{x}/\partial V_{n,m}, \ldots\}$. These are then used in a least-squares adjustment of the linearized model relating observed positions to parameters,

$$\delta \boldsymbol{x} = \mathbf{A}\, \delta \boldsymbol{p} + \varepsilon, \tag{6.142}$$

where $\delta \boldsymbol{x}$ and $\delta \boldsymbol{p}$ are differences with respect to previous estimates, and ε represents errors (Tapley 1973).

Tracking satellites from the ground generally cannot resolve the field with sufficient accuracy to better than maximum degree, $n_{\max} = 70$, due to the signal-to-noise ratio that is limited by the radial attenuation of the field and because of the irregular distribution of tracking stations around the world. However, so-called dedicated satellite gravity missions, in the planning stages since the 1970s were finally realized in the last two decades. The first such significant satellite mission, GRACE (Gravity Research and Climate Experiment, Tapley et al. 2004), launched in 2002 and expected to continue until 2017, yields measures of the relative gravitational perturbations using precision radar tracking between two satellites circling the Earth in tandem, in near polar orbits, and at low altitude. The resolution capability for the average static field is as high as harmonic degree 180 (Mayer-Gürr et al. 2010). The success of this mission in identifying and monitoring the temporal variations in the gravitational field due to mass fluxes has ensured follow-on missions using intersatellite tracking with radar, as well as more precise lasers.

A second type of mission in which the spatial gravitational gradients, equation (6.67), were directly sensed on orbit was GOCE (Gravity field and steady-state Ocean Circulation Explorer), with a four year lifetime, 2009–2013 (Floberghagen et al. 2011). The *in situ* type of measurement system, again in near polar orbit, yielded directly a global grid of data that could be analyzed by the space-wise approach. Due to the lower orbit of the satellite and sensing higher-order gradients of the potential, the resulting resolution of the models typically was greater (e.g., $n_{\max} = 230$, Yi et al. 2013).

Yet, the spatial resolution in geopotential field measurements offered by satellites, whether tracked or carrying sensors, is fundamentally limited by the orbital speed (~ 7 km/s) that acts as a low-pass filter on any system with finite sampling interval (or, *integration time*), typically 1–10 s. On the other hand, radar *altimetry* from satellites yields a significantly higher spatial resolution of a few km, or better, due to a combination of sufficiently narrow beamwidth and short-duration pulses (Chelton et al. 2001). The missions, Seasat, TOPEX/Poseidon, ERS, Jason, among many others and successor missions (Benveniste 2011), have yielded a detailed global grid of ocean heights that translate with corrections for the deviations from the geoid (sea surface topography) to geoid undulations (Figure 6.5) and have provided a valuable addition to global geopotential models. Equations (6.105) together with (6.62) show how these geoid undulations can be transformed to gravity anomalies that, combined with ever expanding land gravity data, form a global set that supplements satellite tracking and sensor missions to enable very-high-resolution gravitational models.

Kaula (1961) introduced the methodology, based on a statistical interpretation (the degree-variance), that optimally combines terrestrial gravimetric data with satellite tracking data (his model had maximum degree, $n_{\max} = 8$). Subsequent models by NASA (the Goddard Earth Models, GEM), by the University of Texas (Texas Earth Gravity, TEG, models), by the Ohio State University (OSU models), and by a German-French consortium (GRIM models), among others, were based on a combination of satellite tracking and terrestrial gravity data (Bouman 1997), in addition to their "satellite-only" models. The International Center for Global Earth Models (ICGEM, Barthelmes and Köhler 2012) is a digital repository for a large archival collection of past and current models. As of 2016 the ICGEM lists 157 models dating back to 1966. Many of the most recent models rely for the harmonic degrees 3 up to about $n = 180$ and $n = 280$ on the dedicated gravity mapping missions, GRACE and GOCE, respectively; and, the very low-degree harmonics up to degree 2 are still best observed with satellite laser tracking.

The Fourier-Legendre spectrum of the gravitational field is often portrayed, for convenience, in terms of degree-variances, equation (5.313), even though the low to medium frequencies are now rarely considered within the realm of the stochastic interpretation (Sections 5.6 and 5.7). Figure 6.8 shows the square roots of the degree-variances,

$$c_n\left(v_g\right) = \left(\frac{GM}{R}\right)^2 \sum_{m=-n}^{n} C_{n,m}^2, \tag{6.143}$$

of several representative global gravitational field models developed during the 50-year period 1960–2010, where

$$GM = 3.986004415 \times 10^{14} \text{ m}^3/\text{s}^2, \quad R = 6378136.3 \text{ m}, \tag{6.144}$$

and $C_{n,m}$ is the unit-less spectrum of the total field, equation (6.75). The depicted models are briefly summarized in Table 6.2. The spatial resolution corresponds to the cell size over which data are averaged, or for which predictions are made based primarily on topography. Where and when available, satellite altimetry data are the source for the estimates of the gravity anomalies on the oceans. Satellite tracking data are obtained on various satellites and used to estimate the spectrum of the gravitational potential directly up to degree, $n_{\max}^{(s)}$. The early models exhibited considerable variation in the low-degree spectrum, due primarily to the inhomogeneous distribution of global data. The high-degree model, EGM96 is also seen to be smoothed significantly at the very high degrees ($n > 230$) as compared to the presumably more accurate model, EGM2008.

The square roots of the degree-variances of the differences of the models in Table 6.2 with respect to the EGM2008 model,

$$c_n\left(v_g^{(\text{model})} - v_g^{(\text{EGM2008})}\right) = \left(\frac{GM}{R}\right)^2 \sum_{m=-n}^{n} \left(C_{n,m}^{(\text{model})} - C_{n,m}^{(\text{EGM2008})}\right)^2, \tag{6.145}$$

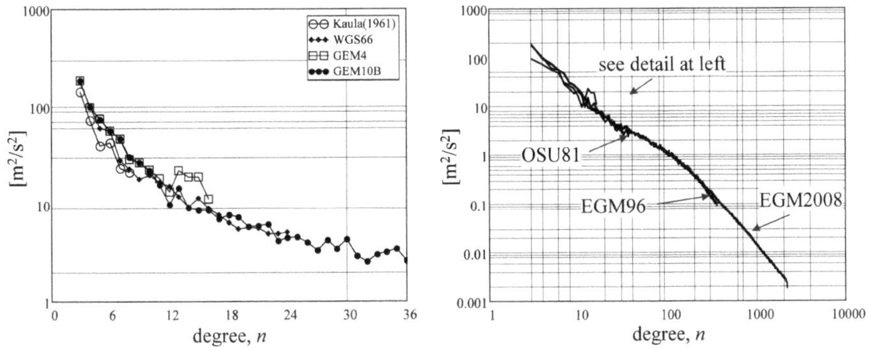

Figure 6.8: Square root of the degree-variance, equation (6.143), of Earth's gravitational potential according to various historical models. The left panel is a detail of some representative early models of the 1960s and 1970s (see Table 6.2 for references). The right panel shows these compared to the high-resolution models, OSU81 (barely distinguishable), EGM96 and EGM2008 (see text).

Table 6.2: A few spherical harmonic models of the Earth's gravitational field.

Name (year)	n_{max}	Satellite altimetry	Satellite tracking, $n_{\text{max}}^{(s)}$	Terrestrial data (incl. altimetry)	Reference
Kaula (1961)	8	no	8	$10^\circ \times 10^\circ$	Kaula (1961)
WGS66 (1966)	24	no	none	$5^\circ \times 5^\circ$	WGS Committee (1974)
GEM4 (1972)*	16	no	12	$5^\circ \times 5^\circ$	Lerch et al. (1972)
GEM10B (1978)	36	yes	30	$5^\circ \times 5^\circ$	Lerch et al. (1978)
OSU81 (1981)	180	yes	36	$1^\circ \times 1^\circ$	Rapp (1981)
EGM96 (1996)	360	yes	70	$30' \times 30'$	Lemoine et al. (1998)
EGM2008 (2008)†	2159	yes	180	$5' \times 5'$	Pavlis et al. (2012)

*with zonal coefficients up to $n = 22$. †with coefficients up to $n = 2190$ and $m = 2159$.

are shown in Figure 6.9. Also shown is EGM2008, itself, and its formal standard deviation (again, per degree). The extremely high accuracy at the low degrees results from the GRACE model on which EGM2008 is based. Assuming that these standard deviations are reasonable, the degree variances of the differences indicate the errors in the previous models, and their improvement over the years.

In summary, most spherical harmonic models of Earth's gravitational potential are constructed from a combination of data—multiple satellite tracking sets, as well as satellite gravity gradiometry, and terrestrial data comprising ground data, satellite altimetry, and data from airborne measurement systems. Such data are heterogeneous in accuracy and resolution, and come from sensors at different altitudes, all of which implies that care must be exercised in calibrating the weights assigned to the component data. Moreover, the optimal estimation of the spectrum involves the inversion of matrices that are likely ill-conditioned and require some form of regularization (Koch and Kusche 2002). In many cases such regularization and error calibration is highly tuned to the specific model

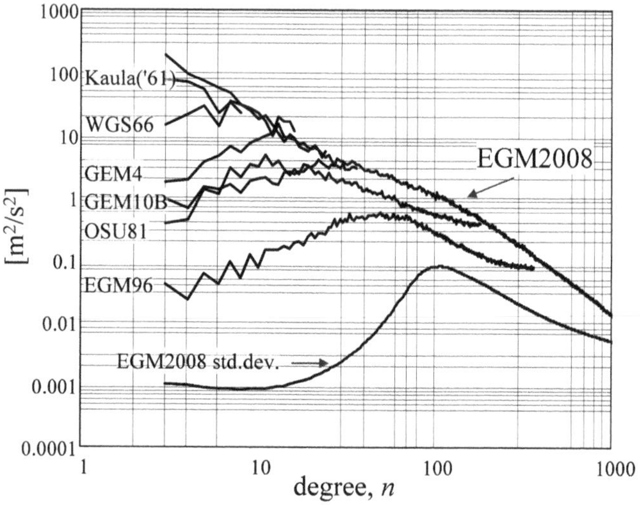

Figure 6.9: Square root of the degree-variances, equation (6.145), of the *differences* between the listed gravitation potential spectra and EGM2008. Also shown are the square roots of the degree-variances of EGM2008 and its standard deviations per degree.

under construction and is, therefore, not elaborated in this text. An excellent study in this respect is the development document for EGM96 (Lemoine et al. 1998).

With the availability of very high resolution topographic elevations from the Shuttle Radar Topography Mission (SRTM, Farr et al. 2007) and extensive ocean depth data from sonar soundings and estimates from satellite altimetry derived gravity anomalies, attention has been drawn to extend the spherical spectrum of the potential field to even higher degree under the premise that the very short wavelengths are caused primarily by the irregular geometric interface between Earth's solid surface and ocean or atmosphere. Notable in this respect is the spherical harmonic expansion up to degree and order, $n_{\max} = 10800$ based on the $1' \times 1'$ grid of topography and ocean depth, ETOPO1 (Amante and Eakins 2009), calculated by Balmino et al. (2012). Most analyses at such short wavelengths, however, are also adequately performed at the local level using Cartesian domains.

6.3.1.1 *Low-Degree Harmonics as Density Moments*

The global spectrum of the gravitational potential also has a geophysical interpretation as a set of constituent *multipoles*. Substituting the spherical harmonic series for the reciprocal distance, equation (6.13), into equation (6.17),

$$v_g(\theta,\lambda,r) = G\iiint_b \frac{\rho(\theta',\lambda',r')}{r'}\sum_{n=0}^{\infty}\left(\frac{r'}{r}\right)^{n+1}\left(\frac{1}{2n+1}\sum_{m=-n}^{n}\bar{Y}_{n,m}(\theta,\lambda)\bar{Y}_{n,m}(\theta',\lambda')\right)db$$

$$= \sum_{n=0}^{\infty}\sum_{m=-n}^{n}\left(\frac{R}{r}\right)^{n+1}\left(\frac{G}{R^{n+1}(2n+1)}\iiint_b \rho(\theta',\lambda',r')r'^n\bar{Y}_{n,m}(\theta',\lambda')\,db\right)\bar{Y}_{n,m}(\theta,\lambda) \tag{6.146}$$

shows that the spectral components (multipoles) are integrals of the mass-density distribution,

$$V_{n,m} = \frac{G}{R^{n+1}(2n+1)} \iiint_b \rho(\theta',\lambda',r') r'^n \bar{Y}_{n,m}(\theta',\lambda')\, db. \tag{6.147}$$

One may also consider the nth-order *moments of density* (from the statistics of distributions, cf. equation (5.201)) defined by[1]

$$\mu_{\alpha,\beta,\gamma}^{(n)} = \iiint_b (x')^\alpha (y')^\beta (z')^\gamma \rho(\theta',\lambda',r')\, db, \quad n = \alpha+\beta+\gamma. \tag{6.148}$$

The multipoles of degree n and the moments of order n are related; although, not all $(n+1)(n+2)/2$ moments of order n can be determined from the $2n+1$ multipoles of degree n, when $n \geq 2$. This indeterminacy is connected directly to the inability to determine the density distribution uniquely from external measurements of the potential (Chao 2005), which is the classic problem in geophysical inverse theory (Blakely 1995, Ch.10).

Equation (6.147) shows that the zero-degree harmonic coefficient of the gravitational potential, representing a monopole, is coordinate invariant and proportional to the total mass of the Earth,

$$V_{0,0} = \frac{G}{R} \iiint_b \rho(\theta',\lambda',r')\, db = \frac{GM}{R}. \tag{6.149}$$

It is also proportional to the zero-th moment of the density, $\mu_{0,0,0}^{(0)} = M$. The first-degree harmonic coefficients, representing *dipoles*, with equations (1.2), (2.218), and (2.217) (in which $y = \cos\theta$ is not to be confused with the y-coordinate), are

$$V_{1,m} = \frac{G}{\sqrt{3}R^2} \iiint_b \rho(x',y',z') \begin{Bmatrix} y', & m=-1 \\ z', & m=0 \\ x', & m=1 \end{Bmatrix} db. \tag{6.150}$$

Therefore, they are proportional to the first moments of the mass-density distribution and the coordinates of the center of mass,

$$V_{1,m} = \frac{GM}{\sqrt{3}R^2} \begin{Bmatrix} y_{cm}, & m=-1 \\ z_{cm}, & m=0 \\ x_{cm}, & m=1 \end{Bmatrix}. \tag{6.151}$$

where, e.g., $x_{cm} = \frac{1}{M} \int_M x'\, dM$ and $dM = \rho db$. It is customary to locate the origin of a global reference system of coordinates at the center of mass; then, the first-degree harmonics of the gravitational potential vanish in such a system.

[1] Clearly risking confusion, deference is given to the common nomenclature of "order" for moments and "degree" for spherical harmonics.

The second-order density moments likewise are related to the second-degree harmonic coefficients (*quadrupoles*). They also define the inertia tensor of the Earth, which is the proportionality factor in the equation that relates its angular momentum vector, $\boldsymbol{H}$, and angular velocity, $\boldsymbol{\omega}_E$,

$$\boldsymbol{H} = \mathbf{I}\boldsymbol{\omega}_E, \tag{6.152}$$

and is denoted by

$$\mathbf{I} = \begin{pmatrix} I_{xx} & I_{xy} & I_{xz} \\ I_{yx} & I_{yy} & I_{yz} \\ I_{zx} & I_{zy} & I_{zz} \end{pmatrix}. \tag{6.153}$$

The inertial tensor comprises the *moments of inertia* on the diagonal,

$$\begin{aligned} I_{xx} &= \iiint_b \rho(x',y',z')\left(y'^2 + z'^2\right) db, \\ I_{yy} &= \iiint_b \rho(x',y',z')\left(z'^2 + x'^2\right) db, \\ I_{zz} &= \iiint_b \rho(x',y',z')\left(x'^2 + y'^2\right) db; \end{aligned} \tag{6.154}$$

and the *products of inertia* off the diagonal,

$$\begin{aligned} I_{xy} &= I_{yx} = -\iiint_b \rho(x',y',z')\,x'y'db, \\ I_{xz} &= I_{zx} = -\iiint_b \rho(x',y',z')\,x'z'db, \\ I_{yz} &= I_{zy} = -\iiint_b \rho(x',y',z')\,y'z'db. \end{aligned} \tag{6.155}$$

From equation (6.148),

$$I_{xx} = \mu_{0,2,0}^{(2)} + \mu_{0,0,2}^{(2)}, \quad I_{xy} = -\mu_{1,1,0}^{(2)}, \quad etc. \tag{6.156}$$

and there are as many (6) independent tensor components as second-order density moments. The expressions for the second-degree spherical harmonic coefficients (multipoles), $V_{2,m}$, in terms of the inertia tensor elements, are *MacCullagh's* formulas,

$$\begin{aligned} V_{2,-2} &= -\frac{\sqrt{15}G}{5R^3} I_{xy}, \quad V_{2,-1} = -\frac{\sqrt{15}G}{5R^3} I_{yz}, \quad V_{2,1} = -\frac{\sqrt{15}G}{5R^3} I_{xz}, \\ V_{2,0} &= \frac{\sqrt{5}G}{10R^3}\left(I_{xx} + I_{yy} - 2I_{zz}\right), \quad V_{2,2} = \frac{\sqrt{15}G}{10R^3}\left(I_{yy} - I_{xx}\right). \end{aligned} \tag{6.157}$$

Not all density moments (or, moments of inertia) can be determined from these five harmonic coefficients. Higher-degree spectral components represent higher-degree "poles", e.g., octupoles (n=3), etc.

If the coordinate axes are chosen so as to diagonalize the inertia tensor, which can always be done, and the products of inertia are zero, then these axes are known as *principal axes of inertia*, or also *figure axes*. For the Earth the z-figure axis is very close to the spin axis (within several meters on average at the coordinate pole). Because of the proximity of the figure axis to the defined reference z-axis, I_{xz} and I_{yz}, and, consequently, the second-degree, first-order harmonic coefficients of the gravitational potential, $V_{2,-1}$ and $V_{2,1}$, are relatively small. On the other hand, because I_{zz} is very much larger than $I_{xx} + I_{yy}$ for the roughly ellipsoidal shape of the Earth, the second zonal harmonic, $V_{2,0}$, is the second largest spectral component after, $V_{0,0}$.

6.3.2 Magnetic Field Models

The global magnetic field potential of the Earth, like the gravitational potential, is modeled as a series of solid spherical harmonics, equation (6.52), on the basis of Laplace's equation (6.22). And, like the gravitational potential, the overall model, due to internal sources, is a combination of a main field and a residual field. Also, external sources are modeled separately, including, as in the case of the gravitational tides, their small indirect effect on the internal sources. The "internal field" is much larger than the "external field" and consists of the core, or main, field and the lithospheric field. The main field is due to the dynamo effect of the convection currents of the highly conductive fluid outer core of the Earth. This field is temporally varying on the scale of several years. Corresponding models are tagged with an epoch of validity and include rates of change in the low-degree harmonic coefficients. The substantially smaller, but still significant lithospheric field is due to the magnetization of the lithosphere (solid crust and upper mantle) where the temperature is less than the Curie point (e.g., 580°C for magnetite) that permits magnetization by the main field. This magnetization of all ferromagnetic minerals includes both the magnetization induced by the present main field and a relatively smaller remanent component created as crustal rocks cooled from a liquid state in the ambient main field (Blakely 1995, Ch.5).

The external field, due to electric currents in the ionosphere and magnetosphere, is highly variable in time, depending on the influence of solar radiation that also changes daily with Earth's rotation. In addition, temporal variations are induced as conducting ocean water flows through Earth's main field. These variable fields and consequent indirect effects are removed from magnetic field data when constructing the core and lithospheric models (e.g., Maus et al. 2006).

The model for the potential of the main field is a spherical harmonic expansion with significantly more components than the few zonal harmonics of the (ellipsoidal) normal gravitational field. Typical maximum degree and order is $\bar{n}$ = 13 at which degree the spectrum begins to intersect that of the field due to the lithospheric magnetization. Both fields, due to internal sources and valid in the exterior space, may be combined in the expression for the potential up to maximum degree and order, n_{max},

$$v_{\mathrm{m}}(\theta,\lambda,r;t)=\sum_{n=1}^{n_{\max}}\sum_{m=-n}^{n}\left(\frac{R}{r}\right)^{n+1} V_{n,m}^{(\mathrm{m})}(t)\bar{Y}_{n,m}(\theta,\lambda), \tag{6.158}$$

where $V_{n,m}^{(\mathrm{m})}(t)$ is the Fourier-Legendre spectrum of the potential whose modeled temporal variation usually is limited to the low degrees (say, up to $n = 8$). Note that the series starts with first-degree harmonics since the magnetic field has no monopole. Equation (6.158) is *not* the conventional expression for the spherical harmonic series of the magnetic potential. Indeed, convention has arranged for the harmonic coefficients to have SI units (Taylor and Thompson 2008, p.25) corresponding to the magnetic field, $\boldsymbol{B}$, namely, tesla [T]. Considering the basic equation (6.10), $\boldsymbol{B} = -\boldsymbol{\nabla} v_m$, this means that a factor of distance must be included explicitly; and, the conventional expression is

$$v_{\mathrm{m}}(\theta,\lambda,r;t)=R\sum_{n=1}^{n_{\max}}\sum_{m=-n}^{n}\left(\frac{R}{r}\right)^{n+1} g_{n,m}(t)\breve{Y}_{n,m}(\theta,\lambda), \tag{6.159}$$

where R is a mean Earth radius, and where the coefficients include a linear temporal variation,

$$g_{n,m}(t)=g_{n,m}(t_0)+\dot{g}_{n,m}(t_0)(t-t_0). \tag{6.160}$$

This usual notation for the harmonic coefficients is adopted here, as well, and should not cause confusion with the general notations used in other sections and chapters. Both the coefficients and the secular rate in the coefficients (for the lower-degree harmonics) are specified for a particular epoch, t_0. In many texts, the coefficients, as in the gravitational case, are separated by the cosine and sine components and denoted, respectively, by $g_{n,m}$ and $h_{n,m}$. The abbreviation that makes this distinction by the sign of the order, m, is continued here for the sake of convenience. In addition, however, another surface spherical harmonic function is introduced in equation (6.159), differing from $\bar{Y}_{n,m}(\theta, \lambda)$ in normalization. While the functions, $\bar{Y}_{n,m}(\theta, \lambda)$, are orthonormal, as in equation (2.219), the functions, $\breve{Y}_{n,m}(\theta, \lambda)$, are only orthogonal and based on the *Schmidt quasi-normalized* (or, semi-normalized) associated Legendre functions,

$$\breve{P}_{n,m}(\cos\theta)=\sqrt{\frac{1}{\varepsilon_m}\frac{(n-m)!}{(n+m)!}}P_{n,m}(\cos\theta)=\frac{1}{\sqrt{2n+1}}\bar{P}_{n,m}(\cos\theta), \tag{6.161}$$

where ε_m is given by equation (2.212). Thus, $\breve{Y}_{n,m}(\theta, \lambda) = \bar{Y}_{n,m}(\theta, \lambda)/\sqrt{2n+1}$, and

$$\frac{1}{4\pi}\iint_{\Omega}\breve{Y}_{n,m}(\theta,\lambda)\breve{Y}_{n',m'}(\theta,\lambda)\,d\Omega=\delta_{n-n'}\delta_{m-m'}/(2n+1). \tag{6.162}$$

The coefficients, $g_{n,m}$, are called *Gauss coefficients* and their values are commonly given in units of nano-tesla [nT = 10^{-9} T]; the secular variations are in units of nT/yr.

As noted in the discussion after equation (6.14), the coefficients of first degree, $g_{1,m}$, are proportional to the Cartesian components of the dipole direction at Earth's center, similar to the center-of-mass coordinates, equation (6.151),

$$\begin{pmatrix} g_{1,-1} \\ g_{1,0} \\ g_{1,1} \end{pmatrix} \sim \begin{pmatrix} \bar{Y}_{1,-1}(\theta',\lambda') \\ \bar{Y}_{1,0}(\theta',\lambda') \\ \bar{Y}_{1,1}(\theta',\lambda') \end{pmatrix} \sim \begin{pmatrix} \sin\theta'\sin\lambda' \\ \cos\theta' \\ \sin\theta'\cos\lambda' \end{pmatrix} = \begin{pmatrix} e_{y'} \\ e_{z'} \\ e_{x'} \end{pmatrix}, \tag{6.163}$$

where $\boldsymbol{e} = (e_{x'}\, e_{y'}\, e_{z'})^{\mathrm{T}}$ is a unit vector whose direction is given by the spherical latitude, $\phi' = 90° - \theta'$, and longitude, λ'. From the values of the IGRF-12 model for 2010 (see below), $g_{1,0} = -2.95 \times 10^4$ nT, $g_{1,1} = -1.586 \times 10^3$ nT, $g_{1,-1} = 4.944 \times 10^3$ nT, the Earth's magnetic dipole is directed (in 2010) toward $\phi' = -80.02°$, $\lambda' = +107.79°$. By convention, since it is near the geographic south pole, this direction identifies the *south geomagnetic pole* and its antipodal point is the *north geomagnetic pole* ($\phi' = +80.02°$, $\lambda' = -72.21°$). The actual magnetic poles on the Earth differ slightly from these due to the additional multipoles and local magnetic anomalies. The dipole accounts for approximately 90% of the total magnetic field, since the second-degree harmonics are already about an order of magnitude less than $g_{1,0}$.

Analogous to the radial component of the gravitation vector, with spectrum, $GM(n + 1)/R^2$ (cf. equation (6.81)), the Fourier-Legendre spectrum of the radial component of the magnetic field, $B_r = -\partial v_m/\partial r$, on the sphere of radius, $r = R$, and with respect to the basis functions, $\underset{\sim}{Y}_{n,m}(\theta, \lambda)$, is given by the coefficients, $(n + 1)g_{n,m}$, (and $(n + 1)g_{n,m}/\sqrt{2n+1}$ with respect to the functions, $\bar{Y}_{n,m}(\theta, \lambda)$). It is common to express and display the spectrum of the field in terms of an average power spectrum, defined by a spectral decomposition of the magnitude of $\boldsymbol{B}$,

$$\left|\boldsymbol{B}(\theta,\lambda,R)\right| = \left|\nabla v_{\mathrm{m}}(\theta,\lambda,r)_{r=R}\right| = \sqrt{\left(\frac{1}{R}\frac{\partial v_{\mathrm{m}}}{\partial\theta}\bigg|_{r=R}\right)^2 + \left(\frac{1}{R\sin\theta}\frac{\partial v_{\mathrm{m}}}{\partial\lambda}\bigg|_{r=R}\right)^2 + \left(\frac{\partial v_{\mathrm{m}}}{\partial r}\bigg|_{r=R}\right)^2}. \tag{6.164}$$

As in Section 5.5, consider the correlation function of the magnitude of $\boldsymbol{B}$ at the origin, equations (5.157) and (5.158),

$$\phi_{|\boldsymbol{B}|,|\boldsymbol{B}|}(0,R;t) = \frac{1}{4\pi}\iint_{\Omega}\left(\left(\frac{1}{R}\frac{\partial v_{\mathrm{m}}}{\partial\theta}\bigg|_{r=R}\right)^2 + \left(\frac{1}{R\sin\theta}\frac{\partial v_{\mathrm{m}}}{\partial\lambda}\bigg|_{r=R}\right)^2 + \left(\frac{\partial v_{\mathrm{m}}}{\partial r}\bigg|_{r=R}\right)^2\right)d\Omega$$

$$= \sum_{n=1}^{\infty}(n+1)\sum_{m=-n}^{n} g_{n,m}^2(t) \tag{6.165}$$

using the relationship, $V_{n,m}^{(\mathrm{m})}(t) = Rg_{n,m}(t)/\sqrt{2n+1}$. This is the total power approximately at the Earth's surface ($r = R$) of the magnitude of the internal-source field. The radial dependence in the spherical harmonic series, equation (6.159), for all the first-order derivatives of v_m is of the form, $(R/r)^{n+2}$. Hence, it is not difficult to see that at greater distances from Earth's center the total power is

$$\phi_{|B|,|B|}(0,r;t)=\sum_{n=1}^{\infty}(n+1)\left(\frac{R}{r}\right)^{2n+4}\sum_{m=-n}^{n}g_{n,m}^{2}(t). \tag{6.166}$$

The degree-variances, equation (5.138),

$$c_n(|\boldsymbol{B}|)=(n+1)\sum_{m=-n}^{n}g_{n,m}^{2}(t), \tag{6.167}$$

are often identified with the "power spectrum" (e.g., Schmitz et al. 1989) of the field, but the proper power spectrum (as defined by equation (5.137); see also (Maus 2008)) is the degree-and-order-variance,

$$\left(\Phi_{|B|,|B|}\right)_n=\frac{n+1}{2n+1}\sum_{m=-n}^{n}g_{n,m}^{2}(t), \tag{6.168}$$

although even this is an approximation to the spectrum of the correlation function of $|\boldsymbol{B}|$, as noted in connection with equation (5.159).

The first spherical harmonic expansion of the magnetic field is attributed to C.F. Gauss (Garland 1979) who in 1838 computed the eponymous coefficients, $g_{n,m}$, to degree and order $\bar{n}$ = 4 from worldwide measurements then available along several parallels of latitude. Many subsequent spherical harmonic analyses benefitted from improved distribution and accuracy in the field measurements, including the secular variations (already known in Gauss's time). The first global maps of the magnetic field from satellite data came during the period 1965–1970 from NASA's POGO (Polar Orbiting Geophysical Observatories) satellites equipped with a scalar magnetometer. This was followed by Magsat (1979–1980) and Denmark's Ørsted satellite (1999–present), both carrying also a vector magnetometer for improved resolution at higher latitudes. More recent satellites include Germany's CHAMP (2002–present) and ESA's SWARM (2013–present) satellites. They contribute to a steady improvement in resolution and accuracy of the global magnetic field models. The standard methodology for estimating the Gauss coefficients is analogous to the space-wise least-squares process for spherical harmonic modeling of the gravitational field (Section 4.5.2) (Mandea et al. 2011), although as in the gravitational case, equation (6.137), also numerical quadrature techniques have been investigated (Schmitz et al. 1989).

A number of centers, institutes, and agencies develop magnetic field models, either primarily the core model, or lithospheric models with these core models as reference, or a combination of core and lithospheric models. The International Association of Geomagnetism supports and guides the collaboration among the corresponding world-wide scientific community to develop and maintain the International Geomagnetic Reference Field (IGRF), which is revised every 5 years. The model, IGRF-12, released in 2014, is definitive for 2010 and includes a predictive model, that is subject to change, for 2015–2020 (Thébault et al. 2015); its maximum degree and order is $\bar{n}$ = 13 with secular components up to degree and order, n = 8. Another core model is the World Magnetic Model (WMM), jointly sponsored by the National Geospatial-Intelligence Agency (NGA) of the U.S. Department of

Defense (DoD) and the U.K.'s Defense Geographic Centre (DGC) and developed by the National Centers for Environmental Information (NCEI, Boulder, CO) and the British Geologic Survey (Edinburgh, Scotland) (Chulliat et al. 2015). Also, updated at 5-year intervals, the maximum degree and order of the 2015 WWM is $\bar{n} = 12$ for both the epoch-2015 coefficients and the secular variation coefficients. The reference radius for both IGRF and WMM is $R = 6371.2$ km.

Lithospheric models and combination models are now developed to high degree and order, enabled by the dedicated magnetic satellite missions, particularly CHAMP and SWARM. For example, NCEI's EMM2015 (Enhanced Magnetic Model, NOAA 2016) is a spherical harmonic expansion to degree and order $n_{\text{max}} = 720$, offering spatial resolution (half-wavelength) at the Earth's surface of about 28 km. Figure 6.10 shows the power spectra of the models, IGRF-12 and EMM2015, both in terms of the degree variances, equation (6.167), and the degree-and-order variances, equation (6.168). The different character of the global magnetic field compared to the gravitational field is apparent. The core and lithospheric spectra are clearly separated by the knee in the graph of EMM2015.

Section 6.4.2.2 shows that under some general assumptions the magnetic field is proportional to the spatial derivative of the gravitational acceleration, equation (6.231). Figure 6.11 compares the degree variances of $|\boldsymbol{B}|$ according to EMM2015 and of $\Gamma_{3,3}$, equation (6.73), according to EGM2008 (Section 6.3.1). Since the spectrum of the gravitational field is given on a sphere of radius, $r = 6378136.3$ m, the corresponding magnetic degree-variances are computed as implied by equation (6.166). Taking square roots of the degree-variances, the figure shows that at higher degrees the proportionality is approximately 10 nT/E.

6.3.3 Topographic and Isostatic Models

An obvious geophysical function suitable for spherical harmonic analysis is the height, h, of the Earth's solid surface, including both land and sea floor topography, with respect to the geoid, the level geopotential surface that approximates mean sea level (Figure 6.5). The spherical harmonic representation is given by equation (2.221),

$$h(\theta,\lambda) \sum_{n=}^{\infty}\sum_{m=-n} H_{n,m}\bar{Y}_{n,m}(\theta,\lambda). \tag{6.169}$$

Early spherical harmonic analyses of Earth's topography are those of Prey (1922) ($n_{\text{max}} = 16$), Lee and Kaula (1967) ($n_{\text{max}} = 36$, corrected by Balmino et al. 1973), and Rapp (1982) ($n_{\text{max}} = 180$) based on contemporaneous maps and data of land elevation and sea floor soundings (bathymetry), where the required data resolution, especially over the oceans, was not always adequate. Bathymetric data generally are much sparser and less accurate than land elevation data, being obtained with acoustic sounders along ship tracks, although modern systems utilize multibeam echosounders that yield swath-type surveys of the ocean bottom (Smith 1993). More uniform global maps of the sea floor topography are inferred indirectly from

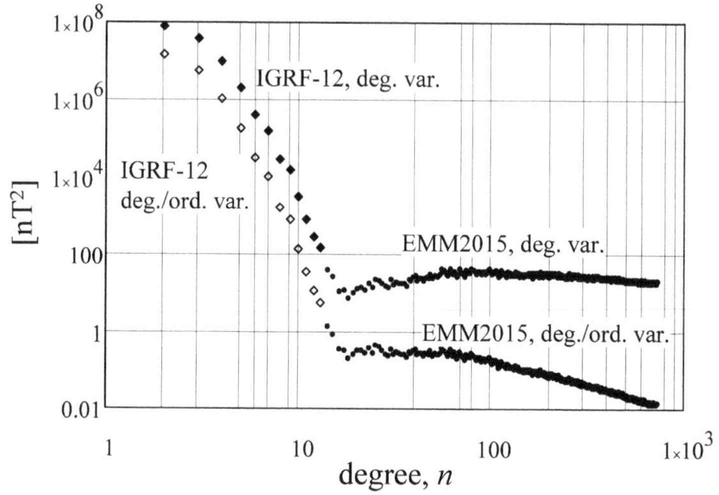

Figure 6.10: Degree variances and degree-and-order variances of the IGRF-12 and EMM2015 magnetic models on the sphere of radius, $R = 6371200$ m. EMM2015 is indistinguishable from IGRF-12 for $n \leq 13$ on this plot.

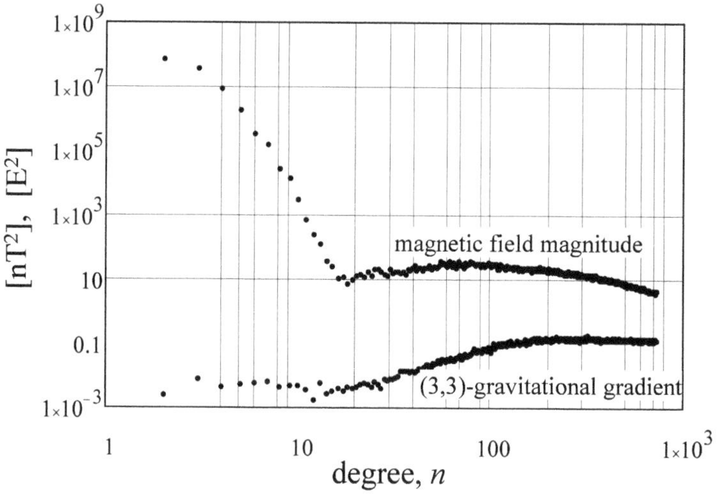

Figure 6.11: Degree variances of EMM2015 and the radial gravitational gradient according to EGM2008, both on a sphere of radius, $r = 6378136.3$ m.

satellite altimetry that determines the shape of the ocean surface (approximately the geoid), which is determined, in part, by the gravitational influence of the masses of the varying sea floor topography (Smith et al. 2005); see equation (6.224) (Section 6.4.2). Topographic heights for much of the Earth's land areas were also obtained from radar altimetry in 2000 with the Shuttle Radar Topographic Mission (SRTM, Farr et al. 2007).

Higher-resolution spherical harmonic models of the solid Earth topography were developed from global geodetic data bases in conjunction with the gravitational potential models, which facilitated the development of the latter using gravity values inferred from land topography in places where actual gravimetry was lacking. The topographic spectra constructed in association with the gravitational potential models EGM96 and EGM2008 are, respectively, JGP95E (Lemoine et al. 1998, Ch.2) and DTM2006 (Pavlis et al. 2012). The square roots of their degree-variances, equation (5.138), together with those of some of the earlier models are shown in Figure 6.12. The slopes on this log-log plot are approximately –1, which means that the degree-variances of the global topography for $n > 5$ attenuate roughly with the inverse-square of the harmonic degree, as also observed for other rocky planets (Phillips and Lambeck 1980, p.32),

$$c_n(h) \sim \frac{1}{n^2}. \tag{6.170}$$

The degree-and-order variance, equation (5.137), then attenuates with the inverse-cube of harmonic degree, as does the PSD with respect to the azimuthally averaged frequency, $\bar{f} = n/(2\pi R)$, using the relationship between spherical and planar PSDs, equation (5.165),

$$\Phi_{h,h}(\bar{f}) \sim \frac{1}{\bar{f}^3}. \tag{6.171}$$

Indeed, this attenuation is also typical for local topographies, as illustrated in Figure 6.13. The PSD slopes on the log-log plot range between –2.9 and –3.2. Note, however, that the amplitudes of the PSD vary depending on the roughness of the topography.

It is well known that topographic profiles, like a coast line, have a fractal character. That is, they resemble a process that is self similar at different scales (Mandelbrot 1983) and the PSD obeys a power law, such as equation (6.171). The fractal dimension of a profile and, by extension, for an *isotropic* process on the sphere with PSD proportional to $\bar{f}^{-\beta}$ is given by $D = 3 - \alpha/2$ (Turcotte 1987, where Turcotte's power spectral density, $S(\bar{f})$, is related to the PSD, equation (5.99), by $(2\pi\bar{f})\Phi_{h,h}(\bar{f}) = S(\bar{f})$). The slopes in Figure 6.13 imply a fractal dimension of about $D = 1.5$.

While the spherical harmonic expansions of Earth's topography are interesting in their own right, their importance in geophysics and geodesy derives from the frequent (though not universal) high correlation, or coherency, between the gravitational field and the topography at medium to high spatial frequencies (e.g., Figure 5.10). As noted, topographic heights often serve as surrogates for gravity in land areas lacking adequate gravimetry. The computation of gravity derives from the "forward model" based on the density integral, equation (6.17), usually with an assumed constant mass-density. On the other hand, at medium frequencies the topographic load on the elastic lithosphere invalidates a simple model of mere topography on the geoid. That is, the actual existence of topography under a long-term trend toward hydrostatic equilibrium for the Earth's lithosphere and liquid mantle requires a more elaborate explanation of how the sub-geoid lithospheric structure can support that topography.

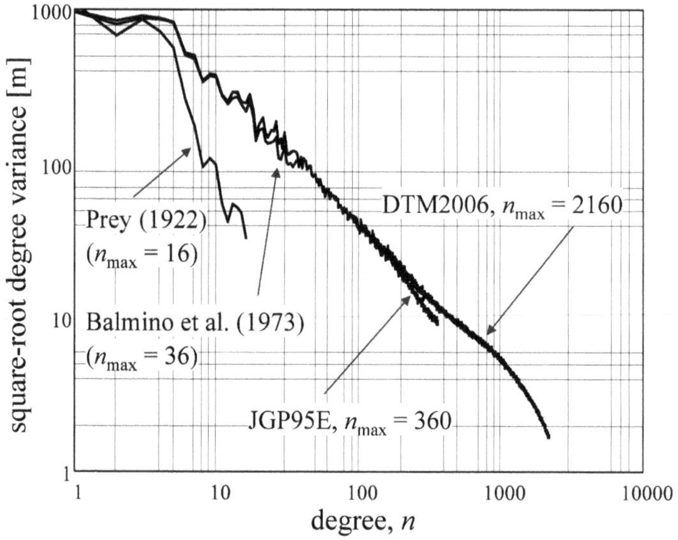

Figure 6.12: Square-roots of the degree-variances of the solid Earth's topography (land and sea floor) according to various global models. The zero-degree harmonic coefficient is approximately, $H_{0,0} \approx -2400$ m; its absolute value is not shown here. The text explains the labels.

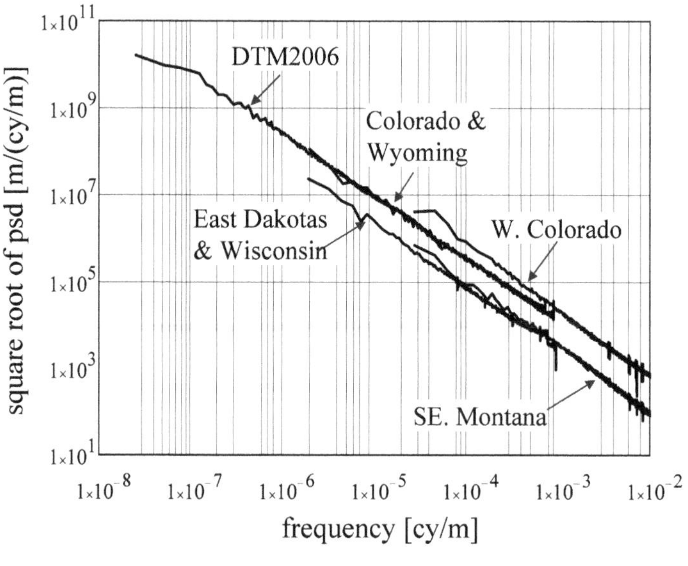

Figure 6.13: Square-roots of the azimuthally averaged PSDs of local topographies, as indicated.

This explanation is the theory of isostasy that can offer an improved forward model for the gravitational field associated with the mass-density structure of the Earth's lithosphere. Indeed, the concept of isostasy was motivated and developed from the disagreement in the mid 1800s between gravity measurements and computed mass attraction due to just the visible topography. Conversely, isostatic models and their parameters are inferred from extensive surface gravity observations.

Early isostasy models assumed only local support for topographic loads, without considering lateral stresses. For example, the model proposed by G.B. Airy in 1855 (Watts 2001) posits that each unit column of topography, with constant crust density, ρ_c, is floating in the denser mantle ($\rho_m > \rho_c$) according to the law of buoyancy. It implies that each column extends into the mantle with a "root" whose depth is proportional to the topographic height (Figure 6.14),

$$w = \frac{\rho_c}{\Delta\rho_{mc}} h, \tag{6.172}$$

where $\Delta\rho_{mc} = \rho_m - \rho_c$ and w is positive downward.

More complicated models account for the fact that the lithosphere is elastic and bends in a regional setting depending on the weight of the topographic loads and the mechanical parameters of this flexure. Assuming that the lithosphere is an elastic shell of constant thickness, T, radius R, and enclosing a highly viscous fluid mantle (Figure 6.14), the equation that describes the vertical deflection, w, due to flexure in response to a load, q, is given by (Kraus 1967, eq.6.56a; revised by Beuthe 2008, eq.88)

$$\eta \frac{D}{R^4}\left(\nabla^6_{(\theta,\lambda)} + 4\nabla^4_{(\theta,\lambda)} + 4\nabla^2_{(\theta,\lambda)}\right) w + \frac{1}{\alpha R^2}\left(\nabla^2_{(\theta,\lambda)} + 2\right) w = \left(\nabla^2_{(\theta,\lambda)} + 1 - \nu\right) q, \tag{6.173}$$

where tangential (horizontal) loads are neglected and the Laplace-Beltrami operator, $\nabla^2_{(\theta,\lambda)}$, is given by equation (2.240). The parameters of this equation are

$$\eta = \frac{\psi}{1+\psi}, \quad \psi = 12\frac{R^2}{T^2}, \quad \alpha = \frac{1}{ET}, \quad D = \frac{ET^3}{12\left(1-\nu^2\right)}, \tag{6.174}$$

where E is Young's modulus, ν is Poisson's ratio (both parameters associated with the material strength of the lithosphere), and D is known as the flexural rigidity. All terms of equation (6.173) have units of pressure (force per unit area). Again, the direction of w is positive downward.

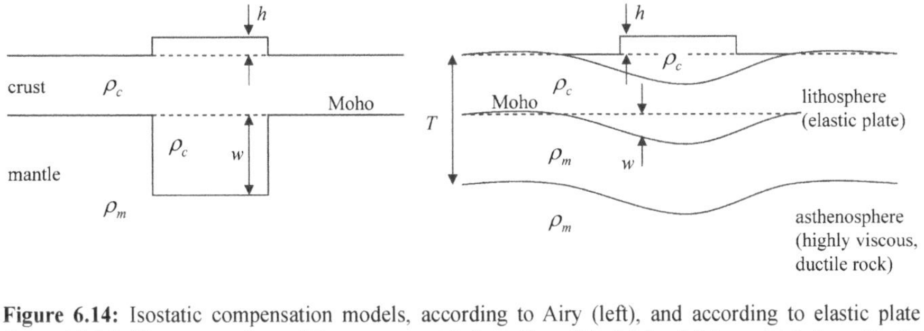

Figure 6.14: Isostatic compensation models, according to Airy (left), and according to elastic plate theory (right). The topography and the consequent deformation of the Moho (Mohorovicic) discontinuity together are responsible for the local gravity anomaly observed at the surface.

The load function, q, includes the buoyancy condition and in terms of a pressure is the resultant of the weight and the counteracting buoyancy. For the continental topography,

$$q = g_0 \rho_c h - g \rho_{mc} w, \tag{6.175}$$

where $g_0 = 9.8$ m/s^2 is an average value of gravity. For the oceans, it is convenient to replace the load of the oceans by that of an "equivalent rock layer" whose gravitational potential approximately equals that of the ocean masses (Figure 6.15). If both the equivalent rock layer, with variable thickness, b_r, and the ocean with variable depth, b, are viewed as infinitesimally thin layers of density proportional to their respective thicknesses, then the condition of equal gravitational potential implies that (see equation (6.206))

$$b_r = \frac{\rho_w}{\rho_c} b, \tag{6.176}$$

where ρ_w is the density of ocean water. The "equivalent rock topography" for the oceans is then $h_r = -b + b_r$. Defining a global equivalent topography,

$$\bar{h} = \begin{cases} h, & \text{land areas} \\ -\left(1 - \dfrac{\rho_w}{\rho_c}\right) b, & \text{ocean areas, depth } = b \geq 0 \end{cases} \tag{6.177}$$

positive upward, the load function is given by

$$q = g_0 \rho_c \bar{h} - g_0 \rho_{mc} w. \tag{6.178}$$

The gradient operator, $\nabla^{2k}_{(\theta,\lambda)}$, appearing in equation (6.173) is the kth product of the Laplace-Beltrami operator. With equation (2.241), there is

$$\nabla^{2k}_{(\theta,\lambda)} \bar{Y}_{n,m}(\theta,\lambda) = (-1)^k n^k (n+1)^k \bar{Y}_{n,m}(\theta,\lambda). \tag{6.179}$$

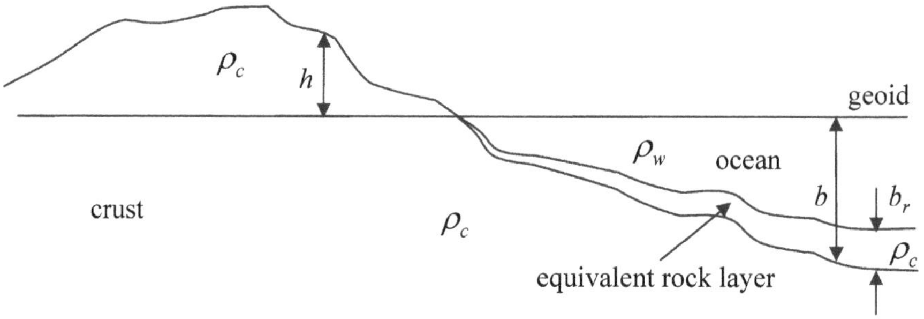

Figure 6.15: Definition of equivalent rock layer of density, ρ_c, replacing the ocean with density, ρ_w. The height of the equivalent rock topography relative to the geoid is $h_r = -b + b_r \leq 0$.

If the deflection, w, is also expanded in spherical harmonics,

$$w(\theta,\lambda)=\sum_{n=0}^{\infty}\sum_{m=-n}^{n}W_{n,m}\bar{Y}_{n,m}(\theta,\lambda), \tag{6.180}$$

then the Fourier-Legendre transform of the combination of equations (6.173) and (6.175) yields

$$W_{n,m}=\frac{g_0\rho_c\left(-n(n+1)+1-\nu\right)\bar{H}_{n,m}}{\eta\frac{D}{R^4}\left(-n^3(n+1)^3+4n^2(n+1)^2\right)-\left(\frac{1}{\alpha R^2}+g_0\Delta\rho_{mc}+4\eta\frac{D}{R^4}\right)n(n+1)+\left(\frac{2}{\alpha R^2}+g_0\Delta\rho_{mc}(1-\nu)\right)} \tag{6.181}$$

where $\bar{H}_{n,m}$ is the Fourier-Legendre transform of the global equivalent topography, $\bar{h}$, defined by equation (6.177). Typical numerical values of the parameters for the Earth's lithosphere are (Turcotte and Schubert 2002, p.123),

$$T=25\text{ km},\;\; E=70\times10^9\text{ N/m}^2,\;\; \nu=0.25,\;\; \rho_c=2800\text{ kg/m}^3,\;\; \rho_m=3300\text{ kg/m}^3. \tag{6.182}$$

With $R=6371$ km, it is easily verified that $\eta\simeq1$ and $D/R^4=5.9\times10^{-5}$ N/m^3. Hence, for large harmonic degrees, n, the numerator and denominator in equation (6.181) are dominated by the first parts in each term, and one may approximate,

$$W_{n,m}=\frac{g_0\rho_c}{\frac{D}{R^4}n^2(n+1)^2+\left(\frac{1}{\alpha R^2}+\Delta\rho_{mc}g_0\right)}\bar{H}_{n,m}. \tag{6.183}$$

Taking the inverse Fourier-Legendre transform then leads to the approximation given by Brotchie and Sylvester (1969),

$$\frac{D}{R^4}\nabla^4_{(\theta,\lambda)}w+\frac{1}{\alpha R^2}w=q. \tag{6.184}$$

If $D=0$ (the lithosphere has no flexural rigidity), then also $1/\alpha\to0$ and equation (6.173) reverts to the Airy model, equation (6.172). Using the parameter values of equations (6.182), the flexural frequency response, $W_{n,m}/\bar{H}_{n,m}$, is practically indistinguishable between equations (6.181) and (6.183), and is shown in Figure 6.16 together with the Airy response. The latter reasonably well approximates the isostatic compensation of topography at low degrees, say $n\le60$, or wavelengths greater than 600 km. For short wavelengths, the response tapers to zero as the lithosphere is able to support such loads with little or no flexure. The figure also shows the isostatic compensation according to these models for an illustrative topographic profile. The characterization of the elastic plate as a low-pass filter of the isostatic compensation is evident.

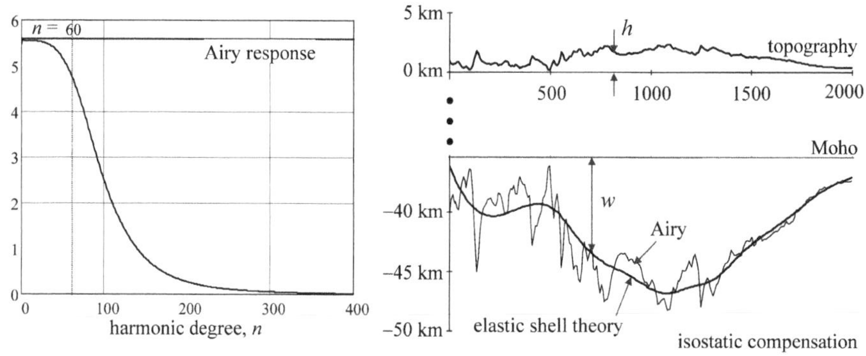

Figure 6.16: Left: Flexural frequency response for continental loads on a thin elastic lithospheric shell (parameter values given by equations (6.182)). Right: Illustrative topography and isostatic compensation according to the Airy and elastic shell theories (w is positive downward).

In the Cartesian limit as the lithospheric shell is approximated by a horizontal plate, the differential operator in equation (6.184) becomes $\nabla^4_{(\theta,\lambda)}/R^4 \rightarrow \nabla^4_{(x_1,x_2)}$, and $1/(\alpha R^2) \rightarrow 0$ as $R \rightarrow \infty$. With $\bar{f} = \sqrt{n(n+1)}/(2\pi R)$, equation (2.330), the isostatic compensation response, equation (6.183), is then

$$W\left(\bar{f}\right) = \frac{\rho_c}{\Delta\rho_{mc}}\left(1 + \frac{16\pi^4 D\bar{f}^4}{\Delta\rho_{mc}\, g}\right)^{-1} \bar{H}\left(\bar{f}\right). \tag{6.185}$$

6.4 Local Spectral Analysis

The importance of local spectral analysis in geophysics and geodesy often lies in obtaining a reasonable covariance model for further estimation in the spatial domain using such methods as kriging and least-squares collocation (Section 6.6.1). Section 5.7 presents non-parametric methods to estimate the power spectral density and covariance function from data given on a discrete domain. Unlike the global models where the physical correlations are usually understood in a non-stochastic sense, being essentially derived directly from the Fourier-Legendre spectrum (however, see also Section 5.7.3), here we may be more inclined to follow a geospatial statistical interpretation since local fields vary significantly from region to region in their high-frequency content. The following Section 6.4.1 considers some analytic models of the covariance function and its transform for the gravitational field that are constructed with an appropriate choice of parameters. Analytic models are particularly useful in propagating covariances and PSDs between linear functionals, particularly spatial derivatives, of the potential. This becomes imperative when combining heterogeneous data in least-squares collocation that depends critically on their physical cross-correlations.

Other applications of local spectral analysis include ascertaining signal correlations that may be physically indefinite in the spatial domain, but significant over a limited spectral domain. For example, Figure 5.11 shows the coherency between gravity anomalies and topography that varies considerably with respect to frequency. Similarly, within certain spectral bands one may correlate magnetic anomalies with gravitational gradients (Figures 5.13, 6.21). A spectral analysis of these signals in a region may indicate the part of the spectrum where this correlation is evident, thus leading to the appropriate manner in which such data could be combined for further geophysical or geodetic estimation or interpretation. These types of analyses are explored in Sections 6.4.2 and 6.4.3.

6.4.1 Power Laws and PSD/Covariance Models

Numerous models of the covariance function for the gravitational field have been proposed in the geodetic and geophysical literature, many from the viewpoint of the spherical spectral domain. Perhaps the most famous is a simple model for the degree variances of Earth's gravitational field, proposed by Kaula (1966, p.98) on the basis of estimated spectral components,

$$c_n = \sum_{m=-n}^{n} \delta C_{n,m}^2 = 10^{-10} \frac{2n+1}{n^4}, \tag{6.186}$$

for the unit-less coefficients, $\delta C_{n,m}$, defined in equations (6.74). This model, now known as Kaula's rule (or law), and based largely on terrestrial gravimetry up to harmonic degree, $n = 32$ (Kaula 1959), holds remarkably well even compared to the most recent spherical harmonic models, such as EGM2008 (Section 6.3.1) to maximum degree in excess of $n = 2000$ ($\bar{f} \approx 5 \times 10^{-5}$ cy/m), as shown in Figure 6.17. Several other more elaborate models have ensued based on fits to available spherical spectra of the gravitational field, usually with the intent to construct closed expressions for the covariance function of the potential and its derivatives (Tscherning 1976). One is the often-used Tscherning-Rapp model (Tscherning and Rapp 1974),

$$c_n = 4.41 \times 10^{-10} \frac{1}{(n-1)(n-2)(n+24)} (0.999617)^{n+1}, \quad n \ge 3, \tag{6.187}$$

which has a better fit for degrees, $80 \le n \le 180$; and another is an extension by H. Moritz, adapted by Rapp (1979),

$$c_n = 3.53 \times 10^{-12} \frac{(0.999617)^{n+1}}{(n-1)(n+1)} + 1.45 \times 10^{-10} \frac{(0.914232)^{n+1}}{(n-1)(n-2)(n+2)}, \quad n \ge 3, \tag{6.188}$$

which also fits better the degrees, $n \le 80$, but still rather poorly beyond degree, $n = 180$ (Figure 6.17). However, such models have lost relevance, except for rudimentary analyses or simple spectral weighting schemes, since the long-wavelength gravitational field is now quite well known from extensive satellite tracking and *in situ* satellite gradiometry (the spherical harmonic models derived from the GOCE mission, Section 6.3.1). Corresponding empirical models based on equation (5.137) suffice, and modern computers easily sum the Legendre series, equation (5.132), for the covariance function to even large finite degrees.

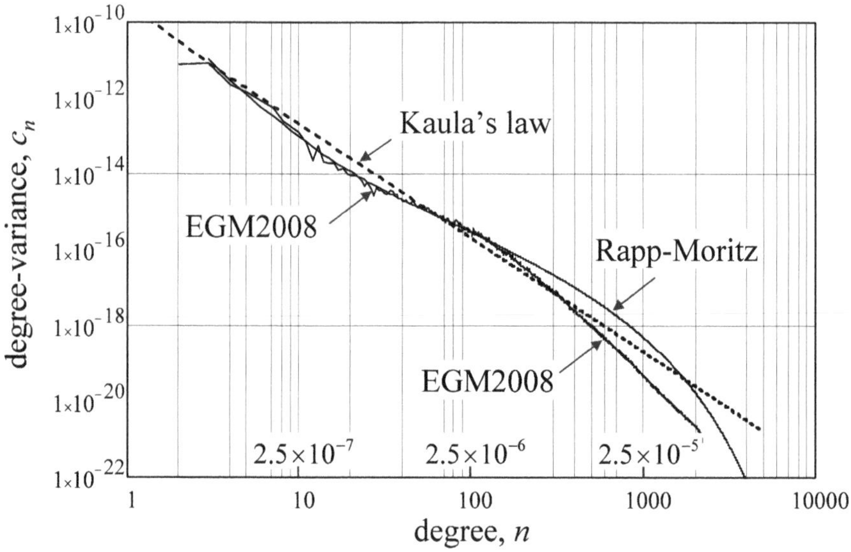

Figure 6.17: Degree variances, c_n, of EGM2008, Kaula's law, and the Rapp-Moritz model, equation (6.188), for the unit-less gravitational potential. Equivalent frequencies, $\bar{f} = n/(2\pi R)$, equation (2.331), in units of cy/m are indicated above the abscissa.

Moreover, local PSDs of the field differ significantly in magnitude from region to region, in large part as a consequence of the local topographic PSDs (Figure 6.13). Thus, for practical purposes, they cannot be simply obtained by high-pass filtering an existing global model. Since modeling the covariance function or PSD is particularly important for the shorter wavelengths (e.g., in least-squares collocation, Section 6.6), they must be designed for the particular region of interest.

A covariance model cannot be an arbitrary function—it must satisfy certain properties (Section 5.6.2) in accordance with the stochastic assumptions of the field. Specifically, for fields that are assumed to be stationary on the plane (or sphere) the covariance function depends only on the coordinate differences between points. Generally, one may also assume isotropy (directional invariance), which is virtually required for covariance models on the sphere, but not for planar approximations. The focus here is on Cartesian domains, although one could consider also spherical radial basis functions (Freeden et al. 1998) and spherical harmonic models fitted to local data are sometimes used. A particularly desirable property is the ability to express both the covariance function and its Fourier transform as closed-form analytic functions, thus more accurately enabling the propagation of covariances and PSDs to all derivatives of the field. Models in Cartesian coordinates, that is, in the planar approximation suitable for local applications, are particularly flexible in this regard.

The auto-covariance function at its origin point must be positive (indicating a variance). All auto-covariances for positive lag distances are less than the variance and the model must attenuate to zero in the limit for large lag distances. Most important is the requirement that its Fourier transform, i.e., the (auto-) PSD of the field on the plane (or sphere) is positive for all frequencies. This is another way of saying that the (auto-) covariance model must be a positive definite function. If the application of interest goes beyond estimation (interpolation or extrapolation) of a single type of data on the plane, then in terms of the potential it must also satisfy Laplace's equation (with respect to both points) if the domain extends into free space and/or the covariance propagation involves vertical derivatives.

From a geophysical perspective, Heller and Jordan (1979) defined an isotropic model for the disturbing potential, T (Section 6.2.4), based on the correlation that is generated by a white noise process for the gravitational potential on a layer at depth, b, when upward continued according to the harmonicity of the potential. The covariance function of a white noise process on a sphere is given by the delta function, equation (5.298), with some scaling factor, σ_0^2,

$$c_{w,w}(\psi) = \sigma_0^2 \sum_{n=0}^{\infty} (2n+1) P_n(\cos\psi), \tag{6.189}$$

where, from equation (5.132), its power spectrum is $(C_{w,w})_n = \sigma_0^2$, for all n. By equations (6.54) and (5.132), its upward continuation for r_1, $r_2 \geq R - b = R_b$ then models the covariance between T on two spheres of radii, r_1 and r_2, respectively,

$$c_{T,T}(\psi;r_1,r_2) = \sigma_0^2 \sum_{n=0}^{\infty} (2n+1)\left(\frac{R_b^2}{r_1 r_2}\right)^{n+1} P_n(\cos\psi), \tag{6.190}$$

with two parameters, σ_0^2 and b. The particular dependence on r_1 and r_2 immediately verifies that $c_{T,T}(\psi; r_1,r_2)$ is harmonic (satisfies Laplace's equation) with respect to either point in free space (see equation (6.51)). From Poisson's kernel, equations (6.77) and (6.78), with the substitutions, $R \rightarrow R_b^2$, $r \rightarrow r_1 r_2$, a closed expression is

$$c_{T,T}(\psi;r_1,r_2) = \sigma_0^2 \frac{R_b^2\left(r_1^2 r_2^2 - R_b^4\right)}{\left(r_1^2 r_2^2 + R_b^4 - 2 r_1 r_2 R_b^2 \cos\psi\right)^{3/2}}. \tag{6.191}$$

Let the variance on the sphere of radius, $r = R$, be $\sigma_T^2 = c_{T,T}(0;R,R)$. Then with $\rho_b = R_b^2/R^2$ and $\rho = R_b^2/(r_1 r_2)$, it is easily shown that

$$c_{T,T}(\psi;r_1,r_2) = \sigma_T^2 \frac{(1-\rho_b)^2}{\rho_b(1+\rho_b)} \frac{\rho(1-\rho^2)}{\left(1+\rho^2 - 2\rho\cos\psi\right)^{3/2}}. \tag{6.192}$$

The PSD of the disturbing potential on a sphere of radius, $r_1 = r_2 = R_1 \geq R$ is simply

$$\left(C_{T,T}\right)_n = \sigma_T^2 \frac{(1-\rho_b)^2}{\rho_b(1+\rho_b)}\left(\frac{R_b^2}{R_1^2}\right)^{n+1}. \tag{6.193}$$

This model, called the *attenuated white noise* (AWN) (or, the *Poisson*) model, when differentiated according to the laws of propagation of correlation (equations (5.146) and (6.55)), also yields models for the covariance functions of the components of the gravity disturbance vector. The corresponding power spectra follow the propagation laws, equations (6.61) and (5.155), for the PSDs of the derivatives of the potential.

The planar approximation is obtained by setting $z_1 = r_1 - R$, $z_2 = r_2 - R$, and $s^2 \approx 2R^2(1 - \cos\psi)$, where the distance, s, is given by

$$s = \sqrt{\left(x_1 - x_1'\right)^2 + \left(x_2 - x_2'\right)^2}. \tag{6.194}$$

Moritz (1980, p.183) shows that by neglecting terms of order $(z_1/R)^2$ and $(b/R)^2$,

$$\sqrt{1+\rho^2 - 2\rho\cos\psi} \approx \sqrt{s^2 + \left(2b + z_1 + z_2\right)^2} \Big/ R. \tag{6.195}$$

With the same level of approximation, it then follows (Exercise 6.7) that

$$c_{T,T}\left(s; z_1, z_2\right) \approx \sigma_T^2 \frac{4b^2\left(2b + z_1 + z_2\right)}{\left(s^2 + \left(2b + z_1 + z_2\right)^2\right)^{3/2}}. \tag{6.196}$$

The corresponding PSD is, from equation (2.334),

$$C_{T,T}\left(\bar{f}; z_1, z_2\right) = 8\pi\sigma_T^2 b^2 e^{-2\pi\bar{f}\left(2b + z_1 + z_2\right)}, \tag{6.197}$$

which shows that the covariance model is a positive definite function. The laws of propagation of covariances and PSDs follow the laws for correlations, equations (5.90), (6.47) and (6.48), and may be applied, respectively, to $c_{T,T}(s; z_1, z_2)$ and $C_{T,T}(\bar{f}; z_1, z_2)$, with due consideration of signs for the coordinates of the first and second functions that are correlated, as in equation (6.194).

Another isotropic model for the disturbing potential is the *reciprocal distance* (RD) model (Moritz 1980), given in spherical form by

$$c_{T,T}\left(\psi; r_1, r_2\right) = \frac{\sigma_T^2\left(1 - \rho_b\right)\rho/\rho_b}{\sqrt{1 + \rho^2 - 2\rho\cos\psi}}, \tag{6.198}$$

with the same notation as for equation (6.192). There is no particular geophysical genesis for this model, but it does satisfy all the required properties, such as positive definiteness and harmonicity. The Legendre transform, or power spectrum, in view of the Coulomb expansion, equation (6.12), is

$$\left(C_{T,T}\right)_n = \frac{\sigma_T^2\left(1 - \rho_b\right)}{\left(2n + 1\right)\rho_b}\rho^{n+1}. \tag{6.199}$$

The planar approximation, obtained as above, evokes the reciprocal of a distance more clearly,

$$c_{T,T}\left(s;z_1,z_2\right)=\frac{\sigma_T^2}{\sqrt{\alpha^2 s^2+\left(1+\alpha\left(z_1+z_2\right)\right)^2}}, \tag{6.200}$$

where $1/\alpha = 2(R - R_b)$. The corresponding PSD is given by equation (2.337),

$$C_{T,T}\left(\bar{f};z_1,z_2\right)=\frac{\sigma_T^2}{\alpha \bar{f}}e^{-2\pi\bar{f}\left(z_1+z_2+1/\alpha\right)}, \tag{6.201}$$

which holds for $\bar{f} \neq 0$ (recall from Section 6.2.2 that the gravitational potential in planar approximation cannot possess a zero-frequency component).

The parameter, b, for the AWN model, or α for the RD model, relates to the correlation distance of the respective model. This and the variance parameter could be used to fit the model to an empirically determined covariance function or PSD. However, the PSDs of both the AWN and the RD models for the disturbing potential, and more importantly, for the gravity disturbance (which multiplies $C_{T,T}$ by $(2\pi\bar{f})^2$, see equation (6.48)), attenuate much more rapidly than suggested empirically by actual gravimetric data. This modeling problem is solved by summing several such models, each with its own two parameters, thus obtaining a better fit to the empirical PSD. Jordan (1978) recognized this from the geophysical viewpoint by identifying the depths of major density contrasts within the Earth and implanting corresponding white noise shells using the radius parameter, R_b. However, the PSD at very high frequencies, representing shallow density contrasts, is difficult to model on the basis of a geophysical interpretation.

A different physical characterization for the gravitational PSD at high frequencies comes from the theory of fractals, as suggested by Turcotte (1987) (Section 6.3.3), which posits that the PSDs of such signals obey a "power-law", viz., $\sim\bar{f}^{-\beta}$, $\beta > 0$, that is, an attenuation in frequency that is linear on a logarithmic scale. Figure 6.13 gives evidence that topography, at least for a reasonably broad band of frequencies, indeed is a fractal; and, since under the condensation approximation (Section 6.4.2), the spectra of the gravity disturbance (at high frequencies) and of topography are linearly related, also the gravitational PSD is expected to behave like a power law. Kaula's rule, equation (6.186), is a power-law model; and, global empirical models, such as EGM2008 (Figure 6.17), confirm the power-law attenuation of the gravitational field in the domain of moderate to high frequencies. Figure 6.18 illustrates the reciprocal distance models for two local gravitational fields in terms of the gravity anomaly, approximated by $\Delta g \approx -\partial T/\partial z$. Each model, comprising 13 components, is given according to the law of propagation of correlations in the frequency domain, equation (6.48), by

$$C_{\Delta g,\Delta g}\left(\bar{f};0,0\right)=\left(2\pi\right)^2\bar{f}\sum_{j=1}^{13}\frac{\left(\sigma_T^2\right)_j}{\alpha_j}e^{-2\pi\bar{f}/\alpha_j}, \tag{6.202}$$

with parameter values for σ_T^2 and α shown in Table 6.4.

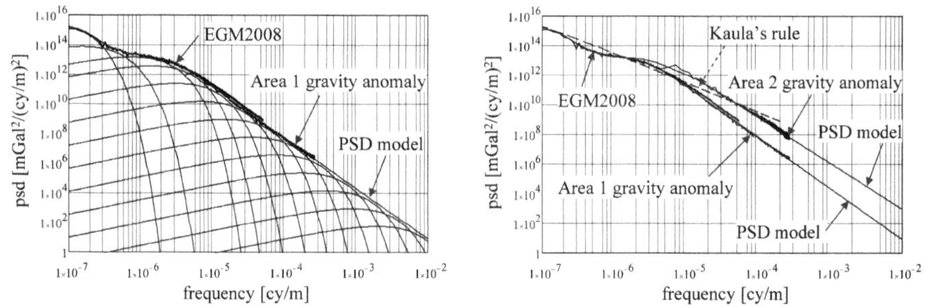

Figure 6.18: PSD models for the gravity anomaly in two areas, fitted to EGM2008 at low frequencies, empirical PSDs based on gravity anomaly data at the medium frequencies, and extended to high frequencies with an assumed power-law attenuation. Area 1, in the region of Wisconsin and the eastern Dakotas in the U.S., has relatively low topography and moderately valued gravity anomalies, and Area 2 in the region of Colorado and Wyoming has rugged topography and correspondingly strong gravity anomalies. The graph on the left illustrates how multiple components of the reciprocal distance model combine to yield the final model. The graph on the right compares the final models in the two areas with their empirical PSDs, as well as Kaula's model (dashed line).

Table 6.4: Parameter values for reciprocal distance PSD model components, equation (6.201), for the gravity anomaly in two areas of the U.S. (see Figure 6.18).

Area 1						Area 2					
j	σ^2 [m⁴/s⁴]	α [1/m]	j	σ^2 [m⁴/s⁴]	α [1/m]	j	σ^2 [m⁴/s⁴]	α [1/m]	j	σ^2 [m⁴/s⁴]	α [1/m]
1	10^5	3×10^{-7}	8	2.53×10^{-3}	2.25×10^{-4}	1	10^5	3×10^{-7}	8	4.74×10^{-2}	2.44×10^{-4}
2	3300	9.69×10^{-7}	9	1.59×10^{-4}	5.03×10^{-4}	2	3300	9.69×10^{-7}	9	3.98×10^{-3}	5.47×10^{-4}
3	650	4.76×10^{-6}	10	9.97×10^{-6}	1.13×10^{-3}	3	640	7.56×10^{-6}	10	3.34×10^{-4}	1.23×10^{-3}
4	162	8.94×10^{-6}	11	6.26×10^{-7}	2.52×10^{-3}	4	951	9.73×10^{-6}	11	2.81×10^{-5}	2.74×10^{-3}
5	10.2	2.00×10^{-5}	12	3.93×10^{-8}	5.64×10^{-3}	5	79.9	2.18×10^{-5}	12	2.36×10^{-6}	6.14×10^{-3}
6	0.641	$4.48\times10^{\times5}$	13	2.47×10^{-9}	1.26×10^{-2}	6	6.71	4.88×10^{-5}	13	1.98×10^{-7}	1.37×10^{-2}
7	4.02×10^{-2}	1.00×10^{-4}				7	0.564	1.09×10^{-4}			

The models described above satisfy all the properties of auto-covariance functions, have closed expressions for the covariance functions as well as the PSD, and are extended to the third dimension on the basis of the harmonicity of the potential. They are ideally suited to provide self-consistent covariances among all the derivatives of the field as required in least-squares collocation that optimally estimates the field in three dimensions from a possibly heterogeneous combination of derivatives of the potential (Section 6.6).

The reciprocal distance model can also be adapted to along-track data as might be collected with an aircraft, yielding closed expressions for corresponding PSDs. For example, suppose the local Cartesian coordinate system is chosen with the x_1-axis parallel to a straight-line profile of a geophysical signal, represented by the function, $g(x_1,x_2)$. The along-track PSD is the one-dimensional Fourier transform of the covariance function with respect to x_1, evaluated at $x_2 = 0$,

$$S_{g,g}(f_1;x_2) = \int_{-\infty}^{\infty} c_{g,g}(x_1,x_2)e^{-i2\pi f_1 x_1}\,dx_1. \tag{6.203}$$

For general values of x_2, $S_{g,g}(f_1;\, x_2)$ is a hybrid covariance/PSD, i.e., a covariance function in the cross-track direction and a PSD in the along-track direction. Substituting the RD model for the potential, equation (6.200), these functions may be expressed in analytic form for any of its derivatives, including both auto- and cross-hybrid covariance/PSDs (Jekeli 2010).

Other less complicated local PSD and covariance function models are also used where extension to the third dimension is not required. For example, the models for Gauss-Markov stochastic processes (Gelb 1974) often are applied in order to give a general idea of variance and correlation distance of the gravity field in inertial gravimetry (Jekeli 2000).

6.4.2 Gravity and Topography

Since the variations of the Earth's surface with respect to the geoid, whether topographic heights on the continents or depths on the oceans, are small relative to the Earth's mean radius ($R = 6371$ km), the consequent gravitational variations, at least from a global perspective, may be approximated as those due to an equivalent, infinitesimally thin density layer. Indeed, from potential theory, it can be shown using Green's identities that the potential, v, due to a three-dimensional mass-density distribution is representable by a mass-layer of density, $-(\partial v/\partial n)_s/(4\pi G)$, on any of its (external) surfaces, S, of constant potential, where the normal derivative is evaluated on this surface. It is called Chasles's Theorem in (Heiskanen and Moritz 1967, p.13) and Green's equivalent layer theorem in (Blakely 1995, p.61).

From a more proletarian viewpoint, the existence of such a density layer is not inconceivable since the mass-density distribution that generates a given potential is not unique. The equivalent density layer has no particular relationship to the three dimensional distribution other than that it generates the same potential, which for the sole purpose of modeling the *external* potential is of no physical consequence. But it simplifies the modeling problem from the boundary-value perspective by changing a triple integral to a double (surface) integral. In fact, this artifice is the foundational premise in solving Molodensky's problem in geodesy (Moritz 1980, Klees 1997).

However, for the present derivation of admittance, or frequency response, between topography and gravitation, a definite relationship between the density layer and the three-dimensional source masses is formulated. Specifically, the density layer is approximated by Helmert's condensation (Martinec 1998, p.28) whereby the layer on a sphere of radius, R, consistent with the usual spherical approximation for the geoid, has the two-dimensional density, κ, defined so as to preserve the total mass,

$$\kappa(\theta,\lambda)=\frac{1}{R^2}\int_{R}^{R+\bar{h}(\theta,\lambda)}\rho(\theta,\lambda,r)r^2dr\approx\rho_c\bar{h}(\theta,\lambda), \tag{6.204}$$

where $\bar{h}$ is the generalized topography, equation (6.177), and the last approximation assumes a constant crust density, ρ_c, and $r\approx R$. Thus, the topographic masses with density, $\rho(\theta,\lambda,r)=\rho_c$, are "condensed" onto a density layer on the geoid, where it is noted that $\bar{h}<0$ implies a deficiency in crustal matter. The gravitational potential of the condensed topography, analogous to equation (6.17), is

$$v^{(h)}(\theta,\lambda,r)=GR^2\rho_c\iint_{\Omega'}\frac{\bar{h}(\theta',\lambda')}{d}d\Omega', \tag{6.205}$$

where d is given by equation (6.79). With equation (6.169) and $r'=R$ in equation (6.12) the potential becomes (Exercise 6.8)

$$v^{(h)}(\theta,\lambda,r)=4\pi G\rho_c R\sum_{n=0}^{\infty}\sum_{m=-n}^{n}\left(\frac{R}{r}\right)^{n+1}\frac{1}{2n+1}\bar{H}_{n,m}\bar{Y}_{n,m}(\theta,\lambda), \tag{6.206}$$

where $\bar{H}_{n,m}$ is the spectrum of the global equivalent topography, equation (6.177). The gravitational effect (defined by the negative radial derivative of the potential, consistent with equation (6.63)) is then given by

$$\delta g^{(h)}(\theta,\lambda,r)=-\frac{\partial v^{(h)}}{\partial r}=4\pi G\rho_c\sum_{n=0}^{\infty}\sum_{m=-n}^{n}\left(\frac{R}{r}\right)^{n+2}\frac{n+1}{2n+1}\bar{H}_{n,m}\bar{Y}_{n,m}(\theta,\lambda). \tag{6.207}$$

The gravitational attraction of the topographic masses, $\delta g^{(h)}$, is called the (complete) Bouguer effect after the eighteenth century French geophysicist, Pierre Bouguer, who studied gravity in relation to topography and introduced the reduction of gravity measurements to sea level. *Bouguer gravity anomalies* are free-air gravity anomalies with this effect due to the equivalent topography, $\bar{h}$, removed (Hinze et al. 2013, p.149),

$$\Delta g_{\text{Bouguer}}=\Delta g-\delta g^{(h)}. \tag{6.208}$$

At $r=R$, the global model, equation (6.207), for $\delta g^{(h)}$ may be approximated by

$$\begin{aligned}\delta g^{(h)}(\theta,\lambda,R)&=4\pi G\rho_c\left(\bar{H}_{0,0}+\sum_{n=1}^{\infty}\sum_{m=-n}^{n}\frac{n+1}{2n+1}\bar{H}_{n,m}\bar{Y}_{nm}(\theta,\lambda)\right)\\&\approx4\pi G\rho_c\left(\bar{H}_{0,0}+\frac{1}{2}\sum_{n=1}^{\infty}\sum_{m=-n}^{n}\bar{H}_{n,m}\bar{Y}_{nm}(\theta,\lambda)\right)\\&=4\pi G\rho_c\left(\bar{H}_{0,0}+\frac{1}{2}\left(\bar{h}(\theta,\lambda)-\bar{H}_{0,0}\right)\right)\\&=2\pi G\rho_c\left(\bar{h}(\theta,\lambda)+\bar{H}_{0,0}\right)\end{aligned} \tag{6.209}$$

The first part of the final result is the traditional *simple* Bouguer effect due to an infinite plate of density, ρ_c, and thickness, $\bar{h}$, as in equation (6.33). The global average height, $\bar{H}_{0,0} = -1378$ m, adds a significant constant bias to local evaluations, requiring care as in all local applications of global models, especially since the Bouguer effect locally is predominantly positive on land and negative on the oceans.

An expression similar to the global Bouguer effect is easily obtained for the isostatic compensation according to the elastic lithospheric plate hypothesis (Section 6.3.3). Assuming that the material of the deflection, $w \geq 0$ (Figure 6.14), is condensed as a layer onto the bottom of the Moho at depth, b_M, the density of this layer, being a deficiency in mass relative to the mantle, is $-\Delta\rho_{mc} w(\theta', \lambda')$. The generated gravitational potential at the surface is

$$v^{(w)}(\theta,\lambda,r) = -G\Delta\rho_{mc} R^2 \iint_{\Omega'} \frac{w(\theta',\lambda')}{d_M} d\Omega', \tag{6.210}$$

where the distance from the computation point to the Moho is given by

$$d_M = \sqrt{r^2 + (R-b_M)^2 - 2r(R-b_M)\cos\psi}, \tag{6.211}$$

and where $b_M = 35000$ m is the average depth of the Moho. Using the expansion (6.180) for w and $r' = R - b_M$ in equation (6.12),

$$v^{(w)}(\theta,\lambda,r) = -4\pi G\Delta\rho_{mc} \frac{R^2}{R-b_M} \sum_{n=0}^{\infty}\sum_{m=-n}^{n} \left(\frac{R-b_M}{r}\right)^{n+1} \frac{1}{2n+1} W_{n,m} \bar{Y}_{n,m}(\theta,\lambda), \tag{6.212}$$

with corresponding gravitational effect ($\delta g^w = -\partial v^w/\partial r$),

$$\delta g^{(w)}(\theta,\lambda,r) = -4\pi G\Delta\rho_{mc} \left(\frac{R}{R-b_M}\right)^2 \sum_{n=0}^{\infty}\sum_{m=-n}^{n} \left(\frac{R-b_M}{r}\right)^{n+2} \frac{n+1}{2n+1} W_{n,m} \bar{Y}_{n,m}(\theta,\lambda). \tag{6.213}$$

If the free-air gravity anomaly is solely responsible for the effects of the topography and the isostatic compensation (i.e., the isostatic model is correct), then with equation (6.208),

$$\Delta g = \delta g^{(h)} + \delta g^{(w)} \quad \Rightarrow \quad \Delta g_{\text{Bouguer}} = \delta g^{(w)}. \tag{6.214}$$

Let $G_{n,m}^{(\text{Bouguer})}$ denote the Fourier-Legendre spectrum of $\Delta g_{\text{Bouguer}}$. With the layer approximations and the condition given by equations (6.214), the spectra of the Bouguer anomaly and lithospheric deflection, from (6.213), are related by

$$G_{n,m}^{(\text{Bouguer})} = -4\pi G\Delta\rho_{mc} \left(\frac{R-b_M}{R}\right)^n \frac{n+1}{2n+1} W_{n,m}. \tag{6.215}$$

Furthermore, the admittance, or frequency response between the Bouguer anomaly and the equivalent topography, from equation (6.183), is then expressed by

$$\frac{G_{n,m}^{(\text{Bouguer})}}{\bar{H}_{n,m}} = -4\pi G \Delta\rho_{mc} \left(\frac{R-b_M}{R}\right)^n \frac{n+1}{2n+1} \frac{g_0 \rho_c}{\frac{D}{R^4}\left(n^2(n+1)^2\right) + \left(\frac{1}{\alpha R^2} + \Delta\rho_{mc} g_0\right)}. \tag{6.216}$$

In planar approximation, for large n, $R \to \infty$, and frequency given by equation (2.331),

$$\frac{G^{(\text{Bouguer})}(f_1, f_2)}{\bar{H}(f_1, f_2)} = -\frac{2\pi G \rho_c}{1 + \frac{D}{\Delta\rho_{mc} g_0}\left(2\pi \bar{f}\right)^4} e^{-2\pi b_M \bar{f}}, \tag{6.217}$$

since, using the property of the exponential, $e^{-1} = \lim_{p \to 0} (1-p)^{\frac{1}{p}}$,

$$\lim_{R\to\infty}\left(\frac{R-b_M}{R}\right)^n = \lim_{R\to\infty}\left(1 - \frac{b_M}{R}\right)^{\frac{R}{b_M} 2\pi b_M \bar{f}} = e^{-2\pi b_M \bar{f}}. \tag{6.218}$$

For the Airy isostatic model, $D = 0$, and equation (6.217) says that the admittance between the Bouguer gravity anomaly and topography is then simply

$$\frac{G^{(\text{Bouguer})}(f_1, f_2)}{H(f_1, f_2)} \approx -2\pi G \rho_c e^{-2\pi b_M \bar{f}}. \tag{6.219}$$

As noted, these admittances assume the condition (6.214) and a particular isostatic compensation model, but without specific parameter values. Thus, a spectral analysis of Bouguer anomalies and topographic heights can be used to infer these values (conversely, the validity of a model, such as the Airy model, can be tested against alternatives suggested, for example, by seismic data; e.g., Hansen et al. 2013).

The admittance between the free-air gravity anomaly and topography follows by combining equations (6.208), (6.207) with $r = R$, and (6.216),

$$\frac{G_{n,m}}{\bar{H}_{n,m}} = 4\pi G \rho_c \frac{n+1}{2n+1}\left(1 - \frac{\Delta\rho_{mc}\left(\frac{R-b_M}{R}\right)^n g_0}{\frac{D}{R^4} n^2 (n+1)^2 + \left(\frac{1}{\alpha R^2} + \Delta\rho_{mc} g_0\right)}\right), \tag{6.220}$$

where $G_{n,m}$ is the Fourier-Legendre spectrum of Δg. Applying the same limiting procedure as for equation (6.217) the planar approximation is

$$\frac{G(f_1, f_2)}{\bar{H}(f_1, f_2)} = 2\pi G \rho_c \left(1 - \frac{e^{-2\pi b_M \bar{f}}}{1 + \frac{D}{\Delta\rho_{mc} g_0}\left(2\pi \bar{f}\right)^4}\right). \tag{6.221}$$

Formulated specifically for local applications, the equivalent topography, $\bar{h}$, in this relationship may be reverted to depth or height, as appropriate.

For the oceans, from equation (6.177), $\bar{h} = -(1 - \rho_w/\rho_c)b$. Also, in this case, instead of condensing the mass deficiency to the geoid, it is more representative for the gravity anomaly at the ocean surface to condense the mass excess of the sea floor topography (relative to the ocean mass-density) to a level surface at the mean depth, $\bar{b}$, in the region. This introduces an attenuation in the gravitational effect given by equation (6.218). Thus, writing equation (6.221) as

$$G(f_1, f_2) = Z(\bar{f})B(f_1, f_2), \tag{6.222}$$

where $B(f_1, f_1)$ is the Fourier transform of the depth to the sea floor, the admittance is

$$Z(\bar{f}) = -2\pi G(\rho_c - \rho_w)\left(1 - \frac{e^{-2\pi b_M \bar{f}}}{1 + \dfrac{D}{\Delta\rho_{mc} g_0}(2\pi\bar{f})^4}\right) e^{-2\pi\bar{b}\bar{f}}. \tag{6.223}$$

Figure 6.19 shows that the admittance defines a "spectral window" wherein surface gravity anomalies are maximally correlated with sea floor topography. The flexure of the lithosphere is responsible for the attenuation at low frequencies—long-wavelength loads are not well supported and are isostatically compensated thus negating the gravitational effect of the topography. Short-wavelength loads are more easily supported and not isostatically compensated. The attenuation at the high end of the spectrum is due to the upward continuation of the gravitational attraction of the sea-floor topography and is governed by the factor, $e^{-2\pi\bar{b}\bar{f}}$, as demonstrated in the right panel of the figure. It is also evident in the figure that the isostatic compensation is negligible for $\bar{f} > 0.01$ cy/km, or wavelengths shorter than 100 km.

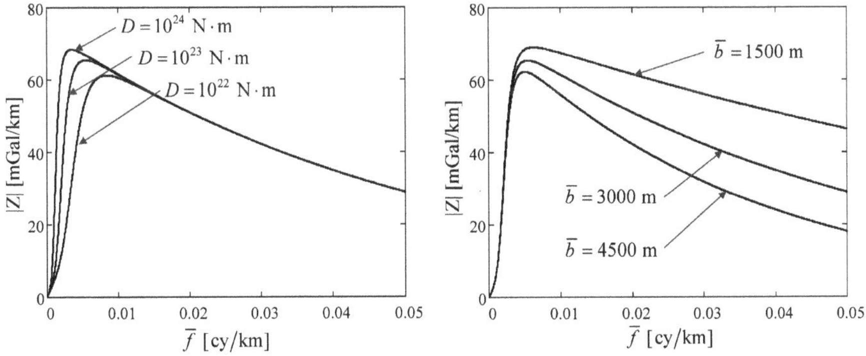

Figure 6.19: Absolute value of admittance, equation (6.223), between sea floor topography (depth) and surface gravity anomaly, assuming a mean depth of $\bar{b} = 3000$ m on the left and a mean flexural rigidity of $D = 10^{23}$ N·m on the right. Other parameter values are as in equations (6.182).

The inverse admittance has been used to advantage in mapping the sea floor using gravity anomalies inferred from satellite altimetry (Smith and Sandwell 1994). Satellite altimetry over the oceans first determines the geoid undulation, N, from the altimeter measurement, h_{alt} (height of the satellite above the sea surface), relative to the computed height, h_{sat}, of the satellite above a given ellipsoid (Shum et al. 1995, Chelton et al. 2001),

$$N = h_{\text{sat}} - \left(h_{\text{alt}} + \delta h_{\text{alt}}\right), \tag{6.224}$$

where δh_{alt} includes various corrections to the altimeter measurement in order to obtain the geoid undulation. Then, from Bruns's formula (6.62) and equation (6.123), the local spectrum of the gravity anomaly follows directly from the altimeter-derived geoid undulation,

$$G(f_1, f_2) = 2\pi \bar{f} \gamma \mathcal{F}(N). \tag{6.225}$$

Alternatively, the computed geoid slopes, $\eta = -\partial N/\partial x_1$, $\xi = -\partial N/\partial x_2$ (equations (6.62) and (6.125)), also lead to the gravity anomaly spectrum by way of equation (6.132) for $x_3 = 0$,

$$G(f_1, f_2) = i \frac{}{} \left(f_1 \mathcal{F}(\eta) + f_2 \mathcal{F}(\xi)\right), \tag{6.226}$$

which, because of the differentiation of N, reduces long-wavelength errors, such as orbit error, in equation (6.224) (McAdoo 1990, Olgiati et al. 1995).

Since the inverse admittance, Z^{-1}, tends to infinity at high and low frequencies, a band-pass filter is required to extract a spectral band of the topography from the estimated gravity anomalies. The design of the filter depends in particular on the mean ocean depth and on the flexural rigidity, D, of the ocean lithosphere, which, unlike the continental lithosphere, varies significantly due to the large range of ages (Watts 2001); see, for example, (Smith and Sandwell 1994).

6.4.3 Poisson's Relationship in the Frequency Domain

If one can assume a physical connection in crustal material between the induced magnetization and the mass-density (certain rock types have a particular magnetic susceptibility), then it is possible to relate the magnetic anomaly and spatial gradients of the gravity anomaly. If the mass-density is constant, $\rho = \rho_0$, then the mass-density integral, equation (6.17), in Cartesian coordinates is simply

$$v_g(x_1, x_2, x_3) = G\rho_0 \iiint_b \frac{1}{d}\, db. \tag{6.227}$$

where the distance, d, is given by equation (6.24) and $db = dx_1'\, dx_2'\, dx_3'$. Similarly, suppose that the magnetization of a body is constant in magnitude *and* direction, $\boldsymbol{\eta}(x_1', x_2', x_3') = \eta_0 \boldsymbol{\kappa}_0$, where $\boldsymbol{\kappa}_0$ is a unit vector of constant direction, $\boldsymbol{\kappa}_0 = (e_1^{(k)}\, e_2^{(k)}\, e_3^{(k)})^{\text{T}}$, and $(e_1^{(k)})^2 + (e_2^{(k)})^2 + (e_3^{(k)})^2 = 1$. Then, the directional derivative in equation (6.18) is

$$\boldsymbol{\kappa}_0 \cdot \boldsymbol{\nabla} = \frac{\partial}{\partial \kappa_0} = e_1^{(\kappa)} \frac{\partial}{\partial x_1} + e_2^{(\kappa)} \frac{\partial}{\partial x_2} + e_3^{(\kappa)} \frac{\partial}{\partial x_3}; \tag{6.228}$$

and, the scalar magnetic potential in Cartesian coordinates becomes

$$v_{\mathrm{m}}\left(x_1, x_2, x_3\right) = -\frac{\mu_0 \eta_0}{4\pi} \frac{\partial}{\partial \kappa_0} \iiint_b \frac{1}{d}\, db. \tag{6.229}$$

Comparing equations (6.229) and (6.227) under these conditions shows that the scalar magnetic potential is related to the directional derivative of the gravitational potential, in Cartesian coordinates,

$$v_{\mathrm{m}}\left(x_1, x_2, x_3\right) = -\frac{\mu_0 \eta_0}{4\pi G \rho_0} \frac{\partial}{\partial \kappa_0} v_{\mathrm{g}}\left(x_1, x_2, x_3\right). \tag{6.230}$$

This is known as *Poisson's relationship*.

In terms of their gradients,

$$\Delta \boldsymbol{B}\left(x_1, x_2, x_3\right) = \frac{\mu_0 \eta_0}{4\pi G \rho_0} \frac{\partial}{\partial \kappa_0} \Delta \boldsymbol{g}\left(x_1, x_2, x_3\right), \tag{6.231}$$

where $\Delta \boldsymbol{B} = -\boldsymbol{\nabla} \boldsymbol{v}_{\mathrm{m}}$ is the vector magnetic anomaly due to the magnetization and $\Delta \boldsymbol{g} = \boldsymbol{\nabla} \boldsymbol{v}_{\mathrm{g}}$ is the *vector* gravitational anomaly (defined as the positive gradient of the potential). Being relevant only to the magnetic field of the magnetized lithospheric and with the assumptions of constant magnetization and mass-density, it is a relationship that may hold, if at all, primarily locally. Thus, indeed, it is appropriate to consider a planar approximation (Cartesian coordinates) and interpret $\Delta \boldsymbol{B}$ and $\Delta \boldsymbol{g}$ as magnetic and gravitational vector *anomalies* relative to their respective reference fields. The right side of equation (6.231) is also called the pseudo-magnetic anomaly. In view of the directional derivative, equation (6.228), it may be written as

$$\Delta \boldsymbol{B}\left(x_1, x_2, x_3\right) = \frac{\mu_0 \eta_0}{4\pi G \rho_0} \boldsymbol{\nabla} \Delta \boldsymbol{g}^{\mathrm{T}}\left(x_1, x_2, x_3\right) \boldsymbol{\kappa}_0. \tag{6.232}$$

The scalar magnetic anomaly is defined as

$$\Delta B\left(x_1, x_2, x_3\right) = \left|\boldsymbol{B}\left(x_1, x_2, x_3\right)\right| - \left|\boldsymbol{B}_0\left(x_1, x_2, x_3\right)\right|, \tag{6.233}$$

where $\boldsymbol{B}$ is the total field and $\boldsymbol{B}_0$ is the main field (Section 6.3.2). Since $|\Delta \boldsymbol{B}| \ll |\boldsymbol{B}|$, $\boldsymbol{B}$ and $\boldsymbol{B}_0$ are nearly parallel and Figure 6.20 shows that the scalar magnetic anomaly is then the projection of the vector anomaly onto the direction of $\boldsymbol{B}_0$, i.e., $\Delta B \approx \boldsymbol{\kappa}_0 \cdot \Delta \boldsymbol{B}$. Therefore,

$$\Delta B\left(x_1, x_2, x_3\right) = \frac{\mu_0 \eta_0}{4\pi G \rho_0} \boldsymbol{\kappa}_0^{\mathrm{T}} \boldsymbol{\Gamma}\left(x_1, x_2, x_3\right) \boldsymbol{\kappa}_0. \tag{6.234}$$

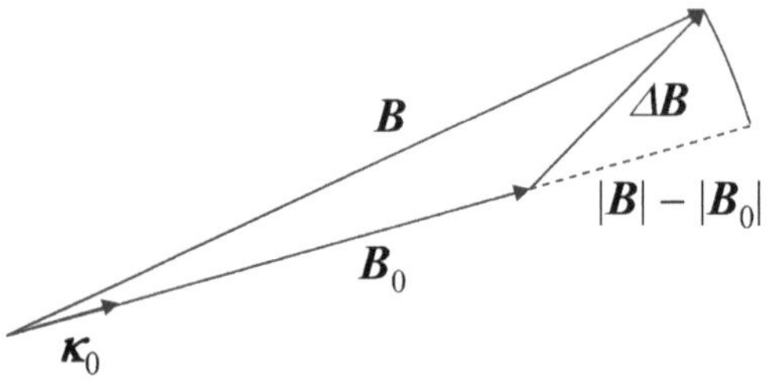

Figure 6.20: The relationship between the total magnetic field, the main magnetic field, and the magnetic anomaly, not to scale.

where $\boldsymbol{\Gamma} = \boldsymbol{\nabla} \boldsymbol{\Delta g}^{\mathrm{T}}$ is the tensor of (anomalous) gravitational gradients, equation (6.67). For local comparisons of the gravitational gradient anomaly and magnetic anomaly an additional scaling is necessary since the respective reference fields do not correspond in the same way.

The assumption on the magnetization, η, may be relaxed slightly by supposing that it has only constant direction, $\boldsymbol{\eta}(x_1', x_2', x_3') = \eta(x_1', x_2', x_3')\boldsymbol{\kappa}_0$, and considering the *condensation* of the three-dimensional (scalar) densities onto a single layer at level, $x_3^{(0)} \le 0$,

$$\bar{\eta}_{x_3^{(0)}}(x_1', x_2') = \int_{x_3'} \eta(x_1', x_2', x_3')\, dx_3', \quad \bar{\rho}_{x_3^{(0)}}(x_1', x_2') = \int_{x_3'} \rho(x_1', x_2', x_3')\, dx_3'. \tag{6.235}$$

Then, the gravitational and magnetic residual potentials, equations (6.227) and (6.229), are convolutions (assuming formally that the area of integration is $E = (-\infty,\infty) \times (-\infty,\infty)$),

$$v_{\mathrm{g}}(x_1, x_2, x_3) = G \iint_E \frac{\bar{\rho}_{x_3^{(0)}}(x_1', x_2')}{d_{x_3^{(0)}}}\, dx_1' dx_2', \tag{6.236}$$

$$v_{\mathrm{m}}(x_1, x_2, x_3) = -\frac{\mu_0}{4\pi} \frac{\partial}{\partial \kappa_0} \iint_E \frac{\bar{\eta}_{x_3^{(0)}}(x_1', x_2')}{d_{x_3^{(0)}}}\, dx_1' dx_2', \tag{6.237}$$

where

$$d_{x_3^{(0)}} = \sqrt{(x_1 - x_1')^2 + (x_2 - x_2')^2 + \left(x_3 - x_3^{(0)}\right)^2}. \tag{6.238}$$

Now, if these condensed densities have the same proportion to each other for all (x_1', x_2'),

$$\bar{\eta}_{x_3^{(0)}}(x_1', x_2') = \frac{\eta_0}{\rho_0} \bar{\rho}_{x_3^{(0)}}(x_1', x_2'), \tag{6.239}$$

then, again, one obtains equation (6.230).

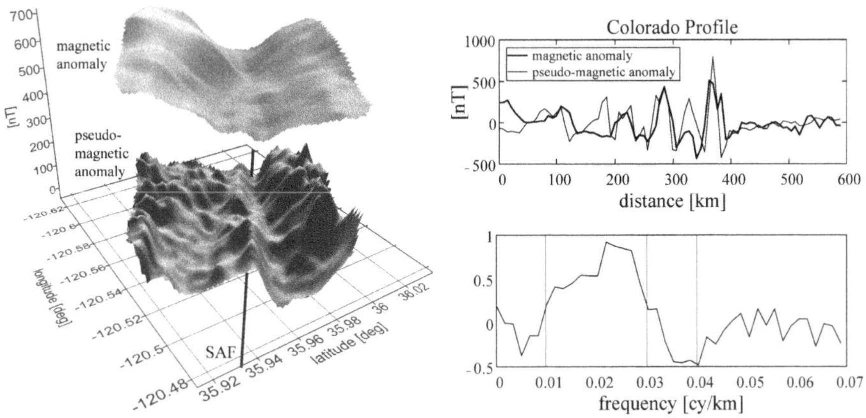

Figure 6.21: Two contrasting examples of Poisson's relationship. The maps on the left straddle the San Andreas Fault (SAF) near Parkfield, CA. The Colorado profiles on the right characterize the relationship in both the spatial domain (top, same as Figure 5.13) and frequency domain (bottom), the latter in terms of coherency. See the text for further explanations.

Poisson's relationship in the frequency domain follows by first noting that the convolution in equation (6.237) is also a harmonic function in free space, since it is a Newtonian type of potential (see equation (6.21)). Hence, the transforms of its derivatives are given by the appropriate transforms (6.46). Specifically, the two-dimensional Fourier transform of the magnetic potential, equation (6.230), with equation (6.228), is (Gunn 1975)

$$V_{\mathrm{m}}(f_1, f_2; x_3) = -\frac{\mu_0 \eta_0}{2G\rho_0}\left(i\left(e_1^{(\kappa)} f_1 + e_2^{(\kappa)} f_2\right) - e_3^{(\kappa)} \bar{f}\right) V_{\mathrm{g}}(f_1, f_2; x_3). \tag{6.240}$$

Similarly, the Fourier transform of the magnetic anomaly, equation (6.234), leads to

$$\mathcal{F}(\Delta B) = \frac{\mu_0 \eta_0}{4\pi G \rho_0} \boldsymbol{\kappa}_0^{\mathrm{T}} \mathcal{F}(\Gamma) \boldsymbol{\kappa}_0. \tag{6.241}$$

Various combinations of derivatives of the anomalous magnetic and gravitational potentials thus can be constructed in the frequency domain under the assumption given by equation (6.239) and using the transforms of derivatives of potentials, equation (6.46).

The validity of Poisson's relationship is illustrated in Figure 6.21, demonstrating both the coherency between gravitational gradients and magnetic anomalies, already indicated in Figure 5.13, as well as its reliance on the underlying assumptions. On the left, both the actual magnetic anomaly (top map) and pseudo-magnetic anomaly (bottom map) generally indicate the magnetic highs on either side of the San Andreas Fault (SAF) near Parkfield, CA. However, the pseudo-magnetic anomaly, in addition, reflects a significant ridge of gravity gradient along the fault that is due to a non-magnetic granitic body between the magnetic bodies (McPhee et al. 2004). The right side (bottom) of Figure 6.21 shows the spectral coherency in a different region between the magnetic anomaly and the pseudo-magnetic anomaly derived from gravity gradients (refer to the explanation of Figure 5.13, which is repeated on the right side (top) for comparison). The coherency is greatest (> 0.5) for frequencies, 0.016 cy/km $\le f \le$ 0.028 cy/km, corresponding to wavelengths, 36 km to 63 km, while virtually no coherency exists at shorter wavelengths. The coherency shown here is computed on the basis of the phase coherency, equation (5.338), and low-pass filtered by a moving average in the frequency domain to highlight the trends.

6.5 Convolutions by FFT

The convolutions described in Section 6.2.3 and their spectral equivalents offer two methods to relate the derivatives of the gravitational potential, either in the spatial domain or the spectral domain. Since an accurate high-degree global spherical harmonic model (or, even an ellipsoidal harmonic model) is now available from satellite gravity missions, such as GOCE (Section 6.2.1), these relationships are also easily formulated by the derivatives of such a harmonic model in the spatial domain. However, detail in the field that is finer than implied by the global model comes only from local higher-resolution data and can supplement the global model either by spectral analysis or by convolution. The classic example in physical geodesy is the computation of the geoid undulation from gravity anomalies (Stokes's integral, equation (6.96)) using a *remove-restore* technique, whereby the global, long-wavelength model, with harmonic coefficients, $\delta C_{n,m}$, $n \le n_{\max}$, is first removed from the anomaly and then restored in the geoid undulation. Just as equation (3.191) is equivalent to equation (3.195), equation (6.96) is readily shown to be equivalent to (Exercise 6.9)

$$N(\theta,\lambda)=\frac{R}{4\pi\gamma}\iint\limits_{\Omega'}\left(\Delta g(\theta',\lambda',R)-\Delta g_M(\theta',\lambda',R)\right)S(\psi)\,d\Omega'+N_M(\theta,\lambda), \tag{6.242}$$

where from equation (6.82)

$$\Delta g_M(\theta,\lambda,R)=\gamma\sum_{n=2}^{n_{\max}}\sum_{m=-n}^{n}(n-1)\,\delta C_{n,m}\bar{Y}_{n,m}(\theta,\lambda), \tag{6.243}$$

and, hence, by the convolution theorem (3.60),

$$N_M(\theta,\lambda,r)=\frac{R}{4\pi\gamma}\iint\limits_{\Omega'}\Delta g_M(\theta',\lambda',R)S(\psi)\,d\Omega'=R\sum_{n=2}^{n_{\max}}\sum_{m=-n}^{n}\delta C_{n,m}\bar{Y}_{n,m}(\theta,\lambda). \tag{6.244}$$

Since the residual, $\delta\Delta g(\theta,\lambda,R)=\Delta g(\theta',\lambda',R)-\Delta g_M(\theta',\lambda',R)$, only contains wavelengths for harmonic degree greater than $n_{\max}$, one could, in theory, also replace Stokes's function by

$$\delta S(\psi) = S(\psi) - \sum_{n=2}^{n_{max}} \frac{2n+1}{n-1} P_n(\cos\psi), \tag{6.245}$$

which, by the orthogonality of spherical harmonics, does not change the integral in equation (6.242). Figure 6.22 shows an example of the involved signals if the global model has maximum harmonic degree and order, $n_{max} = 180$.

Similar procedures may be implemented for any of the other convolutions among the derivatives of the potential in Section 6.2.4. Invariably, the kernel, whether in its original form or modified by removal of its long-wavelength components, exhibits rapid attenuation away from the evaluation point of the integral (Figures 6.2, 6.6, 6.7). A truncation of the integral to the immediate neighborhood of the evaluation point thus is justified from the viewpoint of numerical accuracy; and, indeed, a planar approximation often suffices, as well. Further numerical efficiency may be obtained using FFTs if the data are regularly gridded. Modifications other than equation (6.245) are considered in Section 3.6.3 in order to reduce the error in neglecting the integration outside the data region, i.e., due to data windowing; see also (Featherstone 2013) for a review of many similar modifications. However, in applying the remove-restore procedure with high-degree spherical harmonic models, these have become less important in some applications (e.g., Sanso and Sideris (2013, p.459) suggest using unmodified Stokes's kernel in geoid computations). Thus, it is also assumed here that all kernels are unmodified. The following section reviews the procedures for utilizing the FFT for various types of integrals encountered in physical geodesy, whether they are already convolutions or must be suitably reconstructed. As such these methods process only local short-wavelength information of the field and are formulated strictly in Cartesian coordinates. Although all quantities in these integrals refer to residuals under the remove-restore procedure, the notation for the

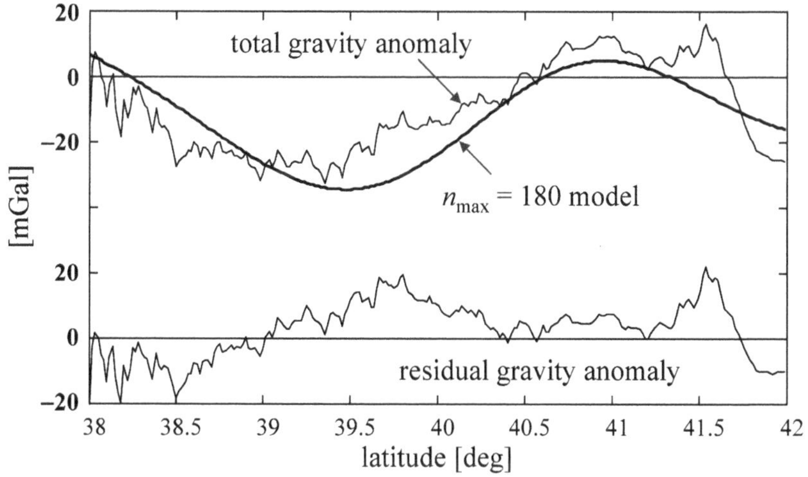

Figure 6.22: A north-south profile of the gravity anomaly in Ohio contrasting the long-wavelength reference model and the high-frequency residual anomaly.

gravitational quantities, T, Δg, η, ξ, is retained with the understanding that they refer to a high-degree model.

6.5.1 Integrals in Physical Geodesy

Section 4.4 shows in general how a discrete planar convolution may be approximated by a discrete cyclic convolution using FFTs. This is investigated here more specifically in terms of the approximations associated with the discretization of the kernel function and its spectrum. In addition, the kernels have singularities at the origin and these must be treated appropriately.

The convolutions of Section 6.2.4 are one of two basic types, those whose kernel attenuates essentially as the inverse of distance, or as a higher power of the inverse distance, where the distance in the planar approximation of the convolution is given by equation (6.24). Stokes's kernel is of the former type, while the Vening-Meinesz and radial derivative kernels belong to the more rapidly decaying type. Both types have a singularity at $d = 0$ when $x_3 = 0$. In either case, one must compute the integral separately over the immediate area around the evaluation point. In the first type, representing a *weakly singular* integral, the result is a straightforward evaluation of the integral by standard numerical approximation. The Vening-Meinesz kernel, depending on the inverse squared distance, represents a *strongly singular integral*; and, the radial derivative is an example of a *hypersingular* integral, where the kernel depends on the inverse cubed distance. It is still possible to compute these strongly singular and hypersingular integrals, assuming a sufficiently smooth data function, using a Cauchy principal value. The idea is to regularize the integrals near the singularity, thus yielding weakly singular integrals that can then be numerically integrated by standard methods (Klees and Lehman 1998). Hofmann-Wellenhof and Moritz (2005, pp.125–127) derive the Cauchy principal values for the Vening-Meinesz and radial derivative integrals.

The planar approximation of the Pizzetti integral, equation (6.86), with approximate kernel, $\bar{S}(x_1, x_2, x_3)$, equation (6.124), and $R^2 d\Omega' \to dx_1' dx_2'$ is given by

$$T(x_1, x_2, x_3) = \frac{1}{2\pi} \iint_E \frac{\Delta g(x_1', x_2')}{\sqrt{(x_1 - x_1')^2 + (x_2 - x_2')^2 + x_3^2}} dx_1' dx_2', \tag{6.246}$$

which is also the planar approximation of Hotine's integral with the gravity disturbance as input function. The integration, or data region, E, is a sub-domain of the infinite plane and also defines the limits of the evaluation points, (x_1, x_2). For present purposes, E is finite and rectangular with sides parallel to the coordinate axes, representing the data and evaluation domains with the coordinate origin at a corner of E, as illustrated in Figure 6.23 (a data region centered on the origin could also be considered with appropriate changes in notation). Equation (6.246) in the spectral domain is equivalent to (as in equation (6.123))

$$\mathcal{F}(T(x_1, x_2, x_3)) = \frac{1}{2\pi \bar{f}} e^{-2\pi x_3 \bar{f}} \mathcal{F}(\Delta g(x_1, x_2, 0)). \tag{6.247}$$

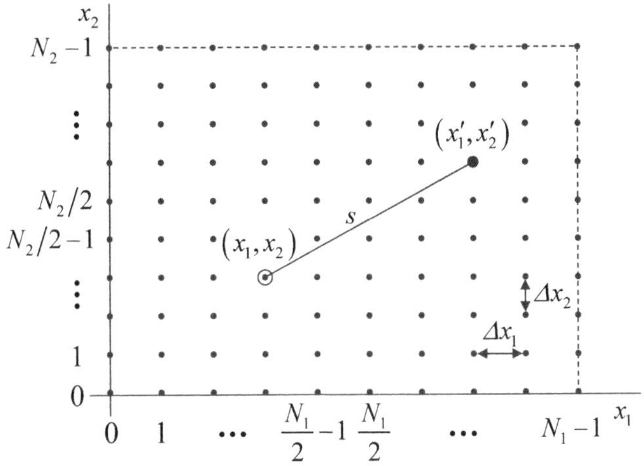

Figure 6.23: Local Cartesian data domain, E, with integration point, (x_1', x_2'), relative to the evaluation point, (x_1, x_2). The grid points are $(n_1\Delta x_1, n_2\Delta x_2)$, $n_1 = 0,\ldots, N_1 - 1$, $n_2 = 0,\ldots, N_2 - 1$ with even N_1 and N_2.

Clearly, equation (6.246) is a two-dimensional convolution as in equation (3.24) (if E is the infinite plane), where the Fourier (Hankel) transform of the kernel function has an analytic form for non-zero frequencies, equation (6.27). The altitude, x_3, serves here strictly as a constant parameter. In order to apply the FFT (DFT) method for computing the convolution the data samples must be distributed on a regular grid on the plane,

$$\Delta\tilde{g}_{n_1,n_2} = \Delta g\left(n_1\Delta x_1, n_2\Delta x_2\right), \quad n_1 = 0,\ldots,N_1-1, \quad n_2 = 0,\ldots,N_2-1; \tag{6.248}$$

where Δx_1 and Δx_2 are constant (not necessarily equal) sampling intervals. The discrete domain is finite in extent, indicated by integers N_1 and N_2. As prescribed for the DFT (Sections 4.3.3 and 4.3.4), it is assumed that the data are periodically extended over the entire plane. The corresponding cyclic convolution error can be eliminated with appropriate zero-padding, as shown in Section 4.4.

Let $\tilde{G}_{k_1,k_2}$ be the DFT of $\Delta\tilde{g}_{n_1,n_2}$, defined by equation (4.90). Computationally one can now proceed along two alternative numerical paths—discretize either the convolution, equation (6.246), or its spectral counterpart, equation (6.247); that is, discretize either the kernel or its spectrum. In either case, the discretization must properly convolve or filter the data with the recognition that both the input, $\Delta\tilde{g}_{n_1,n_2}$, and the output are real-valued arrays and that the DFT assumes periodicity in both the input array *and* the kernel.

6.5.1.1 Weakly Singular Integral

The practical implementation of the Fourier transform of the Pizzetti convolution, equation (6.247), requires that the frequency response, $e^{-2\pi x_3 \bar{f}}/(2\pi\bar{f})$, is discretized in the same way as the transform of the gravity anomalies, $\tilde{G}_{k_1,k_2}$, defined here for wave numbers, $k_{1,2} = 0,\ldots,N_{1,2} - 1$. Since both Δg and T are real-valued, their discrete spectra satisfy the conjugate symmetry property (4.94), $\tilde{G}_{N_1-k_1,N_2-k_2} = \tilde{G}^*_{k_1,k_2}$, for all k_1 and k_2, and similarly for the transform of T. This means that $e^{-2\pi x_3 \bar{f}}/(2\pi\bar{f})$ must be discretized in a way that preserves this property. Moreover, its singularity at $\bar{f} = 0$ (which exists even if $x_3 > 0$) can be avoided by setting the spectrum to zero at zero frequency, meaning that with this method the constant in T for the region, E, is not determined from the data. This is no great loss since local analysis, whether in the space domain or the frequency domain, is unable to solve for the long wavelengths of the field. Thus, let

$$\tilde{C}^{(x_3)}_{k_1,k_2} = \begin{cases} \dfrac{1}{2\pi\bar{f}_{k_1,k_2}} e^{-2\pi x_3 \bar{f}_{k_1,k_2}}, & k_{1,2} = -N_{1,2}/2,\ldots,N_{1,2}/2-1, \quad |k_1|+|k_2| \neq 0 \\ 0, & |k_1|+|k_2| = 0 \end{cases} \tag{6.249}$$

with imposed periodicity, $\tilde{C}^{(x_3)}_{k_1,k_2} = \tilde{C}^{(x_3)}_{k_1+q_1N_1,k_2+q_2N_2}$ for any integers, q_1, q_2, and where

$$\bar{f}_{k_1,k_2} = \sqrt{\left(\frac{k_1}{N_1\Delta x_1}\right)^2 + \left(\frac{k_2}{N_2\Delta x_2}\right)^2}. \tag{6.250}$$

Using its periodicity define $\tilde{C}^{(x_3)}_{k_1,k_2}$ on the frequency domain,

$$\tilde{C}^{(x_3)}_{k_1,k_2} = \begin{cases} \tilde{C}^{(x_3)}_{k_1-N_1,k_2}, & \dfrac{N_1}{2} \le k_1 \le N_1 - 1, \quad 0 \le k_2 \le \dfrac{N_2}{2} - 1 \\ \tilde{C}^{(x_3)}_{k_1,k_2-N_2}, & 0 \le k_1 \le \dfrac{N_1}{2} - 1, \quad \dfrac{N_2}{2} \le k_2 \le N_2 - 1 \\ \tilde{C}^{(x_3)}_{k_1-N_1,k_2-N_2}, & \dfrac{N_1}{2} \le k_1 \le N_1 - 1, \quad \dfrac{N_2}{2} \le k_2 \le N_2 - 1 \end{cases} \tag{6.251}$$

Then, since $\tilde{C}^{(x_3)}_{k_1,k_2}$ is real-valued, $\tilde{G}_{k_1,k_2}\tilde{C}^{(x_3)}_{k_1,k_2}$ preserves the required symmetries; and, the convolution, approximated by

$$\tilde{T}^{(x_3)}_{\ell_1,\ell_2} = \mathrm{DFT}^{-1}\left(\tilde{G}_{k_1,k_2}\tilde{C}^{(x_3)}_{k_1,k_2}\right)_{\ell_1,\ell_2}, \tag{6.252}$$

for $\ell_1 = 0,\ldots,N_1 - 1$ and $\ell_2 = 0,\ldots,N_2 - 1$, yields real values, $\tilde{T}^{(x_3)}_{\ell_1,\ell_2}$, on the data grid.

The alternative approach starts with a discretization of the integral in equation (6.246), in essence, by the rectangle rule for numerical integration,

$$T(\ell_1\Delta x_1,\ell_2\Delta x_2,x_3) = \frac{\Delta x_1\Delta x_2}{2\pi}\sum_{n_1=-\infty}^{\infty}\sum_{n_2=-\infty}^{\infty}\frac{\Delta g(n_1\Delta x_1,n_2\Delta x_2)}{\sqrt{(\ell_1-n_1)^2\Delta x_1^2+(\ell_2-n_2)\Delta x_2^2+x_3^2}}. \tag{6.253}$$

where, for the moment, E is the entire plane. If $x_3 = 0$, the singularity is circumvented by extracting the integral over a small neighborhood of the evaluation point; for example,

$$\delta T(x_1,x_2,0)=\frac{1}{2\pi}\int_{\omega=0}^{2\pi}\int_{s=0}^{s_0}\Delta g(x_1',x_2')\frac{1}{s}sdsd\omega\approx\Delta g(x_1,x_2)s_0 \tag{6.254}$$

where s, ω are polar coordinates (Figure 2.12) and the distance, s_0, is the radius of the neighborhood, approximated by the radius of the circle with area equal to that of the grid cell,

$$s_0\approx\sqrt{\frac{\Delta x_1\Delta x_2}{\pi}}. \tag{6.255}$$

The singularity of the integral in effect disappears, hence the designation of the integral as weakly singular.

Accounting for this separate contribution when $x_3 = 0$, and truncating the convolution, equation (6.253), to the finite area, E, its cyclic approximation is

$$\tilde{T}_{\ell_1,\ell_2}^{(x_3)}=\Delta x_1\Delta x_2\sum_{n_1=0}^{N_1-1}\sum_{n_2=0}^{N_2-1}\Delta\tilde{g}_{n_1,n_2}\tilde{u}_{\ell_1-n_1,\ell_2-n_2}+\begin{cases}\sqrt{\dfrac{\Delta x_1\Delta x_2}{\pi}}\Delta\tilde{g}_{\ell_1,\ell_2}, & x_3=0\\ 0, & x_3>0\end{cases} \tag{6.256}$$

where the discrete, periodic kernel is defined by

$$\tilde{u}_{m_1,m_2}^{(x_3)}=\begin{cases}\dfrac{1}{2\pi\sqrt{m_1^2\Delta x_1^2+m_2^2\Delta x_2^2+x_3^2}}, & m_{1,2}=-\dfrac{N_{1,2}}{2},\ldots\dfrac{N_{1,2}}{2}-1,\ |m_1|+|m_2|\neq 0,\text{ any } x_3\geq 0\\ \dfrac{1}{2\pi x_3}, & |m_1|+|m_2|=0,\text{ and } x_3>0\\ 0, & |m_1|+|m_2|=0,\text{ and } x_3=0\end{cases} \tag{6.257}$$

with $\tilde{u}_{m_1+p_1N_1,m_2+p_2N_2}^{(x_3)}=\tilde{u}_{m_1,m_2}^{(x_3)}$ for any integers, p_1, p_2, so that on the data domain (Figure 6.24),

$$\tilde{u}_{m_1,m_2}^{(x_3)}=\begin{cases}\tilde{u}_{m_1-N_1,m_2}^{(x_3)}, & \dfrac{N_1}{2}\leq m_1\leq N_1-1,\quad 0\leq m_2\leq\dfrac{N_2}{2}-1\\ \tilde{u}_{m_1,m_2-N_2}^{(x_3)}, & 0\leq m_1\leq\dfrac{N_1}{2}-1,\quad \dfrac{N_2}{2}\leq m_2\leq N_2-1\\ \tilde{u}_{m_1-N_1,m_2-N_2}^{(x_3)}, & \dfrac{N_1}{2}\leq m_1\leq N_1-1,\quad \dfrac{N_2}{2}\leq m_2\leq N_2-1\end{cases} \tag{6.258}$$

In this way, the kernel, extended periodically in both directions over the entire plane, correctly represents the weights assigned to the data near the evaluation point. If $\tilde{U}_{k_1,k_2}^{(x_3)}$ is the discrete Fourier transforms of $\tilde{u}_{m_1,m_2}^{(x_3)}$, defined by equation (4.90), then equation (4.115) yields for the convolution

$$\tilde{T}^{(x_3)}_{\ell_1,\ell_2} = \mathrm{DFT}^{-1}\left(\tilde{G}_{k_1,k_2}\tilde{U}^{(x_3)}_{k_1,k_2}\right)_{\ell_1,\ell_2} + \begin{cases} \sqrt{\dfrac{\Delta x_1 \Delta x_2}{\pi}}\Delta\tilde{g}_{\ell_1,\ell_2}, & x_3 = 0 \\ 0, & x_3 > 0 \end{cases} \tag{6.259}$$

Corresponding formulas for equations (6.252) and (6.259) using the FFT and its inverse are obtained by relating these to the DFT with appropriate scale factors.

It is noted that the transform of this discretized kernel does not equal the discretized kernel transform, $\tilde{U}^{(x_3)}_{k_1,k_2} \neq \tilde{C}^{(x_3)}_{k_1,k_2}$; and, therefore, equations (6.252) and (6.259) are not the same approximation to the convolution. Figure 6.25 compares the discretized kernel, equation (6.257), and the inverse DFT of its spectrum, equation (6.249), as well as the discretized spectrum and the DFT of the kernel, $\tilde{U}^{(x_3)}_{k_1,k_2}$. In each case the discrete Fourier transform deviates from the true function, especially

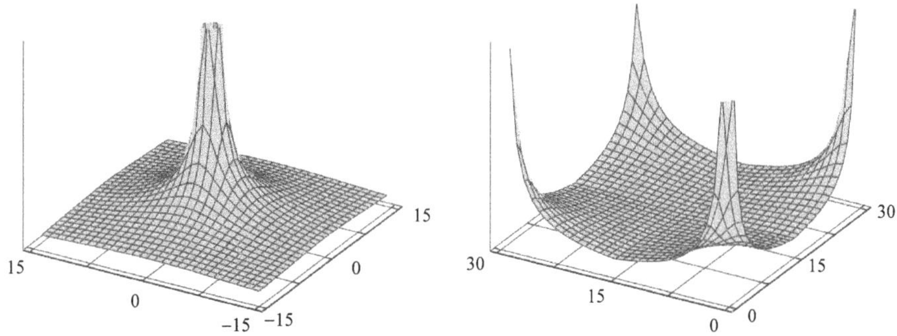

Figure 6.24: Stokes's kernel discretized for the cyclic convolution if the data grid is centered on the coordinate origin (left) and if the data grid occupies the first quadrant (right), as in Figure 6.23. The discretization omits the singularity of the kernel at the origin.

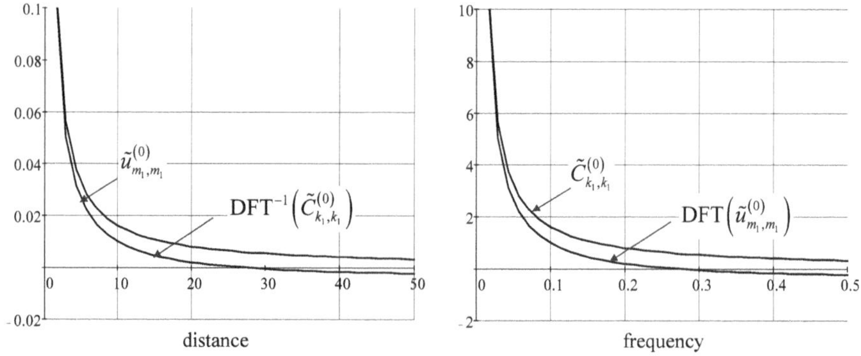

Figure 6.25: Left: The discretized kernel, $\tilde{u}^{(0)}_{m_1,m_1}$ compared to the inverse DFT of its analytic spectrum, $\mathrm{DFT}^{-1}(\tilde{C}^{(0)}_{k_1,k_1})$. Right: The discretized spectrum, $\tilde{C}^{(0)}_{k_1,k_1}$, compared to the DFT of the kernel, $\mathrm{DFT}(\tilde{u}^{(0)}_{m_1,m_1})$. Discrete points are connected for clarity. $N_1 = N_2 = 100$, $\Delta x_1 = \Delta x_2 = 1$, $x_3 = 0$.

in giving erroneous negative values. However, empirical evidence supports the conjecture that discretizing the kernel, which is most significant in a neighborhood of its origin, yields better numerical results than discretizing its analytical frequency response that represents the entire spectral domain (Sideris and Li 1993). That is, in seeking an accurate yet efficient computation of the convolution, the equation (6.259), which initially transforms space-domain functions, is preferred to equation (6.252).

6.5.1.2 *Strongly Singular Integral*

An example of a strongly singular integral is the Vening-Meinesz formula for the deflection of the vertical, equation (6.108). In planar approximation and view of the Vening-Meinesz kernels, equations (6.127), the integrals for $x_3 = 0$ are

$$\begin{Bmatrix} \eta(x_1, x_2) \\ \xi(x_1, x_2) \end{Bmatrix} = \frac{1}{2\pi\gamma} \iint_E \frac{\Delta g(x_1', x_2')}{\left((x_1 - x_1')^2 + (x_2 - x_2')^2\right)^{3/2}} \begin{Bmatrix} x_1 - x_1' \\ x_2 - x_2' \end{Bmatrix} dx_1' dx_2', \tag{6.260}$$

where, with s given by equation (6.194) and $x_1 - x_1' = s\cos\omega$, $x_2 - x_2' = s\sin\omega$, the kernel varies inversely with the squared distance. As in the case of the Pizzetti integral, one may consider two methods for an efficient calculation of the convolution. For the first, the discretized, periodic versions of the Fourier transforms of the kernels are

$$\tilde{D}^{(\eta)}_{k_1,k_2} = \begin{cases} -i\dfrac{\dfrac{k_1}{N_1 \Delta x_1}}{\bar{f}_{k_1,k_2}}, & k_{1,2} = -N_{1,2}/2, \ldots, N_{1,2}/2 - 1, \quad |k_1| + |k_2| \neq 0, \\ 0, & |k_1| + |k_2| = 0 \end{cases} \tag{6.261}$$

$$\tilde{D}^{(\xi)}_{k_1,k_2} = \begin{cases} -i\dfrac{\dfrac{k_2}{N_2 \Delta x_2}}{\bar{f}_{k_1,k_2}}, & k_{1,2} = -N_{1,2}/2, \ldots, N_{1,2}/2 - 1, \quad |k_1| + |k_2| \neq 0, \\ 0, & |k_1| + |k_2| = 0 \end{cases} \tag{6.262}$$

where $f_{k_{1,2}} = k_{1,2} / \left(N_{1,2} \Delta x_{1,2}\right)$ and $\bar{f}_{k1,k2}$ is given by equation (6.250). Nulling the spectra at zero frequency forces a zero average for both the gravity anomaly and the vertical deflection components in the integration region; a non-zero average must be analyzed separately. Also, periodicity implies that

$$\tilde{D}^{(\eta,\xi)}_{k_1+q_1N_1, k_2+q_2N_2} = \tilde{D}^{(\eta,\xi)}_{k_1,k_2}, \quad \text{for all integers } q_1, q_2. \tag{6.263}$$

The required conjugate symmetry given by equations (4.94) through (4.97) additionally constrains the definition of these discrete periodic spectra,

$$\tilde{D}^{(\eta,\xi)}_{-k_1,0} = \left(\tilde{D}^{(\eta,\xi)}_{k_1,0}\right)^*, \quad \tilde{D}^{(\eta,\xi)}_{-k_1,-N_2/2} = \left(\tilde{D}^{(\eta,\xi)}_{k_1,-N_2/2}\right)^*, \quad k_1 = 0, \ldots, \frac{N_1}{2} - 1, \tag{6.264}$$

$$\tilde{D}_{0,-k_2}^{(\eta,\xi)} = \left(\tilde{D}_{0,k_2}^{(\eta,\xi)}\right)^*, \quad \tilde{D}_{-N_1/2,-k_2}^{(\eta,\xi)} = \left(\tilde{D}_{-N_1/2,k_2}^{(\eta,\xi)}\right)^*, \quad k_2 = 0,\ldots,\frac{N_2}{2}-1, \tag{6.265}$$

in addition to the requirement of real values for $\tilde{D}_{0,0}^{(\eta,\xi)}, \tilde{D}_{0,-N_2/2}^{(\eta,\xi)}, \tilde{D}_{-N_1/2,0}^{(\eta,\xi)}, \tilde{D}_{-N_1/2,-N_2/2}^{(\eta,\xi)}$, obtained, e.g., by setting them to their absolute value. Having thus defined the discrete Fourier transform of the Vening-Meinesz kernels, one may work also in the first quadrant of the coordinate system, as in equation (6.251). Then, by the convolution theorem, equation (4.115), the deflection components are approximated by

$$\begin{Bmatrix} \tilde{\eta}_{\ell_1,\ell_2} \\ \tilde{\xi}_{\ell_1,\ell_2} \end{Bmatrix} = \frac{1}{\gamma} \begin{Bmatrix} \mathrm{DFT}^{-1}\left(\tilde{G}_{k_1,k_2} \tilde{D}_{k_1,k_2}^{(\eta)}\right) \\ \mathrm{DFT}^{-1}\left(\tilde{G}_{k_1,k_2} \tilde{D}_{k_1,k_2}^{(\xi)}\right) \end{Bmatrix}_{\ell_1,\ell_2}, \tag{6.266}$$

where the constraints imposed on $\tilde{D}_{k_1,k_2}^{(\eta,\xi)}$ ensure real values for the computed deflection components on the data grid.

The second approach, which discretizes the integral in equation (6.260), must account for its strong singularity at the evaluation point. With equation (6.255), the contribution to the deflection components in its immediate neighborhood may be approximated by (Hofmann-Wellenhof and Moritz 2005, pp.126-127)

$$\begin{Bmatrix} \delta\eta(x_1,x_2) \\ \delta\xi(x_1,x_2) \end{Bmatrix} \simeq -\frac{1}{2\gamma}\sqrt{\frac{\Delta x_1 \Delta x_2}{\pi}} \begin{Bmatrix} \dfrac{\partial}{\partial x_1}\Delta g(x_1,x_2) \\ \dfrac{\partial}{\partial x_2}\Delta g(x_1,x_2) \end{Bmatrix}. \tag{6.267}$$

The discretization of equation (6.260) then is given by

$$\begin{Bmatrix} \tilde{\eta}_{\ell_1,\ell_2} \\ \tilde{\xi}_{\ell_1,\ell_2} \end{Bmatrix} = \begin{Bmatrix} \delta\eta_{\ell_1,\ell_2} \\ \delta\xi_{\ell_1,\ell_2} \end{Bmatrix} + \frac{\Delta x_1 \Delta x_2}{\gamma} \sum_{n_1=0}^{N_1-1}\sum_{n_2=0}^{N_2-1} \Delta\tilde{g}_{n_1,n_2} \begin{Bmatrix} \tilde{w}_{\ell_1-n_1,\ell_2-n_2}^{(\eta)} \\ \tilde{w}_{\ell_1-n_1,\ell_2-n_2}^{(\xi)} \end{Bmatrix}, \quad \ell_{1,2} = 0,\ldots,N_{1,2}-1, \tag{6.268}$$

where (Figure 6.26)

$$\begin{Bmatrix} \tilde{w}_{m_1,m_2}^{(\eta)} \\ \tilde{w}_{m_1,m_2}^{(\xi)} \end{Bmatrix} = \begin{cases} \dfrac{1}{2\pi\left(m_1^2 \Delta x_1^2 + m_2^2 \Delta x_2^2\right)^{3/2}} \begin{Bmatrix} m_1 \Delta x_1 \\ m_2 \Delta x_2 \end{Bmatrix}, & 0 \le m_{1,2} \le \dfrac{N_{1,2}}{2} - 1, \quad |m_1| + |m_2| \ne 0 \\ 0, & |m_1| + |m_2| = 0 \end{cases} \tag{6.269}$$

with

$$\tilde{w}_{m_1,m_2}^{(\eta,\xi)} = \begin{cases} \tilde{w}_{m_1-N_1,m_2}^{(\eta,\xi)}, & \dfrac{N_1}{2} \le m_1 \le N_1 - 1, \quad 0 \le m_2 \le \dfrac{N_2}{2} - 1 \\ \tilde{w}_{m_1,m_2-N_2}^{(\eta,\xi)}, & 0 \le m_1 \le \dfrac{N_1}{2} - 1, \quad \dfrac{N_2}{2} \le m_2 \le N_2 - 1 \\ \tilde{w}_{m_1-N_1,m_2-N_2}^{(\eta,\xi)}, & \dfrac{N_1}{2} \le m_1 \le N_1 - 1, \quad \dfrac{N_2}{2} \le m_2 \le N_2 - 1 \end{cases} \tag{6.270}$$

Now, if $\tilde{W}^{(\eta,\xi)}_{k_1,k_2} = \mathrm{DFT}\left(\tilde{w}^{(\eta,\xi)}_{m_1,m_2}\right)_{k_1,k_2}$, then by equation (4.115),

$$\begin{Bmatrix} \tilde{\eta}_{\ell_1,\ell_2} \\ \tilde{\xi}_{\ell_1,\ell_2} \end{Bmatrix} = \begin{Bmatrix} \delta\tilde{\eta}_{\ell_1,\ell_2} \\ \delta\tilde{\xi}_{\ell_1,\ell_2} \end{Bmatrix} + \frac{1}{\gamma}\mathrm{DFT}^{-1}\left(\tilde{G}_{k_1,k_2}\begin{Bmatrix} \tilde{W}^{(\eta)}_{k_1,k_2} \\ \tilde{W}^{(\xi)}_{k_1,k_2} \end{Bmatrix}\right)_{\ell_1,\ell_2}. \tag{6.271}$$

The first term on the right side can be an appreciable amount. For example, if the average horizontal gradient of the gravity anomaly in a grid cell is 10 E (= 1 Eötvös = 10^{-9} s^{-2}) and $\Delta x_1 = \Delta x_2 = \sqrt{\pi}$ km $\approx$ 1.8 km, then the contribution is about 0.1 arcsec.

The DFTs, $\tilde{W}^{(\eta,\xi)}_{k_1,k_2}$, of the Vening-Meinesz kernels only approximate their (discretized) true spectra, $\tilde{D}^{(\eta,\xi)}_{k_1,k_2}$, as shown for the deflection, ξ, in Figure 6.27. From equation (6.262), the limiting values of the Fourier transform, $\tilde{D}^{(\xi)}_{k_1,k_2}$, for $k_1 = 0$ and $\mp k_2 > 0$ are $\pm i$. The large difference in $\tilde{D}^{(\xi)}_{k_1,k_2}$ between domains, $k_1 < 0$ and $k_2 > 0$, is due to the singularity of the kernel, which when integrated over x_1 is

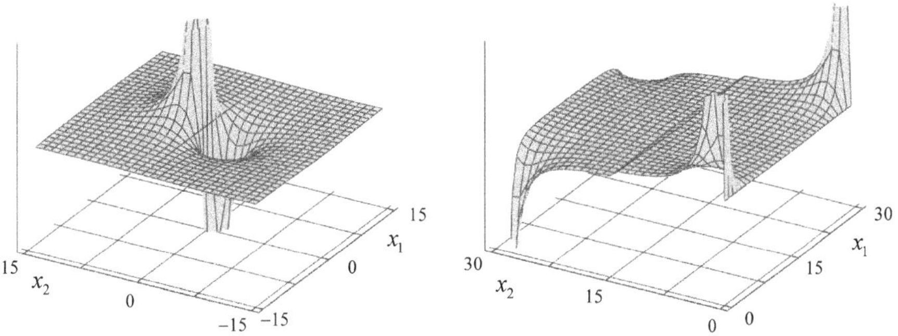

Figure 6.26: Vening-Meinesz kernel (north-south deflection, ξ) discretized for the cyclic convolution if the data grid is centered on the coordinate origin (left) and if the data grid occupies the first quadrant (right), as in Figure 6.23. The apparent discontinuity for $x_2 = 15$ on the right is due to the opposite signs for the kernel on either side of the $x_2 = 0$ line (x_1-axis) in the left graph.

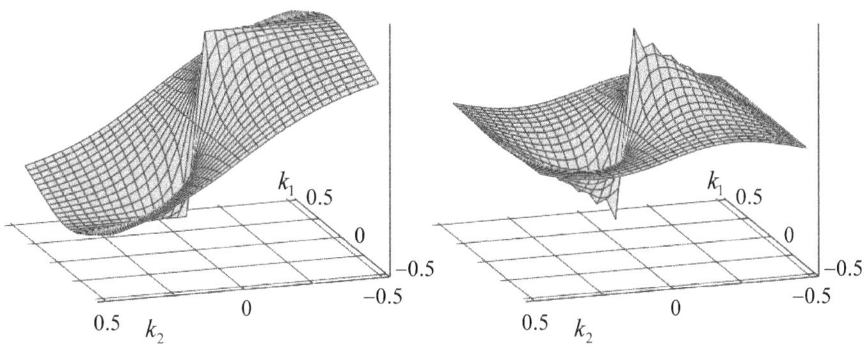

Figure 6.27: The true discretized spectrum, $\mathrm{Im}(\tilde{D}^{(\xi)}_{k1,k2})$ (left), and the DFT of the kernel for the deflection, ξ, $\mathrm{Im}(\tilde{W}^{(\xi)}_{k1,k2})$ (right). The discrete points are displayed as a surface for clarity. Only the imaginary parts are shown, as the real parts are negligible or not important to the comparison.

$$\int_{-\infty}^{\infty} \frac{x_2 dx_1}{2\pi\left(x_1^2 + x_2^2\right)^{3/2}} = \frac{x_2}{\pi} \int_0^{\infty} \frac{dx_1}{\left(x_1^2 + x_2^2\right)^{3/2}} = \frac{x_1}{\pi x_2 \sqrt{x_1^2 + x_2^2}}\Bigg|_{x_1=0}^{\infty} = \frac{1}{\pi x_2}. \tag{6.272}$$

The Fourier transform of this is the step function, $\mathcal{F}(1/(\pi x_2)) = -i\,\mathrm{sgn}(f_2)$, equation (2.104). This causes Gibbs's effect (Figure 2.5) if the discretized spectrum, $\tilde{D}^{(\xi)}_{k_1,k_2}$, is used instead of $\tilde{W}^{(\eta,\xi)}_{k_1,k_2}$, as shown in Figure 6.28. Although $\tilde{W}^{(\eta,\xi)}_{k_1,k_2}$ is a rather poor approximation of the true spectrum, it does attenuate (i.e., smooth) the high frequencies; and, again, for this reason this second approach would be the recommended procedure.

The analysis of these two approaches holds as well for the inverse Vening-Meinesz formulas since the kernels are identical except in sign. Analogous to equations (6.266), (6.271) and in view of equations (6.132), (6.133), one has

$$\Delta\tilde{g}_{\ell_1,\ell_2} = -\gamma\,\mathrm{DFT}^{-1}\left(\tilde{H}_{k_1,k_2}\tilde{D}^{(\eta)}_{k_1,k_2} + \tilde{X}_{k_1,k_2}\tilde{D}^{(\xi)}_{k_1,k_2}\right)_{\ell_1,\ell_2}, \tag{6.273}$$

by discretizing the analytic spectra of the kernels, and,

$$\Delta\breve{g}_{\ell_1,\ell_2} = \delta\Delta\tilde{g}_{\ell_1,\ell_2} - \gamma\,\mathrm{DFT}^{-1}\left(\tilde{H}_{k_1,k_2}\tilde{W}^{(\eta)}_{k_1,k_2} + \tilde{X}_{k_1,k_2}\tilde{W}^{(\xi)}_{k_1,k_2}\right)_{\ell_1,\ell_2}, \tag{6.274}$$

when discretizing the integral (hence the kernels, themselves), where the DFTs of the data are

$$\tilde{X}_{k_1,k_2} = \mathrm{DFT}\left(\tilde{\xi}_{\ell_1,\ell_2}\right)_{k_1,k_2}, \quad \tilde{H}_{k_1,k_2} = \mathrm{DFT}\left(\tilde{\eta}_{\ell_1,\ell_2}\right)_{k_1,k_2}, \tag{6.275}$$

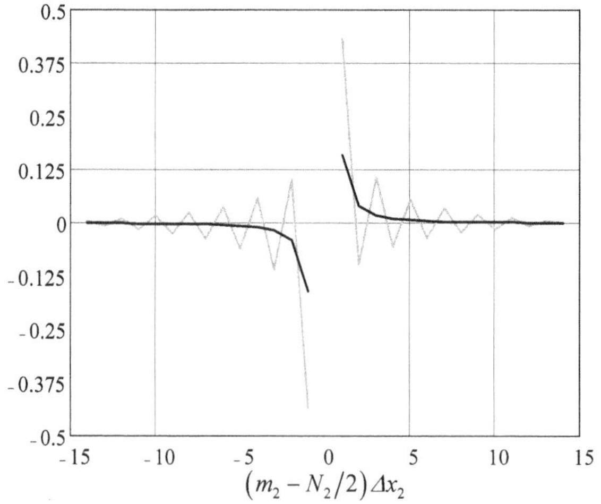

Figure 6.28: The discretized Vening-Meinesz kernel, $\tilde{w}^{(\xi)}_{0,m_2}$ (dark line), for the deflection, ξ, and $m_1 = 0$, compared to the inverse DFT of its analytic spectrum, $\mathrm{DFT}^{-1}(\tilde{D}^{(\xi)}_{0,k_2})$ (light line). Discrete points are connected by a line for clarity.

and the Cauchy principal value of the integral over the immediate neighborhood of the evaluation point is

$$\delta\Delta g\left(x_1, x_2, 0\right) = \frac{\gamma}{2}\sqrt{\frac{\Delta x_1 \Delta x_2}{\pi}}\left(\frac{\partial \eta}{\partial x_1} + \frac{\partial \xi}{\partial x_2}\right). \tag{6.276}$$

An alternative to equation (6.273) is

$$\Delta \tilde{g}_{\ell_1,\ell_2} = \gamma\,\mathrm{DFT}^{-1}\left(\frac{\tilde{X}_{k_1,k_2} + \tilde{H}_{k_1,k_2}}{\tilde{D}^{(\eta)}_{k_1,k_2} + \tilde{D}^{(\xi)}_{k_1,k_2}}\right), \tag{6.277}$$

that is obtained directly from equations (6.266) and is equivalent to equation (6.273) (Exercise 6.10) by noting that $\partial^2 T/(\partial x_1 \partial x_2) = \partial^2 T/(\partial x_2 \partial x_1)$, hence

$$\frac{k_1}{N_1 \Delta x_1}\tilde{X}_{k_1,k_2} = \frac{k_2}{N_2 \Delta x_2}\tilde{H}_{k_1,k_2}. \tag{6.278}$$

Since the deflection components and the gravity anomaly are real signals, the conjugate symmetry of their DFTs, equations (4.94), must hold. Indeed, if $\tilde{X}_{-k_1,k_2} = \tilde{X}^*_{k_1,-k_2}$, and $\tilde{H}_{-k_1,k_2} = \tilde{H}^*_{k_1,-k_2}$, then it is easily verified that the Fourier transforms of equations (6.273) or (6.277) have this conjugate symmetry. However, they do not necessarily give real values at wave numbers, 0, $N_{1,2}/2$, as required by equation (4.95),

$$\tilde{G}_{0,0} = \tilde{G}^*_{0,0},\quad \tilde{G}_{0,\frac{N_2}{2}} = \tilde{G}^*_{0,\frac{N_2}{2}},\quad \tilde{G}_{\frac{N_1}{2},0} = \tilde{G}^*_{\frac{N_1}{2},0},\quad \tilde{G}_{\frac{N_1}{2},\frac{N_2}{2}} = \tilde{G}^*_{\frac{N_1}{2},\frac{N_2}{2}}. \tag{6.279}$$

which must, therefore, be defined. For example, these can be set to zero, to the real part of the complex spectral value, or to the absolute value, depending on which introduces the least error.

6.5.1.3 *Hypersingular Integral*

The kernels for the radial derivative, equation (6.106), and of the inverse Stokes formula, equation (6.105) (with $r = R$), attenuate with the reciprocal of the cubed distance. In planar approximation and for $x_3 = 0$ the radial derivative of the gravity anomaly is given by

$$\left.\frac{\partial}{\partial x_3}\Delta g\left(x_1, x_2, x_3\right)\right|_{x_3=0} = \frac{1}{2\pi}\iint\limits_E \frac{\Delta g\left(x_1', x_2'\right) - \Delta g\left(x_1, x_2\right)}{\left(\left(x_1 - x_1'\right)^2 + \left(x_2 - x_2'\right)^2\right)^{3/2}}\,dx_1' dx_2'. \tag{6.280}$$

Noteworthy is the exceptional situation in which the numerator of the integrand approaches zero as fast as $s^2 = (x_1 - x_1')^2 + (x_2 - x_2')^2$ when $(x_1', x_2') \to (x_1, x_2)$. For then, the integral is not hypersingular nor even strongly singular, but rather weakly singular. The following development is for the general case when this cannot be guaranteed. Only the discretization of the integral, hence the kernel, is considered here since the previous sections already show that discretizing the analytic form of a kernel's spectrum is a numerically inferior procedure. Hofmann-Wellenhof and Moritz (2005, pp.126-127) derive an approximation of the contribution to the innermost zone around the evaluation point, (x_1, x_2),

$$\delta L(x_1, x_2) = \frac{1}{4}\sqrt{\frac{\Delta x_1 \Delta x_2}{\pi}}\left(\frac{\partial^2 \Delta g}{\partial x_1^2} + \frac{\partial^2 \Delta g}{\partial x_2^2}\right), \tag{6.281}$$

where equation (6.255) is used and the second-order derivatives refer to the evaluation point of the convolution. This term is not often applied, since the given gridded data usually do not support an accurate determination of the second-order horizontal derivatives, which in any case are relatively small.

With the cyclic, discretized approximation of the integral in equation (6.280), the vertical derivatives for $\ell_1 = 0,\ldots, N_1 - 1$ and $\ell_2 = 0,\ldots, N_2 - 1$ are estimated by

$$\left(\frac{\partial \Delta \tilde{g}}{\partial x_3}\right)_{\ell_1,\ell_2} = \delta L_{\ell_1,\ell_2} + \frac{\Delta x_1 \Delta x_2}{2\pi}\sum_{n_1=0}^{N_1-1}\sum_{n_2=0}^{N_2-1}\left(\Delta \tilde{g}_{n_1,n_2} - \Delta \tilde{g}_{\ell_1,\ell_2}\right)\tilde{v}_{\ell_1-n_1,\ell_2-n_2}, \tag{6.282}$$

where

$$\tilde{v}_{m_1,m_2} = \begin{cases} \dfrac{1}{\left(m_1^2 \Delta x_1^2 + m_2^2 \Delta x_2^2\right)^{3/2}}, & m_{1,2} = -\dfrac{N_{1,2}}{2},\ldots \dfrac{N_{1,2}}{2} - 1,\ |m_1| + |m_2| \neq 0 \\ 0, & |m_1| + |m_2| = 0 \end{cases} \tag{6.283}$$

with $\tilde{v}_{m_1+p_1N_1,m_2+p_2N_2} = \tilde{v}_{m_1,m_2}$ for any integers, p_1, p_2. Defined over the data domain, the discretized kernel is

$$\tilde{v}_{m_1,m_2} = \begin{cases} \tilde{v}_{m_1-N_1,m_2}, & \dfrac{N_1}{2} \le m_1 \le N_1 - 1,\quad 0 \le m_2 \le \dfrac{N_2}{2} - 1 \\ \tilde{v}_{m_1,m_2-N_2}, & 0 \le m_1 \le \dfrac{N_1}{2} - 1,\quad \dfrac{N_2}{2} \le m_2 \le N_2 - 1 \\ \tilde{v}_{m_1-N_1,m_2-N_2}, & \dfrac{N_1}{2} \le m_1 \le N_1 - 1,\quad \dfrac{N_2}{2} \le m_2 \le N_2 - 1 \end{cases} \tag{6.284}$$

Equation (6.282) then becomes

$$\left(\frac{\partial \Delta \tilde{g}}{\partial x_3}\right)_{\ell_1,\ell_2} = (\delta L)_{\ell_1,\ell_2} + \frac{\Delta x_1 \Delta x_2}{2\pi}\left(\Delta \tilde{g} \,\#\, \tilde{v}\right)_{\ell_1,\ell_2} - \Delta \tilde{g}_{\ell_1,\ell_2}\frac{\Delta x_1 \Delta x_2}{2\pi}\sum_{n_1=0}^{N_1-1}\sum_{n_2=0}^{N_2-1}\tilde{v}_{\ell_1-n_1,\ell_2-n_2} \tag{6.285}$$

By the defined periodicity of $\tilde{v}_{m_1,m_2}$, the last double sum is independent of ℓ_1, ℓ_2 and thus is equal to the zero-frequency term of $\tilde{V}_{k_1,k_2} = \mathrm{DFT}\left(\tilde{v}_{m_1,m_2}\right)_{k_1,k_2}$. As in equation (6.259) let the discrete Fourier transform of the data be $\tilde{G}_{k_1,k_2}$. Then by equation (4.115),

$$\left(\frac{\partial \Delta\tilde{g}}{\partial x_3}\right)_{\ell_1,\ell_2} = (\delta L)_{\ell_1,\ell_2} + \frac{1}{2\pi}\mathrm{DFT}^{-1}\left(\tilde{G}_{k_1,k_2}\tilde{V}_{k_1,k_2}\right)_{\ell_1,\ell_2} - \Delta\tilde{g}_{\ell_1,\ell_2}\frac{\tilde{V}_{0,0}}{2\pi}. \tag{6.286}$$

The planar approximation of the inverse Stokes formula for $x_3 = 0$ follows immediately in view of equations (6.105) and (6.130),

$$\Delta g(x_1,x_2) = \frac{1}{2\pi}\iint_E \frac{T(x_1',x_2') - T(x_1,x_2)}{\left((x_1-x_1')^2 + (x_2-x_2')^2\right)^{3/2}}\,dx_1'dx_2'\;; \tag{6.287}$$

and, hence, analogous to equation (6.286),

$$\Delta\tilde{g}_{\ell_1,\ell_2} = \delta\Gamma_{\ell_1,\ell_2} + \frac{1}{2\pi}\mathrm{DFT}^{-1}\left(\tilde{\tau}_{k_1,k_2}\tilde{V}_{k_1,k_2}\right)_{\ell_1,\ell_2} - \tilde{T}_{\ell_1,\ell_2}\frac{\tilde{V}_{0,0}}{2\pi}, \tag{6.288}$$

where $\tilde{\tau}_{k_1,k_2} = \mathrm{DFT}\left(\tilde{T}_{\ell_1,\ell_2}\right)$ and

$$\delta\Gamma_{\ell_1\ell_2} = \frac{1}{4}\sqrt{\frac{\Delta x_1\Delta x_2}{\pi}}\left(\frac{\partial^2 T}{\partial x_1^2} + \frac{\partial^2 T}{\partial x_2^2}\right)_{\substack{x_1=\ell_1\Delta x_1\\ x_2=\ell_2\Delta x_2}}. \tag{6.289}$$

By Laplace's equation, the parenthetical term is also the vertical gradient of the gravity anomaly. If its mean value in a grid cell is about 20 E and $\Delta x_1 = \Delta x_2 = \sqrt{\pi}$ km ≈ 1.8 km, then $\delta\Gamma_{\ell_1,\ell_2} = 0.5$ mGal, which is not insignificant.

6.5.2 Forward Models

The previous Section 6.5.1 expresses convolutions in the spectral domain as derived primarily from solutions to boundary-value problems, that is, solutions for the potential via Laplace's equation. The mass-density function does not enter the solution, rather the potential function or its derivatives are given on a boundary (the plane or sphere) and functionally related quantities are determined from their spectral interdependencies. Forward models of the potential, that is, models based on an assumed density distribution for a three-dimensional source, similarly may be expressed in terms of spectral transforms with some approximations, where the third dimension of the source is integrated specifically in the space domain.

Assuming constant density, $\rho(x_1', x_2', x_3') = \rho_0$, in the density integral, equation (6.17), as well as the planar approximation, the gravitational potential due to the topographic mass over a finite area is

$$v^{(h)}(x_1,x_2,x_3) = G\rho_0 \int_{-T_1/2}^{T_1/2}\int_{-T_2/2}^{T_2/2}\int_{0}^{h(x_1',x_2')} \frac{dx_3'}{d}\,dx_1'dx_2', \tag{6.290}$$

where the distance, d, is given by equation (6.24), $x_3>0$, and $h(x_1',x_2')$ is the elevation of the terrain. With a finite area of integration, $E=\{(x_1,x_2)\,|\,-T_1/2\le x_1\le T_1/2,-T_2/2\le x_2\le T_2/2\}$, it is presumed that the topographic mass at a certain distance from the evaluation points of interest has negligible gravitational effect. Using $\partial(1/d)/\partial x_3=-\partial(1/d)/\partial x_3'$, the vertical derivative of the potential, $g_3^{(h)}=\partial v^{(h)}/\partial x_3$, easily reduces the integral to just two dimensions,

$$g_3^{(h)}(x_1,x_2,x_3)=G\rho_0\iint\limits_{E'}\left(\frac{1}{d_0}-\frac{1}{d_h}\right)dE', \tag{6.291}$$

where $dE'=dx_1'dx_2'$ and

$$d_0=\sqrt{(x_1-x_1')^2+(x_2-x_2')^2+x_3^2}, \tag{6.292}$$

$$d_h=\sqrt{(x_1-x_1')^2+(x_2-x_2')^2+(x_3-h(x_1',x_2'))^2}. \tag{6.293}$$

However, this is not in the form of a convolution and thus not amenable to fast computation using FFTs. A general method to obtain convolutions for such integrals is to expand the reciprocal distance into a series.

The Taylor expansion of $1/d$ as a function of x_3' with respect to the point, $x_3'=0$, is given by

$$\frac{1}{d}=\sum_{n=0}^{\infty}\frac{1}{n!}\frac{\partial^n(1/d)}{(\partial x_3')^n}\bigg|_{x_3'=0}(x_3')^n. \tag{6.294}$$

It is easy to show by the symmetry of d with respect to the primed and un-primed coordinates that (Exercise 6.11)

$$\frac{\partial^n(1/d)}{(\partial x_3')^n}\bigg|_{x_3'=0}=(-1)^n\frac{\partial^n(1/d_0)}{(\partial x_3)^n}. \tag{6.295}$$

Now, substitute equation (6.294) with this into equation (6.290) and interchange summation and integration to obtain the potential as the series,

$$\begin{aligned}v^{(h)}(x_1,x_2,x_3)&=G\rho_0\iint\limits_{E'}\sum_{n=0}^{\infty}\frac{(-1)^n}{n!}\frac{\partial^n(1/d_0)}{(\partial x_3)^n}\left(\int_0^{h(x_1',x_2')}(x_3')^n\,dx_3'\right)dE'\\&=G\rho_0\sum_{n=0}^{\infty}\frac{(-1)^n}{(n+1)!}\iint\limits_{E'}\frac{\partial^n(1/d_0)}{(\partial x_3)^n}\left(h(x_1',x_2')\right)^{n+1}dE'\end{aligned} \tag{6.296}$$

Replacing h by

$$h_T\left(x_1',x_2'\right)=\begin{cases} h\left(x_1',x_2'\right), & -\frac{T_1}{2}\le x_1'\le\frac{T_1}{2}\text{ and }-\frac{T_2}{2}\le x_2'\le\frac{T_2}{2} \\ 0, & \text{otherwise}\end{cases} \tag{6.297}$$

the integrals in the series may be extended over the infinite plane and are *convolutions* of two bounded functions. Since $1/d_0$ is like a potential function (see equation (6.6)), equations (6.45) shows that

$$\mathcal{F}\left(\frac{\partial^n\left(\dfrac{1}{\sqrt{x_1^2+x_2^2+x_3^2}}\right)}{\partial x_3^n}\right)=\left(-2\pi\bar{f}\right)^n\mathcal{F}\left(\frac{1}{\sqrt{x_1^2+x_2^2+x_3^2}}\right)=\left(-2\pi\bar{f}\right)^n\frac{1}{\bar{f}}e^{-2\pi x_3\bar{f}}, \tag{6.298}$$

using the transform (2.337). By the convolution theorem (3.26),

$$\mathcal{F}\left(v^{(h)}\left(x_1,x_2,x_3\right)\right)=G\rho_0\frac{e^{-2\pi x_3\bar{f}}}{2\pi\bar{f}^2}\sum_{n=1}^{\infty}\frac{\left(2\pi\bar{f}\right)^n}{n!}\mathcal{F}\left(\left(h_T\left(x_1',x_2'\right)\right)^n\right), \tag{6.299}$$

where, again, as in equation (6.27), one must exclude the zero frequency.

This equation was derived by Parker (1972) who also showed that the series converges as fast as a series with terms, $(\bar{h}/x_3)^n$, where $\bar{h}=\max_{E'} h(x_1', x_2')$. Indeed, from inequality (2.67) the Fourier transforms on the right side of equation (6.299) are bounded,

$$\left|\mathcal{F}\left(\left(h_T\left(x_1',x_2'\right)\right)^n\right)\right|<A_E\bar{h}^n, \tag{6.300}$$

where $A_E = T_1T_2$ is the area of E. From the uniformly converging exponential series, one has

$$e^{2\pi x_3\bar{f}}=1+\ldots+\frac{1}{n!}\left(2\pi x_3\bar{f}\right)^n+\ldots\quad\Rightarrow\quad\frac{\left(2\pi\bar{f}\right)^n}{n!}<\frac{e^{2\pi x_3\bar{f}}}{x_3^n}; \tag{6.301}$$

and, therefore, the terms of the series in equation (6.299) are bounded by

$$\frac{\left(2\pi\bar{f}\right)^n}{n!}\mathcal{F}\left(\left(h_T\left(x_1',x_2'\right)\right)^n\right)<A_E\frac{\left(2\pi\bar{f}\right)^n}{n!}\bar{h}^n<A_E e^{2\pi x_3\bar{f}}\left(\frac{\bar{h}}{x_3}\right)^n. \tag{6.302}$$

These are terms of a uniformly converging series if $\bar{h}/x_3<1$, or $x_3>\max_{E'} h\left(x_1',x_2'\right)$. That is, the plane on which the potential is computed by Fourier transform must be above all the terrain. The rate of convergence of the series in equation (6.299) is governed primarily by the ratio, $\bar{h}/x_3$—the smaller the ratio, the more rapid the convergence.

This ratio can be minimized by adjusting the origin of the x_3-coordinate, which is the same as adjusting the zero-frequency component of the topographic height. It may be set to zero without affecting equation (6.299) since the zero-frequency is excluded, except that h is now replaced by $h–h_{\text{avg}}$, where h_{avg}, a constant, is the average height within the integration area.

For a *residual* topographic height relative to a reference terrain, defined by $h_{\text{ref}}(x'_1, x'_2)$, such as a global model of spherical harmonics, equation (6.169) truncated at some finite degree, it is easy to see that the lower limit on the x'_3-integral in equations (6.290) and (6.296) is then $h_{\text{ref}}(x'_1, x'_2)$. The corresponding Fourier transform of the residual topographic potential in this case is given by

$$\mathcal{F}\left(\delta v^{(h)}\left(x_1,x_2,x_3\right)\right)=G\rho_0\frac{e^{-2\pi x_3\bar{f}}}{2\pi\bar{f}^2}\sum_{n=1}^{\infty}\frac{\left(2\pi\bar{f}\right)^n}{n!}\left(\mathcal{F}\left(\delta\left(h_T\left(x'_1,x'_2\right)\right)^n\right)\right), \tag{6.303}$$

where the difference in powers of the height is

$$\delta\left(h_T\left(x'_1,x'_2\right)\right)^n=\left(h_T\left(x'_1,x'_2\right)\right)^n-\left(\left(h_{\text{ref}}\right)_T\left(x'_1,x'_2\right)\right)^n. \tag{6.304}$$

Fourier transforms of the derivatives of the topographic potential make use of the spectral relationships given by the transform pairs (6.46). For example, the Fourier transform of the vertical gravitational acceleration, $g_3^{(h)} = \partial v^{(h)}/\partial x_3$, due to the total terrain and on a plane, $x_3 > \max_{E'} h\left(x'_1,x'_2\right)$, is given by

$$\mathcal{F}\left(g_3^{(h)}\left(x_1,x_2,x_3\right)\right)=-G\rho_0\frac{e^{-2\pi x_3\bar{f}}}{\bar{f}}\sum_{n=1}^{\infty}\frac{\left(2\pi\bar{f}\right)^n}{n!}\left(\mathcal{F}\left(\left(h_T\left(x'_1,x'_2\right)\right)^n\right)\right). \tag{6.305}$$

If horizontal variations of the mass-density are known, but $\partial\rho/\partial x'_3 = 0$, i.e., $\rho = \rho(x'_1, x'_2)$, then it is straightforward to modify the previous derivation of equation (6.299) and obtain (Exercise 6.12)

$$\mathcal{F}\left(v^{(h)}\left(x_1,x_2,x_3\right)\right)=G\frac{e^{-2\pi x_3\bar{f}}}{2\pi\bar{f}^2}\sum_{n=1}^{\infty}\frac{\left(2\pi\bar{f}\right)^n}{n!}\mathcal{F}\left(\rho\left(x'_1,x'_2\right)\left(h_T\left(x'_1,x'_2\right)\right)^n\right). \tag{6.306}$$

In practice, the infinite sum in equation (6.303) is terminated at some finite order, often already at the first term, with the assumption that higher-order terms are negligible. For the potential, or even its first derivative this may be justified, but for the second and higher derivatives, whose spectral content is concentrated toward the high frequencies, these higher-order terms become significant.

A slightly different approach to transform the forward model into the spectral domain, given by Forsberg (1985), expresses the kernel functions of the integrals for the potential and its derivatives as Taylor series in the spatial domain before applying the Fourier transform. For example, for the gravitation vector components, $g_j^{(h)} = \partial v^{(h)}/\partial x_j$, $j = 1,2,3$, equation (6.296) leads to

$$g_j^{(h)}(x_1,x_2,x_3)=G\rho_0\sum_{n=0}^{\infty}\frac{(-1)^n}{(n+1)!}\iint_{E'}q_j^{(n)}(x_1-x_1',x_2-x_2',x_3)\left(h(x_1',x_2')\right)^{n+1}dE', \tag{6.307}$$

where

$$q_j^{(n)}(x_1,x_2,x_3)=\frac{\partial^{n+1}}{\partial x_j(\partial x_3)^n}\left(\frac{1}{\sqrt{x_1^2+x_2^2+x_3^2}}\right) \tag{6.308}$$

is readily derived for each $n\geq 0$. Subsequently, replacing the finite integrals by infinite integrals using equation (6.297), the Fourier transform of the sum of convolutions in equation (6.307) yields (by the convolution theorem)

$$\mathcal{F}\left(g_j^{(h)}(x_1,x_2,x_3)\right)=G\rho_0\sum_{n=0}^{\infty}\frac{(-1)^n}{(n+1)!}\mathcal{F}\left(q_j^{(n)}(x_1,x_2,x_3)\right)\mathcal{F}\left(\left(h_T(x_1,x_2)\right)^{n+1}\right). \tag{6.309}$$

Similar expressions hold for the potential and higher derivatives. In fact, equations (6.305) and (6.309) (for $j=3$) are equivalent (Exercise 6.13), as are corresponding expressions for other derivatives of the potential. Thus, the same convergence criteria hold for equation (6.309). And, as for Parker's formulation, the partial derivatives, $q_j^{(n)}$, typically are not needed beyond the first few orders, n, but more are required for higher-order derivatives of the potential on the left side of equation (6.307).

The essential difference between the practical application of these two methods is in the calculation of the Fourier transforms of the derivatives in the Taylor series. That is, one either discretizes the analytic Fourier transform,

$$\mathcal{F}\left(q_3^{(n)}(x_1,x_2,x_3)\right)=\left(-2\pi\bar{f}_{k_1,k_2}\right)^{n+1}\frac{1}{\bar{f}_{k_1,k_2}}e^{-2\pi x_3\bar{f}_{k1,k2}}, \tag{6.310}$$

as in the numerical implementation of equation (6.305), where equation (6.250) gives $\bar{f}_{k_1,k_2}$; or, one computes the DFT of the discretization of $q_3^{(n)}(x_1,x_2,x_3)$,

$$\mathcal{F}\left(q_3^{(n)}(x_1,x_2,x_3)\right)\approx DFT\left(q_3^{(n)}(\ell_1\Delta x_1,\ell_2\Delta x_2;x_3)\right)_{k_1,k_2}, \tag{6.311}$$

as implied by equation (6.309). Figure 6.29 compares these two options assuming a 100×100 computation grid (only the amplitude spectra are shown for $f_1=f_2$). As in the case of the Vening-Meinesz formulas, there is considerable difference between the analytic and DFT-based responses. Numerical tests have shown that in some cases the latter (i.e., the numerical realization of equation (6.309) rather than (6.305)) is, in fact, more accurate in representing the local convolution (Jekeli and Zhu 2006), which is consistent with similar remarks made in Section 6.5.1.

When determining the residual Bouguer effect relative to that of an infinite plate ($2\pi G\rho_0 h(x_1,x_2)$) at points, (x_1, x_2, $x_3=h(x_1,x_2)$), directly on the topographic surface, the vertical derivative of equation (6.290), $-\partial v^{(\Delta h)}/\partial x_3$, is (note the different sign convention in this application)

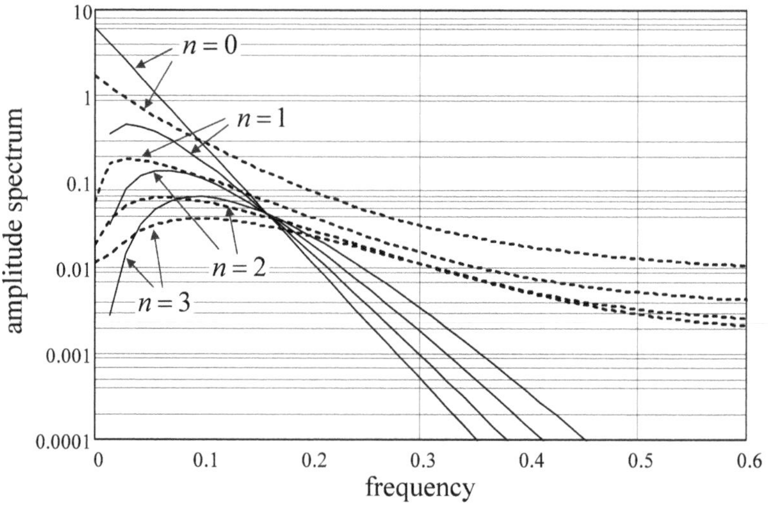

Figure 6.29: Analytic (solid lines) and DFT-based (dashed lines) Fourier transforms of the vertical derivatives, $q_3^{(n)}(x_1, x_2, x_3)$, for $x_3 = 5$ and $n = 0,1,2,3$. The size of the sampling grid for the DFT-based transform is 100×100 with unit sampling intervals. The curves represent the amplitude spectra for $f_1 = f_2$ in the frequency domain.

$$
\begin{aligned}
g_3^{(\Delta h)}\left(x_1, x_2, h(x_1, x_2)\right) &= G\rho_0 \iint\limits_{E'} \int\limits_{h(x_1,x_2)}^{h(x_1',x_2')} \frac{\partial}{\partial x_3}\left(\frac{1}{d}\right)_{x_3 = h(x_1,x_2)} dx_3' \, dE' \\
&= G\rho_0 \iint\limits_{E'} \left(\frac{1}{s} - \frac{1}{d_h}\right) dE'
\end{aligned}
\tag{6.312}
$$

where equation (6.194) defines the distance, s, and d_h is equation (6.293) with $x_3 = h(x_1, x_2)$ substituted. This gravitational effect is due to the mass *excess or deficiency* of the residual topography, Δh, that is bounded by a horizontal plane through the point, $(x_1, x_2, h(x_1, x_2))$. Denoting $h \equiv h(x_1, x_2)$ and $h' \equiv h(x_1', x_2')$, a binomial series expansion of the integrand yields

$$
\frac{1}{s} - \frac{1}{d_h} = \frac{1}{s} - \frac{1}{s} \frac{1}{\sqrt{1 + \frac{(h-h')^2}{s^2}}} = \frac{1}{s} \sum_{n=1}^{\infty} \frac{(-1)^{n-1}}{2^{2n}} \binom{2n}{n} \left(\frac{h-h'}{s}\right)^{2n},
\tag{6.313}
$$

which converges if a *Lipschitz condition* (Davis 1975, p.8) holds,

$$
\left|\frac{h-h'}{s}\right| < 1.
\tag{6.314}
$$

Thus, at the computation point, the terrain slope must be less than 45°. The gravitational effect of the residual terrain is then given by

$$g_3^{(\Delta h)}\left(x_1,x_2,h(x_1,x_2)\right)=G\rho_0\sum_{n=1}^{\infty}\frac{(-1)^{n-1}}{2^{2n}}\binom{2n}{n}\iint_{E'}\frac{\left(h(x_1,x_2)-h(x_1',x_2')\right)^{2n}}{s^{2n+1}}dE' \quad (6.315)$$

By the Lipschitz condition, $(h-h')^{2n}/s^{2n}$ is bounded for all integration points and each integral in the sum is of the weakly singular type (Section 6.5.1.1), where the integrand essentially behaves like $1/s$.

Retaining only the first term in equation (6.315), $g_3^{(\Delta h)}$ is approximated by

$$g_3^{(\Delta h)}\left(x_1,x_2,h(x_1,x_2)\right)\simeq\frac{G\rho_0}{2}\iint_{E'}\frac{\left(h(x_1,x_2)-h(x_1',x_2')\right)^{2}}{s^{3}}dE'. \quad (6.316)$$

It is not a convolution in this form; but expanding the square,

$$g_3^{(\Delta h)}\left(x_1,x_2,h(x_1,x_2)\right)\simeq\frac{G\rho_0}{2}\left(\begin{array}{l}h^2(x_1,x_2)\iint_{E'}\frac{dE'}{s^3}-2h(x_1,x_2)\iint_{E'}\frac{h(x_1',x_2')}{s^3}dE'\\ \qquad+\iint_{E'}\frac{h^2(x_1',x_2')}{s^3}dE'\end{array}\right), \quad (6.317)$$

it becomes the sum of three convolutions, respectively, of 1, h, and, h^2 with $1/s^3$. However, each convolution is now hypersingular (Section 6.5.1.3). That is, a weakly singular integral, equation (6.316), is traded for hypersingular integrals in order to accelerate practical computations using the discrete Fourier transform. With an appropriate discretization applied to h^2 and to h, the practical computation, analogous to equation (6.286), is

$$\left(\tilde{g}_3^{(\Delta h)}\right)_{\ell_1,\ell_2}=\frac{G\rho_0}{2}\left(\tilde{h}^2_{\ell_1,\ell_2}\tilde{V}_{0,0}-2\tilde{h}_{\ell_1,\ell_2}\,\mathrm{DFT}^{-1}\left(\tilde{H}_{k_1,k_2}\tilde{V}_{k_1,k_2}\right)_{\ell_1,\ell_2}+\mathrm{DFT}^{-1}\left(\tilde{K}_{k_1,k_2}\tilde{V}_{k_1,k_2}\right)_{\ell_1,\ell_2}\right), \quad (6.318)$$

where $\tilde{K}_{k_1,k_2}=\mathrm{DFT}\left(\tilde{h}^2_{\ell_1,\ell_2}\right)_{k_1,k_2}$ and $\tilde{H}_{k_1,k_2}=\mathrm{DFT}\left(\tilde{h}_{\ell_1,\ell_2}\right)_{k_1,k_2}$, and where the near-zone contributions, analogous to equation (6.281), have been neglected under the assumption that the slopes of the terrain and of its square are practically constant near the computation point (then the second-order gradients are zero). This method is commonly used when computing *terrain corrections* in gravimetry (Schwarz et al. 1990, p.502; Sansò and Sideris 2013, Chapter 10).

The preceding applications of the Fourier transform to forward models of the gravitational potential and its derivatives is motivated primarily by the opportunity to obtain efficient computational formulas via the convolution theorem. As seen by the various examples, the price to be paid often includes approximations and restrictions, such as the planar approximation and a constant altitude for evaluations, which may or may not be worth the savings in computation time. The Fourier transforms here are two-dimensional reflecting the typically horizontal distribution of data; and, in forward modeling this necessitates an assumption at least of vertical homogeneity in the mass-density function. However, a three-dimensional model, such as equation

(6.17) in Cartesian coordinates, is also a three-dimensional convolution of a density function and the inverse distance. In this case, no series expansions are required for the height coordinate and the three-dimensional Fourier transform could be applied directly, given a three dimensional density function (e.g., Caratori Tontini et al. 2009). This method is outside the present scope.

6.6 Least-Squares Collocation

The physical models of the previous sections usually serve well as scaffolding for geodetic and geophysical measurements. However, they typically require more data than is theoretically possible; for example, the convolution models of Section 6.5 assume continuous input over the entire Earth. Besides being discrete and of limited spatial extent, our data also have measurement errors. The fact that the Earth's surface, on which measurements are made, is not a sphere (nor a plane) further complicates the situation as the model itself then is not strictly valid. In many cases all these imperfections are acceptable approximations and one proceeds with the given measurements and appropriate reductions and corrections to accommodate the model.

An alternative approach builds specifically on the premise that the data are discrete and noisy. This kind of *operational approach* was developed in physical geodesy and is known as *least-squares collocation* (Moritz 1978, 1980). In its most general form it is an optimal estimation method that assumes a stochastic interpretation (Section 5.6) for a (residual) potential and combines any finite, discrete set of linear functionals of the potential using mutually consistent covariance models (Section 6.4.1). The estimation minimizes the effect of random error and puts no *a priori* restrictions on the domains of the data and estimation points (other than being discrete and on or above the Earth's surface). In geophysics, a similar methodology is known as kriging, though the application is usually one of interpolation or *prediction* of a like quantity (Cressie 1991). However, under special conditions mixed data types can also be considered (*cokriging*). Similarly, *collocation* refers to the option of combining different data types (different functionals of the potential) and, in fact, it is done in an optimal way. Indeed, *least-squares* implies that the potential is estimated with minimum mean square error.

Least-square collocation has found particular application where discrete data are irregularly distributed and sparse, two conditions that seriously encumber the use of the physical models. This situation is becoming less common as ground, airborne, and satellite measurement systems fill gaps in the regional and global landscape. In that sense, least-squares collocation (or kriging) in the frequency domain, the main topic of this section, represents a philosophical contradiction. A transformation into the spectral domain assumes a sufficiently high-resolution and uniform distribution of data; whereas, the estimation method was developed precisely on the premise of data scarcity. On the other hand, the technique employs a physically stochastic interpretation of the observable, and, as such, its analysis in the spectral domain leads to a better quantification of errors associated with the finite resolution of the observables and consequent aliasing errors, as well as the finite data extent. Moreover, there are situations that call for least-squares collocation if the data are sparse in one dimension if not in the other, as in the case of data trajectories obtained by airborne measurement systems. The transformations into the spectral domain then also lead to computational efficiency. It is from these viewpoints that the spectral techniques are applied to least-squares collocation.

6.6.1 Theoretical Setup

A brief introduction of the theory of least-squares collocation prepares for the frequency domain formulation, which is given only with respect to a Cartesian domain since the applications for the most part are local in extent. The theory of least-squares collocation is based on the fundamental hypothesis that the observed and predicted or estimated signals are stochastic processes. Moreover, it is assumed that these processes are stationary in order to apply corresponding concepts in the frequency domain, such as power spectral density. Ordinary kriging is developed under somewhat more general assumptions on the variability of the underlying process and stationarity is not a prerequisite (Section 5.6.3). However, by assuming stationary processes with zero mean, *simple* kriging is identical to least-squares collocation; and, the latter, therefore, subsumes the former in this case and is the subject of the following development.

Thus, consider a stationary, random process, g, on the plane, that is sampled with measurements, $\tilde{g}_{n}$, at a finite number of points, $p_n' = \left((x_1)_n, (x_2)_n\right)$, $n = -N/2, \ldots, N/2-1$,

$$\tilde{g}_n = g(p_n') + \varepsilon_g(p_n'), \tag{6.319}$$

where ε_g is the measurement error. It is assumed that the mean value of the process is zero, $\mathcal{E}(g) = 0$, and that also $\mathcal{E}(\varepsilon_g) = 0$. In the absence of these conditions systematic biases in the data and in the errors can be estimated under a more general formulation (Moritz 1980). The covariance function of g is presumed known and is given formally by equation (5.251),

$$c_{g,g}(\xi_1, \xi_2) = \mathcal{E}\left(g(x_1, x_2)\, g(x_1 + \xi_1, x_2 + \xi_2)\right). \tag{6.320}$$

By convention, the covariance is evaluated at lag distances, $\xi_1 = (x_1)_{\text{second quantity}} - (x_1)_{\text{first quantity}}$ and $\xi_2 = (x_2)_{\text{second quantity}} - (x_2)_{\text{first quantity}}$, where "first" and "second" quantity refer to the order of the subscripts on the covariance function. Similarly, $c_{\varepsilon_g,\varepsilon_g}(\xi_1, \xi_2)$ is the covariance function of the error, and it is reasonable to assume the physical signal, g, and the measurement error are independent, so that $c_{g,\varepsilon_g}(\xi_1, \xi_2) = 0$.

At a point, $p = (x_1, x_2)$, one seeks an estimate, $\hat{u}$, of the realization of a zero-mean, stationary, random process, u, that is related to g through a linear operator, typically a derivative. For example, g, may be the gravity anomaly and u the disturbing potential, related according to equation (6.64). Derivatives of stochastic processes and the laws of propagation of covariances are assumed to exist under the interpretations discussed in Section 5.6.4 (e.g., limits in the mean, equation (5.248)). The covariance functions, $c_{g,g}$, $c_{u,g}$, and $c_{u,u}$, thus are mutually consistent and all depend only on the differences between coordinates of u and g. In accordance with the motivation for the operational approach, the estimator of u at p is assumed to be a linear combination of the measurements, having the form,

$$\hat{u}(p) = \boldsymbol{h}^{\mathrm{T}}(p, \boldsymbol{p}')\hat{\boldsymbol{g}}(\boldsymbol{p}'), \tag{6.321}$$

where the $n \times 1$ vector, $\hat{\boldsymbol{g}}(\boldsymbol{p}')$, collects the measurements and $\boldsymbol{p}'$ denotes the collection of measurement points. The $1 \times n$ vector, $\boldsymbol{h}^{\mathrm{T}}(p, \boldsymbol{p}')$, is a non-random quantity to be determined such that the variance of the estimation error is minimized. It is noted that $\boldsymbol{h}^{\mathrm{T}}(p, \boldsymbol{p}')$, representing weights applied to the measurements, does not generally depend on coordinate differences between p and $\boldsymbol{p}'$. For example, if a uniformly spaced sequence of measurements is finite, then one should not expect the same set of weights when the estimation point is near the ends of the data sequence as when it is near the middle of the sequence.

Let the estimation error be denoted by $\varepsilon_u = \hat{u} - u$. With the zero-mean assumption for the processes, $\mathcal{E}(\varepsilon_u) = 0$; and, the variance of the error is

$$\begin{aligned} \sigma_{\varepsilon_u}^2 &= \mathcal{E}\left(\left(\boldsymbol{h}^{\mathrm{T}}(p, \boldsymbol{p}')\hat{\boldsymbol{g}}(\boldsymbol{p}') - u(p)\right)^2\right) \\ &= \boldsymbol{h}^{\mathrm{T}}\mathcal{E}\left(\hat{\boldsymbol{g}}\hat{\boldsymbol{g}}^{\mathrm{T}}\right)\boldsymbol{h} - 2\mathcal{E}\left(u\hat{\boldsymbol{g}}^{\mathrm{T}}\right)\boldsymbol{h} + \mathcal{E}\left(u^2\right) \end{aligned} \tag{6.322}$$

where the dependence on the point coordinates is omitted for easier notation. With further parsimony in notation, the auto- and cross-covariance matrices of the observations and estimated quantity are, respectively,

$$\overline{\mathbf{c}}_{g,g} = \mathcal{E}\left(\hat{\boldsymbol{g}}\hat{\boldsymbol{g}}^{\mathrm{T}}\right) = \mathcal{E}\left(\boldsymbol{g}\boldsymbol{g}^{\mathrm{T}}\right) + \mathcal{E}\left(\boldsymbol{\varepsilon}_g\boldsymbol{\varepsilon}_g^{\mathrm{T}}\right) = \mathbf{c}_{g,g} + \mathbf{c}_{\varepsilon_g,\varepsilon_g}, \tag{6.323}$$

$$\boldsymbol{c}_{u,g} = \mathcal{E}\left(u\hat{\boldsymbol{g}}^{\mathrm{T}}\right) = \mathcal{E}\left(u\boldsymbol{g}^{\mathrm{T}}\right), \tag{6.324}$$

again, noting the zero means and the independence of the measurement error with respect to g or u. Thus,

$$\sigma_{\varepsilon_u}^2 = \boldsymbol{h}^{\mathrm{T}}\overline{\mathbf{c}}_{g,g}\boldsymbol{h} - 2\boldsymbol{c}_{u,g}\boldsymbol{h} + \sigma_u^2, \tag{6.325}$$

where $\sigma_u^2 = \mathcal{E}(u^2)$, and, explicitly,

$$\mathbf{c}_{g,g}(\boldsymbol{p}', \boldsymbol{p}') = \begin{pmatrix} c_{g,g}(\boldsymbol{\xi}_{1,1}) & \cdots & c_{g,g}(\boldsymbol{\xi}_{n,1}) \\ \vdots & \ddots & \vdots \\ c_{g,g}(\boldsymbol{\xi}_{1,n}) & \cdots & c_{g,g}(\boldsymbol{\xi}_{n,n}) \end{pmatrix}, \tag{6.326}$$

$$\boldsymbol{c}_{u,g}(p, \boldsymbol{p}') = \begin{pmatrix} c_{u,g}(\boldsymbol{\xi}_1) & \cdots & c_{u,g}(\boldsymbol{\xi}_n) \end{pmatrix}. \tag{6.327}$$

In these matrices, $\boldsymbol{\xi}_{j,k} = \begin{pmatrix} x_1^{(k)} - x_1^{(j)} & x_2^{(k)} - x_2^{(j)} \end{pmatrix}^{\mathrm{T}}$ is the two-element vector of the differences in coordinates between measurement points, p_j and p_k, and

$\xi_j = \left(x_1^{(j)} - x_1 \quad x_2^{(j)} - x_2\right)^{\mathrm{T}}$ represents the corresponding difference between the jth measurement point and the estimation point.

As an aside, it is noted that the covariance matrix, $\mathbf{c}_{g,g}(\boldsymbol{p}', \boldsymbol{p}')$, is symmetric and positive definite. The symmetry is obvious from the analogue of equation (5.89), $c_{g,g}\left(\xi_{j,k}\right) = c_{g,g}\left(-\xi_{j,k}\right) = c_{g,g}\left(\xi_{k,j}\right)$. Positive definiteness of the matrix follows from the corresponding property of the auto-covariance function (Section 5.6.4); or, also because for an arbitrary non-zero vector, $\boldsymbol{v}$, the quadratic form, $\boldsymbol{v}^{\mathrm{T}}\mathbf{c}_{g,g}\boldsymbol{v}$, is positive,

$$\boldsymbol{v}^{\mathrm{T}}\mathbf{c}_{g,g}\boldsymbol{v} = \boldsymbol{v}^{\mathrm{T}}\,\mathcal{E}\left(\boldsymbol{g}\boldsymbol{g}^{\mathrm{T}}\right)\boldsymbol{v} = \mathcal{E}\left(\boldsymbol{v}^{\mathrm{T}}\boldsymbol{g}\boldsymbol{g}^{\mathrm{T}}\boldsymbol{v}\right) = \mathcal{E}\left(s^2\right) > 0. \tag{6.328}$$

where $s = \boldsymbol{v}^{\mathrm{T}}\mathbf{g}$.

Re-writing the error variance, equation (6.325), as

$$\sigma_{\varepsilon_u}^2 = \sigma_u^2 - c_{u,g}\overline{\mathbf{c}}_{g,g}^{-1}c_{u,g}^{\mathrm{T}} + \left(\boldsymbol{h}^{\mathrm{T}} - c_{u,g}\overline{\mathbf{c}}_{g,g}^{-1}\right)\overline{\mathbf{c}}_{g,g}\left(\boldsymbol{h}^{\mathrm{T}} - c_{u,g}\overline{\mathbf{c}}_{g,g}^{-1}\right)^{\mathrm{T}}, \tag{6.329}$$

which is easily verified, only the last term depends on $\boldsymbol{h}$ and by equation (6.328) it is never negative. Therefore, a necessary and sufficient condition for minimum error variance is obtained by setting the last term to zero with the choice,

$$\boldsymbol{h}^{\mathrm{T}} = c_{u,g}\overline{\mathbf{c}}_{g,g}^{-1}. \tag{6.330}$$

The optimal, i.e., minimum error variance, estimator of u, equation (6.321), is then given by

$$\hat{u}(p) = c_{u,g}(p, \boldsymbol{p}')\left(\overline{\mathbf{c}}_{g,g}(\boldsymbol{p}', \boldsymbol{p}')\right)^{-1}\overline{\boldsymbol{g}}(\boldsymbol{p}'); \tag{6.331}$$

and the error variance, equation (6.329), is

$$\sigma_{\varepsilon_u}^2(p) = \sigma_u^2(p) - c_{u,g}(p, \boldsymbol{p}')\left(\overline{\mathbf{c}}_{g,g}(\boldsymbol{p}', \boldsymbol{p}')\right)^{-1}c_{u,g}^{\mathrm{T}}(p, \boldsymbol{p}'). \tag{6.332}$$

The estimation point on the right side of equation (6.331) enters only in the row-vector, $\mathbf{c}_{u,g}(p, \boldsymbol{p}')$. Thus, one may collect estimates, based on the given $\overline{\boldsymbol{g}}(\boldsymbol{p}')$, at L points in a vector, $\hat{\boldsymbol{u}}(\boldsymbol{p})$, and then expand $\boldsymbol{c}_{u,g}(p, \boldsymbol{p}')$ into an $L \times N$ matrix, or a column of row-vectors. The vector of minimum error variance estimators is

$$\hat{\boldsymbol{u}}(\boldsymbol{p}) = \mathbf{c}_{u,g}(\boldsymbol{p}, \boldsymbol{p}')\left(\overline{\mathbf{c}}_{g,g}(\boldsymbol{p}', \boldsymbol{p}')\right)^{-1}\overline{\boldsymbol{g}}(\boldsymbol{p}'), \tag{6.333}$$

where

$$\mathbf{c}_{u,g}(\boldsymbol{p}, \boldsymbol{p}') = \begin{pmatrix} c_{u,g}(p_1, \boldsymbol{p}') \\ \vdots \\ c_{u,g}(p_L, \boldsymbol{p}') \end{pmatrix}; \tag{6.334}$$

and, the error covariance matrix is

$$\mathbf{c}_{\varepsilon_u,\varepsilon_u}(p,p) = \mathbf{c}_{u,u}(p,p) - \mathbf{c}_{u,g}(p,p')\left(\overline{\mathbf{c}}_{g,g}(p',p')\right)^{-1}\mathbf{c}_{u,g}^{\mathrm{T}}(p',p). \tag{6.335}$$

If the covariance matrix of measurement errors, $\mathbf{c}_{\varepsilon_g,\varepsilon_g}$, is zero (i.e., there is no measurement error), then $\overline{\mathbf{c}}_{g,g} = \mathbf{c}_{g,g}$; and, further, if the process, u, is the same process, g (i.e., the estimator is a simple *interpolation*), then an estimate at a measurement point, p'_n, is exactly the measurement,

$$\hat{g}(p'_n) = c_{g,g}(p'_n,p')\left(\mathbf{c}_{g,g}(p',p')\right)^{-1}\tilde{g}(p') = \tilde{g}(p'_n), \tag{6.336}$$

since $\mathbf{c}_{g,g}(p'_n,p')$ is now a row of $\mathbf{c}_{g,g}(p',p')$ and $\mathbf{c}_{g,g}(p'_n,p')\,(\mathbf{c}_{g,g}(p',p'))^{-1} = (0 \cdots 1 \cdots 0)$, where the 1 is at the nth location. The estimator for this special case thus has the *reproducing property*. The estimated function, $\hat{g}(x_1, x_2)$ (a realization of g), in this case passes through all values of g at the measurement points.

Because $\overline{\mathbf{c}}_{g,g}^{-1}$ is positive definite, the second term on the right side of equation (6.332) is always positive (or, zero), and the variance of the estimation error satisfies

$$0 \le \mathrm{var}\left(\varepsilon_{\hat{u}}(p)\right) \le \sigma_u^2. \tag{6.337}$$

That is, in principle the estimation (even using just a single measurement!) is no worse than what is already known about u from its variance (or auto-covariance function).

A further generalization allows multiple measurements of different types at each measurement point, as well as multiple and different types of quantities to be estimated at each estimation point. For example, measurements of the gravity anomaly and of its spatial gradients could be used to estimate the geoid undulation and the deflection of the vertical. In this case, each of the entries in the covariance matrices in equations (6.326) and (6.327) or (6.334) is itself a covariance matrix that quantifies the correlations between *vectors* of quantities at two points. The only requirement is that all the covariances are mutually consistent, which is possible by the propagation of covariances if the quantities are related through linear functionals (e.g., derivatives) of the disturbing potential.

In principle, least-squares collocation depends solely on the data and their locations, as well as the covariance function that characterizes with certain assumptions the randomness of the underlying processes. It yields the best possible estimates from the available data and their distribution. Thus, being free of all other physical models, it is ideally suited to analyze errors associated with the lack of data, that is, errors due to finite extent and resolution. The following spectral analysis is concerned with the latter since resolution error implies high-frequency spectral omission error.

6.6.2 Frequency Domain Formulation

Consider a finite set of long, straight, and parallel tracks of measurements, as might be acquired with an airborne sensor system. The sensor measures the realization of a random process, g, at regularly distributed points along the tracks, $p'_{n,j} = \left(n\Delta x_1, j\Delta x_2\right)$, where Δx_1 is the along-track spacing, $-N/2 < n < N/2 - 1$, and Δx_2 is the spacing between tracks, $j = 0,\ldots, J-1$. The $J \times 1$ vectors, $\hat{\boldsymbol{g}}_n = \left(\hat{g}\left(p'_{n,0}\right) \;\cdots\; \hat{g}\left(p'_{n,J-1}\right)\right)^{\mathrm{T}}$, include the measurements for the nth point along *all* tracks (Figure 6.30). It is desired to estimate the realization of a related random process, u, at points, $p_\ell = \left(\ell \Delta x_1 + b_1, b_2\right)$, $-N/2 < \ell < N/2 - 1$, on a parallel track as shown in the figure. The estimation points have the same along-track spacing, Δx_1, but are offset from the $j = 0$ track and from the data along the measurement tracks by the vector $\boldsymbol{b} = (b_1 \;\; b_2)^{\mathrm{T}}$. The processes, u and g, are related as discussed in the previous section and have mutually consistent covariance functions.

For N sufficiently large it may be assumed that the tracks are infinitely long (i.e., let $N \to \infty$). Then, the covariance matrices, $\mathbf{c}_{u,g}(\boldsymbol{p}, \boldsymbol{p}')$ and $\overline{\mathbf{c}}_{g,g}(\boldsymbol{p}', \boldsymbol{p}')$ are also infinitely dimensioned,

$$\mathbf{c}_{u,g}\left(\boldsymbol{p},\boldsymbol{p}'\right)=\begin{pmatrix} \ddots & \vdots & \vdots & \iddots \\ \cdots & \mathbf{c}_{u,g}\left(p_\ell,\boldsymbol{p}'_n\right) & \mathbf{c}_{u,g}\left(p_\ell,\boldsymbol{p}'_{n+1}\right) & \cdots \\ \cdots & \mathbf{c}_{u,g}\left(p_{\ell+1},\boldsymbol{p}'_n\right) & \mathbf{c}_{u,g}\left(p_{\ell+1},\boldsymbol{p}'_{n+1}\right) & \cdots \\ \iddots & \vdots & \vdots & \ddots \end{pmatrix}, \tag{6.338}$$

$$\mathbf{c}_{g,g}\left(\boldsymbol{p}',\boldsymbol{p}'\right)=\begin{pmatrix} \ddots & \vdots & \vdots & \iddots \\ \cdots & \mathbf{c}_{g,g}\left(\boldsymbol{p}'_m,\boldsymbol{p}'_n\right) & \mathbf{c}_{g,g}\left(\boldsymbol{p}'_m,\boldsymbol{p}'_{n+1}\right) & \cdots \\ \cdots & \mathbf{c}_{g,g}\left(\boldsymbol{p}'_{m+1},\boldsymbol{p}'_n\right) & \mathbf{c}_{g,g}\left(\boldsymbol{p}'_{m+1},\boldsymbol{p}'_{n+1}\right) & \cdots \\ \iddots & \vdots & \vdots & \ddots \end{pmatrix}. \tag{6.339}$$

The elements, $\mathbf{c}_{u,g}(p_\ell, \boldsymbol{p}'_n)$, in equation (6.338) are $1 \times J$ row-vectors of covariances between u at $\boldsymbol{p}_\ell$ and the elements of $\boldsymbol{g}_n$ at the nth point along the tracks, represented by $\boldsymbol{p}'_n$,

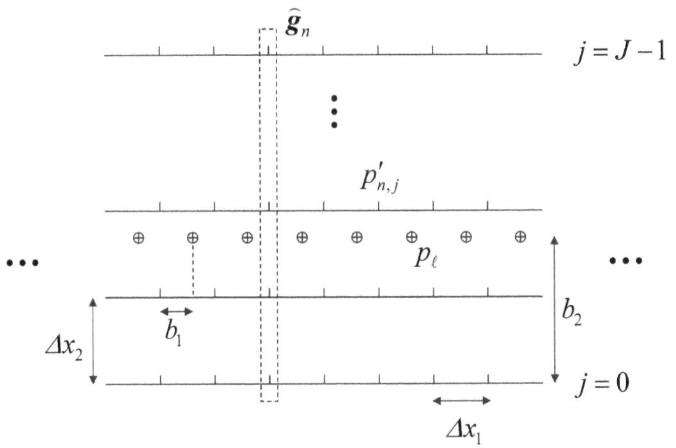

Figure 6.30: Geometry of the data points (ticks on the horizontal lines) and the estimation points (circled crosses) for the spectral analysis of least-squares collocation. The nth measurement vector, $\hat{\boldsymbol{g}}_n$, comprises the J measurements at the nth point on each track.

$$c_{u,g}\left(p_\ell, p'_n\right) = \left(c_{u,g}\left(\xi_{n-\ell,0} - b\right) \quad \cdots \quad c_{u,g}\left(\xi_{n-\ell,J-1} - b\right)\right). \tag{6.340}$$

Similarly, the elements of the matrix, $\mathbf{c}_{g,g}(p', p')$, are $J \times J$ matrices of covariances among the elements of the vectors, g_m and g_n, at track points, p'_m and p'_n,

$$\mathbf{c}_{g,g}\left(p'_m, p'_n\right) = \begin{pmatrix} c_{g,g}\left(\xi'_{n-m,0}\right) & \cdots & c_{g,g}\left(\xi'_{n-m,J-1}\right) \\ \vdots & \ddots & \vdots \\ c_{g,g}\left(\xi'_{n-m,J-1}\right) & \cdots & c_{g,g}\left(\xi'_{n-m,0}\right) \end{pmatrix}; \tag{6.341}$$

and $\bar{\mathbf{c}}_{g,g}\left(p', p'\right) = \mathbf{c}_{g,g}\left(p', p'\right) + \mathbf{c}_{\varepsilon_g,\varepsilon_g}\left(p', p'\right)$. The indicated coordinate vectors are

$$\xi_{n-\ell,j} = \left(\left(n-\ell\right)\Delta x_1 \quad j\Delta x_2\right)^{\mathrm{T}}. \tag{6.342}$$

Re-writing equation (6.330),

$$\mathbf{c}_{u,g}\left(p, p'\right) = \mathbf{h}^{\mathrm{T}}\left(p, p'\right)\bar{\mathbf{c}}_{g,g}\left(p', p'\right), \tag{6.343}$$

the nth element, a $1 \times J$ vector, of the ℓth row on the left side (see equation (6.338)) is a linear combination of elements of the nth column of $\bar{\mathbf{c}}_{g,g}(p', p')$ (see equation (6.339)),

$$c_{u,g}\left(p_\ell, p'_n\right) = \sum_{m=-\infty}^{\infty} h^{\mathrm{T}}\left(p_\ell, p'_m\right)\bar{\mathbf{c}}_{g,g}\left(p'_m, p'_n\right), \quad -\infty < \ell, n < \infty, \tag{6.344}$$

where $h^{\mathrm{T}}(p_\ell, p'_m)$ is the mth $1 \times J$ vector within the infinitely dimensioned row vector, $h^{\mathrm{T}}(p_\ell, p'_m)$, being a row of the matrix, $\mathbf{h}^{\mathrm{T}}(p, p')$. The vector, $h^{\mathrm{T}}(p_\ell, p'_m)$, defines the optimal weights given to the measurements at the mth along-track location (on all tracks). For infinitely long tracks, *it can depend only on the difference in the along-track coordinates with respect to the estimation point* since the measured and estimated processes are stationary and there is no change in the *relative* geometry between the measurement and estimation points along the tracks. That is, the geometric distribution of the measurements is x_1-translation invariant and, therefore, so are the weight vectors, $h^{\mathrm{T}}(p_\ell, p')$, which, in fact, must be identical for all estimation points, i.e., for all ℓ. Each depends on the coordinate difference vectors, $\xi_{m-\ell,j} - b = \left(\left(m-\ell\right)\Delta x_1 - b_1 \quad j\Delta x_2 - b_2\right)^{\mathrm{T}}$. For finite tracks, the rows of $\mathbf{h}^{\mathrm{T}}$ would differ, especially for estimation points near the endpoints of the tracks.

Continuing with the assumption of infinitely long tracks, one needs to consider only a single row of $\mathbf{c}_{u,g}(p, p')$; and, equation (6.344) with $\ell = 0$ is fully representative of elements of this row,

$$c_{u,g}\left(p_0, p'_n\right) = \sum_{m=-\infty}^{\infty} h^{\mathrm{T}}\left(p_0, p'_m\right)\bar{\mathbf{c}}_{g,g}\left(p'_m, p'_n\right), \quad -\infty < n < \infty. \tag{6.345}$$

With equation (6.340) for the left side, this sum is recognized as a $1 \times J$ vector of sums of convolutions with respect to the coordinates, $m\Delta x_1$, for each track, $j = 0,\dots, J-1$. Specifically, the kth element of the $1 \times J$ vector on the left is

$$\begin{aligned} c_{u,g}\left(\boldsymbol{\xi}_{n,k} - \boldsymbol{b}\right) &= \sum_{m=-\infty}^{\infty}\sum_{j=0}^{J-1} h\left(\boldsymbol{\xi}_{m,j} - \boldsymbol{b}\right)\overline{c}_{g,g}\left(\xi_{n-m,k-j}\right) \\ &= \sum_{j=0}^{J-1}\sum_{m=-\infty}^{\infty} h\left(\boldsymbol{\xi}_{m,j} - \boldsymbol{b}\right)\overline{c}_{g,g}\left(\xi_{n-m,k-j}\right) \end{aligned} \tag{6.346}$$

For any k, this is a sum of J convolutions, equation (4.35); or, in view of the first of these equations, it is a convolution of vectors, defined as an infinite sum of the inner product of $\boldsymbol{h}(p_0, \boldsymbol{p}'_m)$ and $\overline{\mathbf{c}}_{g,g}(\boldsymbol{p}'_m, \boldsymbol{p}'_n)$ for a given n.

Let the Fourier transforms of $\boldsymbol{c}_{u,g}(p_0, \boldsymbol{p}'_n)$, $\overline{\mathbf{c}}_{g,g}(\boldsymbol{p}'_m, \boldsymbol{p}'_n)$, and $\boldsymbol{h}^{\mathrm{T}}(p_0, \boldsymbol{p}'_m)$, with respect to $n\Delta x_1$ be denoted, respectively, by

$$\begin{aligned} \tilde{\boldsymbol{S}}_{u,g}\left(f_1;\boldsymbol{b}\right) &= \mathcal{F}\left(c_{u,g}\left(\boldsymbol{\xi}_{n,0} - \boldsymbol{b}\right) \quad \cdots \quad c_{u,g}\left(\boldsymbol{\xi}_{n,J-1} - \boldsymbol{b}\right)\right) \\ &= \Delta x_1 \left(\sum_{n=-\infty}^{\infty} c_{u,g}\left(\boldsymbol{\xi}_{n,0} - \boldsymbol{b}\right)e^{-i2\pi n\Delta x_1 f_1} \quad \cdots \quad \sum_{n=-\infty}^{\infty} c_{u,g}\left(\boldsymbol{\xi}_{n,J-1} - \boldsymbol{b}\right)e^{-i2\pi n\Delta x_1 f_1}\right) \end{aligned} \tag{6.347}$$

$$\tilde{\mathbf{S}}_{\tilde{g},\tilde{g}}\left(f_1\right) = \mathcal{F}\begin{pmatrix} \overline{c}_{g,g}\left(\xi_{n,0}\right) & \cdots & \overline{c}_{g,g}\left(\xi_{n,J-1}\right) \\ \vdots & \ddots & \vdots \\ \overline{c}_{g,g}\left(\xi_{n,-J+1}\right) & \cdots & \overline{c}_{g,g}\left(\xi_{n,0}\right) \end{pmatrix}, \tag{6.348}$$

$$\tilde{\boldsymbol{H}}^{\mathrm{T}}\left(f_1;\boldsymbol{b}\right) = \mathcal{F}\left(h\left(\boldsymbol{\xi}_{m,0} - \boldsymbol{b}\right) \quad \cdots \quad h\left(\boldsymbol{\xi}_{m,J-1} - \boldsymbol{b}\right)\right), \tag{6.349}$$

where the details of the last two equations are similar to equation (6.347). Elements of $\tilde{\boldsymbol{S}}_{u,g}$ and $\tilde{\mathbf{S}}_{\tilde{g},\tilde{g}}$ are *hybrid PSD-covariances*, equation (6.203); that is, they are a PSD with respect to x_1 and covariance with respect to x_2. Being transforms of sequences on the infinite domain, they are periodic with period, $1/\Delta x_1$, in continuous frequency, f_1. The dependence on $j\Delta x_2$ is omitted for convenience in notation and is embedded in the vectors and matrices. Then, by the convolution theorem (4.36), the transform of equation (6.346) is

$$\tilde{\boldsymbol{S}}_{u,g}\left(f_1;\boldsymbol{b}\right) = \frac{1}{\Delta x_1}\tilde{\boldsymbol{H}}^{\mathrm{T}}\left(f_1;\boldsymbol{b}\right)\tilde{\mathbf{S}}_{\tilde{g},\tilde{g}}\left(f_1\right). \tag{6.350}$$

Solving equation (6.350) for $\tilde{\boldsymbol{H}}^{\mathrm{T}}$,

$$\tilde{\boldsymbol{H}}^{\mathrm{T}}\left(f_1;\boldsymbol{b}\right) = \Delta x_1 \tilde{\boldsymbol{S}}_{u,g}\left(f_1;\boldsymbol{b}\right)\left(\tilde{\mathbf{S}}_{\tilde{g},\tilde{g}}\left(f_1\right)\right)^{-1}. \tag{6.351}$$

The estimator for any point, p_ℓ, from equation (6.321), is given by

$$\hat{u}(p_\ell) = \sum_{n=-\infty}^{\infty} \boldsymbol{h}^{\mathrm{T}}(p_\ell, \boldsymbol{p}'_n)\,\hat{\boldsymbol{g}}_n, \tag{6.352}$$

which is, as in equation (6.345), a convolution of vectors. However, since $\boldsymbol{h}^{\mathrm{T}}(p_\ell, \boldsymbol{p}'_n)$ depends on the difference, $(n-\ell)\Delta x_1$, it is the convolution of $\hat{\boldsymbol{g}}_n$ and $\boldsymbol{h}^{\mathrm{T}}_{-n}$. Since $\boldsymbol{h}^{\mathrm{T}}_{-n}$ is a real vector, its Fourier transform is $\tilde{\boldsymbol{H}}^{\mathrm{T}*}(f_1;\boldsymbol{b})$, by the duality of transforms applied to the symmetry property (2.22) and the equivalence relation (2.28). The convolution theorem (4.36) then gives the transform of equation (6.352) as

$$\tilde{\hat{U}}(f_1;\boldsymbol{b}) = \frac{1}{\Delta x_1}\tilde{\boldsymbol{H}}^{\mathrm{T}*}(f_1;\boldsymbol{b})\tilde{\hat{\boldsymbol{G}}}(f_1). \tag{6.353}$$

An element of the covariance matrix of the estimation error at points, p_{ℓ_1} and p_{ℓ_2}, is given by equation (6.335),

$$c_{\varepsilon_u,\varepsilon_u}\left(p_{\ell_1}, p_{\ell_2}\right) = c_{u,u}\left(p_{\ell_1}, p_{\ell_2}\right) - \boldsymbol{c}_{u,g}\left(p_{\ell_1}, \boldsymbol{p}'\right)\left(\overline{\mathbf{c}}_{g,g}(\boldsymbol{p}',\boldsymbol{p}')\right)^{-1}\boldsymbol{c}^{\mathrm{T}}_{u,g}\left(p_{\ell_2}, \boldsymbol{p}'\right), \tag{6.354}$$

which, upon substituting equation (6.343), becomes

$$\begin{aligned} c_{\varepsilon_u,\varepsilon_u}\left(p_{\ell_1}, p_{\ell_2}\right) &= c_{u,u}\left(p_{\ell_1}, p_{\ell_2}\right) - \boldsymbol{h}^{\mathrm{T}}\left(p_{\ell_1}, \boldsymbol{p}'\right)\boldsymbol{c}^{\mathrm{T}}_{u,g}\left(p_{\ell_2}, \boldsymbol{p}'\right) \\ &= c_{u,u}\left(p_{\ell_1}, p_{\ell_2}\right) - \sum_{n=-\infty}^{\infty}\boldsymbol{h}^{\mathrm{T}}\left(p_{\ell_1}, \boldsymbol{p}'_n\right)\boldsymbol{c}^{\mathrm{T}}_{u,g}\left(p_{\ell_2}, \boldsymbol{p}'_n\right) \end{aligned} \tag{6.355}$$

Now, $\boldsymbol{h}^{\mathrm{T}}(p_{\ell_1},\boldsymbol{p}'_n)$ depends on the difference, $(n-\ell_1)\Delta x_1$, and the covariances, $\boldsymbol{c}^{\mathrm{T}}_{u,g}(p_{\ell_1},\boldsymbol{p}'_n)$, depend on the differences, $(n-\ell_2)\Delta x_1$. Because of the along-track invariance, ℓ_1 may be fixed while all values of ℓ_2 define the possible elements of $\mathbf{c}_{\varepsilon_u,\varepsilon_u}$ (i.e., its diagonals, as in equation (4.38)). Then the second term on the right side of equation (6.355) is a convolution of vectors, $\boldsymbol{h}^{\mathrm{T}}_n$ and $(\boldsymbol{c}^{\mathrm{T}}_{u,g})_{-n}$, for fixed ℓ_1; and, the Fourier transform with respect to $\ell_2\Delta x_1$, applied to both sides, together with the convolution theorem yield

$$\tilde{S}_{\varepsilon_u,\varepsilon_u}(f_1;\boldsymbol{b}) = \tilde{S}_{u,u}(f_1) - \frac{1}{\Delta x_1}\tilde{\boldsymbol{H}}^{\mathrm{T}}(f_1;\boldsymbol{b})\tilde{\boldsymbol{S}}^{*\mathrm{T}}_{u,g}(f_1;\boldsymbol{b}). \tag{6.356}$$

As before, $\tilde{S}_{\varepsilon_u,\varepsilon_u}$ and $\tilde{S}_{u,u}$ are along-track hybrid PSD-covariance functions. Substituting equation (6.351), one finally arrives at

$$\tilde{S}_{\varepsilon_u,\varepsilon_u}(f_1;\boldsymbol{b}) = \tilde{S}_{u,u}(f_1) - \tilde{\boldsymbol{S}}_{u,g}(f_1;\boldsymbol{b})\left(\tilde{\mathbf{S}}_{\bar{g},\bar{g}}(f_1)\right)^{-1}\tilde{\boldsymbol{S}}^{*\mathrm{T}}_{u,g}(f_1;\boldsymbol{b}). \tag{6.357}$$

The relationship between the Fourier transforms of sampled and continuous functions, given by equation (4.8), expresses $\tilde{\boldsymbol{S}}_{u,g}(f_1;\boldsymbol{b})$ as

$$\tilde{\boldsymbol{S}}_{u,g}(f_1;\boldsymbol{b}) = \sum_{n=-\infty}^{\infty}\boldsymbol{S}_{u,g}\left(f_1 + 2nf_{\mathcal{N}_1};b_2\right)e^{-i2\pi\left(f_1+2nf_{\mathcal{N}_1}\right)b_1}, \tag{6.358}$$

where $\boldsymbol{S}_{u,g}(f_1;b_2)$ is the (non-periodic) hybrid PSD-covariance function, defined for the PSD on the infinite spectral domain, $-\infty < f_1 < \infty$, and $f_{\mathcal{N}_1}$ is the Nyquist frequency. The exponential factor comes from the translation property (2.73), associated with a shift in the coordinate origin for the cross-covariances relative to the auto-covariances of $\tilde{\boldsymbol{g}}$. Substituting equation (6.358) and the analogous equation for $\tilde{\boldsymbol{S}}_{g,u}$ into equation (6.357) yields

$$\tilde{S}_{\varepsilon_u,\varepsilon_u}(f_1;\boldsymbol{b}) = \tilde{S}_{u,u}(f_1) - \sum_{n=-\infty}^{\infty}\sum_{n'=-\infty}^{\infty} \boldsymbol{S}_{u,g}\left(f_1 + 2nf_{\mathcal{N}_1};b_2\right)\left(\tilde{\mathbf{S}}_{\tilde{g},\tilde{g}}(f_1)\right)^{-1} \boldsymbol{S}_{u,g}^{*\mathrm{T}}\left(f_1 + 2n'f_{\mathcal{N}_1};b_2\right)e^{-i4\pi(n-n')f_{\mathcal{N}_1}b_1} \tag{6.359}$$

To simplify the analysis of the estimation error, consider the average of the PSD over all possible along-track shifts of the estimation grid relative to the measurement grid. The result may be interpreted as the expected error PSD for an estimation point arbitrarily placed along the estimation track. The only part of the error PSD that depends on b_1 is the exponential in equation (6.359), which, with $f_{\mathcal{N}_1} = 1/(2\Delta x_1)$, averages according to equation (2.12) as

$$\frac{1}{\Delta x_1}\int_0^{\Delta x_1} e^{i\frac{2\pi}{\Delta x_1}(n-n')b_1}\,db_1 = \begin{cases} 0, & n \neq n' \\ 1, & n = n' \end{cases} \tag{6.360}$$

Thus, by averaging, the double sum in equation (6.359) reduces to a single sum,

$$\bar{\tilde{S}}_{\varepsilon_u,\varepsilon_u}(f_1;b_2) = \frac{1}{\Delta x_1}\int_0^{\Delta x_1} \tilde{S}_{\varepsilon_u,\varepsilon_u}(f_1;\boldsymbol{b})\,db_1$$
$$= \sum_{n=-\infty}^{\infty}\left(S_{u,u}\left(f_1 + 2nf_{\mathcal{N}_1}\right) - \boldsymbol{S}_{u,g}\left(f_1 + 2nf_{\mathcal{N}_1};b_2\right)\left(\tilde{\mathbf{S}}_{\tilde{g},\tilde{g}}(f_1)\right)^{-1}\boldsymbol{S}_{u,g}^{*\mathrm{T}}\left(f_1 + 2nf_{\mathcal{N}_1};b_2\right)\right) \tag{6.361}$$

where equation (4.8) is used also for $\tilde{S}_{u,u}(f_1)$. Defined for $-f_{\mathcal{N}_1} \le f_1 < f_{\mathcal{N}_1}$, $\bar{\tilde{S}}_{\varepsilon_u \varepsilon_u}$ is the *average* hybrid PSD-covariance of a discrete process, specifically, the error in estimating u at discrete points along the $\oplus$-track (Figure 6.30) that is *arbitrarily* displaced in the direction of the measurement tracks. It is given in terms of the hybrid PSD-covariances of *continuous* domain processes u and g, except for $\tilde{\mathbf{S}}_{\tilde{g}\tilde{g}}$, which can be decomposed into the hybrid PSD-covariances of a continuous process, an aliasing effect, and the measurement noise,

$$\tilde{\mathbf{S}}_{\tilde{g},\tilde{g}}(f_1) = \mathbf{S}_{\tilde{g},\tilde{g}}(f_1) + \mathbf{S}_{g\text{-alias}}(f_1) + \mathbf{S}_{\varepsilon_g,\varepsilon_g}(f_1), \tag{6.362}$$

where

$$\mathbf{S}_{g\text{-alias}}(f_1) = \sum_{\substack{n=-\infty \\ n\neq 0}}^{\infty} \mathbf{S}_{g,g}\left(f_1 + 2n\,f_{\mathcal{N}_1}\right). \tag{6.363}$$

To calculate $\bar{\tilde{S}}_{\varepsilon_u \varepsilon_u}$ one would expect to include only a few terms of the sums in equations (6.361) and (6.363), e.g., $-1 \le n \le 1$, if the cross-PSDs for u and g (and the auto-

PSD for g) attenuate sufficiently rapidly with frequency. The error variance is obtained by integrating the PSD, equation (6.361), over the principal domain, $-f_{N_1} \le f_1 < f_{N_1}$, according to equation (5.343) with $\ell = 0$,

$$\begin{aligned}\sigma_{\varepsilon_u}^2 &= \int_{-f_{N_1}}^{f_{N_1}} \bar{\tilde{S}}_{\varepsilon_u,\varepsilon_u}(f_1';b_2)\,df_1' \\ &= \sum_{n=-\infty}^{\infty} \int_{-f_{N_1}}^{f_{N_1}} \left(S_{u,u}(f_1'+2nf_{N_1}) - \boldsymbol{S}_{u,g}(f_1'+2nf_{N_1};b_2)\left(\tilde{\mathbf{S}}_{\tilde{g},\tilde{g}}(f_1')\right)^{-1} \boldsymbol{S}_{u,g}^{*\mathrm{T}}(f_1'+2nf_{N_1};b_2)\right) df_1'\end{aligned} \tag{6.364}$$

where it is noted that $\tilde{S}_{\tilde{g},\tilde{g}}(f_1')$ is periodic with period, $2f_{N_1}$, so that a change in integration variable, $f_1 = f_1' + 2nf_{N_1}$, yields

$$\tilde{\mathbf{S}}_{\tilde{g}\tilde{g}}(f_1') = \tilde{\mathbf{S}}_{\tilde{g}\tilde{g}}(f_1 - 2nf_{N_1}) = \tilde{\mathbf{S}}_{\tilde{g}\tilde{g}}(f_1), \tag{6.365}$$

and thus,

$$\begin{aligned}\sigma_{\varepsilon_u}^2 &= \sum_{n=-\infty}^{\infty} \int_{(2n-1)f_{N_1}}^{(2n+1)f_{N_1}} \left(S_{u,u}(f_1) - \boldsymbol{S}_{u,g}(f_1;b_2)\left(\tilde{\mathbf{S}}_{\tilde{g},\tilde{g}}(f_1)\right)^{-1} \boldsymbol{S}_{u,g}^{*\mathrm{T}}(f_1;b_2)\right) df_1 \\ &= \int_{-\infty}^{\infty} \left(S_{u,u}(f_1) - \boldsymbol{S}_{u,g}(f_1;b_2)\left(\tilde{\mathbf{S}}_{\tilde{g},\tilde{g}}(f_1)\right)^{-1} \boldsymbol{S}_{u,g}^{*\mathrm{T}}(f_1;b_2)\right) df_1\end{aligned} \tag{6.366}$$

where

$$\tilde{\mathbf{S}}_{g,g}(f_1) = \sum_{m=-\infty}^{\infty} \mathbf{S}_{g,g}(f_1 + 2mf_{N_1}), \quad -f_{N_1} \le f_1 < f_{N_1}. \tag{6.367}$$

Even though the integrand in equation (6.364) looks like (and is) the PSD of a continuous process, it is formally the *average* of a PSD (with respect to b_1); hence, $\sigma_{\varepsilon_u}^2$ only approximates the variance of ε_u.

The development above supposes a relatively small number of long parallel tracks with many along-track measurements of a single data type and a single type of quantity estimated not necessarily on any track, but parallel to them. Various generalizations of this scenario are possible. For example, the number of parallel tracks could be large, which would suggest two-dimensional Fourier transforms. The measurements at any point along a track could include multiple data types, for example different components of the gravity gradient tensor (airborne gravity gradiometry). The covariance matrices then have a more complicated block structure. One might also consider estimating more than one data type, for example for components of the deflection of the vertical. These generalizations are possible with least-squares collocation and require only more elaborate formulations. They are left to the fearless reader.

6.6.3 Example

For an illustration of least-squares collocation in the frequency domain, consider the estimation of the geoid undulation, N, from a track of gravity anomaly data, $\widehat{\Delta g} = \Delta g + \varepsilon_{\Delta g}$, that include observation errors. The physical model that relates these two functionals of the disturbing potential is Stokes's integral, equation (6.96), where, in principle, data over the entire sphere are required to determine the geoid undulation. In this example of the alternative operational approach, a single track of data is available and the geoid undulation is estimated along this track. Figure 6.22 shows the data, as well as the model, EGM2008 ($n_{\max} = 180$), that serves as reference. That is, the geoid undulation is estimated as in equation (6.242) with Stokes's integral replaced by the least-squares collocation estimator.

The data domain is a relatively short 462 km track of $K = 240$ points, and the estimation may be performed with a planar approximation. For a single track of data, $J = 1$ in Figure 6.30. Also, $\Delta x_1 = R\Delta\phi = 1.9$ km, corresponding to the 1 arcmin grid spacing of the gravity anomaly data. (Although it is a south-to-north profile, the spatial coordinate, x_1, is used for consistency with the previous section.) Furthermore, the displacement of the estimation profile from the data profile is zero, $\boldsymbol{b} = \boldsymbol{0}$. Equation (6.353) with equation (6.351) gives the Fourier transform of the estimator,

$$\tilde{\hat{U}}(f_1;\boldsymbol{0}) = \frac{\tilde{S}_{u,g}(f_1;\boldsymbol{0})}{\tilde{S}_{\hat{g},\hat{g}}(f_1)}\tilde{\hat{G}}(f_1), \tag{6.368}$$

where the residual geoid undulation is $u \equiv \delta N = N - N_{\mathrm{EGM2008}(n_{\max}=180)}$, and the residual gravity anomaly is $\hat{g} \equiv \delta\widehat{\Delta g} = \widehat{\Delta g} - \Delta g_{\mathrm{EGM2008}(n_{\max}=180)}$. The covariance of the data is the sum of the covariance of the observable, $\delta\Delta g$, and the covariance of the observation error, presumed independent of the observable. Moreover, it is assumed that the errors among the observations are uncorrelated and have equal variance (the value, 1 mGal^2, is assigned). Since both covariance functions, $c_{\hat{g},\hat{g}}$ and $c_{u,g}$, are symmetric (and real), the along-track PSDs, $\tilde{S}_{\hat{g},\hat{g}}(f_1)$ and $\tilde{S}_{u,g}(f_1)$, are also symmetric and real (equivalence (4.87)).

The model for the covariances is based on the reciprocal distance model, equation (6.200), for the residual disturbing potential, δT, where $c_{u,g}$ and $c_{g,g}$ are derived by the law of propagation of covariances, equation (6.47), applied to $\delta N = \delta T/\gamma$ and the planar approximation, $\delta\Delta g = -\partial(\delta T)/\partial r$. Figure 6.31 shows the two-dimensional PSD of the total and residual gravity anomalies according to the analytical expression derived from the reciprocal distance covariance model (Jekeli 2010). The residual PSD model varies smoothly across the cutoff frequency of the reference field. Using a model that instead abruptly truncates to zero for $\bar{f} < \bar{f}_0$ creates significant oscillations in the estimates (Gibbs's effect). The PSD models are functions of continuous frequency and thus do not account for the aliasing error associated with the finite along-track Nyquist frequency. Although it makes very little difference for a small sampling interval, as in this case, rather than using analytic expressions for the along-track hybrid PSD/covariances, they are computed according to

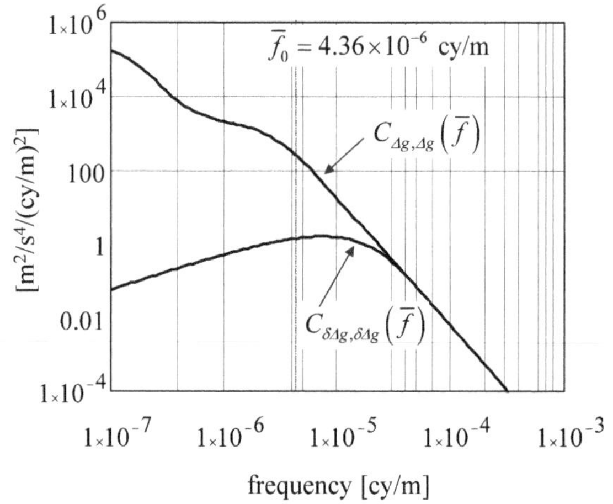

Figure 6.31: Two-dimensional PSDs of the total and residual gravity anomalies based on the model of Table 6.4 (Area 1). The PSD for the residual anomaly is the sum of the components, $j = 6,\ldots,9$. The dotted vertical line indicates the cutoff frequency, $\bar{f}_0$, of the reference field, corresponding to harmonic degree, $n_{\max} = 180$.

$$\tilde{S}_{\tilde{g},\tilde{g}}\left(\left(f_1\right)_k\right) = \mathrm{DFT}\left(c_{\tilde{g},\tilde{g}}\left(\ell\Delta x_1\right)\right)_k, \quad \tilde{S}_{u,g}\left(\left(f_1\right)_k\right) = \mathrm{DFT}\left(c_{u,g}\left(\ell\Delta x_1\right)\right)_k, \tag{6.369}$$

for $(f_1)_k = k/(K\Delta x_1)$, $k = 0,\ldots, K-1$, where the symmetries, $c_{\tilde{g}\tilde{g}}(\ell\Delta x_1) = c_{\tilde{g}\tilde{g}}((K/2-\ell)\Delta x_1)$ and $c_{u,g}(\ell\Delta x_1) = c_{u,g}((K/2-\ell)\Delta x_1)$, $\ell = 1,\ldots, K/2$, are enforced.

Additional tracks of data might be included for a more accurate estimation. For three tracks ($J = 3$), that is, one additional track on either side of the estimation track, equations (6.351) and (6.353) combine to yield

$$\tilde{\hat{U}} = \begin{pmatrix} \tilde{S}_{u,g}^{(-)} & \tilde{S}_{u,g}^{(0)} & \tilde{S}_{u,g}^{(+)} \end{pmatrix} \begin{pmatrix} \tilde{S}_{\tilde{g},\tilde{g}}^{(0)} & \tilde{S}_{\tilde{g},\tilde{g}}^{(+)} & \tilde{S}_{\tilde{g},\tilde{g}}^{(++)} \\ \tilde{S}_{\tilde{g},\tilde{g}}^{(+)} & \tilde{S}_{\tilde{g},\tilde{g}}^{(0)} & \tilde{S}_{\tilde{g},\tilde{g}}^{(+)} \\ \tilde{S}_{\tilde{g},\tilde{g}}^{(++)} & \tilde{S}_{\tilde{g},\tilde{g}}^{(+)} & \tilde{S}_{\tilde{g},\tilde{g}}^{(0)} \end{pmatrix}^{-1} \begin{pmatrix} \tilde{\tilde{G}}^{(-)} \\ \tilde{\tilde{G}}^{(0)} \\ \tilde{\tilde{G}}^{(+)} \end{pmatrix}. \tag{6.370}$$

Every element on the left and right sides of this equation is a function of the frequency, f_1. The along-track Fourier transforms of the residual data are $\tilde{\tilde{G}}^{(-,0,+)}$ for the three tracks; and, $\tilde{S}_{u,g}^{(-,0,+)}$ are the along-track PSD/cross-covariances between the residual geoid undulation on the central track and the residual gravity anomalies on the three data tracks. The matrix elements, $\tilde{S}_{\tilde{g},\tilde{g}}^{(0,+,++)}$, are the along-track PSD/auto-covariances among the residual data on any two tracks, respectively, separated by $b_2 = 0\Delta x_2$, $+\Delta x_2$, or $+\Delta x_2+\Delta x_2$, and Δx_2 is the distance between tracks. Figure 6.32 shows the geoid undulation estimates along the (central) data track for one and three data tracks. As expected, the estimation improves with an increase in the number of tracks, which comes closer to the true physical model, equation (6.242), i.e., a surface integral. Since there are K data per track, the spatial-domain least-squares collocation estimator, equations (6.321) and (6.330), would require the inversion of

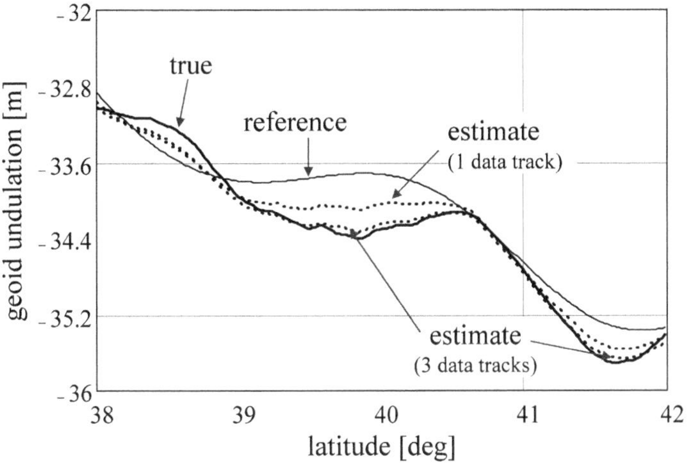

Figure 6.32: Geoid undulation estimated by frequency-domain least-squares collocation along the track, $38° \le \phi \le 42°$, at $\lambda = 277.27°$, from gravity anomaly data on this track (Figure 6.22) or on this and two additional tracks at $\lambda = 277.17°$ and $\lambda = 277.37°$. A reference field, EGM2008($n_{max} = 180$) is first removed from the gravity anomalies and subsequently added back in the form of the geoid undulation (shown). The covariance model used in the estimation is shown in Figure 6.31. The "true" geoid undulation is obtained from EGM2008($n_{max} = 2160$).

a $K \times K$ matrix, which may be unstable, the matrix, $\mathbf{c}_{g,g}$, being nearly singular due to closely spaced data. For the frequency-domain method, the estimation requires four DFTs and no matrix inversion for the single data track, and ten DFTs and $K/2$ 3×3 matrix inversions (due to the conjugate symmetry of $\tilde{U}$) for three data tracks.

For the single-track case in this example, the average error spectrum given by the final integrand of equation (6.366) with $\delta N = \delta T/\gamma$ and $\delta\Delta g = -\partial(\delta T)/\partial r$ is

$$\overline{S}_{\varepsilon_{\delta N},\varepsilon_{\delta N}}(f_1;0) = S_{\delta N,\delta N}(f_1) - S_{\delta N,\delta\Delta g}(f_1;0)\left(\tilde{S}_{\widehat{\delta\Delta g},\widehat{\delta\Delta g}}(f_1)\right)^{-1} S_{\delta N,\delta\Delta g}(f_1;0), \quad (6.371)$$

where the averaging refers to the possible locations of the estimation points along the track. With the transfer functions between the spectra of the geoid undulation and gravity anomaly, $S_{\delta N,\delta\Delta g}(f_1) = S_{\delta\Delta g,\delta\Delta g}(f_1)/(2\pi f_1\gamma) = (2\pi f_1\gamma) S_{\delta N,\delta N}(f_1)$, this leads to

$$\overline{S}_{\varepsilon_{\delta N},\varepsilon_{\delta N}}(f_1) = S_{\delta N,\delta N}(f_1)\left(1 - \frac{1}{1 + \dfrac{S_{\delta\Delta g\text{-alias}}(f_1) + S_{\varepsilon_{\delta\Delta g},\varepsilon_{\delta\Delta g}}(f_1)}{S_{\delta\Delta g,\delta\Delta g}(f_1)}}\right) \quad (6.372)$$

in view of equation (6.362), and where all PSDs are for quantities along the track. As aliasing increases due to a larger sampling interval, Δx_1, and/or as the observation error PSD increases, the estimation error PSD, $\overline{S}_{\varepsilon_{\delta N},\varepsilon_{\delta N}}(f_1)$, approaches $S_{\delta N,\delta N}(f_1)$— the estimation is never worse, in the statistical sense, than the *a priori* PSD of the residual geoid undulation. Note that the hybrid aliasing error PSD/covariance function accounts also for the lack of data in the cross-track direction. A formula similar to equation (6.372) holds in the two-dimensional case that represents a large number of long parallel data tracks. The PSD of the aliasing error does not vanish until data occupy the entire plane (in this planar approximation). Only then and with vanishing observation error does the PSD of the estimation error also go to zero.

* * *

EXERCISES

Exercise 6.1

Show that the reciprocal distance, $1/d$, is harmonic (satisfies Laplace's equation) for $d \neq 0$, where $d = \sqrt{\left(x_1 - x_1'\right)^2 + \left(x_2 - x_2'\right)^2 + \left(x_3 - x_3'\right)^2}$.

Exercise 6.2

Show that the Laplace operator in spherical coordinates, equation (6.49), can be written as

$$\nabla^2 = \frac{1}{r^2}\frac{\partial}{\partial r}\left(r^2 \frac{\partial}{\partial r}\right) + \frac{1}{r^2}\nabla^2_{(\theta,\lambda)}, \tag{E6.1}$$

where the Laplace-Beltrami operator, $\nabla^2_{(\theta,\lambda)}$, is given by equation (2.240). Then show that $\left(\frac{\partial}{\partial r}\left(r^2 \frac{\partial}{\partial r}\right)\right) r^n = r^n n(n+1)$ and prove equation (6.50) using equation (2.241). Similarly prove equation (6.51).

Exercise 6.3

Show that for $r > R$ and $d = \sqrt{r^2 + R^2 - 2rR\cos\psi}$,

$$\sum_{n=0}^{\infty}(2n+1)\left(\frac{R}{r}\right)^{n+1} P_n(\cos\psi) = \frac{R\left(r^2 - R^2\right)}{d^3}. \tag{E6.2}$$

Hint: Consider the series expression for $-2Rr\frac{\partial}{\partial r}\left(\frac{1}{d}\right) - \frac{R}{d}$ using the Coulomb expansion (6.12).

Exercise 6.4

For r and d defined in Problem 6.3, show that

$$\sum_{n=1}^{\infty}(2n+1)n\left(\frac{R}{r}\right)^{n+2} P_n(\cos\psi) = \frac{R^3}{d^3}\left(5\cos\psi - \frac{r}{R} + \frac{d^2}{rR} - 6\frac{rR}{d^2}\sin^2\psi\right). \tag{E6.3}$$

Hint: Write $(2n + 1)n = (n + 1)(n + 2) + n(n + 1) - 3(n + 1) + 1$ and consider the series expression for $2R^2 r\dfrac{\partial^2}{\partial r^2}\left(\dfrac{1}{d}\right)+5R^2\dfrac{\partial}{\partial r}\left(\dfrac{1}{d}\right)+\dfrac{R^2}{rd}$ using the Coulomb expansion (6.12).

Exercise 6.5

With reference to Figure 2.13 and equation (2.196), derive the following derivatives,

$$\frac{\partial\psi}{\partial\lambda} = -\sin\theta\sin\zeta, \quad \frac{\partial\psi}{\partial\theta} = -\cos\zeta, \tag{E6.4}$$

making use of equations (2.198) and (2.199). With these in hand, verify equation (6.109).

Exercise 6.6

Derive equation (6.138) starting with an approximation as in equation (6.137) and using $\Delta\lambda = 2\pi/M$, $\lambda_\ell = \left(\ell + \dfrac{1}{2}\right)\Delta\lambda$ to derive the integral of the exponential (see also equation (4.227)).

Exercise 6.7

With reference to equation (6.195) let $L^2 = 1 + \rho^2 - 2\rho\cos\psi$ and show to first-order approximation that $\rho = R_b^2/(r_1 r_2) \approx 1 - \dfrac{2b + z_1 + z_2}{R}$ and $L^2 \approx (1 - \rho)^2 + \rho s^2/R^2$. Then, prove equation (6.196) together with $\sigma_T^2 = c_{T,T}(0,0,0) \approx \sigma_0^2 R^2/(2b^2)$.

Exercise 6.8

Use the Coulomb expansion for the reciprocal distance, the addition theorem for spherical harmonics, and their orthogonality, to derive equation (6.206) from equation (6.205).

Exercise 6.9

Prove equations (6.242) through (6.244) by making use of the spectral and spatial relationships between the disturbing potential and the gravity anomaly in Section 6.2.4.

Exercise 6.10

Derive equation (6.277) starting with

$$\left(\tilde{\xi}_{\ell_1,\ell_2}, \tilde{\eta}_{\ell_1,\ell_2}\right) = \frac{1}{\gamma}\mathrm{DFT}^{-1}\left(\tilde{G}_{k_1,k_2}\tilde{D}^{(\xi,\eta)}_{k_1,k_2}\right)_{\ell_1,\ell_2}, \tag{E6.5}$$

obtained by the convolution theorem, and which implies that

$$\tilde{H}_{k_1,k_2} + \tilde{X}_{k_1,k_2} = \frac{1}{\gamma}\tilde{G}_{k_1,k_2}\left(\tilde{D}^{(\xi)}_{k_1,k_2} + \tilde{D}^{(\eta)}_{k_1,k_2}\right) = -i\frac{1}{\gamma}\frac{\tilde{G}_{k_1,k_2}}{\bar{f}_{k_1,k_2}}\left(\frac{k_2}{N_2\Delta x_2} + \frac{k_1}{N_1\Delta x_1}\right). \tag{E6.6}$$

Solve for $\tilde{G}_{k_1,k_2}$ and simplify using equations (6.261) and (6.262). Then show that equations (6.273) and (6.277) are equivalent using equation (6.278).

Exercise 6.11

Prove equation (6.295) by setting $u_j = x_j - x'_j$, $j = 1,2,3$ and noting that $\partial^n/(\partial x'_j)^n = (\partial u_j/\partial x'_j)^n \partial^n/(\partial u_j)^n$ and $\partial^n/(\partial u_j)^n = (\partial x_j/\partial u_j)^n \partial^n/(\partial x_j)^n$.

Exercise 6.12

Derive equation (6.306) under the assumption that $\rho(x'_1,x'_2,x'_3) = \rho(x'_1,x'_2)$ instead of $\rho(x'_1,x'_2,x'_3) = \rho_0$ in equation (6.290).

Exercise 6.13

Show that the Parker and Forsberg expressions are equivalent for the Fourier transform of the gravitational effect of the topography, equations (6.305) and (6.309) (for $j = 3$), starting with the Fourier transform of the reciprocal distance, from equation (2.337),

$$\mathcal{F}\left(\frac{1}{\sqrt{x_1^2 + x_2^2 + x_3^2}}\right) = \frac{1}{\bar{f}}e^{-2\pi x_3 \bar{f}}, \quad \bar{f} \neq 0, \quad x_3 > 0. \tag{E6.7}$$

References

Abramowitz, M. and Stegun, I.A. (1972): Handbook of Mathematical Functions. Dover Publications, Inc., New York.

Albertella, A., Sansò, F. and Sneeuw, N. (1999): Band-limited functions on a bounded spherical domain: the Slepian problem on the sphere. Journal of Geodesy, 73: 436–447.

Amante, C. and Eakins, B.W. (2009): ETOPO1 1 arc-minute global relief model: procedures, data sources and analysis. Technical Memorandum NESDIS NGDC-24, National Geophysical Data Center, Marine Geology and Geophysics Division, Boulder, CO.

Arfken, G. (1970): Mathematical Methods for Physicists. Academic Press, New York.

Aster, R.C., Borchers, B. and Thurber, C.H. (2005): Parameter Estimation and Inverse Problems. Elsevier Academic Press, Burlington, Massachusetts.

Balmino, G., Lambeck, K. and Kaula, W.M. (1973): A spherical harmonic analysis of the Earth's topography. Journal of Geophysical Research, 78(2): 478–481.

Balmino, G., Vales, N., Bonvalot, S. and Briais, A. (2012): Spherical harmonic modelling to ultra-high degree of Bouguer and isostatic anomalies. Journal of Geodesy, 86: 499–520.

Barrera, R.G., Estévez, G.A. and Giraldo, J. (1985): Vector spherical harmonics and their application to magnetostatics. Eur. J. Phys., 6: 287–294.

Barthelmes, F. and Köhler, W. (2012): International Centre for Global Earth Models (ICGEM). Journal of Geodesy, 86(10): 932–934.

Bath, M. (1974): Spectral Analysis in Geophysics. Elsevier Scientific Publishing Co.

Bendat, J.S. and Piersol, A.G. (1986): Random data, Analysis and Measurement Procedures, Second Edition. John Wiley and Sons, New York.

Benveniste, J. (2011): Radar Altimetry: Past, present and future. pp. 1–17. *In*: Vignudelli, S. (ed.). Coastal Altimetry, Springer-Verlag Berlin.

Berberian, S.K. (1999): Fundamentals of Real Analysis. Springer-Verlag, New York.

Beuthe, M. (2008): Thin elastic shells with variable thickness for lithospheric flexure of one-plate planets. Geophysical Journal International, 172(2): 817–841.

Blakely, R.J. (1995): Potential Theory in Gravity and Magnetic Applications. Cambridge University Press, Cambridge, UK.

Boas, M.L. (1966): Mathematical Methods in the Physical Sciences. John Wiley & Sons, Inc., New York.

Bölling, K. and Grafarend, E.W. (2005): Ellipsoidal spectral properties of the Earth's gravitational potential and its first and second derivatives. Journal of Geodesy, 79: 300–330.

Bouman, J. (1997): A survey of global gravity models. DEOS Report no. 97.1, Delft Institute for Earth-Oriented Space Research, Delft University of Technology, Delft, The Netherlands.

Brigham, E.O. (1988): The Fast Fourier Transform and its Applications, Prentice Hall, Engelwood, New Jersey.

Brillinger, D.R. (2001): Time Series—Data analysis and Theory. Society for Industrial and Applied Mathematics, Philadelphia.

Brotchie, J.F. and Sylvester, R. (1969): On crustal flexure. Journal of Geophysical Research, 74(22): 5240–5252.

Buttkus, B. (2000): Spectral Analysis and Filter Theory in Applied Geophysics. Springer-Verlag, Berlin.

Butzer, P.L. and Nessel, R.J. (1971): Fourier Analysis and Approximation, vol.1—One-Dimensional Theory. Academic Press, New York.

Cappelari, J.O., Velez, C.E. and Fuchs, A.J. (eds.) (1976): Mathematical theory of the Goddard Trajectory Determination System. GSFC Document X-582-76-77, Goddard Space Flight Center, Greenbelt, Maryland.

Caratori Tontini, F., Cocchi, L. and Carmisciano, C. (2009): Rapid 3-D forward model of potential fields with application to the Palinuro Seamount magnetic anomaly (southern Tyrrhenian Sea, Italy). Journal of Geophysical Research, 114: B02103, doi:10.1029/2008JB005907.

Chapman, Bartels (1940): Geomagnetism, vol.1 and 2. Oxford University Press, London.

Carter, B. and Carter, M.S. (2002): Latitude: How American Astronomers Solved the Mystery of Variation. U.S. Naval Institute Press.

Chao, B.F. (2005): On inversion for mass distribution form global (time-variable) gravity field. Journal of Geodynamics, 29: 223–230.

Chelton, D.B., Ries, J.B., Haines, B.J., Fu, L.L. and Callahan, P.S. (2001): Satellite altimetry. *In*: Fu, L.L. and Cazenave, A. (eds.). Satellite Altimetry and Earth Sciences, A Handbook of Techniques and Applications. Academic Press.

Chui, C.K. (1992): An Introduction to Wavelets. Academic Press, San Diego.

Chulliat, A., Macmillan, S., Alken, P., Beggan, C., Nair, M., Hamilton, B., Woods, A., Ridley, V., Maus, S. and Thomson, A. (2015): The US/UK World Magnetic Model for 2015–2020. Technical Report, National Geophysical Data Center, NOAA. doi:10.7289/V5TB14V7.

Churchill, R.V. and Dolph, C.L. (1954): Inverse transforms of products of Legendre transforms. Proceedings of the American Mathematical Society, 5(1): 93–100.

Colombo, O.L. (1980): Numerical analysis for harmonic analysis on the sphere. Report no. 310, Department of Geodetic Science, Ohio State University, Columbus, Ohio.

Courant, R. and Hilbert, D. (1966): Methods of Mathematical Physics, Volume I. Interscience Publishers, Inc., New York.

Cressie, N.A.C. (1991): Statistics for Spatial Data. John Wiley & Sons, New York.

Cross, R.S. (2000): The excitation of the Chandler wobble. Geophysical Research Letters, 27(15): 2329–2332.

Cushing, J.T. (1975): Applied Analytical Mathematics for Physical Scientists. John Wiley & Sons, Inc., New York.

Dahlen, F.A. and Simons, F.J. (2008): Spectral estimation on a sphere in geophysics and cosmology. Geophysical Journal International, 174: 774–807.

Daubechies, I. (1992): Ten Lectures on Wavelets. SIAM Press, Philadelphia, PA.

Davis, P.J. (1975): Interpolation and Approximation. Dover Publications.

De Finetti, B. (1970): Logical foundations and measurement of subjective probability. Acta Psychologica, 34: 129–145.

Edmonds, A.R. (1957): Angular Momentum in Quantum Mechanics. Princeton University Press, Princeton, New Jersey.

Farr, T.G., Rosen, P.A., Caro, E., Crippen, R., Duren, R., Hensley, S., Kobrick, M., Paller, M., Rodriguez, E., Roth, L., Seal, D., Shaffer, S., Joanne, J., Umland, J., Werner, M., Oskin, M., Burbank, D. and Alsdorf, D. (2007): The Shuttle Radar Topography Mission. Rev. Geophys., 45, RG2004, doi:10.1029/2005RG000183.

Featherstone, W.E. (2013): Deterministic, stochastic, hybrid and band-limited modifications of Hotine's integral. Journal of Geodesy, 87: 487–500.

Floberghagen, R., Fehringer, M., Lamarre, D., Muzi, D., Frommknecht, B., Steiger, C., Piñeiro, J. and da Costa, A. (2011): Mission design, operation and exploitation of the gravity field and steady-state ocean circulation explorer mission. Journal of Geodesy, 85: 749–758.

Forsberg, R. (1985): Gravity field terrain effect computations by FFT. Bulletin Géodésique, 59: 342–360.

Foster, M.R. and Guinzy, N.J. (1967): The coefficient of coherence: its estimation and use in geophysical data processing. Geophysics, 32(4): 602–616.

Freeden, W., Gervens, T. and Schreiner, M. (1994): Tensor spherical harmonics and tensor spherical splines. Manuscripta Geodaetica, 19(2): 70–100.

Freeden, W., Gervens, T. and Schreiner, M. (1998): Constructive Approximation on the Sphere. Clarendon Press, Oxford University Press, New York.

Fukushima, T. (2012a): Numerical computation of spherical harmonics of arbitrary degree and order by extending exponent of floating point numbers: II first-, second-, and third-order derivative. Journal of Geodesy, DOI 10.1007/s00190-012-0561-8.

Fukushima, T. (2012b): Recursive computation of finite difference of associated Legendre functions. Journal of Geodesy, 86: 745–754.

Garland, G.D. (1979): The contributions of Carl Friedrich Gauss to geomagnetism. Historia Mathematica, 6: 5–29.

Garmier, R. and Barriot, J.P. (2001): Ellipsoidal expansions of the gravitational potential: theory and application. Celestial Mechanics and Dynamical Astronomy, 79: 235–275.

Gaunt, J.A. (1929): The triplets of helium. Philosophical Transactions of the Royal Society, London, A, 228: 151–196.

Gelb, A. (ed.) (1974): Applied Optimal Estimation. The M.I.T. Press, Cambridge, Massachusetts.

Gleason, D.M. (1985). Partial sums of Legendre series via Clenshaw summation. Manuscripta Geodaetica, 10: 115–130.

Goldberg, R.R. (1964): Methods of Real Analysis. Xerox College Publishing, Lexington, Massachusetts.

Golub, G.H. and Van Loan, C.F. (1996): Matrix Computation, Third Edition. The Johns Hopkins University Press, Baltimore, Maryland.

Gradshteyn, I.S. and Ryzhik, I.M. (1980): Table of Integrals, Series, and Products. Academic Press, New York.

Grafarend, E.W. (2001): The spherical horizontal and spherical vertical boundary value problem—vertical deflections and geoidal undulations—the completed Meissl diagram. Journal of Geodesy, 75: 363–390.

Gruenbacher, D.M. and Hummels, D.R. (1994): A simple algorithm for generating discrete prolate spheroidal sequences. IEEE Transactions on Signal Processing, 42(11): 3276–3278.

Gunn, P.J. (1975): Linear transformations of gravity and magnetic fields. Geophysical Prospecting, 23: 300–312.

Haagmans, R., De Min, E. and Van Gelderen, M. (1993): Fast evaluation of convolution integrals on the sphere using 1D FFT, and a comparison with existing methods for Stokes' integral. Manuscripta Geodaetica, 18(5): 227–241.

Hagiwara, Y. (1976): A new formula for evaluating the truncation error coefficient. Bulletin Géodésique, 50: 31–35.

Han, S.C., Shum, C.K., Jekeli, C., Kuo, C.Y., Wilson, C. and Seo, K.W. (2005): Non-isotropic filtering of GRACE inferred time-variable gravity fields and verification with independent geophysical models and data. Geophysical Journal International, 63: 18–25.

Hansen, S.M., Dueker, K.G., Stachnik, J.C., Aster, R.C. and Karlstrom, K.E. (2013): A rootless rockies—Support and lithospheric structure of the Colorado Rocky Mountains inferred from CREST and TA seismic data. Geochemistry, Geophysics, Geosystems, 14: 2670–2695, doi:10.1002/ggge.20143.

Harris, F.J. (1978): On the use of windows for harmonic analysis with the discrete Fourier transform. Proceedings of the IEEE, 66(1): 51–83.

Heiskanen, W.A. (1938): Investigations on the gravity formula. Ann. Acad. Sci. Fennicae A, 51(8); reprinted in Publ. Isostatic Inst. of the IAG, no.1, Helsinki.

Heiskanen, W.A. and Moritz, H. (1967): Physical Geodesy. W.H. Freeman and Co., San Francisco.

Heller, W.G. and Jordan, S.K. (1979): Attenuated white noise statistical gravity model. Journal of Geophysical Research, 84(B9): 4680–4688.

Hill, E.L. (1954): The theory of vector spherical harmonics. Am. J. Phys., 22: 211–214; doi:10.1119/1.1933682.

Hinze, W.J., von Frese, R.R.B. and Saad, A.F. (2013): Gravity and Magnetic Exploration, Principles, Practices, and Applications. Cambridge University Press, Cambridge, U.K.

Hobson, E.W. (1965): The Theory of Spherical and Ellipsoidal Harmonics. Chelsea Publishing Company, New York.

Hoel, P.G. (1971): Introduction to Mathematical Statistics, 4th Edition. John Wiley and Sons, Inc., New York.

Hofmann-Wellenhof, B. and Moritz, H. (2005): Physical Geodesy. Springer-Velag, Berlin.

Holmes, S.A. and Featherstone, W.E. (2002): A simple and stable approach to high degree and order spherical harmonic synthesis. pp. 259–264. *In*: Adam, J. and Schwarz, K.P. (eds.). Vistas for Geodesy in the New Millennium, Proceedings of the IAG 2001 Scientific Assembly, Budapest, Hungary September 2–7, 2001, Springer, Berlin. doi:10.1007/978-3-662-04709-5_43.

Hotine, M. (1969): Mathematical Geodesy. Monograph No. 2, Environmental Sciences and Services Administration (ESSA), U.S. Department of Commerce, Washington, DC.
Hwang, C. (1998): Inverse Vening Meinesz formula and deflection-geoid formula—applications to the predictions of gravity and geoid over the South China Sea. Journal of Geodesy, 72: 304–312.
Jackson, J.D. (1999): Classical Electrodynamics, Third Edition. John Wiley and Sons, Inc., New York.
Jeffreys, H. (1943): The determination of the Earth's gravitational field. Mon. Not. Roy. Astron. Soc., Geophys. Suppl., 5: 55–66.
Jeffreys, H. (1955): Two properties of spherical harmonics. Quart. J. Mech. Appl. Math., 8(4): 448–451.
Jekeli, C. (1981a): Alternative methods to smooth the Earth's gravity field. Report no. 327, Department of Geodetic Science and Surveying, Ohio State University.
Jekeli, C. (1981b): Modifying Stokes' function to reduce the error of geoid undulation computations. Journal of Geophysical Research, 86(B6): 6985–6990.
Jekeli, C. (1988): The exact transformation between ellipsoidal and spherical harmonic expansions. Manuscripta Geodaetica, 13: 106–113.
Jekeli, C. (1996): Spherical harmonic analysis, aliasing, and filtering. Journal of Geodesy, 70(4): 214–223.
Jekeli, C. (1998): Error Analysis of Padding Schemes for DFTs of Convolutions and Derivatives. Report no. 446, Department of Civil and Environmental Engineering and Geodetic Science, Ohio State University.
Jekeli, C. (2000): Inertial Navigation Systems with Geodetic Applications. Walter deGruyter, Inc., Berlin.
Jekeli, C. (2010): Correlation Modeling of the Geopotential Field in Classical Geodesy. pp. 834–863. *In*: Freeden, W. et al. (eds.). Handbook of Geomathematics, Springer-Verlag, Berlin.
Jekeli, C. (2015): Potential theory and static gravity field of the Earth. Treatise on Geophysics, 2nd edition, G. Schubert (ed.), vol.3, p. 9–35. Elsevier Publ., Oxford.
Jekeli, C. and Zhu, L. (2006): Comparison of methods to model the gravitational gradients from topographic data bases. Geophysical Journal International, 166: 999–1014.
Jekeli, C., Lee, J.K. and Kwon, J.H. (2007): On the computation of ultra-high-degree spherical harmonic series. Journal of Geodesy, 81(9): 603–615.
Jones, M.N. (1985): Spherical Harmonics and Tensors for Classical Field Theory. John Wiley and Sons, Inc., New York.
Jordan, S.K. (1978): Statistical model for gravity, topography, and density contrasts in the Earth. Journal of geophysical Research, 83(B4): 1816–1824.
Kaplan, W. (1973): Advanced Calculus. Addison-Wesley Publ. Co., Reading, MA.
Kaula, W.M. (1959): Statistical and harmonic analysis of gravity. Journal of Geophysical Research, 54(1): 2401–2421.
Kaula, W.M. (1961): A geoid and world geodetic system based on a combination of gravimetric, astrogeodetic, and satellite data. J. Geophys. Res., 66: 1799–1812.
Kaula, W.M. (1966): Theory of Satellite Geodesy. Blaisdell Publ. Co., London.
Kellogg, O.D. (1953): Foundations in Potential Theory. Dover Publications, Inc., New York.
Klees, R. (1997): Topics on Boundary Element Methods. pp. 482–531. *In*: Sanso, F. and Rummel, R. (eds.). Geodetic Boundary Value Problems in View of the One Centimeter Geoid, Lecture Notes in Earth Sciences, vol.65, Springer-Verlag, Berlin.
Klees, R. and Lehmann, R. (1998): Calculation of strongly singular and hypersingular surface integrals. Journal of Geodesy, 72: 530–546.
Koch, K.R. and Kusche, J. (2002): Regularization of geopotential determination from satellite data by variance components. Journal of Geodesy, 76: 259–268.
Kolmogorov, A.N. (1956): Foundations of the Theory of Probability (English translation). Chelsea Publ., Inc., New York.
Kraus, H. (1967): Thin Elastic Shells, An Introduction to the Theoretical Foundations and the Analysis of Their Static and Dynamic Behavior. John Wiley and Sons, Inc., New York.
Lakatos, I. (1976): Proofs and Refutations, The Logic of Mathematical Discovery. Edited by J. Worrall and E. Zahar, Cambridge University Press, Cambridge.
Lambeck, K. (1988): Geophysical geodesy—The Slow Deformations of the Earth. Clarendon Press, Oxford, United Kingdom.
Lee, W.H.K. and Kaula, W.M. (1967): A spherical harmonic analysis of the Earth's topography. Journal of Geophysical Research, 72(2): 753–758.
Lemoine, F.G., Kenyon, S.C., Factor, J.K., Trimmer, R.G., Pavlis, N.K., Chinn, D.S., Cox, C.M., Klosko, S.M., Luthcke, S.B., Torrence, M.H., Wang, Y.M., Williamson, R.G., Pavlis, E.C., Rapp, R.H.

and Olson, T.R. (1998): The development of the joint NASA GSFC and the National Imagery and Mapping Agency (NIMA) geopotential model EGM96, NASA Technical Paper NASA/TP-1998-206861, Goddard Space Flight Center, Greenbelt.
Lerch, F., Wagner, C., Putney, B., Sandson, M., Brownd, J., Richardson, J. and Taylor, W. (1972): Gravitational field models GEM 3 and 4. Goddard Space Flight Center Document X-592-72-476, Greenbelt, Maryland.
Lerch, F.J., Wagner, C.A., Klosko, S.M., Belott, R.P., Laubscher, R.E. and Taylor, W.A. (1978): Gravity model improvement using GEOS-3 altimetry (GEM 10A and 10B). Report of the Goddard Space Flight Center presented at the 1978 Spring Annual Meeting of the American Geophysical Union, Miami, FL.
Lighthill, M.J. (1958): An Introduction to Fourier Analysis and Generalized Functions. Cambridge University Press, Cambridge, U.K.
Lorenz, E.N. (1979): Forced and free variations of weather and climate. Journal of Atmospheric Sciences, 36(8): 1367–1376.
Lovisolo, L. and da Silva, E.A.B. (2001): Uniform distribution of points on a hypersphere with applications to vector bit-plane encoding. IEE Proceedings—Visual, Image, and Signal Processing, 148(3): 187–193.
Mandea, M., Gaina, C. and Lesur, V. (2011): Magnetic Modeling, theory and computation. pp. 781–792. *In*: Gupta, H.K. (ed.). Encyclopedia of Solid Earth Geophysics, Springer, Dordrecht, The Netherlands.
Mandelbrot, B. (1983): The Fractal Geometry of Nature. Freeman and Co., San Francisco.
Marple, S.L. (1987): Digital Spectral Analysis with Applications. Prentice-Hall, Inc., Englewood Cliffs, New Jersey.
Martinec, Z. (1998): Boundary-Value Problems for Gravimetric Determination of a Precise Geoid. Springer-Verlag, Berlin.
Marvasti, F. (ed.) (2001): Nonuniform Sampling, Theory and Practice. Kluwer Academic/Plenum Publ., New York.
Maus, S. (2008): The geomagnetic power spectrum. Geophysical Journal International, 174: 135–142.
Maus, S., Rother, M., Hemant, K., Stolle, C., Lühr, H., Kuvshinov, A. and Olsen, N. (2006): Earth's lithospheric magnetic field determined to spherical harmonic degree 90 from CHAMP satellite measurements. Geophysical Journal International, 164: 319–330.
Maus, S. and 22 other authors (2009): EMAG2: A 2–arc min resolution Earth Magnetic Anomaly Grid compiled from satellite, airborne, and marine magnetic measurements. Geochemistry, Geophysics, Geosystems, 10(8):doi:10.1029/2009GC002471.
Maybeck, P.S. (1979): Stochastic Models, Estimation, and Control, Volume I. Academic Press, New York.
Mayer-Gürr, T., Kurtenbach, E. and Eicker, A. (2010): ITG-Grace2010 gravity field model, available at: http://www.igg.uni-bonn.de/apmg/index.php?id=itg-grace2010.
McAdoo, D.C. (1990): Gravity field of the Southeast Central Pacific from GEOSAT exact repeat mission data. Journal of Geophysical Research, 95(C3): 3041–3047.
McPhee, D.K., Jachens, R.C. and Wentworth, C.M. (2004): Crustal structure across the San Andreas Fault at the SAFOD site from potential field and geologic studies. Geophysical Research Letters, 32, L12S03.
Meissl, P. (1971): A study of covariance functions related to the Earth's disturbing potential. Report 151, Department of Geodetic Science, Ohio State University, Columbus, Ohio.
Meskó, A. (1984): Digital Filtering: Applications in Geophysical Exploration for Oil. Akadé miai Kiadó, Budapest.
Mohr, P.J., Newell, D.B. and Taylor, B.N. (2016): CODATA recommended values of the fundamental physical constants: 2014. Reviews of Modern Physics, 88, DOI: 10.1103/RevModPhys.88.035009.
Molodensky, M.S. (1958): Grundbegriffe der Geodätischen Gravimetrie. VEB Verlag Technik, Berlin.
Moore, G.E. (1965): Cramming More Components onto Integrated Circuits. Electronics, 19 April 1965, pp. 114–117; reprinted in Proceedings of the IEEE, 86(1): 82–85, January 1998.
Moore, I.C. and Cada, M. (2004): Prolate spheroidal wave functions, an introduction to the Slepian series and its properties. Applied and Computational Harmonic Analysis, 16: 208–230.
Moritz, H. (1972): Advanced least-squares methods. Report no. 175, Department of Geodetic Science, Ohio State University, Columbus, Ohio.
Moritz, H. (1978): Least-square collocation: Reviews of Geophysics, 16(3): 421–430.
Moritz, H. (1980): Advanced Physical Geodesy. Herbert Wichmann Verlag, Karlruhe.
Moritz, H. and Mueller, I.I. (1987): Earth Rotation, Theory and Observation. Ungar Publishing Company, New York.

Morse, P.H. and Feshbach, H. (1953): Methods of Theoretical Physics, Parts I and II. McGraw-Hill Book Co., Inc., New York.

Müller, C. (1966): Spherical Harmonics. Lecture Notes in Mathematics, 17, Springer-Verlag, Berlin.

NOAA (2016): Enhanced Magnetic Model (2015). https://www.ngdc.noaa.gov/geomag/EMM/.

Oberhettinger, F. (1957): Tabellen zur Fourier Transformation. Springer-Verlag, Berlin.

Olgiati, A., Balmino, G., Sarrailh, M. and Green, C.M. (1995): Gravity anomalies from satellite altimetry: comparison between computation via geoid heights and via deflections of the vertical. Bulletin Géodésique, 69: 252–260.

Osipov, A. and Rokhlin, V. (2014): On the evaluation of prolate spheroidal wave functions and associated quadrature rules. Applied and Computational Harmonic Analysis, 36: 108–142.

Papoulis, A. (1977): Signal Analysis. McGraw Hill Book Co., New York.

Papoulis, A. (1991): Probability, Random Variables, and Stochastic Processes; Third Edition. McGraw-Hill, Inc., New York.

Parke, M.E., Stewart, R.H., Farless, D.L. and Cartwright, D.E. (1987): On the choice of orbits for an altimetric satellite to study ocean circulation and tides. Journal of Geophysical Research, 92(C11): 11,693-11,707.

Parker, R.L. (1972): The rapid calculation of potential anomalies. Geophysical Journal of the Royal Astronomical Society, 31: 447–455.

Parzen, E. (1960): Modern Probability Theory and Its Applications. John Wiley and Sons, Inc., New York.

Paul, M.K. (1978): Recurrence relations for integrals of associated Legendre functions. Bulletin Géodésique, 52: 177–190.

Pavlis, D.E., Moore, D., Luo, S., McCarthy, J.J. and Luthcke, S.B. (1999): GEODYN Operations Manual: 5 volumes. Raytheon ITSS, Greenbelt, Maryland.

Pavlis, N.K., Holmes, S.A., Kenyon, S.C. and Factor, J.F. (2012): The development and evaluation of Earth Gravitational Model (EGM2008). Journal of Geophysical Research, 117, B04406, doi:10.1029/2011JB008916.

Pavlis, N.K., Holmes, S.A., Kenyon, S.C. and Factor, J.F. (2013): Correction to the development and evaluation of the Earth Gravitational Model 2008 (EGM2008). Journal of Geophysical Research, 118, 2633, doi:10.1002/jgrb.50167.

Percival, D.B. and Walden, A.T. (1993): Spectral Analysis for Physical Applications. Cambridge University Press, Cambridge, U.K.

Phillips, R.J. and Lambeck, K. (1980): Gravity fields of the terrestrial planets: long-wavelength anomalies and tectonics. Reviews of Geophysics and Space Physics, 81(1): 27–76.

Pollack, H.N., Hurter, S.J. and Johnson, J.R. (1993): Heat flow from the Earth's interior: analysis of the global data set. Reviews of Geophysics, 31(3): 267–280.

Prey, A. (1922): Darstellung der Höhen- und Tiefenverhältnisse der Erde durch eine Entwicklung nach Kugelfunktionen bis zur 16 Ordnung. Abhandl. Konig. Ges. Wiss. Göttingen, Math-Physik neue Folge, 11(1).

Priestley, M.B. (1981): Spectral Analysis and Time Series, Vols. 1 and 2. Academic Press, London.

Rapp, R.H. (1969): Analytical and numerical differences between two methods for the combination of gravimetric and satellite data. Bollettino di Geofisica Teorica ed Applicata, XI(41-42): 108–118.

Rapp, R.H. (1979): Potential coefficient and anomaly degree variance modeling revisited. Report no. 293, Department of Geodetic Science, Ohio State University, Columbus, Ohio.

Rapp, R.H. (1981): The earth's gravity field to degree and order 180 using Seasat altimeter data, terrestrial gravity data, and other data. Report no. 322, Department of Geodetic Science, Ohio State University, Columbus, Ohio.

Rapp, R.H. (1982): Degree-variances of the Earth'spotential, topography and its isostatic compensation. Bulletin Géodésique, 56: 84–94.

Rapp, R.H. (1998): Past and future developments in geopotential modeling. pp. 58–78. *In*: Forsberg, R., Feissel, M. and Dietrich, R. (eds.). Geodesy on the Move. Springer-Verlag, Berlin.

Rapp, R.H. and Pavlis, N.K. (1990): The development and analysis of geopotential coefficient models to spherical harmonic degree 360. Journal of Geophysical Research, 95(B13): 21,885–21,911.

Reed, G.B. (1973): Application of kinematical geodesy for determining the short wave length components of the gravity field by satellite gradiometry. Report no. 201, Department of Geodetic Science, Ohio state University, Columbus, Ohio.

Rummel, R. (1997): Spherical spectral properties of the Earth's gravitational potential and its first and second derivatives. pp. 359–404. *In*: Sanso, F. and Rummel, R. (eds.). Geodetic Boundary Value

Problems in View of the One Centimeter Geoid. Lecture Notes in Earth Sciences, vol. 65, Springer-Verlag, Berlin.
Rummel, R. and Van Gelderen, M. (1992): Spectral analysis of the full gravity tensor. Geophysical Journal International, 111: 159–169.
Rummel, R. and Van Gelderen, M. (1995): Meissl scheme—spectral characteristics of physical geodesy. Manuscripta Geodaetica, 20: 379–385.
Sansò, F. and Sideris, M.G. (eds.) (2013): Geoid Determination, Theory and Methods. Lecture Notes in the Earth Sciences, Springer-Verlag, Berlin.
Schmitz, D.R. and Cain, J.C. (1983): Geomagnetic spherical harmonic analyses—1. Techniques. Journal of Geophysical Research, 88(B2): 1222–1228.
Schmitz, D.R., Meyer, J. and Cain, J.C. (1989): Modelling the Earth's geomagnetic field to high degree and order. Geophysical Journal, 97: 421–430.
Schumaker, L.L. and Traas, C. (1991): Fitting scattered data on sphere-like surfaces using tensor products of trigonometric and polynomial splines. Numer. Math., 60: 133–144.
Schwarz, K.P., Sideris, M.G. and Forsberg, R. (1990): The use of FFT techniques in physical geodesy. Geophysical Journal International, 100: 485–514.
Seeber, G. (2003): Satellite geodesy: Foundations, Methods, and Applications, 2nd ed. Walter de Gruyter, Berlin.
Shum, C.K., Ries, J.C. and Tapley, B.D. (1995): The accuracy and applications of satellite altimetry. Geophysical Journal International, 121: 321–336.
Sideris, M.G. and Li, Y.C. (1993): Gravity field convolutions without windowing and edge effects. Bulletin Géodésique, 67(2): 107–118.
Simons, F.J. (2010): Slepian functions and their use in signal estimation and spectral analysis. pp. 891–923. *In*: Freeden, W., Nashed, M.Z. and Sonar, T. (eds.). Handbook of Geomathematics, chap.30, Springer, Heidelberg (2010). doi:10.1007/978-3-642-01546-5_30.
Simons, F.J., Dahlen, F.A. and Wieczorek, M.A. (2006): Spatiospectral concentration on a sphere. SIAM Review, 48(3): 504–536.
Simons, F.J. and Dahlen, F.A. (2007): A spatiospectral localization approach to estimating potential fields on the surface of a sphere from noisy, incomplete data taken at satellite altitudes. pp. 670117. *In*: Van de Ville, D., Goyal, V.K. and Papadakis, M. (eds.). Wavelets XII, Vol. 6701, doi:10.1117/12.732406, Proc. SPIE.
Simons, F.J. and Wang, D.V. (2011): Spatiospectral concentration in the Cartesian plane. International Journal on Geomathematics, 2(1): 1–36.
Singleton, R.C. (1969): An algorithm for computing the mixed radix Fast Fourier Transform. IEEE Transactions on Audio and Electroacoustics, AU-17(2): 93–103.
Slepian, D. (1976): On bandwidth. Proceedings of the IEEE, 64(3): 292–300.
Slepian, D. (1983): Some comments on Fourier analysis, uncertainty, and modeling. SIAM Review, 25(3): 379–393.
Slepian, D. and Pollak, H.O. (1961): Prolate spheroidal wave functions, Fourier analysis and Uncertainty—I. Bell Syst. Tech. J., 40(1): 43–63.
Smith, W.H.F. (1993): On the accuracy of digital bathymetric data. Journal of Geophysical Research, 98(B6): 9591–9603.
Smith, W.H.F. and Sandwell, D.T. (1994): Bathymetric prediction from dense satellite altimetry and sparse shipboard bathymetry. Journal of Geophysical Research, 99(Bll): 21,803–21,824.
Smith, W.H.F., Sandwell, D.T. and Raney, R.K. (2005): Bathymetry from satellite altimetry: Present and future, in OCEANS, 2005. Proceedings of MTS/IEEE, pp. 2586–2589, New York.
Sneddon, I.N. (1961): Fourier Series. Dover Publications, Inc., New York.
Sneeuw, N. and Van Gelderen, M. (1997): The polar gap. pp. 559–568. *In*: Sanso, F. and Rummel, R. (eds.). Geodetic Boundary Value Problems in View of the One Centimeter Geoid. Lecture Notes in Earth Sciences, vol. 65, Springer-Verlag, Berlin.
Sra, S. (2012): A short note on parameter approximation for von Mises-Fisher distributions: and a fast implementation of Is(x). Computational Statistics, 27: 177–190.
Stein, E.M. and Shakarchi, R. (2003): Fourier Analysis, An Introduction. Princeton University Press, Princeton, New Jersey.
Tapley, B.D. (1973): Statistical orbit determination theory. pp. 396–425. *In*: Tapley, B.D. and Szebehely, V. (eds.). Recent Advances in Dynamical Astronomy. D. Reidel Publ. Co., Dordtecht-Holland.
Tapley, B.D., S. Bettadpur, M. Watkins and Ch. Reigber (2004): The Gravity Recovery and Climate Experiment, mission overview and early results. Geophys. Res. Lett., 31(9): 10.1029/2004GL019920.

Taylor, B.N. and Thompson, A. (eds.) (2008): The International System of Units (SI). NIST Special Publication 330, 2008 Edition, National Institute of Standards and Technology, Gaithersburg, MD.
Telford, W.M., Geldart, L.P. and Sheriff, R.E. (1990): Applied Geophysics, Second Edition. Cambridge University Press, Cambridge, U.K.
Thébault, E. et al. (47 other authors) (2015): International Geomagnetic Reference Field: the 12th generation. Earth, Planets and Space, 67(79), 19 pp.
Thomson, D.J. (1982): Spectral estimation and harmonic analysis. Proceedings of the IEEE, 70(9): 1055–1096.
Tikhonov, A.N. and Arsenin, V.Y. (1977): Solutions of Ill-Posed Problems. V.H. Winston and Sons, Washington, D.C.
Titchmarsh, E.C. (1939): The Theory of Functions. Oxford University Press, London.
Titchmarsh, E.C. (1948): Introduction to the Theory of Fourier Integrals. Oxford University Press, London.
Torge, W. and Müller, J. (2012): Geodesy, 4th Edition. de Gruyter, Berlin.
Tscherning, C.C. (1976): Covariance expressions for second and lower order derivatives of the anomalous potential. Report no. 225, Geodetic Science, Ohio State University.
Tscherning, C.C. and Rapp. R.H. (1974): Closed covariance expressions for gravity anomalies, geoid undulations, and deflections of the vertical implied by anomaly degree variance models. Report no.208, Geodetic Science, Ohio State University.
Turcotte, D.L. (1987): A fractal interpretation of topography and geoid spectra on the Earth, Moon, Venus, and Mars. J. Geophys. Res., 92(B4): E597–E601.
Turcotte, D.L. and Schubert, G. (2002): Geodynamics, Second Edition. Cambridge University Press, Cambridge, U.K.
Uotila, U.A. (1962): Harmonic analysis of world-wide gravity material. Publ. of Isost. Inst. of the IAG, No. 39, Helsinki.
Van Gelderen, M. and Rmmuel, R. (2001): The solution of the general geodetic boundary value problem by least-squares. Journal of Geodesy, 75: 1–11.
Ver Hoef, J.M., Cressie, N.A. and Barry, R.P. (2004): Flexible spatial models for kriging and cokriging using moving averages and the fast Fourier Transform (FFT). Journal of Computational and Graphical Statistics, 13(2): 265–282.
Von Frese, R.R.B., Jones, M.B., Kim, J.W. and Kim, J.H. (1997): Analysis of anomaly correlations. Geophysics, 62(1): 342–351.
Wahr, J. (1988): The Earth's rotation. Annual Review of Earth and Planetary Science, 16: 231–249.
Wahr, J., Molenaar, M. and Bryan, F. (1998): Time variability of the Earth's gravity field—hydrological and oceanic effects and their possible detection using GRACE. Journal of Geophysical Research, 103(B12): 30205–30229.
Watts, A.B. (2001): Isostasy and Flexure of the Lithosphere. Cambridge University Press, Cambridge, U.K.
Wenzel, H.G. and Arabelos, D. (1981): Zur Schätzung von Anomalie-Gradvarianzen aus Localen Emperischen Kovarianz-functionen. Zeitschrift für Vermessungswesen, 106(5): 234–243.
Wieczorek, M.A. and Simons, F.J. (2005): Localized spectral analysis on the sphere. Geophysical Journal International, 162: 655–675.
Wiener, N. (1930): Generalized harmonic analysis. Acta Mathematica, 55:117-258.
Wikipedia (2014): http://en.wikipedia.org/wiki/Moore's_law.
Wikipedia (Window function) (2016): https://en.wikipedia.org/wiki/Window_function.
World Geodetic System (WGS) Committee (1974): The Department of Defense World Geodetic System 1972. NTIS AD Accession number 110165, Defense Mapping Agency, Washington, DC.
Xiao, H., Rokhlin, V. and Yarvin, N. (2001): Prolate spheroidal wavefunctions, quadrature and interpolation. Inverse Problems, 17: 805–838.
Yi, W., Rummel, R. and Gruber, T. (2013): Gravity field contribution analysis of GOCE gravitational gradient components. Studia Geophysica et Geodaetica, 57(2): 174–202.
Zerilli, F.J. (1970): Tensor harmonics in canonical form for gravitational radiation and other applications. Journal of Mathematical Physics, 11(7): 2203–2208.
Zhongolovich, I.D. (1952): The external gravitational field of the Earth and the fundamental constants related to it (in Russian). Acad. Sci. Publ. Inst. Teor. Astron., Leningrad; translation to English by Aeronautical Charting and Information Service, gov. acc. no. AD-733840, 1971.
Zia, R.K.P., Redish, E.F. and McKay, S.R. (2009): Making sense of the Legendre transform. American Journal of Physics, 77(7): 614–622.

Index

Q

R

S

T

Milton Keynes UK
Ingram Content Group UK Ltd.
UKHW021833071024
449327UK00021B/1492

9 780367 781828